# PHYSICS WITH ULTRA SLOW ANTIPROTON BEAMS

To learn more about the AIP Conference Proceedings, including the Conference Proceedings Series, please visit the webpage **http://proceedings.aip.org/proceedings**

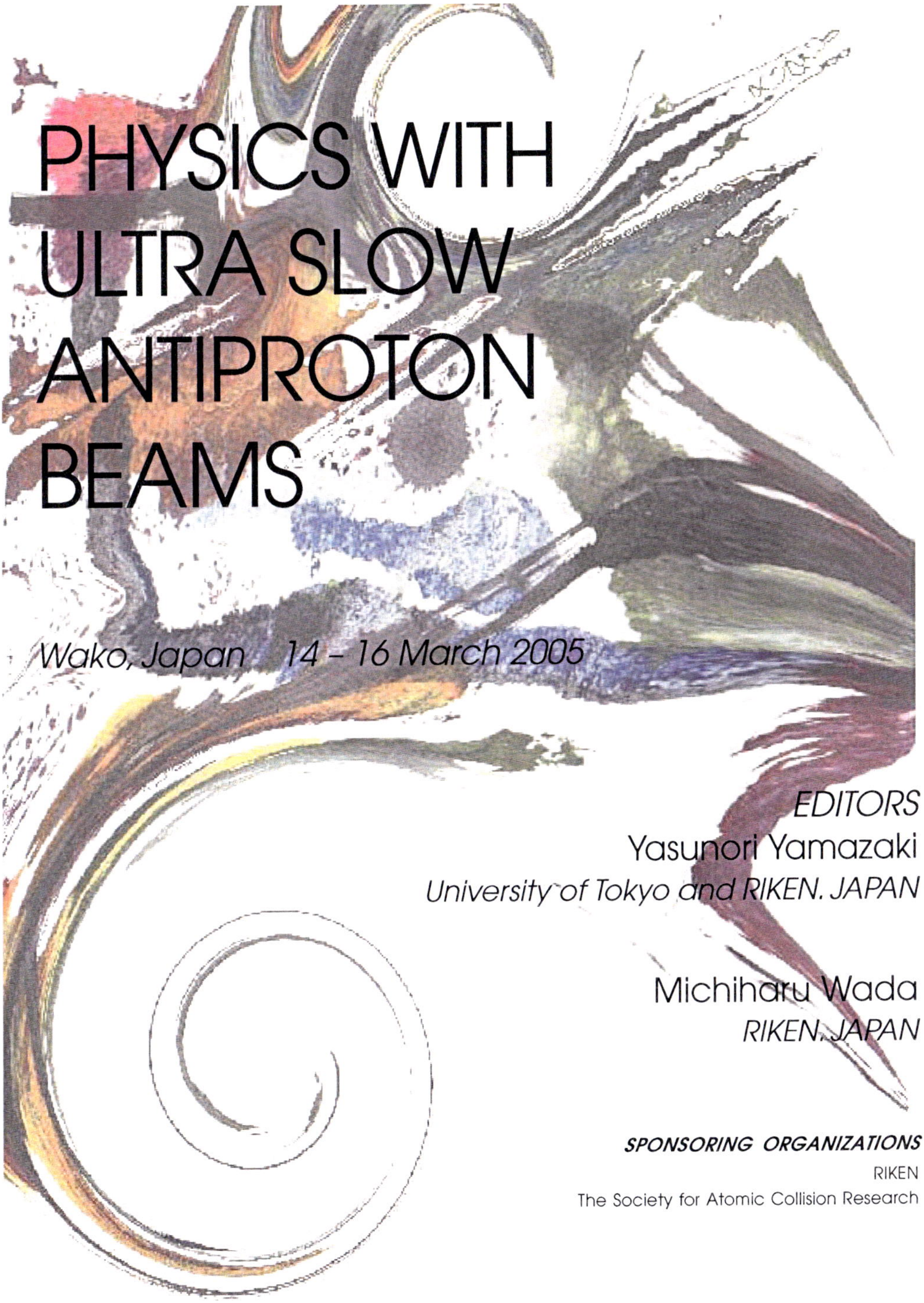

# PHYSICS WITH ULTRA SLOW ANTIPROTON BEAMS

*Wako, Japan 14 – 16 March 2005*

*EDITORS*

Yasunori Yamazaki

*University of Tokyo and RIKEN, JAPAN*

Michiharu Wada

*RIKEN, JAPAN*

***SPONSORING ORGANIZATIONS***

RIKEN

The Society for Atomic Collision Research

AMERICAN INSTITUTE OF PHYSICS

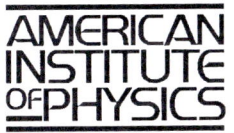

CD-ROM INCLUDED

**Melville, New York, 2005**

**AIP CONFERENCE PROCEEDINGS ■ VOLUME 793**

**Editors:**

Yasunori Yamazaki

University of Tokyo,
3-8-1 Komaba,
Meguro, Tokyo,
153-8902, JAPAN

RIKEN
2-1 Hirosawa, Wako, Saitama,
351-0198, JAPAN


Michiharu Wada

RIKEN
2-1 Hirosawa, Wako, Saitama,
351-0198, JAPAN

E-mail:  yasunori@riken.jp
mw@riken.jp

L.C. Catalog Card No. 2005932870
ISBN 0-7354-0282-5
ISSN 0094-243X
Printed in the United States of America

# CONTENTS

## NON-NEUTRAL PLASMA

# Preface

Just fifty years has passed since the discovery of antiprotons in 1955 by Owen Chamberlain, Emilio Segre, Clyde Wiegand and Tom Ypsilantis. Since then, an extremely diverse range of research topics has developed that involve antiproton beams with kinetic energies of order keV or less. This was the subject of this workshop, the Workshop on Physics with Ultra Slow Antiproton Beams, held 14–16 March 2005 at Wako, Saitama, Japan. Although the announcement was made only two months before the Workshop, it attracted some 70 participants from 12 different countries, and covered subjects ranging from fundamental questions about CPT symmetry and gravitation, to the structure of exotic nuclei, atomic collisions and atomic physics. As is well-known, secondary particles like positrons, muons, kaons, unstable nuclei, and antiprotons are originally produced at high kinetic energies, although interesting physics is often hidden in low energy regions. Among these, positrons and antiprotons are exceptional in the sense that they can be manipulated during a macroscopic time, and accordingly they can really be cooled.

The workshop was motivated by the recent progress in manipulating large numbers of ultra-slow antiprotons that enabled to make a large number of antihydrogen atoms and to make high precision spectroscopy of antiprotonic-helium atoms with the Antiproton Decelerator (AD) at CERN. The latest of these developments was in summer 2004, when the Mono-energetic Ultra-Slow Antiproton Source for High-Precision Investigations (MUSASHI) group of the ASACUSA collaboration first slowed the 5.3 MeV pulsed AD beam in a radio-frequency quadrupole decelerator (RFQD) to some tens of keV, then confined and cooled more than 1 million antiprotons in a large multi-ring trap. This number was approximately 100 times higher in efficiency than every achieved previously. Further, an ultra-slow mono-energetic beam of 10–500 eV antiprotons was also extracted for the first time, which has opened a new possibility in sharing precious antiprotons more effectively for a variety of experiments.

This proceedings contains the latest knowledge of the field presented at the workshop, which could be categorized into three, (1) physics of antiprotons such as CPT symmetry and gravitational interaction including the so-called Standard Model Extension to bridge the Standard Model and general relativity, (2) physics with antiprotons such as atomic collisions, atomic physics, nuclear structure studies, and non-neutral plasma physics, and (3) technological developments in manipulating antiprotons and synthesizing antihydrogen atoms with new schemes.

Regarding the antihydrogen synthesis, the past three years have seen important progress by both the ATHENA and the ATRAP collaborations employing "nested" Penning trap schemes. An unexpected consequence was that these antihydrogen atoms were created before their constituent antiprotons had been fully cooled, with the result that they are themselves too hot to be easily stored and manipulated with existing techniques. Moreover, they are primarily formed in highly excited Rydberg states, while it is the ground and first-excited states that are of most interest for testing CPT symmetry. One of the major subjects was to propose/discuss new ideas in designing traps, going beyond the configuration of the nested electrostatic potential well used so far, such as multipole traps, an RF trap, and a cusp trap. Muonic antihydrogen ($\mu^+ \bar{p}$), the antimatter equivalent of muonic hydrogen ($\mu^- p$), was proposed as an alternative to antihydrogen, which could probe a distance 200 times closer to the antiproton nucleus than positrons.

Atomic collisions with antiprotons were seriously discussed as a potentially important subject to study collision dynamics, which is not well understood when more than three particles are involved. This observation is in good contrast with bound systems like antiprotonic helium, the transition energies of which can be predicted with a precision of one part per billion.

Newly proposed experiments with antiprotons for nuclear structure studies especially for unstable nuclei were discussed. One is to form antiprotonic radioactive nuclear atoms to probe different abundance of protons and neutrons at nuclear surface, and the other is intermediate energy collisions in a collider ring to determine rms radii of protons and neutrons independently. Related pioneering experiments with stable nuclei were reported. Comparison with conventional nuclear collision experiments and these future experiments including muonic atom, proton or electron scattering was also discussed.

It was the pleasure of the organizing committee that the whole period of the Workshop was filled with active and vivid discussions. The Workshop was supported by Special Research Projects for Basic Science of RIKEN.

On behalf of the Organizing committee:   Y.Yamazaki

# Sponsors

RIKEN

The Society for Atomic Collision Research

# Committee Members

Yasunori Yamazaki, University of Tokyo and RIKEN

Michiharu Wada, RIKEN

Naofumi Kuroda, RIKEN

# GENERAL

# The Antiproton and How It Was Discovered

## John Eades

*University of Tokyo*

**Abstract.** The antiproton celebrates its 50th birthday this year. Although its existence had been suspected since the discovery of the positron in 1932, there was still doubt in some quarters that such a companion particle to the proton could exist. I will try to trace the scientific history of the antiproton from that time to the publication of the definitive paper by Chamberlain, Segrè, Wiegand and Ypsilantis in November 1955, with a brief look at what happened next. The narrative will be supplemented with thoughts and opinions of some of the main actors, both at the time and in retrospect.

**Keywords:** Antiproton, Discovery, Dirac
**PACS:** 03.65.P 01.65.+g 11.30.Er 25.75.Dw

We tend to think that the antiproton was discovered in a flash, perhaps during a single night shift at the Berkeley Bevatron in 1955. The real story is much more interesting than that and goes back long before that date. Owen Chamberlain, who got the Nobel prize with Emilio Segrè for the discovery of the antiproton, began his talk at the 1985 Symposium 'Pions to Quarks – Particle Physics in the 1950s' with the words ' *I believe that the antiproton story starts with P.A.M. Dirac, who in 1930 published his paper 'A Theory of Electrons and Protons''.* It is as good a place as any to start.

## The 1930s

At the time, as everyone knows, Dirac was trying to make sense of the negative energy solutions that turned up when he introduced his relativistic wave equation for electrons. Usually the important point is overlooked that this problem occurs even classically, since we get from Einstein the equation :

$$E^2 = p^2c^2 + m^2c^4 \qquad (1)$$

which is going to give negative square roots for the energy $E$ as well as positive ones. Indeed, the transformation of equation 1 into a linear operator equation acting on an electron wave function is where the Dirac equation starts. Classically, however, we can simply ignore the negative roots because discontinuous energy changes like the $2mc^2$ gap that occurs for $p = 0$ between the positive and negative continua do not occur in classical physics.

We cannot do this in a quantum theory, where discontinuous quantum jumps in energy are the name of the game, and where mass energy of particles appears or disappears discontinuously in collisions and radioactive decays. This was the reason why Dirac was so reluctant to reject the infinity of negative energy solutions, and again as everyone knows, proposed that they really existed but were usually all occupied.

## New problems appear

This immediately caused new problems because it gives us an infinite charge density ρeverywhere, so the Maxwell equation:

$$div \; E = -4\pi \; \rho \qquad (2)$$

results in an infinite divergence for $E$, now the electric field vector. Dirac dodged this one by what looks suspiciously like renormalisation in QED : he just redefined the vacuum as that state where all negative energy solutions are occupied by electrons, so that only departures from this infinite charge density distribution contribute to $div \; E$ :

*Let us assume that there are so many electrons in the world that all the states of negative energy are occupied except perhaps a few of small velocity. …. We shall have an infinite number of electrons in negative energy states and indeed an infinite number all over the world, but if their distribution is exactly uniform we should expect them to be unobservable.*

Dirac, 1930 [1]

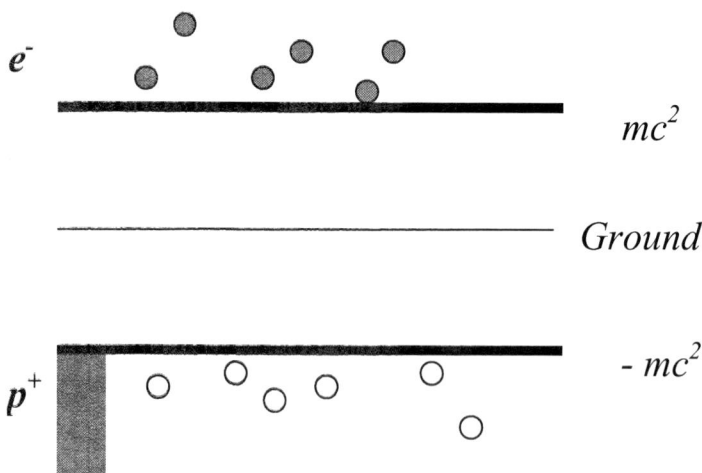

FIGURE 1. Dirac's Unitary Theory of Electrons and protons

It is well attested in his writings that at the time he really did believe in the reality of this kind of Aether of negative energy electrons filling up the entire universe. One anonymous critic said that it all sounded like something an engineer might have dreamed up. Almost no-one now recalls that Dirac's first degree was in Electrical Engineering, and that he is said to have replied *'But I AM an engineer!'*

Accepting both ideas – vacuum 'renormalisation' and fully occupied negative energy states – allows us for all practical purposes to return to the classical physics situation of ignoring the negative energy states, but only as long as every one is occupied by an electron, the exclusion principle for fermions then being sufficient to forbid any $\Delta E \geq 2mc^2$ discontinuous transitions. Strictly speaking, unobservability would mean that these gymnastics would really have no content, but Dirac tried to supply some observable consequences by introducing what he later referred to as a *Unitary Theory of the Nature of Electrons and Protons*, in which the proton, the only other known fundamental particle at the time, was tied to the idea of *unfilled* negative energy states:

*Only the small departures from exact uniformity brought about by some of the negative energy states being unoccupied can we hope to observe... We are led to the assumption that the holes in the distribution of negative energy electrons are the protons.*

Dirac, 1930 [1]


# Contemporary reactions

Even before the 1930 paper, the audacity and counter-intuitiveness of many of the consequences of Dirac's equation made sure that almost nowhere was it accepted with great enthusiasm :

*In spite of the wonderful progress which we owe to Dirac, new difficulties are revealed every day ... A new particularly instructive example has recently been brought to light by Klein.*

Bohr, Letter to C.W. Oseen 5 Nov 1928 [2]

The wonderful progress, was of course the automatic inclusion of spin (which Pauli had, 'without new justification' as he put it, introduced to explain the interactions of the electron with magnetic fields), giving an almost correct value for the electron's magnetic dipole moment, the Thomas factor and so forth. Klein's paradox was that more electrons would be reflected from a sharp potential barrier (one of height $>> mc^2$ over a distance $\sim h/2\pi mc$) than were incident on it. Heisenberg was still more pessimistic:

*The saddest chapter of modern physics is and remains the Dirac theory.*

Heisenberg, Letter to Pauli, 31 Jul 1928 [3]

while after hearing a talk by Dirac, Landau sent a one-word cable to Bohr on September 9, 1930 which said simply *'Quatsch'* (gibberish).

What is striking however in retrospect is the extent to which many of these counter-intuitive ideas were, in some other form, and much later, to become part and parcel of our present ideas of particle physics. Any contemporary who considered the cosmological consequences of Dirac's equation would surely have remarked that the deexcitation of most of the electrons in the universe to negative energy states at some time in the past, must have filled it with radiation. This radiation must now have reached a dynamic equilibrium condition with the particle content of the universe, under which those electrons excited to real positive energy states, together with the real protons left behind, constitutes the material substance of the universe we experience. Secondly there was the redefinition of the vacuum – far from being a place containing no matter, the vacuum now had an infinite density of it everywhere. The non-empty vacuum concept was to reappear within QED in the late 1940s, but it must have been very difficult to accept in 1930. Next, it is important to remember that in 1930 the only known particles of matter were the electron and proton, and that everyone was hoping to include them in what we would call today a Standard Model of particle physics. If Dirac's *Unitary Theory* (that the holes left behind among negative states represent real protons) was correct, why did the masses of protons and electrons differ so much? Dirac suggested a kind of proto-Higgs argument - that the Coulomb interaction between the holes and the electrons might somehow give the protons their mass, although he deferred any attempt at calculation until some unspecified later date. Finally, he was resurrecting a new version of the Aether at a time when no one had any use for it. It was perhaps no wonder that Pauli suggested jokingly at the time that authors be subjected to their own theories, and that Dirac should therefore be annihilated !

## Oppenheimer and Weyl

By March1 1930, Oppenheimer had published a two page letter in Physical Review pointing out several inescapable difficulties with the Unitary Theory, among which was that Thompson scattering (scattering of light by an electron gas) should occur from both positive and negative energy states with the same matrix elements, and that any gaps (as he called Dirac's holes) in the negative 'sea' would be rapidly filled up. He calculated the rate for :

$$p^+ + e^- \rightarrow 2\,\gamma \qquad\qquad (3)$$

and got the result:

$$T = (m + M)^2\, c^3 / 64\ \pi^5\, e^4\, n_p \qquad\qquad (4)$$

m and M being the electron and proton masses, and $n_p$ the local density of protons. He is known to have dropped a factor $(2\pi)^4$ on top. The numerical result is still much too short to keep the universe in existence, and he concludes:

*.... The mean lifetime for ordinary matter will be only about $10^{-10}$ seconds... If we return to the assumption of two independent elementary particles , of opposite charge and dissimilar mass , we can resolve all the difficulties raised in this note, and retain the hypothesis that the reason why no transitions to states of negative energy occur, either for electrons or protons is that all such states are filled.*

Oppenheimer, 1930 [4]

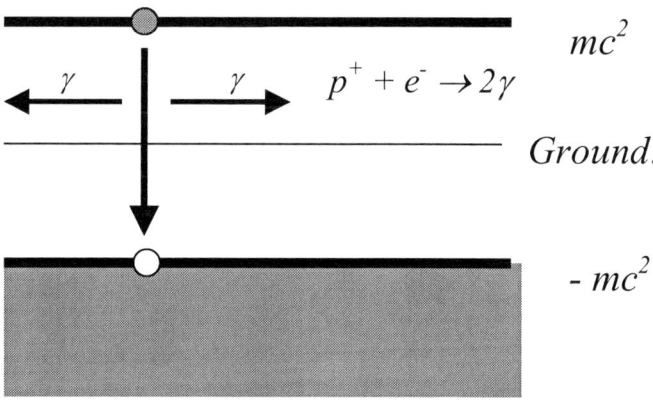

FIGURE 2. Oppenheimer, Dirac - Calculations of annhilation rate .

This is the first hint of the antiproton although it is not explicitly said that there must be some unfilled negative energy proton states.

A few weeks later, probably before he saw this, Dirac sent the Proceedings of the Cambridge Philosophical Society a paper entitled *On the Annihilation of Electrons and Protons*, in which he came to similar conclusions about the instability of matter, for which he calculated a lifetime:

$$T = m^2 c^3 / \pi\, e^4\, n_e \qquad\qquad (5).$$

The question of the mass of the proton having been deferred, Dirac plugged in the average of the electron and proton values and got $10^{-9}$ seconds. By October he had read Oppenheimer's paper and concluded :

*Oppenheimer... does get over these difficulties ...but only at the expense of the unitary theory of the nature of electrons and protons. .....There being now no holes which we can call protons, we must assume that the protons are really independent particles . The proton will now itself have negative energy states, which we must again assume to be all occupied.*

Dirac, 1930 [5]

Shortly afterwards, Weyl also showed that according to Dirac's own theory, the holes would in any case have to have the same mass as the electron. Early in 1931, Dirac finally felt obliged to give up the unitary idea and quoted both Oppenheimer and Weyl as having provided the decisive arguments that the holes were new particles.

*Subsequent investigation has shown that this particle necessarily has the mass of an electron and if it collides with an electron, the two will have a chance of annihilating one another much too great to be consistent with the known stability of matter... We must abandon the identification of holes with protons... A hole, if there were one, would be a new kind of particle, unknown to experimental physics, having the same mass and opposite charge to the electron. We may call such a particle the anti-electron. We should not expect to find any of them in nature ...*

Dirac, 1931 [6]

This paper is also the first in which the antiproton is explicitly mentioned:

*The protons, on the above view, are quite unconnected with with electrons. Presumably, the protons will have their own negative energy states, all of which are normally occupied, an unoccupied one appearing as an anti-proton.*

Dirac, 1931 [6]

It is also, incidentally, the paper in which Dirac first showed that magnetic monopoles are not precluded by quantum mechanics.

## The positron appears

Dirac's cautiousness about finding positrons in nature was based on the assumption that they would have to be produced via the collision of two intense beams of radiation. In fact, when doing his hole-electron annihilation rate calculation, he had not used QED (which he said was unnecessarily complicated) but worked out the stimulated transition rate for this inverse two-photon process, and divided it by the ratio of stimulated to spontaneous emission. Decades before the invention of the laser, he was apparently quite ready to entertain the idea that one day, colliding beams of 512kV radiation might become available to do this (we are still waiting for them). Of course the very next thing was that the positron DID appear in nature, but in cosmic rays, as reported by Anderson, and removed Klein's problem, as the extra electrons reflected from a potential barrier could now be seen to be pair-produced positrons! In the very paper reporting the positron discovery, Anderson himself speculates about antiprotons:

*The greater symmetry, however, between the positive and negative charges revealed by the discovery of the positron should prove a stimulus to search for the evidence of negative protons...   While this paper was in preparation press reports have announced that P.M.S. Blackett and G. Occhialini in an extensive study of cosmic-ray tracks have also obtained evidence for the existence of light positive particles confirming our earlier report.*

Anderson, 1933 [7]

The error on the mass with respect to the electron was about 50%. With regards to the second remark, the Blackett-Occhialini paper may have been accepted as the real discovery, had they been less cautious about publishing. They say in their paper:

*We have recently developed  a method by which the high speed particles associated with penetrating radiation can be made to take their own photographs... It will be shown that it is necessary to come to the same conclusion that has already been drawn by Anderson from similar photographs. This is that some of the tracks must be due to to particles with  a positive charge but whose mass is much less than that of a proton...*

Blackett and Occhialini, 1933 [8]

Whatever the truth,  the Nobel prize was awarded  to Anderson 4 years later,  in 1936.

## The Idea of the Antiproton

Dirac now drew the explicit conclusion that what was true for electrons might also be true for protons:

*One would like to have an equally satisfactory theory for the proton. One might perhaps think that the same theory (as the electron one) could be applied to protons. This would require the possibility of the existence of negatively charged protons forming  a mirror image of the usually positively charged ones.... We must regard it as an accident that the earth (and presumably the whole solar system) contains a preponderance of negative electrons and positive protons. It is quite possible that for some of the stars it is the other way about.... There would be no way of distinguishing them by present astronomical methods.*

Dirac, 1934 [9]

So the hint of the antiproton had now become a serious proposition and the possibility of a baryon-symmetric universe had been seen for the first time.

# The situation in 1955

The fact that Anderson had found positrons in cosmic rays made it seem natural to keep one's eyes open for antiprotons in later cosmic ray exposures of cloud chambers and in emulsions. There had indeed been several  potential annihilation star candidates in cosmic ray experiments as far back as 1947. With the hindsight of knowing what the cosmic ray flux of antiprotons is near sea level, we can  see how weak most of these claims were.  A clear experimental demonstration of the existence of the antiproton was one of the factors considered in deciding the energy of the Bevatron. A parallel emulsion search had been installed by members of the same research team, in the same Bevatron beamline, during the same period in 1955, and it was probably anybody's guess who was going to produce the first evidence.

Or indeed any evidence at all,  since as Tom Ypsilantis often said, there was quite serious doubt in many people's minds as to whether the antiproton existed at all. The main reason given for this was that the proton's large anomalous magnetic moment could be taken to mean that it might not be describable by the Dirac equation, or at least that it was not a simple Dirac particle. Dirac himself had recognized as much in his Nobel Prize lecture:

*There is however some recent experimental evidence obtained by Stern about the spin magnetic moment of the proton, which conflicts with this theory for the proton. As the proton is so much heavier than the electron it is quite likely that it requires some more complicated theory, though one cannot at the present time say what this theory is.*

Dirac, 1934 [9]

Now the electron of course had an anomalous moment too, but its value was a few per mil, not almost two hundred percent, and it was understood in terms of QED. This doubt concerning the antiproton resulted in some now famous bets made in and around Berkeley, which ranged over several orders of magnitude:

*Speculations about the  existence or non-existence of antiprotons were rife during the planning stages of the experiment, and even included a 25 cent bet between Physics Division Head Ed McMillan and Segrè. (By the way 25 cents was a heretofore unimaginably high amount when it came to bets by Segrè). Most of the members of our group were on the pro-antiproton side. The detailed technical planning for the experiment was done primarily by Chamberlain and Wiegand. Tom (Ypsilantis) soon joined the effort and played an active role in all aspects of this work. I was a graduate student who was fortunate enough to be sucked in as well.*

Herbert Steiner, private communication

*As to bets my brother Maurice did indeed bet [Hartland Snyder] $500 against the existence of the antiproton. As I understand his motivation 1) No antiproton galaxies, 2) The proton had an anomalous magnetic moment i.e. not obviously a Dirac particle like the positron. This was long before CP violation and Sakharov's argument.*

Gerson Goldhaber, private communication

One is reminded of similar bets made a year or so later against parity non-conservation, notably by Pauli. An important point about the anomalous moment was that no satisfactory Relativistic Quantum Field Theory existed for hadrons in 1955 to explain the huge anomaly. Nor (see below) would there be one for decades afterwards.

## The experimental discovery of the antiproton

The upshot of these previous inconclusive searches and falsely identified events was the counter experiment method as described in the Physical Review Letter paper announcing the observation of some 60 antiprotons produced by the Berkeley Bevatron . This paper notes in the introductory paragraphs that:

*There have been several experimental events recorded in cosmic ray investigations which might be due to antiprotons although no sure conclusion can be drawn from them at present.*

Chamberlain et al., 1955 [10]

Tom Ypsilantis recalled this inconclusiveness about annihilation searches many years later:

*When we started thinking about an experiment to find the antiproton (1953-1954) we decided to build a spectrometer which could measure both mass and charge rather than trying to observe the annihilation process. This decision... turned out to be crucial ...*

T. Ypsilantis [11]

The emulsion experiment mentioned above was nevertheless thought necessary as a kind of insurance policy and was already well under way using the same beamline:

*I arrived as a postdoc to Berkeley in September 1955 and could follow the progress of the counter experiment. I got deeply involved in the parallel emulsion search and discovery of the annihilations.*

Gösta Ekspong, private communication

The point here is that tracking through a magnetic spectrometer does not give the full record of an emulsion star, which contains ionization information for identifying the annihilation products and therefore for determining the annihilating mass. The counter experiment alone might not therefore have shown unambiguous evidence that the antiproton hypothesis was both necessary and sufficient to explain any candidate events seen. So neither experimental approach seems to have been regarded as having a cast iron guarantee of success.

Furthermore, magnetic tracking does not of course determine the particle mass itself but its radius of curvature, which is proportional to the charge to mass ratio times the velocity (or rather $\beta\gamma$). To ensure that their charge $Q$ was $e$ and not $2e$ or more it was therefore necessary to check that the antiproton candidates had the same pulse height as protons, and also to add a velocity measurement. This was done by measuring the time of flight, between two scintillation counters, *S1* and *S2*, *12 m* apart in the beam. The time of flight for $\pi$ or $K$ mesons (both with $\beta > 0.96$) for this distance was *40 ns*, while for antiprotons ($\beta \sim 0.76$) it was *51 ns*.

This would have been enough were it not for the possibility of mesons interacting in, for example, the air, or a scintillation counter after the magnetic bend, and producing a lower momentum pion that looks like an antiproton. With *$10^5$* mesons per antiproton even very unlikely occurrences of this kind have to be foreseen. So in fact there were three bends in the beamline (including the Bevatron magnet around the target) so that these spurious antiproton events were repeatedly swept out of the beam. Effectively the momentum was measured three times. Still more redundancy was added by measuring the velocity by an independent method:

*Since the antiprotons must be selected from a heavy background of pions it has been necessary to measure the velocity by more than one method.... C2 is a Cerenkov counter that counts particles only within a narrow velocity interval 0.75 < β < 0.78 ... the requirement that the particle be counted in this counter constituted one of the determinations of the velocity of the particle....*

Chamberlain et al., 1955 [10]

In both the text and the figure included in the paper (reproduced as Fig 3 below) the labels C1 and C2 seem to have been interchanged. In spite of all this care taken to reject spurious events, the authors still thought that:

*As outlined so far, the apparatus has some shortcomings ... Accidental coincidences between S1 and S2 cause some mesons to count .... [and] C2 could be actuated by .... [one of these] if the meson suffered a nuclear scattering in the radiator of the counter [C2]. ... Both of these deficiencies have been eliminated by the insertion of the guard counter C1 , which records all particles of β > 0.79. A pulse from C1 indicates a particle (meson) moving too fast to be an antiproton of the selected momentum and indicates that this event should be rejected*

Chamberlain et al., 1955 [10]

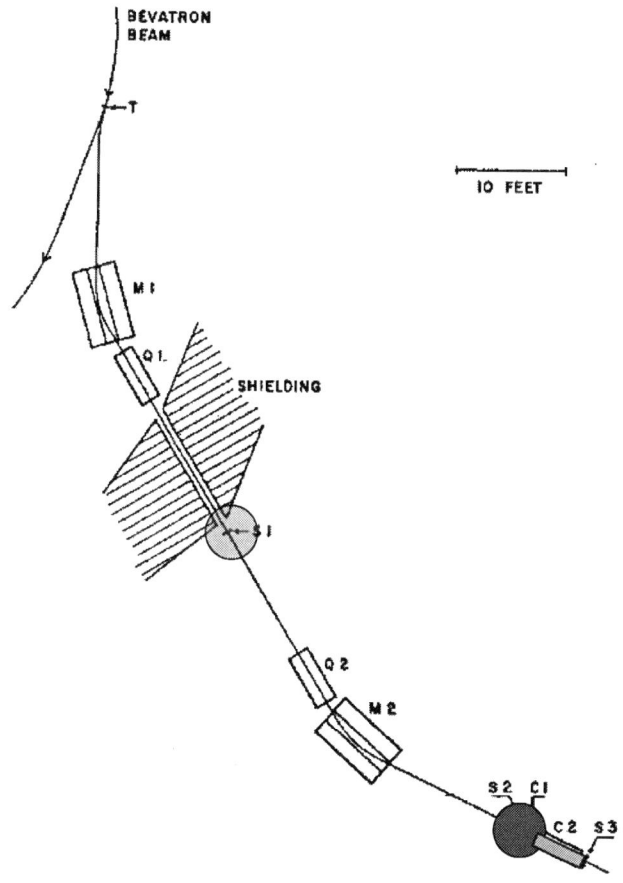

**FIGURE 3.** The Berkeley antiproton spectrometer, Chamberlain et al., Phys. Rev. 100 (1955) p 947.

Even then, the 'golden' annihilation event which came somewhat later from the emulsion experiment [12] seems to have been taken as the conclusive evidence for the antiproton. Indeed the emulsion group might have reported the first antiprotons, because the Bevatron energy – chosen to be 6.2 GeV in order to be above the theoretical six proton-mass threshold for antiproton production from a stationary target – was thought to maximize the beamline yield at 1.19 GeV/c, and this was the momentum to which the beamline was tuned. This was too high to allow the antiprotons to stop in the emulsion, and *132g/cm$^2$* of copper had to be placed in the beam to degrade its energy sufficiently for this. Now the internal Fermi energy within the beryllium target played an important role in both experiments by reducing the threshold for antiproton production to 3.5 Gev/c so the emulsion experiment could have been done at a lower beam momentum without the degrader, with better sensitivity :

*However, the major crime was forgetting Fermi motion. … For protons on a copper target at the Bevatron, the true antiproton threshold was only 3.5 GeV. At 6.2 GeV, the lab antiproton momentum distribution peaked at 0.6 – 0.7 GeV/c, not 1 GeV/c. So no degrader was necessary, just retune of the beam! Of course that was done later. In the first experiment, we could in any case not have reduced the momentum, because the Segrè experiment was designed for 1 GeV/c. You can see therefore that the antiproton "discovery" was a shambles all the way round….*

D. Perkins, private communication

There was some consolation for the emulsion group in the acceptance of the emulsion star as the crucial piece of evidence for the antiproton. Incidentally the event from reference [12], as sketched by Gerson Goldhaber, was reproduced in TIME magazine on Feb 13 of the following year. The overall experimental uncertainty quoted on the equality of the proton and antiproton masses was about 5% (it is now a few parts in 10$^8$). As with the positron experiment, the Nobel prize was only awarded four years later, in 1959.

## Aftermath - P, C, T and CPT

In 1957, quite soon after the antiproton was discovered, parity was shown to be violated in weak interactions by Wu's experiment at Columbia University, New York. It is not widely known that Dirac had always believed that it was unwise to assume either parity or time reversal invariance anyway. Pais asked him in 1959, why parity was never mentioned in his book *Principles of Quantum Mechanics* and he replied '*because I did not believe in it*' , and in 1949 he had written:

*A transformation of this (inhomogeneous Lorenz) type may involve a reflection of the coordinate system in three spatial dimensions and it may involve a time reflection... I do not believe there is any need for physical laws to be invariant under these*

14

*reflections, although the exact laws of nature so far known do have this invariance.*

Dirac, 1949 [13]

Nevertheless it has been said that he was very reluctant to accept the possibility of C-violation (which after the Wu experiment had to be violated to keep CP invariance) until the experimental evidence for this came from an experiment at Liverpool, in 1957. C-invariance, up to then was thought to be the condition that would ensure the equality of particle and antiparticle properties:

*Originally, properties 1-4 (charge, mass, spin, magnetic moment) were derived from C-invariance – a possible physical situation is transformed into another possible physical situation by changing the sign of all electric charges. Since this principle is violated in weak interactions, it is important to point out that it is not necessary to establish the properties listed above, but that the weaker requirement expressed by the invariance under CPT is sufficient ....*

Segrè, 1958 [14]

It has also been said that even after these events Dirac was not convinced of the usefulness of the CPT theorem, which replaced individual C-P- and T- invariance from 1957. If this was true (and I have found no statement by Dirac himself to suggest that it was) the reason may have been that the CPT-theorem is a theorem about relativistic quantum fields, and RQFTs for the strong interaction had more or less been abandoned by then :

*In retrospect we can now see that the reason field theory failed back in the 1950s and 1960s to give an adequate account of the strong interactions was not that it was wrong but that it was misapplied. The fundamental fields of the strong interactions correspond not to the hadronic quanta but rather to the quarks, and the gluons that bind them together. In the mid 1970s theoretical physicists finally invented a successful field theory of the strong interaction – quantum chromodynamics – based on interacting quarks and gluons.*

Pagels, 1983 [15]

It is implied in reference [16] that what did finally put the CPT theorem on firm grounds as far as Dirac was concerned was the appearance of the antideuteron in 1965. As with the positron and to some extent the antiproton, there were two almost simultaneous antideuteron experiments, first at CERN [17] and then at BNL [18] by a Columbia University team. Both were based on the fixed-momentum, long path time-of –flight method used in the antiproton discovery, with now multiple flight paths and momentum bends for additional redundancy since there was now only one antideuteron per $10^8$ mesons, while there had been one antiproton per $10^4$ or $10^5$ of them. Thus the antideuteron experiments had much longer beamlines (*65 m* and *120 m* instead of *24 m*). In the CERN experiment, the beam was partly separated electrostatically, so the added redundancy was not so crucial. In the Brookhaven experiment, there were five bends and the time of flight was measured five times over overlapping flight paths and was supplemented by the same kind of 'velocity-window' Cerenkov counter used in the antiproton experiment. Indeed, the BNL experiment was probably over-redundant for antideuterons because it was looking for heavy, metastable unitary symmetry triplets rather than antideuterons, which were nevertheless expected as a by-product. Now time-of-flight measurements are not a particularly precise way of arriving at a particle's mass, and in both experiments the error on the antideuteron mass was ~ *3%*. Since this far exceeds the deuteron binding energy, neither of them could in fact really be said to constitute a real test of the invariance of nuclear binding expected under CPT, which makes the assertion of reference [16] somewhat difficult to understand.

## Conclusion

A physicist writing about history walks on the same dangerous ground as the historian discussing quantum chromodynamics. I am conscious of the fact that in doing so one is confronted with a confusing mass of ideas, papers and books, without the connections and sequences presented in textbook accounts. All this surrounded by complete mystery as to how much person A knew about what B was doing, and when did he know it. Furthermore, the symbols, terminology, and modes of expression of the time are often now unfamiliar to us. All this makes science history even more difficult than science itself. Nevertheless, I hope I have here conveyed some of the flavour of a long-gone period in particle physics history when the tools of the trade were slide rules rather than computers, and author lists usually took up only a single line.

## ACKNOWLEDGMENTS

It is a pleasure to express my thanks to Gösta Ekspong, Gerson Goldhaber, Don Perkins and Herbert Steiner for sharing their recollections with me of the events surrounding the discovery of the antiproton at Berkeley in 1955. Sadly missing from among us is Tom Ypsilantis, a constant source of ideas and information on the art of experimentation and a friend and colleague, with whom many of us now wish we had discussed the discovery of the antiproton at greater length. For the reasons outlined in the previous paragraph, I would here like to apologize for any errors, omissions and misattributions, sole responsibility for which is mine.

# REFERENCES

1.  Dirac,, P.A.M. , Proc Roy Soc (London) A126 (1930) p 360 .
2.  Reprinted in Niels Bohr, *Collected Works*, North Holland, Amsterdam, 1972.
3.  Reprinted in W. Pauli, *Scientific Correspondence*, vol 1, Springer New York, 1979.
4.  Oppenheimer J.R., Phys. Rev. 35 (1930), p 562.
5.  Dirac, P.A.M.,  Nature vol 126  (1930), p 605.
6   Dirac P.A.M., Proc. Roy. Soc,. A133,  61 (1931) p 60.
7.  Anderson C., Phys. Rev. Vol 43 (1933), p 491.
8.  Blackett P.M.S, and Occhialini G.P.S. , Proc. Roy. Soc. A139 (1933) p 699
9.  Dirac, P.A.M., Nobel Prize lecture (1934)
10. Chamberlain, O., Segrè, E., Wiegand C. and Ypsilantis, T.,  Phys. Rev. 100 (1955) p 947.
11. Ypsilantis T., in *The Discovery of Nuclear Antimatter*, L. Maiani and R.A. Ricci eds., Italian  Physical Society, Bologna 1996, p 37
12. Chamberlain, O, *et al.*,  Phys. Rev, 102 (1956), p. 921.
13. Dirac, P.A.M.,  Rev. Mod. Phys. vol 21 (1949) ,p 392.
14. Segrè, E., , Ann. Rev. Nuc. Sci., vol 8 (1958), p 127
15. Pagels. H., *The Cosmic Code,* Hollen Street Press, Slough 1983, p 295
16. Zichichi, A, in *The Discovery of Nuclear Antimatter, op.cit.*,  p 123
17. T. Massam et al., Il Nuovo Cimento, Serie X, vol 39 (1965), p10
18. D. Dorfan et al., Phys. Rev. Lett. vol 14 (1965) p 1003

# Low Energy Antiproton Experiments – A Review

## Klaus P. Jungmann

*Kernfysisch Versneller Instituut, Rijksuniversiteit Groningen, Zernikelaan 25*
*9747AA Groningen, The Netherlands*

**Abstract.** Low energy antiprotons offer excellent opportunities to study properties of fundamental forces and symmetries in nature. Experiments with them can contribute substantially to deepen our fundamental knowledge in atomic, nuclear and particle physics. Searches for new interactions can be carried out by studying discrete symmetries. Known interactions can be tested precisely and fundamental constants can be extracted from accurate measurements on free antiprotons ($\overline{p}$'s) and bound two- and three-body systems such as antihydrogen ($\overline{H} = \overline{p}e^-$), the antprotonic helium ion ($He^{++}\overline{p}$)$^+$ and the antiprotonic atomcule ($He^{++}\overline{p}e^-$). The trapping of a single $\overline{p}$ in a Penning trap, the formation and precise studies of antiprotonic helium ions and atoms and recently the production of $\overline{H}$ have been among the pioneering experiments. They have led already to precise values for $\overline{p}$ parameters, accurate tests of bound two- and three-body Quantum Electrodynamics (QED), tests of the CPT theorem and a better understanding of atom formation from their constituents. Future experiments promise more precise tests of the standard theory and have a robust potential to discover new physics. Precision experiments with low energy $\overline{p}$'s share the need for intense particle sources and the need for time to develop novel instrumentation with all other experiments, which aim for high precision in exotic fundamental systems. The experimental programs - carried out in the past mostly at the former LEAR facility and at present at the AD facility at CERN - would benefit from intense future sources of low energy $\overline{p}$'s. The highest possible $\overline{p}$ fluxes should be aimed for at new facilities such as the planned FLAIR facility at GSI in order to maximize the potential of delicate precision experiments to influence model building. Examples of key $\overline{p}$ experiments are discussed here and compared with other experiments in the field. Among the central issues is their potential to obtain important information on basic symmetries such as CPT and to gain insights into antiparticle gravitation as well as the possibilities to learn about nuclear neutron distributions.

**Keywords:** antiprotons, antihydrogen, antiprotonic helium, CPT-invariance, gravity
**PACS:** 14.20.-c,11.30.-j,32.10.-f,33.15.-e

## INTRODUCTION

The availability of beams of low energy antiprotons ($\overline{p}$) has led to a number of precision experiments on the properties of free $\overline{p}$'s, of simple two- and three-body bound systems and to studies concerning the interactions of $\overline{p}$'s with matter. Central to the motivation for the precision experiments is the goal to test fundamental forces and symmetries in physics, in particular the CPT invariance [1, 2].

To date we know four fundamental interactions: (i) Electromagnetism, (ii) Weak Interactions, (iii) Strong Interactions, and (iv) Gravitation. These forces are considered fundamental, because all observed dynamical processes in physics can be traced back to one or a combination of them. Together with fundamental symmetries they from a framework on which all physical descriptions ultimately rest. The Electromagnetic, the Weak and many aspects of the Strong Interactions can be described to astounding precision in one single coherent picture, the Standard Model (SM). Gravity is not included in the SM. It is a major goal in modern physics to find a unified quantum field

*CP793 Physics with Ultra Slow Antiproton Beams*
edited by Yasunori Yamazaki and Michiharu Wada
© 2005 American Institute of Physics 0-7354-0282-5/05/$22.50

18

theory which includes all the four known fundamental forces in physics. A satisfactory quantum description of gravity remains yet to be found.

In modern physics - and in particular in the SM - symmetries play an important and central role. Whereas global symmetries relate to conservation laws, local symmetries yield forces [3]. It is rather unsatisfactory that within the SM the physical origin of the observed breaking of discrete symmetries in weak interactions, e.g. of parity (P), of time reversal (T) and of combined charge conjugation and parity (CP), remains unrevealed, although the experimental findings can be well described. Further, there are observed conservation laws which have no status in physics, as there is no symmetry known from which they could be derived, e.g. conservation of baryon and lepton numbers.

Among the intriguing questions in modern physics are the hierarchy of the fundamental fermion masses and the number of fundamental particle generations. Further, the electro-weak SM has a rather large number of some 27 free parameters. All of them need to be extracted from experiments.

The spectrum of speculative models beyond the standard theory, which try to expand the SM, includes such which involve left-right symmetry, fundamental fermion compositeness, new particles, leptoquarks, supersymmetry, supergravity and many more. Interesting candidates for an all encompassing quantum field theory are string or membrane (M) theories which in their low energy limit include supersymmetry.

Searches for effects predicted in speculative models can be carried out in accelerator experiments at the highest at presently possible energies, where, e.g., the direct production of new particles can be looked for. Complementary, at low energies one can look for small deviations from accurate predictions within the SM in high precision experiments. Low energy antiproton experiments are important examples of such precision experiments. In this article we describe precision low energy antiproton experiments and compare them in physics potential and experimental techniques with related low energy precision experiments in other systems.

## ANTIPROTON SOURCES

Today there exists only one facility for low energy antiproton research, the Antiproton Decelerator (AD) at CERN. It succeeded the former LEAR facility and delivers about $10^5$ $\bar{p}$/s and has only pulsed extraction ($3 \times 10^7$ $\bar{p}$ in a 100 ns long pulse every 85 seconds). The AD output energy is 5 MeV. This relatively high kinetic energy is a major disadvantage of the AD . For the experimental programme energies below 100 keV would be much better suited. To achieve this, typically degrader foils are used and one obtains about 2500 $\bar{p}$/s per AD pulse. The AD serves three experimental areas occupied by the ASACUSA, the ATHENA and the ATRAP collaborations. The AD serves a physics program including the spectroscopy of antiprotonic atoms and $\bar{\mathrm{H}}$ and investigations of the $\bar{p}$ interaction with matter.

Recently an ultra slow monoenergetic $\bar{p}$ beam could be extracted with relatively high efficiency. A radiofrequency quadrupole (RFQ) decelerator was employed to achieve $1.2 \times 10^6$ particles per AD pulse could be obtained. When loaded into a multiring trap (MRT) and with electron cooling beams of 10-500eV energy could be made [4]. This new development presents a major step forward and could open new research fields with

AD PROJECT

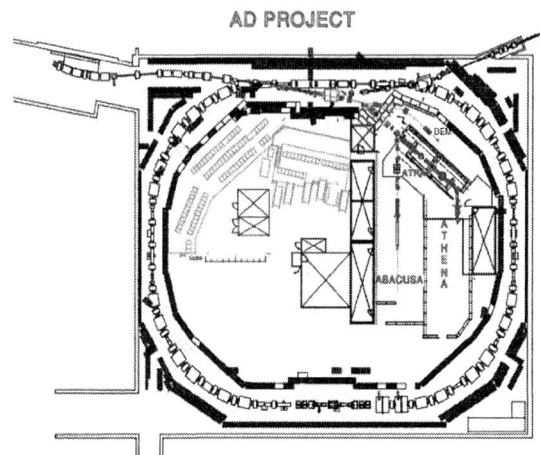

**FIGURE 1.** At present only the Antiproton Decelerator (AD) facility at CERN provides low energy $\bar{p}$'s for precision experiments. A possible upgrade with the ELENA storage ring could increase the $\bar{p}$ flux significantly. In the long term future the FLAIR facility at GSI could provide low energy $\bar{p}$'s for a variety of experiments at somewhat higher rates

ultra slow $\bar{p}$'s, in particular precision tests of fundamental interactions.

At CERN a new Extra Low ENergy Antiproton ring (ELENA) has been proposed, which would be installed behind the AD. One expects 300 ns wide bunches in the energy range 5.3 MeV to 0.1 MeV. Slow extraction appears possible. The space charge limit of the device is about $1.7 \times 10^7$ particles; the longitudinal and transverse temperatures are expected in the 100 eV range.

In connection with the future upgrade plans of GSI, Germany, into a Facility Antiproton and Ion Research (FAIR) also a Facility for Low-energy Antiproton and Ion Research (FLAIR) is foreseen. This promises cooled $\bar{p}$ beams of $1 \times 10^6$ $\bar{p}$/s at 300 keV, and $5 \times 10^5$ $\bar{p}$/s at 20 keV. Emittances of $1\pi$ mm mrad and a momentum spread of $< 10^{-3}$ are possible which is a significant improvement in brilliance over the present AD facility.

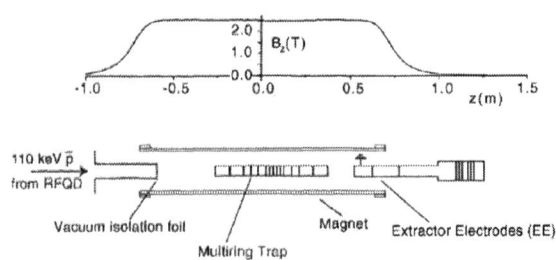

**FIGURE 2.** The multiring trap at the CERN AD facility which achieved 10-500 eV $\bar{p}$ beams [4].

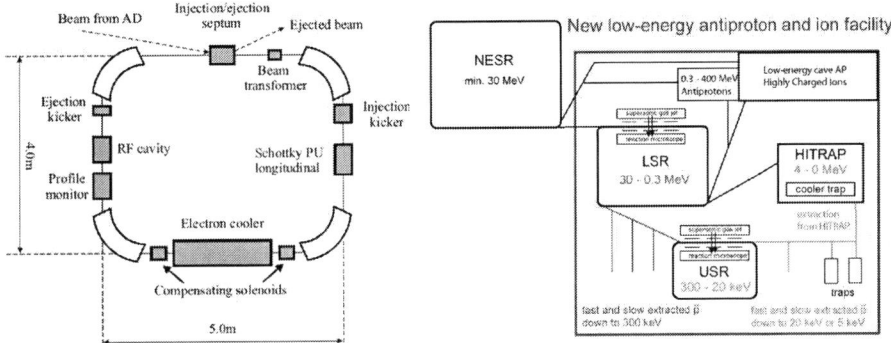

**FIGURE 3.** A possible upgrade of the CERN AD facility with the LENA ring (left) [5] could increase the $\overline{p}$ flux significantly. In the long term future the FLAIR facility (right) at FAIR(GSI) [6] could provide low energy $\overline{p}$'s for a variety of experiments at somewhat higher rates.

# PROPERTIES OF ELEMENTARY PARTICLES – STORED AND TRAPPED LEPTONS AND BARYONS

The properties of elementary particles offer opportunities to test standard theory, and to search for new physics. Particularly the comparisons of particle and antiparticle properties have often been reported as tests of the CPT theorem, which predicts that except for the sign of their electric charge (and charge related quantities) particles and antiparticles should be identical (see Fig. 4).

Trapping and storing of charged particles in combined magnetic and electric fields has been very successfully applied for obtaining properties of the respective species and for determining most accurate values of fundamental constants. Most accurate results were obtained from single trapped and cooled charged particles. The comparison of proton ($p$) and $\overline{p}$ has already reached an impressive level of precision (see Fig. 4).

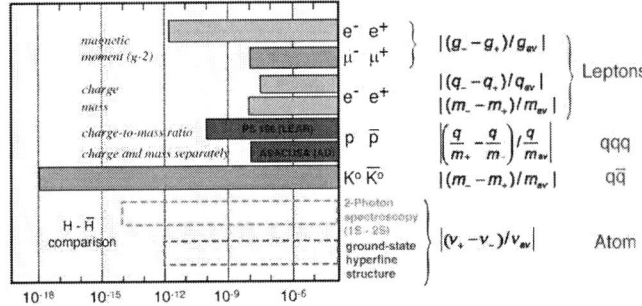

**FIGURE 4.** Tests of the CPT theorem comparing particle and antiparticle properties. Reported is the limit of the fractional differences achieved. Also shown are expectations from $\overline{H}$ experiments [7].

# Leptons

The magnetic anomaly of fermions $a = \frac{1}{2} \cdot (g-2)$ describes the deviation of their magnetic g-factor from the value 2 predicted in the Dirac theory. It could be determined for single electrons and positrons in Penning traps by Dehmelt and his coworkers to 10 ppb [9] by measuring the cyclotron frequency and its difference to the spin precession frequency (g-2 measurement). The good agreement for the magnetic anomaly for electrons and positrons is considered the best CPT test for leptons [2]. Accurate calculations involving almost exclusively the "pure" QED of electron, positron and photon fields allow the most precise determination of the fine structure constant $\alpha$ [10, 11] by comparing experiment and theory for the electron magnetic anomaly in which $\alpha$ appears as an expansion coefficient (see Fig. 5). One order of magnitude improvement appears possible [12] with a new experimental approach (also involving single cooled trapped particles) which aims for reducing the effect of cavity QED, the major systematic contribution to the previous experiment [13].

Muons have been stored in series of measurements at CERN and BNL in magnetic storage rings with weak electrostatic focusing, which is conceptually equivalent to Penning traps. The latest experimental results [14] yield values for positive and negative muons which agree at the 0.7 ppm level in agreement with CPT. The muon is by a factor $(\frac{m_\mu}{m_e})^2$ more sensitive to heavy particles compared to the electron. The muon g-2 measurements are sensitive to new physics involving heavy particles at 40,000 times lower experimental precision. Whether the present experimental results are in agreement with standard theory remains an open question, as not sufficiently accurate values for corrections due to known strong interaction effects exist yet.

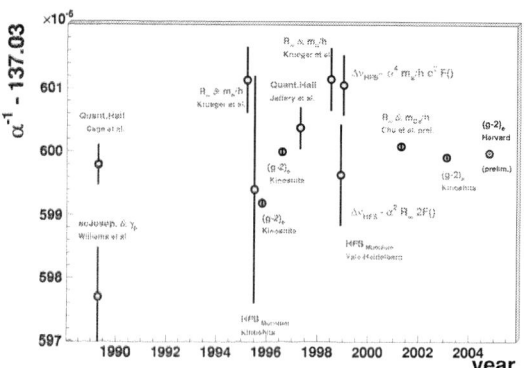

**FIGURE 5.** Measurements of the electron magnetic anomaly using single electrons confined and cooled in a Penning trap yield the best values for the fine structure constant $\alpha$.

# Protons and Antiprotons

After moderation of a $\bar{p}$ beam from LEAR at CERN, $\bar{p}$'s could be trapped in a cylindrical Penning trap for the first time in 1986 [15]. Effective moderation of MeV $\bar{p}$'s and their capture were very important, which has led to detailed studies of the range differences when $p$'s and $\bar{p}$'s are slowed down in matter, known as the Barkas effect[16]. Further electron cooling is essential and could be demonstrated already in the early experiments [17].

In a series of measurements in which the cyclotron frequencies were measured the accuracy of the charge to mass ratio for $\bar{p}$'s could be improved (see Fig. 6) and compared to the proton value. The best results were achieved when a single H$^-$ ion and a single $\bar{p}$ where measured alternatively in the same trap [19]. At present these experiments are interpreted as a CPT test for $p$ and $\bar{p}$ at the level of $9 \times 10^{-11}$. A new experiment has been proposed to measure the magnetic g-factor of the $\bar{p}$ using single particle trapping. Similar to measurements on single electrons and positrons the cyclotron and spin precession frequencies shall be determined. One expects for the comparison of $p$ and $\bar{p}$ g-factors an improvement by a factor of $10^6$ [6].

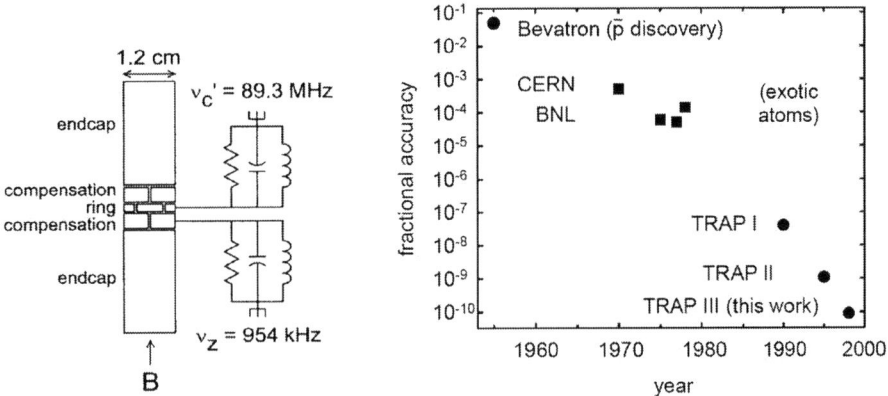

**FIGURE 6.** Left: Cylindrical Penning trap for $\bar{p}$'s [15]. Right: The charge to mass ratio of protons and $\bar{p}$'s could be measured from the cyclotron frequency in a Pennig trap. The fractional uncertainty was constantly improved. The best value was reached, when a single $\bar{p}$ was stored and compared to H$^-$ ions [18, 19].

# HYDROGEN-LIKE ATOMS

Next to single particles, H-like atoms ( see Table 1) are the simplest systems in atomic physics. The electromagnetic part of their binding can be calculated to very high precision in the framework of QED [11, 20]. This allows to study the influence of other known interactions in the Standard Model such as strong and weak forces. Also new yet unknown interactions beyond the Standard Model can be searched for when the precise

**TABLE 1.** The ground state hyperfine structure splitting $\Delta\nu_{HFS}$ and the 1S-2S level separation $\Delta\nu_{1S-2S}$ of H and some exotic H-like systems offer narrow transitions for studying the interactions in Coulomb bound two-body systems. In the exotic systems the natural linewidths $\delta\nu_{HFS}$ and $\delta\nu_{1S-2S}$ have a fundamental lower limit given by the finite lifetime of the systems, because of annihilation, as in the case of positronium and antiprotonic helium, or because of weak muon or pion decays. The very high quality factors (transition frequency divided by the natural linewidth $\Delta\nu/\delta\nu$) in H and other systems with hadronic nuclei can unfortunately not be fully utilized to test the theory, because of the insufficiently known charge distribution and dynamics of the charge carrying constituents within the hadrons. The purely leptonic systems are not affected by this complication.

| | Positronium $e^+e^-$ | Muonium $\mu^+e^-$ | Hydrogen / Antihydrogen $pe^- / \bar{p}e^+$ | Muonic Helium4 $(\alpha\mu^-)e^-$ | Pionium $\pi^+e^-$ | Muonic Hydrogen $p\mu^-$ | Pionic Hydrogen $p\pi^-$ | Antiprotonic Helium4 ion $\alpha\bar{p}$ | Protonium $\bar{p}p$ |
|---|---|---|---|---|---|---|---|---|---|
| $\Delta\nu_{1S-2S}$ [THz] | 1233.6 | 2455.6 | 2466.1 | 2468.5 | 2458.6 | $4.59\times10^5$ | $5.88\,10^5$ | $9\times10^7$ | $2.25\times10^7$ |
| $\delta\nu_{1S-2S}$ [MHz] | $1.28^\dagger$ | .145 | $1.3\times10^{-6}$ | .145 | 12.2 | .176 | $3.5\,10^7$ | $10^6$ | $2.5\times10^5$ |
| $\dfrac{\Delta\nu_{1S-2S}}{\delta\nu_{1S-2S}}$ | $9.5\times10^8$ | $1.7\times10^{10}$ | $1.9\times10^{15}$ | $1.7\times10^{10}$ | $2.0\times10^8$ | $2.6\times10^{12}$ | $1.7\,10^4$ | $10^2$ | $10^2$ |
| $\Delta\nu_{HFS}$ [GHz] | 203.4 | 4.463 | 1.420 | 4.466 | -- | $4.42\times10^7$ | -- | -- | $2.1\times10^{-6}$ |
| $\delta\nu_{HFS}$ [MHz] | 1200 | .145 | $4.5\times10^{-22}$ | .145 | -- | .145 | -- | -- | $2.5\times10^5$ |
| $\dfrac{\Delta\nu_{HFS}}{\delta\nu_{HFS}}$ | $1.7\times10^2$ | $3.1\times10^4$ | $3.2\times10^{24}$ | $3.1\times10^4$ | -- | $3.1\times10^8$ | -- | -- | $8\times10^{-12}$ |

standard theory calculations are confronted with precision experiments. In the past 50 years the chain of three natural isotopes has been significantly expanded with artificially created "exotic" atoms, which often allow to study aspects of fundamental interactions more precisely than the natural atoms.

The H-like atoms fall in three groups: (i) purely leptonic systems such as positronium $(e^+e^-)$ and muonium $(\mu^+e^-)$, (ii) systems containing leptons and hadrons, such as the natural H $(pe^-)$, deuterium $(de^-)$ and tritium $(te^-)$, but also antihydrogen $(\bar{p}e^+)$, pionium $(\pi^+e^-)$, muonic hydrogen $(p\mu^-)$, the muonic helium atom $((He^++\mu^-)e^-)$ and antiprotonic helium atom $(He^{++}\bar{p}e^-)$ and (iii) purely hadronic systems such as pionic hydrogen $(p\pi^-)$, the antiprotonic helium ion $(He^{++}\bar{p})^+$ and protonium $(p\bar{p})$. Although in principle three-body systems, the atoms of muonic helium and antiprotonic helium may be regarded as hydrogen-like, as one may consider the $(He^{++}\mu^-)$ and $(He^{++}\bar{p})$ bound systems as pseudo-nuclei orbited by the lighter $(e^-)$. It should be mentioned that the antiprotonic helium atom with the $\bar{p}$ in an excited state shows both properties of an atom and of a molecule and therefore has been named an "atomcule" [48].

## Positronium

The positronium atom (PS) as a particle-antiparticle bound system is its own antiatom. This causes that in the theoretical description virtual annihilations play an important role and cause significant level shifts. PS has been formed in gases, powders and at certain metal surfaces. The development of sources providing the atoms at thermal velocities in vacuum or in the vacuum of the intergranular regions in powders has taken several decades after the discovery of the atom [21]. This provided a base for later precision experiments. Electromagnetic transitions have been measured in PS with microwave spectroscopy such as the ground state hyperfine splitting and the fine structure transitions in the first excited state of the triplet system. All are in good agreement with standard theory. The $1^3S_1$ and $1^3S_1$ energy difference has been determined with Doppler-free laser spectroscopy to 2.5 ppb. This has been interpreted as the best test of the equality of electron and positron masses [22].

The annihilation of triplet positronium into three gammas has been reported to deviate from SM calculations over the last decade. Recent measurements[23] are in good agreement with theory [24] (see Figure 7). Apparently systematic errors had been underestimated in the past. Sensitive searches for rare positronium decays, e.g. decays into C-parity violating numbers of photons or into new particles such as axions, have not given any result which would be inconsistent with the SM. More sensitive searches are underway and motivated by a number of speculative models [24].

## Muonium

Muonium (M) consists of an antiparticle $(\mu^+)$ and a particle $(e^-)$ [26] and is therefore to be considered in part as antimatter. This fact causes some principle differences in the interactions in the bound state compared to H. Well understood are the differences

25

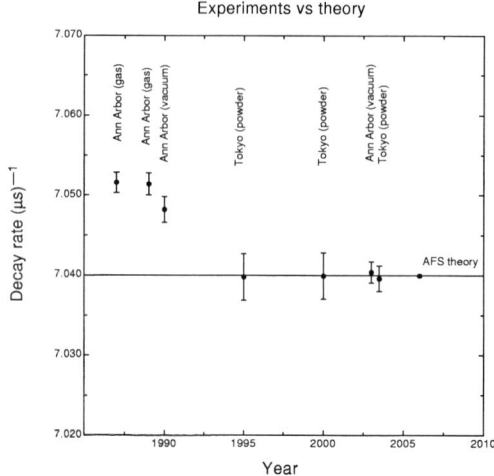

**FIGURE 7.** The history of orthopositronium lifetime measurements (from [24]).

caused (i) by the the different muon and proton masses , which lead to different reduced mass effects, (ii) by the different anomalous magnetic moments reflecting the significantly different muon and proton g-factors due to the proton's internal structure, and (iii) by different signs in the weak interaction contributions to level energies, because the proton is a particle and the muon is an antiparticle.

In the early years muonium research concentrated on measurements that were possible with atoms created by stopping muons in a gas and studying them in this environment [25]. All high precision experiments in M up to date atom have involved the 1s ground state (see Fig.8), in which the atoms can be produced in sufficient quantities. M at thermal velocities in vacuum can be obtained by stopping $\mu^+$ close to the surface of a $SiO_2$ powder target, where the atoms are formed through $e^-$ capture and some of them diffuse through the target surface into the surrounding vacuum, or from hot metal surfaces. This process has an efficiency of a few percent. Only moderate numbers of atoms in the metastable 2s state can be produced with a beam foil technique. Because furthermore these atoms have keV energies due to a velocity resonance in their formation. Electromagnetic transitions in excited states, particularly the n=2 fine structure and Lamb shift splittings could be induced by microwave spectroscopy. However, the experimental accuracy is 1.5 % , which represents not a severe test of theory. The detailed work over 3 decades since the discovery of M in order to understand the atom formation resulted in improvements of the efficiency and quality of M sources. It was indispensable for the success of novel precision experiments [26].

For the M ground state hyperfine structure splitting the constant improvements in experimental techniques resulted in a factor of 10 gain in the accuracy every six years for a period of 20 years after the atom had been discovered. Today the experiments are limited by the available muon fluxes. This is the main reason why the accuracy improvement became slower. The latest measurements of the M hyperfine structure

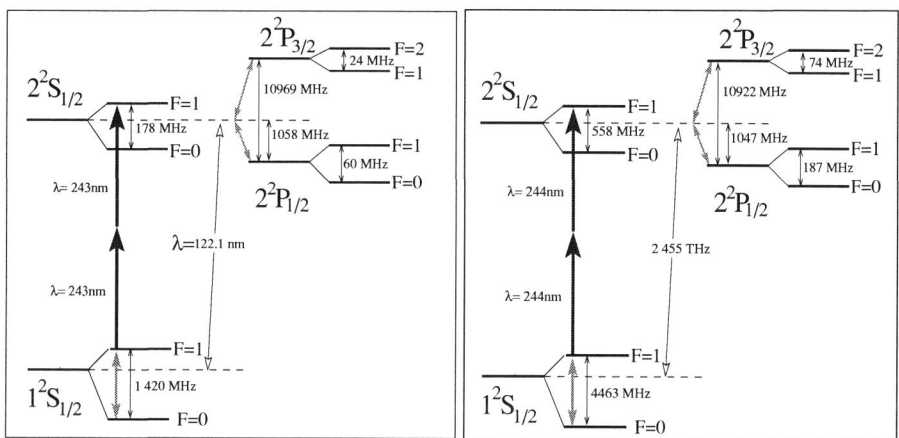

**FIGURE 8.** The energy levels of H and $\overline{\text{H}}$ (right) and M (left) are identical and standard theory predicts exactly the same energy differences. Muonium energy levels differ mainly because of the differences in the proton and muon masses and magnetic moments. At a smaller scale a difference arises from proton internal hadronic structure. New and yet unknown interactions such as CPT or lepton number violation forces could cause additional small energy shifts, which could be revealed in precision experiments.*(not to scale)*

which yield important fundamental constants such as the muon magnetic moment or $\alpha$ were performed at LAMPF, the brightest quasi-pulsed muon source, which delivered $10^8$ $\mu^+$/s. The experiment reached 12 ppb accuracy [27] and was only limited by statistics.

The process of muonium to antimuonium-conversion (M-$\overline{\text{M}}$) violates additive lepton

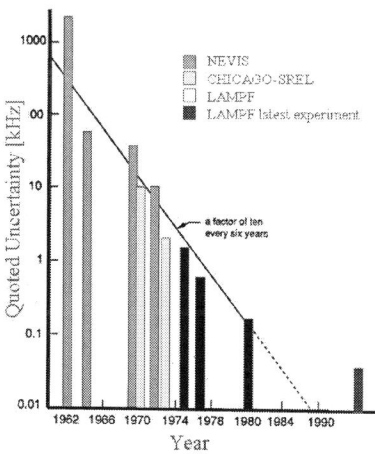

**FIGURE 9.** The accuracy of M $n = 1$ state hyperfine structure measurements over four decades.

family number conservation. It would be an analogy in the lepton sector to the well known $K^0$-$\overline{K^0}$ and $B^0$-$\overline{B^0}$ oscillations in the quark sector. As the oscillations are hindered by gas collisions, the availability of M in vacuum was essential for significant progress. When thermal M atoms in vacuum from a $SiO_2$ powder target, could be observed for their decays in an apparatus covering a large solid angle, three and a half orders of magnitude improvement were possible for the limit on the conversion probability which led to severe restrictions for several speculative models [28]. Again these results are limited by statistics, i.e. the available particle fluxes.

With the availability of thermal M atoms in vacuum from either $SiO_2$ powder or hot metal surfaces, laser spectroscopy became possible. It took one decade from the pioneering Doppler-free 1S-2S two-photon transitions at KEK [29] to the precision experiment at RAL [30] where 4 ppb accuracy was achieved which led to the best test of the equality of muon end electron charges at 2 ppb. Some 15 years after the first laser experiment in M a novel technique succeeded to obtain ultraslow polarized muons through resonant photo ionization of M [31]. Very recently a first muon spin rotation signal with muons from such a source was reported. This development is very important for surface and thin film science.

Precise CPT tests from comparing particle and antiparticle properties (see Fig. 4) neglect in part the fact that symmetry violations ought to be connected with an interaction. Furthermore, the CPT theorem relates to many more intriguing features in physics than particle properties. Recently, generic extensions of the Standard Model, in which both Lorentz invariance and CPT invariance are not assumed, have attracted widespread attention in physics. In particular, new approaches try to quantify CPT violation through additional terms in the Lagrangian of a system [32]. For M diurnal variations of the ratio $(v_{12} - v_{34})/(v_{12} + v_{34})$ could be a sign of Lorentz and CPT violation (see Fig. 10). An upper limit can be set at $2 \cdot 10^{-23}$ GeV for the relevant parameter. In a specific model by Kostelecky and co-workers a dimensionless figure of merit for CPT tests is sought by normalizing this parameter to the particle mass. In this framework $\Delta v_{HFS}$ provides a significantly better test of CPT invariance than electron g-2 and the neutral Kaon oscillations[33].

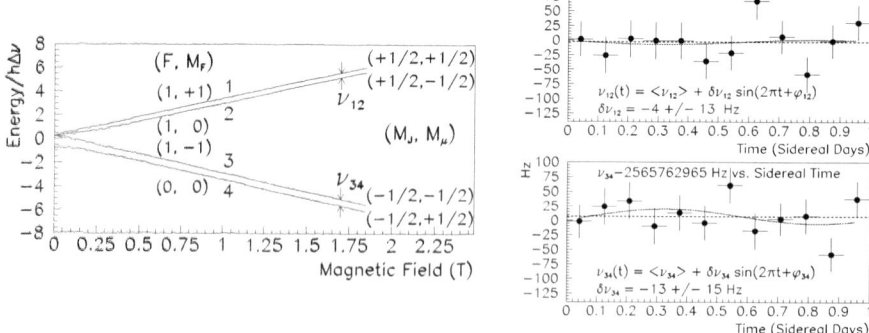

**FIGURE 10.** Left: Breit Rabi diagram of the M ground state. Right: The absence of a significant siderial oscillation in the ratio of transition frequencies between Zeeman levels in a 1.7 T magnetic field confirms CPT invariance at the best level tested for muons.

# Hydrogen

Atomic H has not only enormously contributed to the development of modern physics, in particular to quantum mechanics and QED. Spectroscopic techniques have been pushed forward in attempts to gain ever higher accuracy. Microwave techniques around the H maser and laser techniques and Doppler-free spectroscopic methods are among the examples.

In atomic H the ground state hyperfine structure splitting at 1.42 GHz is used in H masers. It's high quality factor (ratio of transition frequency to natural linewidth) of $3.2 \times 10^{24}$ makes it interesting for a secondary frequency standard. The reproducibility of experimental setups, e.g. H masers, provides the major limitation here to a long term stability of order $10^{-14}$. The transition frequency itself has been measured (compared to the Cs frequency standard) to 0.006 ppb [34]. However, the accuracy of bound state QED calculations is limited to the ppm level due to the proton's internal structure which is not well enough known for calculations of similar accuracy. The main uncertainty arises from the distribution of magnetism within the proton, i.e. its magnetic radius.

Doppler-free two-photon spectroscopy in a cryogenically cooled atomic beam of H yielded the 1S-2S energy interval with a relative uncertainty of $1.8 \times 10^{-14}$ [35]. The coldest H beam one can imagine at this time results from two-photon laser extraction from a H Bose condensate [36, 37]. Typical parameters for the H 1S-2S transition are: a saturation intensity of $I_s = 0.9 W/cm^2$; an excitation rate of $R_e = 4\pi \times 84 \times (I/W/s * cm^2)^2/\Delta v_{exp}/Hz$, where I is the laser intensity and $\Delta v_{exp}$ is the actual observed linewidth of the transition; a 2S state photo-ionization rate $R_p = 9 \times I/W/s * cm^2$; a Zeeman shift of $\delta v_Z = 93 \times B \; kHz/T$, with a magnetic field B; an ac-Stark shift of $\delta v_{ac} = 1.7 \times I \; Hz/W * cm^2$. At 1 mK temperature the average velocity of the atoms is $4 m/s$ causing in a typical 600 $\mu m$ diameter beam a time-of-flight broadening of $\Delta v_{exp} = \Delta v_{TOF} = 3 kHz$. Assuming that a 1S-2S transition is observed with an MCP after field quenching of the 2S state a $10^{-6}$ overall detection efficiency (intrinsic efficiency $\times$ solid angle) must be taken into account. For $10^{11}$ atoms of 30 $\mu K$ in a trap a relative uncertainty for the line center of $\delta v/v_{1s-2s} = 10^{-13}$ has been achieved in 1 s integration time [37]. (It should be noted that at saturation intensity the ionization rate is of order $1.7 \; s^{-1}$ and that the atoms can be used essentially only once, because after the transition the $m_F$ states get equally populated.) In the absence of systematic effects the uncertainty of the line center scales as

$$\delta v = \Delta v_{exp}/(Signal/Noise) = \Delta v_{exp}/\sqrt{N} \; , \tag{1}$$

where $N$ represents the number of observed transitions which is in approximation proportional to the number of atoms in the laser field. The reachable precision depends strongly on the number of available atoms.

A limit on the time variation of $\alpha$ was extracted from two series of repeated measurements of the 1s-2s energy difference in H over a long time and $\frac{\partial \alpha}{\partial t}/\alpha = \frac{\partial}{\partial t}(\ln \alpha) = (-0.9 \pm 2.9) \times 10^{-15} y^{-1}$ could be established [38]. We note, the two series have reduced $\chi^2$ values of 4.2 respectively 9, which may be viewed as a hint that present experiments suffer from not well understood systematic errors. Therefore, presently optical spectroscopy appears to be limited at the $10^{-13}$ level of relative accuracy.

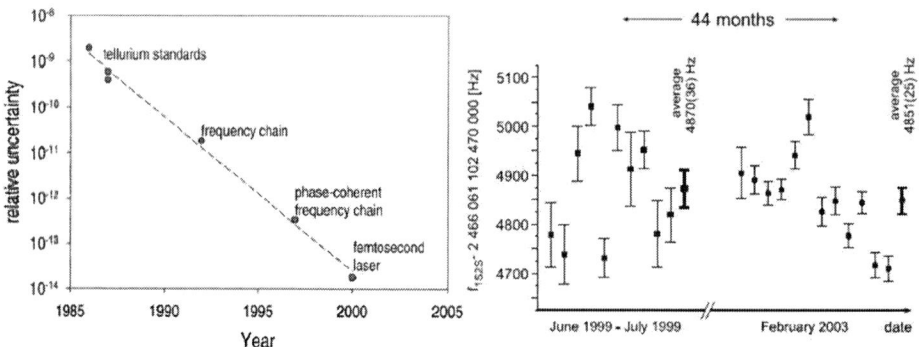

**FIGURE 11.** Right: Measurements of the 1S-2S transition frequency in atomic H have been reported with continuously improving relative uncertainty down to the $10^{-14}$ level due to ever improving technology. Left: Two series of precision measurements of the 1s-1s energy difference in H over a4 years time allow to extract a limit on the time variation of the finestructure constant $\alpha$ [38].

# Antihydrogen

## *Antihydrogen Formation*

$\overline{H}$ was produced first at CERN in 1995 [39] . The atoms were fast as the production mechanism required $e^+e^-$ pair creation when $\overline{p}$'s were passing near heavy nuclei. A small fraction of the $e^+$ form a bound state with the $\overline{p}$. The experiment was an important step forward showing that a few $\overline{H}$ could be produced. Unfortunately, the speed of the atoms does not allow any meaningful spectroscopy. Later a similar experiment was carried out at FERMILAB [40].

The successful production of slow $\overline{H}$ was first reported by the ATHENA collaboration in 2002 [41] and shortly later also by the ATRAP collaboration [42]. Both experiments use combined Penning traps in which first $e^+$ and $\overline{p}$ are stored separately and cooled. The atoms form when both species are brought into contact by proper electric potential switching in the combined traps. The detection in ATHENA relies on diffusion of the neutral atoms out of the interaction volume and the registration of $\pi$'s which appear when the atoms annihilate on contact with matter walls of the container. In ATRAP the hydrogen atoms are re-ionized in an electric field and the $\overline{p}$'s are observed using a capture Penning trap.

Most of the atoms are in excited states (n > 15) which can be seen from the fact that their physical size is above 0.1 $\mu$m[43]. For spectroscopy the atoms need to be in states with low $n$, preferentially the ground state. The production of such states is a major goal of the community for the immediate future. The kinetic energy of the produced $\overline{H}$ atoms has been measured to be of order 200 meV corresponding to a velocity of $6 \times 10^4$ m/s . This is a factor of 400 above the value where neutral atom traps can hold them. Therefore cooling such atoms or identifying a production mechanism for colder $\overline{H}$ are a central topic.

For laser cooling of $\overline{H}$ a continuous laser at the H Lyman-$\alpha$ frequency for $\overline{H}$ cooling has recently been developed [47]. One hopes to achieve the photo-recoil limit of 1.3 mK.

Recently a promising new method was demonstrated to obtain antihydrogen. It uses resonant charge exchange with excited positronium to obtain $\overline{H}$ atoms with essentially the same velocities as the $\overline{p}$'s in the trap which can be made rather low by cooling. [44, 43].

## CPT Tests with Antihydrogen

A main motivation to perform precision spectroscopy on $\overline{H}$ is to test CPT invariance. There are two electromagnetic transitions which offer a high quality factor (see Table 1) and therefore promise high experimental precision when H and $\overline{H}$ are compared: the 1S-2S two-photon transition at frequency $\Delta\nu_{1s-2s}$ and the ground state hyperfine splitting $\Delta\nu_{HFS}$, which both have within the SM in addition to the leading order contributions from QED, nuclear structure, weak and strong interactions,

$$\Delta\nu_{1s-2s} = \frac{3}{4} \times R_\infty + \varepsilon_{QED} + \varepsilon_{nucl} + \varepsilon_{weak} + \varepsilon_{strong} + \varepsilon_{CPT} \qquad (2)$$

$$\Delta\nu_{HFS} = const \times \alpha^2 \times R_\infty + \varepsilon^*_{QED} + \varepsilon^*_{nucl} + \varepsilon^*_{weak} + \varepsilon^*_{strong} + \varepsilon^*_{CPT}. \qquad (3)$$

In eq.(2) and eq.(3) it is assumed that only CPT violating contributions exist from interactions beyond the SM. If one assumes that $\varepsilon_{CPT}$ and $\varepsilon^*_{CPT}$ are of the same order of magnitude, there relative contributions is larger by order $\alpha^{-2}$ for $\Delta\nu_{HFS}$. Further one can speculate that a new interaction may be of short range (contact interaction), which also favors measurements of $\Delta\nu_{HFS}$. Such an experiment has been recently proposed. It utilizes a cold $\overline{H}$ atom beam and has sextupole state selection magnets in a Rabi type atomic beam experiment [45]. For both experiments eq. (1) governs the reachable

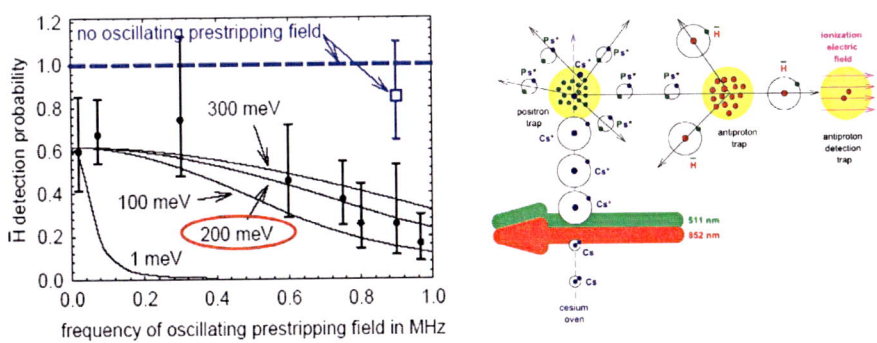

**FIGURE 12.** Right: The $\overline{H}$ atoms produced up to date have typical kinetic energies of order 200meV. Kinetic energies below 0.5 meV are needed to trap the atoms. Left: A new method to produce $\overline{H}$ uses charge exchange with PS atoms. [43]

31

precision, i.e. the atoms should be as cold as possible and one should use as many as possible atoms.

## *Gravitational Force on H̄*

One of the completely open questions in physics concerns the sign of gravitational interaction for antimatter. It can only be answered by experiment. A proposal [46] exists (see Fig. 13) in which the deflection of a horizontal cold beam is measured in the earth's gravitational field. The experiment plans on a number of modern state of the art atomic physics techniques like sympathetic cooling of $\bar{H}^+$ ions by ,e.g. $Be^+$ ions in an ion trap to achieve the neccessary low temperatures of some $20\ \mu K$ . After pulsed laser photo-dissociation of the ion into $\bar{H}$ and a $e^+$ the neutral atoms can then leave the trap. The atom's ballistic path can be measured.

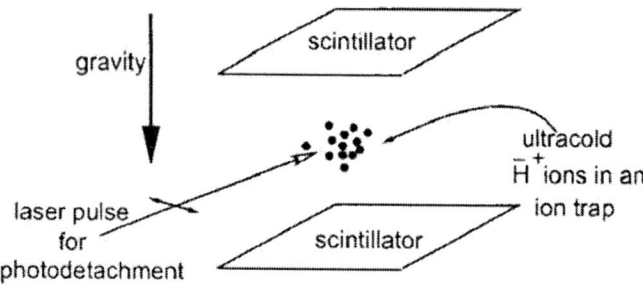

**FIGURE 13.** The sign of the gravitational force on atomic $\bar{H}$ needs to be determined by experiment, as there is no experimental evidence yet that antimatter and matter show identical behavior concerning gravity [46].

## ANTIPROTONIC HELIUM

The potential of antiprotonic helium for precision measurements in the field of fundamental interaction research was realized shortly after it had been discovered that $\bar{p}$'s stopped in liquid or gaseous helium exhibit long lifetimes and do not rapidly annihilate with nucleons in the helium nucleus [48]. This can be explained, if one assumes that the $\bar{p}$'s are captured in metastable states of high principal quantumnumber $n$ and high angular momentum $l$ , with $l \approx n$ (see Fig. 14 [49]. The capture happens at typically at $n \approx \sqrt{M^*/m_e} \approx 38$ , where $M^*$ is the reduced mass of the $(\bar{p}He)$ bound system.

With laser radiation the $\bar{p}$'s in these atoms can be transferred into states where Auger de-excitation can take place. In the resulting H-like system Stark mixing with s-states results in nuclear $\bar{p}$ absorption and annihilation which is signaled by emitted pions. This way a number of transitions could be induced and measured with continuously increasing accuracy over the past decade. A precision of $6 \times 10^{-8}$ has been reached for the transition frequencies [50], which has been stimulating for improving three-body

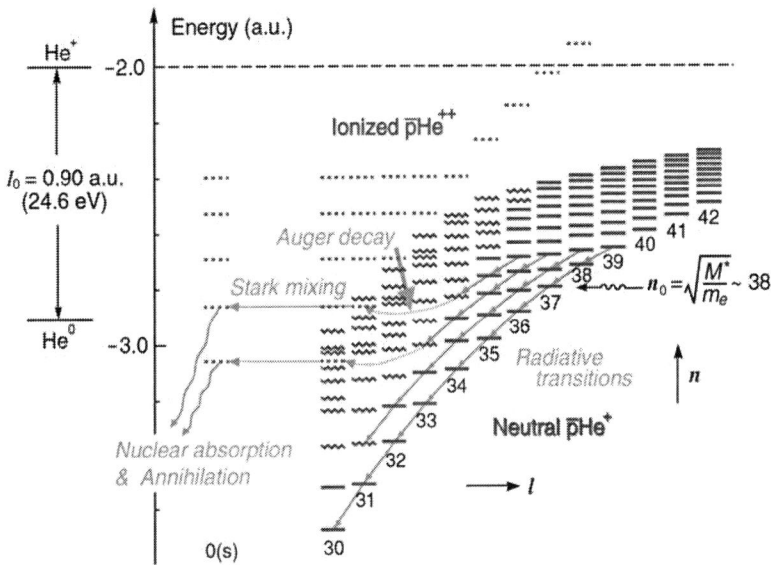

**FIGURE 14.** In the formation process of antiprotonic helium the capture of the $\bar{p}$ into a state with quantum numbers $n \approx 38$ and maximal $l$ is very likely. Such states are rather stable against $\bar{p}$ annihilation [48].

QED calculations. It should be noted that with the high principal quantum numbers for the $\bar{p}$ the system shows also molecular type character [51].

Among the spectroscopic sucesses the laser-microwave double resonance measurements of hyperfine splittings of $\bar{p}$ transitions could be measured (see Fig. 15) [53]. There

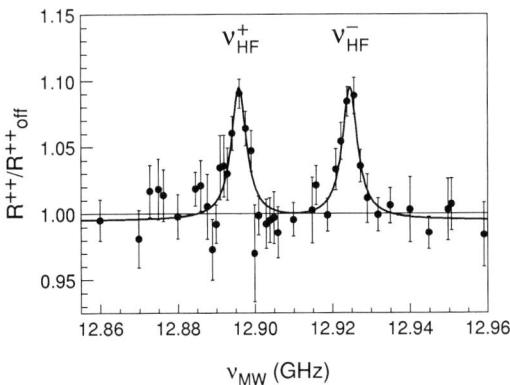

**FIGURE 15.** Example of a line splitting due to hyperfine structure in antiprotonic helium [53].

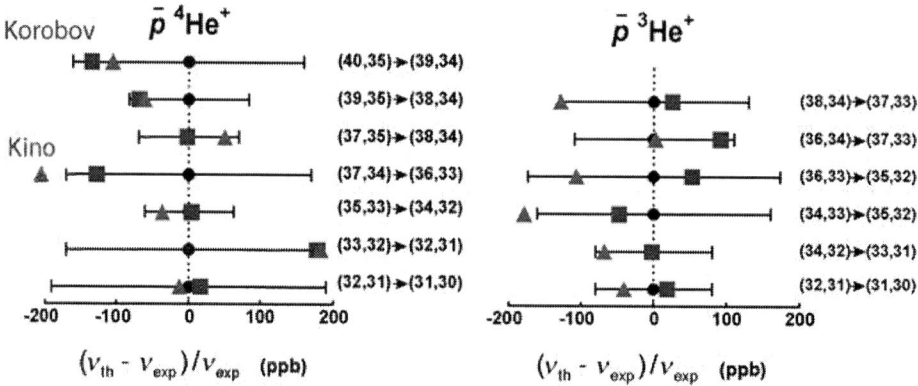

**FIGURE 16.** Accurate QED calculations have been performed for antiprotonic $\bar{P}^4He^+$ and $\bar{P}^3He^+$. The differences between the results from independent calculations are below the accuracy achieved in experiments. These differences may be regarded as a an indication of the size of systematic uncertainties in the theoretical approaches chosen [55].

is greement with QED theory [54] at the $6 \times 10^{-5}$ level which can be interpreted as a measurement of the antiprotonic bound state g-factor to this accuracy. In principle, hyperfinestructure measurements in antipprotonic helium offer the possibility to measure the magnetic moment of the $\bar{p}$.

## CPT tests with Antiprotonic Helium Ions

The very good agreement of the QED calculations with the measurements of several transitions can be exploited to extract a limit on the equality of the charge$^2$ to mass ratio for proton and $\bar{p}$. Combined with the results of cyclotron frequency measurements [19] in Penning traps one can conclude that masses and charges of proton and $\bar{p}$ are equal within $6 \times 10^{-8}$ in full agreement with expectations based on the CPT theorem [55]. The collaboration estimates that a test down to the 10 ppb level should be possible.

## ANTIPROTONIC (RADIOACTIVE) ATOMS

It has been shown at the LEAR facility at CERN that antiprotonic x-rays from atoms in which a $\bar{p}$ has been captured can be utilized to obtain information on the neutron mean square radii of nuclei (see Fig. 17) [56]. The accuracy of the experiments is limited at present by nuclear theory. Neutron distributions are expected to be the limiting factor in the theory of the upcoming round of precision experiments on atomic parity violation. In particular some radioactive nuclei of Francium and Radium isotopes are of interest. At a combined radioactive beam and antiproton facility one can expect experiments to determine the neutron radii with sufficient accuracy for,e.g., the needs

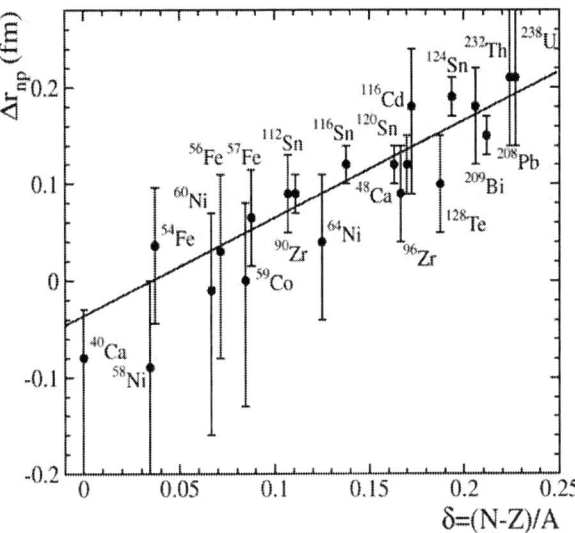

**FIGURE 17.** Difference between the rms radii of the neutron and proton distributions. They were extracted from antiprotonic atom x-ray experiments [56].

of theory to describe improved atomic parity violation experiments including such in heavy radioactive atoms.

# STATUS OF SLOW ANTIPROTON PHYSICS (SPRING 2005) - CONCLUSIONS

Low energy antiproton research has made already a number of important contributions to test fundamental symmetries and to verify precise calculations. With cyclotron frequency measurements of a single trapped $\overline{p}$ and with precision spectroscopy of antiprotonic helium ions stringent CPT tests could be performed on $\overline{p}$ parameters. With precise measurements of $\overline{p}$ in antiprotonic helium atomcules the bound state QED three-body systems could be challenged, which has led to significant advances already. With antiprotonic heavy atoms new input could be provided to obtain neutron radii of nuclei. The differences in proton/ antiproton interactions with matter could expand on similar work with other particle/ antiparticle systems.

$\overline{H}$ atoms have been produced by two independent collaborations. Precision spectroscopy of these atoms will depend on the availability of atoms in the ground state, the successful cooling of the systems to below the 100 $\mu$eV range and their confinement in neutral particle traps. Work in this direction is in progress. A comparison with other exotic atom experimental programs shows that one must allow for sufficient time to develop the necessary understanding of production mechanisms and one must allow time for improving the techniques. The experiments will benefit in their speed of progress and in their ultimate precision from future slow $\bar{p}$ sources of significantly improved particle fluxes and brightness as compared to today's only operational facility.

The ongoing and planned experiments bear a robust discovery potential for new physics, in particular when searching for CPT violation. We can look forward to future precision $\bar{p}$ experiments continuing to deepen insights in fundamental interactions and symmetries, providing important data and parameters with standard theory and providing improved searches for new physics.

## ACKNOWLEDGEMENTS

The author wishes to express his gratitude to the organizers under the leadership of Y. Yamazaki for creating a stimulating atmosphere during the workshop and for their support. The author is further grateful to E. Widmann for numerous discussions on the various subjects covered and to L. Willmann for carefully reading the manuscript. This work was in part supported by the Dutch Stichting voor Fundamenteel Onderzoek (FOM) in the framework of the TRIμP programme.

## REFERENCES

1. J. Schwinger, Phys. Rev. **82**, 914 (1951); G. Lueders, Dansk. Mat. Fys. Medd. **28**, 17 (1954),Ann.Phys. **2**, 1 (1957); W. Pauli, in: *Niels Bohr and the Development of Physics* , McGraw-Hill (1955), Nuovo Cimento **6**, 204 (1957)
2. S. Eidelmann et al., Phys. Lett. **B 592**, 1 (2004)
3. T.D. Lee and C.N. Yang, Phys. Rev. **98**, 1501 (1955)
4. N. Kuroda et al., Phys. Rev. Lett. **94**, 023401 (2005)
5. P. Belochitskii et al., Letter of Intent to CERN, SPSC-I-231, 2005-010 (2005)
6. E. Widmann et al., Letter of Intent to GSI/FAIR (2005) {http://www.oeaw.ac.at/smi/flair/LOI/FLAIR-LOI-resub.pdf}
7. E. Widmann, Nuclear Physics Meeting, Schleching (2005)
8. E. Wildmann et al., Nucl. Instr. Meth. **214**, 31 (2004)
9. R. Van Dyck, Jr.,in *Quantum Electrodynamics*, T. Kinoshita, ed., p. 322, World Scientific (1990)
10. T. Kinoshita and M. Nio, Phys. Rev. **D70** 113001 (2004)
11. T. Kinoshita (ed.), *Quantum Electrodynamics*, World Scientific (1990)
12. G. Gabrielse, priv. comm. (2005)
13. L.S. Brown and G. Gabrielse, Rev. Mod. Phys. **58**, 233 (1986)
14. G.W. Bennett et al., Phys. Rev. Lett. **92**, 161802 (2004)
15. G. Gabrielse et al., Phys. Rev. Lett. **57**, 2504 (1986)
16. G. Gabrielse et al., Phys. Rev. **A 40**, 481 (1989)
17. G. Gabrielse et al., Phys. Rev. Lett.**63**,1360 (1989)
18. G. Gabrielse et al., Phys. Rev. Lett. **74**, 3544 (1995)
19. G. Gabrielse et al., Phys. Rev. Lett. **82**, 3198 (1999);
20. M. Eides, H. Grotch and V.A. Shelyuto, Physics Reports **342**, 63 (2001)
21. M. Deutsch, Phys. Rev. **82**, 455 (1951)
22. M.S. Fee et al., Phys. Rev. **A48**, 192 (1993)
23. R.S. Vallery et al., Phys. Lett. **90** (2003); O. Jinnouchi et al., Phys. Lett. **B572**, 117 (2003)
24. A. Rubbia, Int.J. Mod. Phys. **A 19**, 3961 (2004); S.G. Karshenboim, Int.J. Mod. Phys. **A19**, 3879 (2004)
25. V.W. Hughes et al., Phys. Rev. Lett. **5**, 63 (1960)

26. K. Jungmann, in: *In Memory of Vernon Willard Hughes*, p. 134, World scientific (2004) and nucl-ex/0404013
27. W. Liu et al., Phys. Rev. Lett. **82**, 711 (1999)
28. L. Willmann et al., Phys. Rev. Lett. **82** 249 (1999) ; L. Willmann and K. Jungmann, in: *Lecture Notes in Physics 499*, p. 43, Springer (1997)
29. Steven Chug et al., Phys. Rev. Lett. **60**, 101 (1988) ; see also: K. Danemark et al., Phys. Rev. **A 39**, 6072 (1989)
30. V. Meyer et al., Phys. Rev. Lett. **84**, 1136 (2000)
31. Y. Matsue et al. J. Phys. **G 29**, 2039 (2003)
32. R. Lehár, this volume
33. V.W. Hughes et al., Phys. Rev. Lett. **87**, 111804 (2001); R. Bluhm et al., Phys. Rev. **D 57**, 3932 (1998); R. Bluhm et al., Phys.Rev.Lett. **84**, 1098 (2000)
34. L. Essen et al., Nature **229**, 110 (1971); see also: H. Hellwig et al., IEEE Trans. Instrum. **IM-19**, 200 (1970)
35. M. Niering et al., Phys. Rev. Lett. **84**, 5496 (2000)
36. D. Landhuis et al., Phys. Rev. **A 67**, 022718 (2003) and references therein;
37. L. Willmann, priv. comm. (2004)
38. M. Fischer et al., Lecture Notes in Physics **648**, 209 (2004); see also: M. Fischer et al., Phys. Rev. Lett. **92**, 230802-1 (2004)
39. G. Baur et al., Phys. Lett. **B 368**, 251 (1996)
40. G. Blanford et al., Phys. Rev. Lett. **80**, 3037 (1998)
41. M. Amoretti et al., Nature **419**, 456 (2002)
42. G. Gabrielse et al., Phys. Rev. Lett. **89**, 213401 (2002).
43. G. Gabrielse, Adv. At. Mol. Opt. Phys. **50** (2004); see also: P. Oxley et al, Phys, Lett. **B 395**, 60 (2004); M. Amoretti et al., Physics of Plasmas **10**, 3056 (2003); M. Amoretti et al., Phys. Rev. Lett. **91**, 0055001-1 (2003); M. Amoretti et al., Phys. LEtt. **B 590**, 133 (2004)
44. C.H. Storry et al., Phys. Rev. Lett. **93**, 263401(2004)
45. E. Wildman et al., Nucl. Instrum. Meth.**B214**, 89 (2004)
46. J. Walz and T. Hänsch, Gen. Rel. Grav.**36**, 561 ( 2004)
47. K.S.E. Eikema et al, Phys. Rev. Lett. **86**, 5679 (2001)
48. T. Yamazaki et al., Physics Reports **366**, 183 (2002)
49. S.N. Nakamura et al., Phys. Rev. **A 49**, 4457 (1994)
50. M. Hori et al., Phys. Rev. Lett. **87**, 093401 (2001)
51. T. Yamazaki, in: *The Hydrogen Atom*, S.G. Karshenboim et al. (eds.), Springer, p. 246 (2001)
52. V.I. Korobov, Nucl. Phys. **A689**, 75 (2001)
53. E. Widmann et al., Phys. Rev. Lett. **89**, 243402 (2002)
54. V. Korobov and D. Baklanov, J. Phys. B: At. Mol. Opt. Phys. **34**, L519 (2001) and Phys. Rev. **A 57**, 1662 (1998)
55. M. Hori et al., Phys. Rev. Lett. **91**, 123401 (2003)
56. A. Trzcinska et al., Nucl. Instr. Meth. **B214**, 157 (2004)

# Lorentz and CPT tests involving antiprotons

## Ralf Lehnert

*Department of Physics and Astronomy*
*Vanderbilt University, Nashville, Tennessee, 37235*
*E-mail: ralf.lehnert@vanderbilt.edu*

**Abstract.** Perhaps the largest gap in our understanding of nature at the smallest scales is a consistent quantum theory underlying the Standard Model and General Relativity. Substantial theoretical research has been performed in this context, but observational efforts are hampered by the expected Planck suppression of deviations from conventional physics. However, a variety of candidate models predict minute violations of both Lorentz and CPT invariance. Such effects open a promising avenue for experimental research in this field because these symmetries are amenable to Planck-precision tests.

The low-energy signatures of Lorentz and CPT breaking are described by an effective field theory called the Standard-Model Extension (SME). In addition to the body of established physics (i.e., the Standard Model and General Relativity), this framework incorporates all Lorentz- and CPT-violating corrections compatible with key principles of physics. To date, the SME has provided the basis for the analysis of numerous tests of Lorentz and CPT symmetry involving protons, neutrons, electrons, muons, and photons. Discovery potential exists in neutrino physics.

A particularly promising class of Planck-scale tests involve matter–antimatter comparisons at low temperatures. SME predictions for transition frequencies in such systems include both matter–antimatter differences and sidereal variations. For example, in hydrogen–antihydrogen spectroscopy, leading-order effects in a 1S–2S transition as well as in a 1S Zeeman transition could exist that can be employed to obtain clean constraints. Similarly, tight bounds can be obtained from Penning-trap experiments involving antiprotons.

## MOTIVATIONS

The study of the hydrogen (H) atom is closely associated with two of the most important achievements in 20th-century physics [1]: the explanation of its discrete spectrum provided a great triumph for quantum theory, and many details of the H spectrum serve as a powerful testimonial for Special Relativity in the microscopic world of atoms. The combination of these two groundbreaking theories lies at the foundation of the most successful physical theory to date—the Standard Model of particle physics.

For roughly half a century, theoretical research in fundamental physics has been dominated by various approaches to synthesize the Standard Model and General Relativity into a single unified theory that incorporates, for example, a quantum description of the gravitational interaction. Although a substantial amount of progress has been achieved on the theoretical front, the expected Planck suppression of quantum-gravity effects hampers experimental research in this field [2]: low-energy measurements are likely to require sensitivities of at least one part in $10^{17}$. This talk argues that the recent creation and observation of hot antihydrogen ($\overline{\text{H}}$) [3], the subsequent production of cold $\overline{\text{H}}$ by the ATHENA and ATRAP collaborations [4], and the synthesis of antiprotonic Helium

*CP793 Physics with Ultra Slow Antiproton Beams*
edited by Yasunori Yamazaki and Michiharu Wada
© 2005 American Institute of Physics 0-7354-0282-5/05/$22.50

by the ASACUSA collaboration [5] pave the way for tests that could shed some light on this issue. The basic ideas behind this belief are summarized in the remainder of this section.

The presumed minuscule size of candidate quantum-gravity signatures requires a careful choice of experiments. A promising avenue that one can pursue is testing physical laws that satisfy three primary criteria. First, one should consider presently established fundamental laws that are believed to hold *exactly*. Any measured deviations would then definitely indicate qualitatively new physics. Second, the chances of finding an effect are increased by testing laws that can be *violated* in credible candidate fundamental theories. Third, from a practical viewpoint, this law must be amenable to *ultrahigh-precision* tests.

An example of a physics law that satisfies all of these criteria is CPT invariance. As a brief reminder, this law requires that the physics remains invariant under the combined operations of charge conjugation (C), parity inversion (P), and time reversal (T). Here, the C transformation links particles and antiparticles, P corresponds to a spatial reflection of physics quantities through the coordinate origin, and T reverses a given physical process in time [6]. The Standard Model is CPT-invariant by construction, so that the first criterion is satisfied. With regards to criterion two, we mention that a variety of approaches to fundamental physics can lead to CPT breakdown. Examples include strings [7], spacetime foam [8], nontrivial spacetime topology [9], and cosmologically varying scalars [10]. The third criterion is met as well. Consider, for instance, the conventional figure of merit for CPT conservation in the Kaon system: its value lies currently at $10^{-18}$, as quoted by the Particle Data Group [11].

The CPT transformation relates a particle to its antiparticle. One would therefore expect that CPT invariance implies a symmetry between matter and antimatter. Indeed, one can prove that the magnitude of the mass, charge, decay rate, gyromagnetic ratio, and other intrinsic properties of a particle are exactly equal to those of its antiparticle. This prove can be extended to systems of particles and their dynamics. For example, atoms and anti-atoms must exhibit identical spectra and a particle-reaction process and its CPT-conjugate process must possess the same reaction cross section. It follows that experimental matter–antimatter comparisons can serve as probes for the validity of CPT symmetry. In particular, the extraordinary sensitivities offered by atomic spectroscopy suggest comparative tests with H and $\overline{\text{H}}$ as high-precision tools in this context.

CPT violation in Nature would also lead to other, less obvious effects. The celebrated CPT theorem of Bell, Lüders, and Pauli states that CPT invariance arises under a few mild assumptions through the combination of quantum theory and Lorentz symmetry. If CPT invariance is broken one or more of the assumptions necessary prove the CPT theorem must be false. This leads to the obvious question which one of the fundamental assumptions in the CPT theorem has to be dropped. Since both CPT and Lorentz symmetry involve spacetime transformations, it is natural to suspect that CPT violation implies Lorentz-invariance breakdown. This has recently been confirmed rigorously in Greenberg's "anti-CPT theorem," which roughly states that in any local, unitary, relativistic point-particle field theory CPT violation implies Lorentz violation [12, 13]. Note, however, that the converse of this statement—namely that Lorentz violation implies CPT breaking—is not true in general. *In any case, it follows that CPT tests also probe Lorentz symmetry.* This result offers the possibility for a another class of CPT-violation searches

39

in addition to instantaneous matter–antimatter comparisons: probing for sidereal effects in matter–antimatter and other systems.

This talk gives an overview of Lorentz and CPT violation as a tool in the search for underlying physics—possibly arising at the Planck scale. Section  discusses some forms of Lorentz and CPT violation that could be considered when constructing test models. The Standard-Model Extension (SME), which is currently the standard and most general framework for CPT and Lorentz tests, is reviewed in Sec. . Section  gives two examples of how a Lorentz- and CPT-invariant model can lead to the violation of these symmetries in the ground state generating SME coefficients. In Sec. , some experimental tests of Lorentz and CPT invariance that involve antimatter are discussed. Section  contains a brief summary.

## TYPES OF LORENTZ VIOLATION

The first step in constructing a test model parametrizing the breakdown of Lorentz and CPT symmetry is to determine possible types of Lorentz violation. Additional considerations for CPT breakdown are unnecessary by the anti-CPT theorem because general Lorentz violation will include CPT breaking. It turns out that a clear understanding of the *fundamental principle of coordinate independence* will provide us with a useful, rough classification of types of Lorentz violation. For this reason, it appears appropriate begin with a brief review of this fundamental principle and its implementation.

Coordinate independence is one of the most basic principles in physics. Its need in the presence of Lorentz breaking is well established, and it has served as the basis for the construction of the SME [14, 15]. However, this principle is sometimes not fully appreciated. For example, in some investigations of Lorentz and CPT violation coordinate-*dependent* physics emerges, and occasionally Lorentz-symmetry breakdown is identified with the loss of coordinate independence.

A given labeling scheme for events in space and time is called a coordinate system. Such a labeling typically depends on the observer choosing the coordinates, and it is thus arbitrary to a large extent. In other words, coordinate systems are *mathematical tools* for the measurement, description, and prediction of physical phenomena. But since they are a pure product of human thought, coordinates *lack physical reality*. It follows that the physics should remain unaffected by the choice of a particular coordinate system. This principle of coordinate independence is one of the most fundamental in science. Since it assures that the physics remains independent of the observer, it is also called *observer invariance*. It should therefore be possible to formulate the fundamental laws of physics in a coordinate-free language. For example, this can be achieved mathematically, when spacetime is given a manifold structure and physical quantities are represented by geometric objects, such as tensors or spinors.

The principle of coordinate invariance is more fundamental than Lorentz symmetry. Consider, for instance, Newton's second law of motion in nonrelativistic classical mechanics

$$\vec{F} = m\dot{\vec{v}}, \tag{1}$$

as well as its relativistic version

$$F^\mu = m \frac{dv^\mu}{d\tau}, \tag{2}$$

where $\tau$ denotes the mass' proper time, and $F^\mu$ is the usual relativistic generalization of the force $\vec{F}$. Both laws are coordinate independent. Equation (1) takes the same form in all inertial gallilean frames; it is formulated in the coordinate-free language of 3-vectors. Similarly, Eq. (2) remains of the same form in all inertial Minkowski coordinate systems; it is expressed in terms of 4-vectors. However, only Eq. (2) is Lorentz invariant. We conclude that coordinate independence is more general than Lorentz symmetry because there might very well be laws—such as Eq. (1)—that maintain coordinate invariance but violate Lorentz symmetry.

The above situation becomes even more transparent with the following observation. Lorentz symmetry does not only require coordinate independence, but it also dictates the transformations that relate different inertial frames. Although Eq. (1) is coordinate independent, inertial frames are related by Gallilei instead of Lorentz transformations. Mathematically speaking, both cases allow a coordinate-invariant spacetime-manifold description, but the manifold structure is gallilean in the case of Eq. (1) and lorentzian in the case of Eq. (2). The question which spacetime manifold is realized in Nature must be answered experimentally.

**Lorentz violation through non-lorentzian manifolds.** The above considerations lead to one possible type of Lorentz violation maintaining coordinate independence: the spacetime manifold could be non-lorentzian. Then, the fundamental physics laws have the same form in all inertial frames, but the frames are no longer related by the usual Lorentz transformations. This point of view is taken in the early relativity test model of Robertson and its extension by Mansouri and Sexl, as well as in the so called "doubly special relativities." In the present work, we do not treat this type of Lorentz-symmetry violation separately because it is known that (at least some of) these effects are equivalent to those of another type of Lorentz violation discussed next. Moreover, such frameworks are typically purely kinematical precluding the analysis of atomic level shifts, for example.

**Lorentz violation through a nontrivial vacuum.** Most modern approaches to fundamental physics involve lorentzian manifolds, where inertial frames are related by the usual Lorentz transformations. Such approaches take Lorentz symmetry as a key ingredient, and the issue arises as to whether Lorentz violation can occur in such situations. It turns out that this is indeed the case if the vacuum contains a tensorial background. The primary emphasis in this section is to gain some intuitive understanding of Lorentz breakdown in the presence of such a background. The question of how to generate tensorial backgrounds in a Lorentz-invariant theory is deferred to Sec. .

It is again useful to consider a familiar example from classical physics. Suppose the particle described by Eq. (2) has charge $q$ and is subjected to an *external* electromagnetic field $F^{\mu\nu}$. We remind the reader that the components of $F^{\mu\nu}$ are determined by the usual electric and magnetic fields $\vec{E}$ and $\vec{B}$. The left-hand side of Eq. (2) is now given by the Lorentz force, which reads $qF^{\mu\nu}v_\nu$ in covariant form. The equation of motion

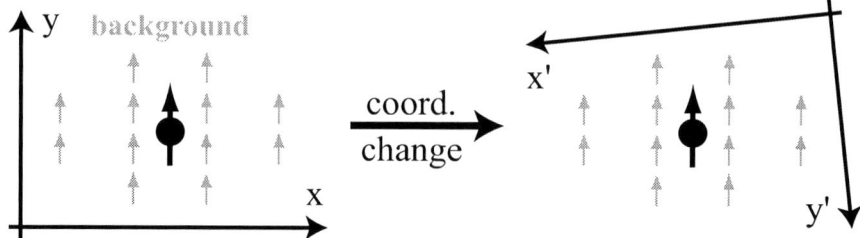

**FIGURE 1.** Coordinate independence. Two experimenters observe identical physical systems represented by the black "particle with spin." Although they may choose to employ different coordinate systems to describe their observations, the outcome of the experiment remains unaffected by this choice. It must therefore be possible to relate observations recorded with respect to different reference frames by appropriate transformations of coordinates. The principle of coordinate independence therefore assures that the physics is independent of the observer.

determining the trajectory of our particle becomes

$$qF^{\mu\nu}v_\nu = m\frac{dv^\mu}{d\tau}.\qquad(3)$$

Note that Eq. (3) remains valid in *all* inertial coordinate systems because it is a tensor equation. Invariance under Lorentz transformations of the coordinate system is therefore maintained. However, the external $F^{\mu\nu}$ background breaks, for example, symmetry under arbitrary rotations of the charge's trajectory. Among the consequences of this rotation-symmetry violation is the non-conservation of the particle's angular momentum. Notice the difference to coordinate changes, which leave unaffected the physics: in the present case, only the trajectory is rotated, so that its orientation with respect to $F^{\mu\nu}$ can change. One then says that particle Lorentz symmetry is violated, despite the presence of Lorentz coordinate independence [14, 15]. Figures 1 and 2 illustrate the difference between particle Lorentz transformations and Lorentz coordinate transformations.

Although the above example captures the main features of Lorentz violation through background vectors or tensors, there is an important difference to situations where these vectors or tensors can be considered as part of the effective vacuum. Our above background $F^{\mu\nu}$ is a local electromagnetic field caused by other 4-currents that can in principle be accessed experimentally. Such backgrounds are therefore *not* a feature of the vacuum, so that these situations cannot be considered as exhibiting fundamental Lorentz violation. This is to be contrasted with situations involving candidate underlying physics, where tensorial backgrounds can extend over the entire Universe and are outside of experimental control. Such backgrounds *must* be viewed as a property of the effective vacuum, which can then be considered as violating Lorentz symmetry (see Sec ).

As a result of the lorentzian structure of the underlying manifold and the usual Lorentz-covariant dynamics at the fundamental level, this approach appears closest to established theories. The physical effects in such Lorentz-breaking vacua are perhaps comparable to those inside certain crystals: the physics remains independent of the chosen coordinate system, but particle propagation, for example, can depend upon the

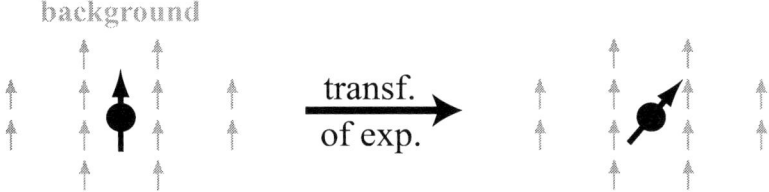

**FIGURE 2.** Particle transformations. Tests of rotational invariance, for example, would *not* be carried out as in Fig. 1: identical experiments with the observer rotated. Instead, one would perform a suitable measurement, repeat it with a rotated apparatus, and then compare the two measurements. Under these types of transformations, which involve localized particles and fields and leave unchanged the background, symmetry can be lost because of the different orientation with respect to the vacuum structure.

direction. An immediate consequence is that one can locally still work with the metric $\eta^{\mu\nu} = \mathrm{diag}(1, -1, -1, -1)$, particle 4-momenta are still additive and still transform in the usual way under coordinate changes, and the conventional tensors and spinors still represent physical quantities.

**Models with coordinate-dependent physics.** We have argued above that coordinate independence is a principle more fundamental than Lorentz symmetry. If one is willing to give up coordinate independence, the loss of Lorentz invariance is unsurprising. Although it seems to be impossible to perform meaningful scientific investigations involving coordinate-*dependent* physics, such approaches to Lorentz breaking have been considered in the literature. More specifically, there have been the following two attempts in the context of neutrino phenomenology: the first one forces the masses of particle and antiparticle to be different [16], while the second one suggests to construct a model from positive-energy eigenspinors only [17]. Both of these approaches are known to involve coordinate-dependent off-shell physics [12, 18]. In what follows, we do not consider these models further.

## THE STANDARD-MODEL EXTENSION

The next step after determining general low-energy manifestations of Lorentz violation is the identification of specific experimental signatures for these effects and the theoretical analysis of Lorentz-violation searches. This task is most conveniently accomplished employing a suitable test model. Many Lorentz tests are motivated and analyzed in purely kinematical frameworks allowing for small violations of Lorentz symmetry. Examples are the aforementioned Robertson's framework and its Mansouri–Sexl extension, as well as the $c^2$ model and phenomenologically constructed modified dispersion relations. However, the implementation of general dynamical features significantly increases the scope of Lorentz tests. For this reason, the SME mentioned in the Sec. has been developed. But the use of dynamics in Lorentz-violation searches has recently been questioned on grounds of test-framework dependence. We disagree with this assertion and begin with a few arguments in favor of a dynamical test model.

The construction of a dynamical test framework is constrained by the requirement that known physics must be recovered in certain limits, despite some residual freedom in introducing dynamical features compatible with a given set of kinematical rules. Moreover, it appears difficult and may even be impossible to develop an effective theory containing the Standard Model with dynamics significantly different from that of the SME. We also point out that kinematical analyses are limited to only a subset of potential Lorentz-violating signatures from fundamental physics. From this viewpoint, it seems to be desirable to explicitly implement dynamical features of sufficient generality into test models for Lorentz and CPT symmetry.

**The generality of the SME.** In order to understand the generality of the SME, we review the main cornerstones of its construction [14]. Starting from the usual Standard-Model lagrangian $\mathscr{L}_{SM}$, Lorentz-violating modifications $\delta\mathscr{L}$ are added:

$$\mathscr{L}_{SME} = \mathscr{L}_{SM} + \delta\mathscr{L} \,, \qquad (4)$$

where the SME lagrangian is denoted by $\mathscr{L}_{SME}$. The correction term $\delta\mathscr{L}$ is obtained by contracting Standard-Model field operators of any dimensionality with Lorentz-violating tensorial coefficients that describe the nontrivial vacuum discussed in the previous section. To guarantee coordinate independence, this contraction must give observer Lorentz scalars. It becomes thus apparent that all possible contributions to $\delta\mathscr{L}$ yield the most general effective dynamical description of Lorentz violation at the level of observer Lorentz-invariant quantum field theory. For simplicity, we have focused on nongravitational physics in the above construction. We mention that the complete SME also contains an extended gravity sector.

Possible Planck-scale features, such as non-pointlike elementary particles or a discrete spacetime, are unlikely to invalidate the above effective-field-theory approach at currently attainable energies. On the contrary, the phenomenologically successful Standard Model is widely believed to be an effective-field-theory approximation for underlying physics. If fundamental physics indeed leads to minute Lorentz-breaking effects, it would seem contrived to consider low-energy effective models outside the framework of effective quantum field theory. We finally note that the necessity for a low-energy description beyond effective field theory is also unlikely to arise in the context of candidate fundamental models with novel Lorentz-*invariant* aspects, such as additional particles, new symmetries, or large extra dimensions. Lorentz-symmetric modifications can therefore be implemented into the SME, if needed [19].

**Advantages of the SME.** The SME allows the identification and direct comparison of virtually all currently feasible experiments searching for Lorentz and CPT breaking. Moreover, certain limits of the SME correspond to classical kinematics test models of relativity (such as the previously mentioned Robertson's framework, its Mansouri-Sexl extension, or the $c^2$ model) [20]. Another advantage of the SME is the possibility of implementing further desirable conditions besides coordinate independence. For example, one can choose to impose spacetime-translation invariance, $SU(3) \times SU(2) \times U(1)$ gauge symmetry, power-counting renormalizability, hermiticity, and pointlike interactions. These requirements further restrict the parameter space for Lorentz violation. One could also adopt simplifying choices, such as a residual rotation symmetry in certain coordinate systems. This latter assumption together with additional simplifications of the

SME has been considered in Ref. [21].

**Analyses performed within the SME.** To date, the flat-spacetime limit of the minimal SME has provided the basis for numerous experimental and theoretical studies of Lorentz and CPT violation involving mesons [22, 23, 24, 25], baryons [26, 27, 28], electrons [29, 30, 31], photons [32, 20], muons [33], and the Higgs sector [34]. We remark that neutrino-oscillation experiments offer the potential for discovery [14, 35, 36]. A number of these studies involve some form of antimatter. CPT and Lorentz tests with antimatter will be discussed further in Sec. .

# SAMPLE MECHANISMS FOR LORENTZ AND CPT VIOLATION

In the previous two sections, we have studied various general *types* of manifestations of Lorentz and CPT breakdown, as well as the *description* of the corresponding effects in a microscopic model, such as the SME. However, the question of *how* exactly a Lorentz- and CPT-invariant candidate theory can lead to the violation of these symmetries has thus far been left unaddressed. The purpose of this section is to provide some intuition about such mechanisms for Lorentz and CPT breaking in underlying physics. Of the various possible mechanisms mentioned in Sec. , we will focus on spontaneous Lorentz violation and Lorentz breakdown through varying scalars.

**Spontaneous Lorentz and CPT violation.** The mechanism of spontaneous symmetry breaking is well established in various subfields of physics, such as the physics of elastic media, condensed-matter physics, and elementary particle theory. From a theoretical perspective, this mechanism is very attractive because the invariance is essentially violated through a non-trivial ground-state solution. The underlying dynamics of the system governed by the Hamiltonian remains completely invariant under the symmetry. To gain intuition about spontaneous Lorentz and CPT breakdown, we will consider three sample systems, whose features will gradually lead us to a better understanding of the effect. An illustration supporting these three examples is given in Fig. 3.

First, let us consider classical electrodynamics. Any electromagnetic-field configuration is associated with an energy density $V(\vec{E}, \vec{B})$ given by

$$V(\vec{E}, \vec{B}) = \frac{1}{2}\left(\vec{E}^2 + \vec{B}^2\right), \tag{5}$$

where we have employed natural units, and $\vec{E}$ and $\vec{B}$ denote the electric and magnetic field, respectively. With Eq. (5), we can determine the field energy of any given solution of the Maxwell equations. Note that if the electric field, or the magnetic field, or both are nonzero somewhere in spacetime, the energy stored in these fields will be strictly positive. The field energy can only be exactly zero when both $\vec{E}$ and $\vec{B}$ vanish everywhere. The vacuum (or ground state) is usually identified with the lowest-energy configuration of a system. We see that in conventional electrodynamics the configuration with the lowest energy is the field-free one, so that the Maxwell vacuum is empty (disregarding quantum fluctuations).

Second, let us look at the Higgs field, which is part of the phenomenologically very successful Standard Model of particle physics. Unlike the electromagnetic field, the

Higgs field is a scalar. In what follows, we can adopt some simplifications without distorting the features important in the present context. The expression for the energy density of our Higgs scalar $\phi$ in situations with spacetime independence is given by

$$V(\phi) = (\phi^2 - \lambda^2)^2, \tag{6}$$

where $\lambda$ is a constant. As in the Maxwell case discussed above, the lowest possible field energy is zero. Note, however, that this configuration *requires* $\phi$ to be nonzero: $\phi = \pm\lambda$. It follows that the vacuum for a system containing a Higgs-type field is not empty; it is, in fact, filled with constant scalar field $\phi_{vac} \equiv \langle\phi\rangle = \pm\lambda$. In quantum theory, the quantity $\langle\phi\rangle$ is called the vacuum expectation value (VEV) of $\phi$. One of the physical effects caused by the VEV of the Higgs is to give masses to many elementary particles. Note, however, that $\langle\phi\rangle$ is a scalar and does *not* select a preferred direction in spacetime.

Third, we consider a vector field $\vec{C}$ (the relativistic generalization is straightforward) not contained in the Standard-Model. Of course, there is no observational evidence for such a field at the present time, but fields like $\vec{C}$ frequently arise in approaches to more fundamental physics. In analogy to Higgs case, we take its expression for energy density in situations with constant $\vec{C}$ to be

$$V(\vec{C}) = (\vec{C}^2 - \lambda^2)^2. \tag{7}$$

As in the previous two examples, the lowest possible energy is exactly zero. As for the Higgs, this lowest energy configuration is attained for nonzero $\vec{C}$. More specifically, we must have $\vec{C}_{vac} \equiv \langle\vec{C}\rangle = \vec{\lambda}$, where $\vec{\lambda}$ is any constant vector satisfying $\vec{\lambda}^2 = \lambda^2$. Again, the vacuum is not empty, but filled with the VEV of our vector field. Since we have only considered constant solutions $\vec{C}$, $\langle\vec{C}\rangle$ is also spacetime independent ($x$ dependence would lead to positive definite derivative terms in Eq. (7) raising the energy density). The true vacuum in our model therefore contains an intrinsic direction determined by $\langle\vec{C}\rangle$ *violating rotation invariance and thus Lorentz symmetry.* We mention that interactions leading to energy densities like (7) are absent in conventional renormalizable gauge theories, but can be found in the context of strings, for example.

**Cosmologically varying scalars.** A spacetime-dependent scalar, regardless of the mechanism driving the variation, typically implies the breaking of spacetime-translation invariance. Since translations and Lorentz transformations are closely linked in the Poincaré group, it is reasonable to expect that the translation-symmetry violation also affects Lorentz invariance.

Consider, for instance, the angular-momentum tensor $J^{\mu\nu}$, which is the generator of Lorentz transformations:

$$J^{\mu\nu} = \int d^3x \left(\theta^{0\mu}x^\nu - \theta^{0\nu}x^\mu\right). \tag{8}$$

Note that this definition contains the energy–momentum tensor $\theta^{\mu\nu}$, which is not conserved when translation invariance is broken. In general, $J^{\mu\nu}$ will possess a nontrivial dependence on time, so that the usual time-independent Lorentz-transformation generators do not exist. As a result, Lorentz and CPT symmetry are no longer assured.

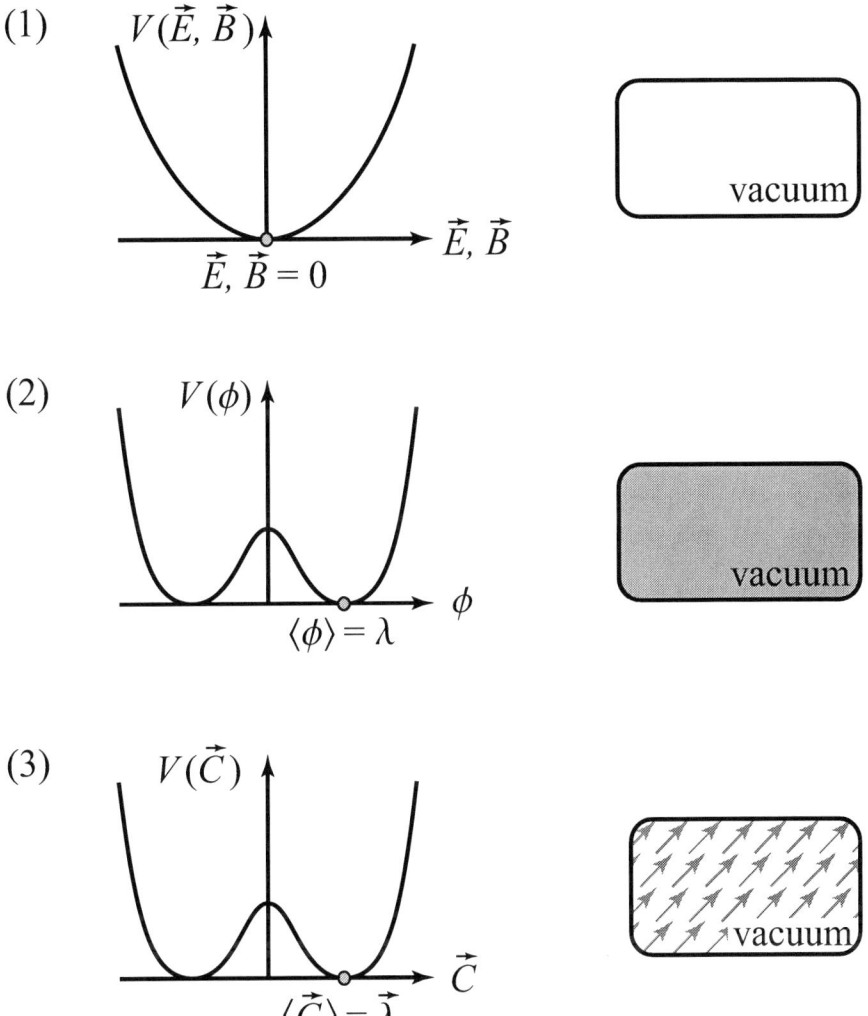

**FIGURE 3.** Spontaneous symmetry breaking. In conventional electrodynamics (1), the lowest-energy state is attained for zero $\vec{E}$ and $\vec{B}$ fields. The vacuum remains essentially empty. For the Higgs-type field (2), interactions lead to an energy density $V(\phi)$ that forces a nonzero value of $\phi$ in the ground state. The vacuum fills with a scalar condensate shown in gray. Lorentz invariance still holds (other, internal symmetries may be violated though). Vector fields occurring, e.g., in string theory (3) can have interactions similar to those of the Higgs requiring a nonzero field value in the lowest-energy state. The VEV of a vector field selects a preferred direction in the vacuum, which has now properties paralleling those of a crystal.

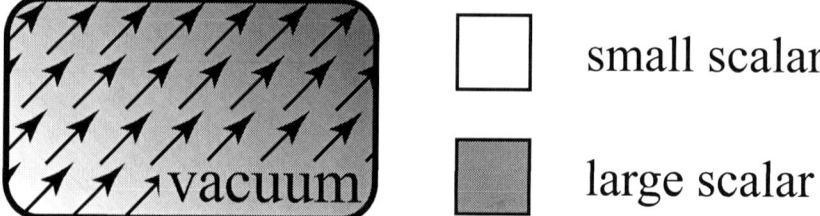

**FIGURE 4.** Lorentz breakdown through varying scalars. The value of the scalar corresponds to the background shade of gray: the darker regions are associated with greater values of the scalar. The gradient represented by the black arrows selects a preferred direction in the vacuum. Although Lorentz coordinate independence is maintained, particle Lorentz invariance is violated.

Intuitively, the violation of Lorentz invariance in the presence of a varying scalar can be understood as follows. The 4-gradient of the scalar must be nonzero in some regions of spacetime. Such a gradient then selects a preferred direction in this region (see Fig. 4). Consider, for example, a particle that interacts with the scalar. Its propagation features might be different in the directions parallel and perpendicular to the gradient. Physically inequivalent directions imply the violation of rotation symmetry. Since rotations are contained in the Lorentz group, Lorentz invariance must be violated.

Lorentz violation induced by varying scalars can also be established at the Lagrangian level. Consider, for instance, a system with a varying coupling $\xi(x)$ and scalar fields $\phi$ and $\Phi$, such that the Lagrangian $\mathscr{L}$ contains a term $\xi(x)\,\partial^\mu \phi\,\partial_\mu \Phi$. The action for this system can be integrated by parts (e.g., with respect to the first partial derivative in the above term) without affecting the equations of motion. An equivalent Lagrangian $\mathscr{L}'$ would then obey

$$\mathscr{L}' \supset -K^\mu \phi\, \partial_\mu \Phi, \tag{9}$$

where $K^\mu \equiv \partial^\mu \xi$ is an external nondynamical 4-vector, which clearly violates Lorentz symmetry. We remark that for variations of $\xi$ on cosmological scales, $K^\mu$ is constant to an excellent approximation locally—say on solar-system scales.

## LORENTZ AND CPT TESTS INVOLVING ANTIMATTER

Numerous Lorentz and CPT tests among those listed in Sec. involve some form of anti-matter. As pointed out earlier, certain matter–antimatter comparisons are extremely sensitive to CPT violations. This is unsurprising because CPT symmetry connects particles and antiparticles. CPT tests with subatomic particles typically involve quantum numbers like mass, charge, spin, etc. Atoms and their anti-atoms possess additional, qualitatively different properties, such as spectra, that can be compared. The possibility of $\bar{\text{H}}$ production combined with the ultrahigh sensitivities attainable in atomic spectroscopy and the simplicity of the two-body problem (antiproton nucleus and orbiting positron) make this anti-atom particularly well suited for such investigations. The determination of SME predictions for such physical systems follows the outline shown in Fig. 5.

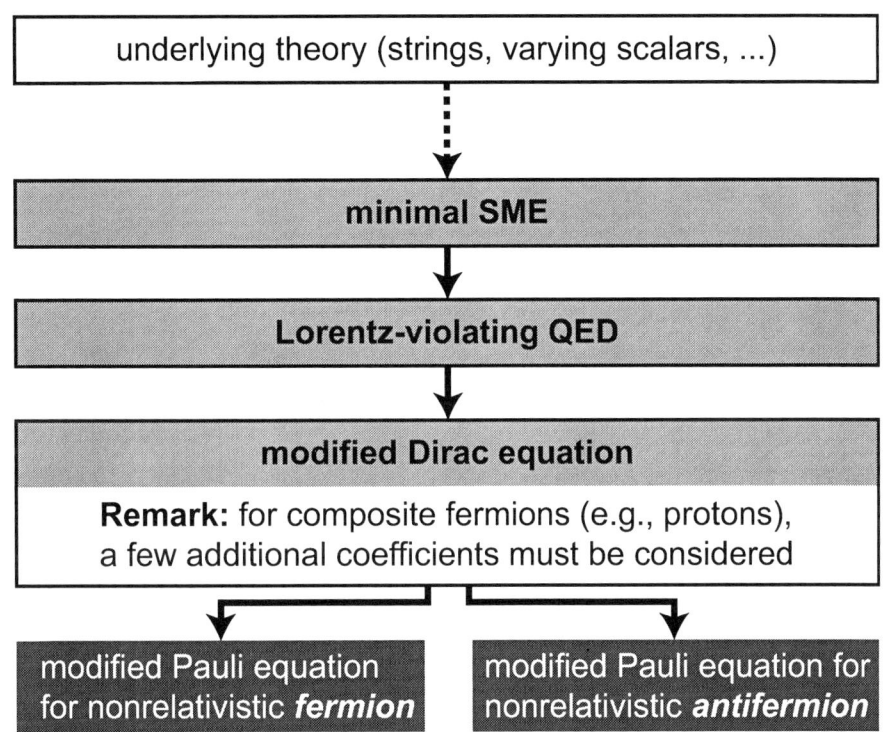

**FIGURE 5.** SME analysis of atomic spectra. The resulting modified Pauli equation for fermions and that for antifermions—which typically differ—are employed to describe the (anti)proton and the orbiting (anti)electron in the H ($\overline{\text{H}}$) system.

**The unmixed 1S–2S transition.** The experimental resolution of the transition involving the unmixed spin states is expected to be about one part in $10^{-18}$. This sensitivity appears promising in light of potential Planck-suppressed quantum-gravity effects. On the other hand, the leading-order SME calculation shows identical shifts for free H or $\overline{\text{H}}$ in the initial and final states with respect to the conventional levels. It follows that this transition is less suitable for the measurement of unsuppressed Lorentz- and CPT-breaking effects. The largest non-trivial contribution to this transition within the SME test framework arises through relativistic corrections, and it involves two additional powers of the fine-structure parameter $\alpha = \frac{1}{137}$. The expected energy shift—already at zeroth order in $\alpha$ anticipated to be minute—comes therefore with an additional suppression by a factor of more than ten thousand. This further exemplifies the need and importance of a viable test model for Lorentz- and CPT-violation searches.

**The spin-mixed 1S–2S transitions.** When H or $\overline{\text{H}}$ is confined with magnetic fields—such as in a Ioffe–Pritchard trap—the 1S and the 2S levels are each split by the Zeeman effect. In the framework of the SME, one can show that in this situation the 1S–2S transition between the spin-mixed states is affected by Lorentz and CPT violation at leading order. A disadvantage from a practical perspective is the field dependence of this

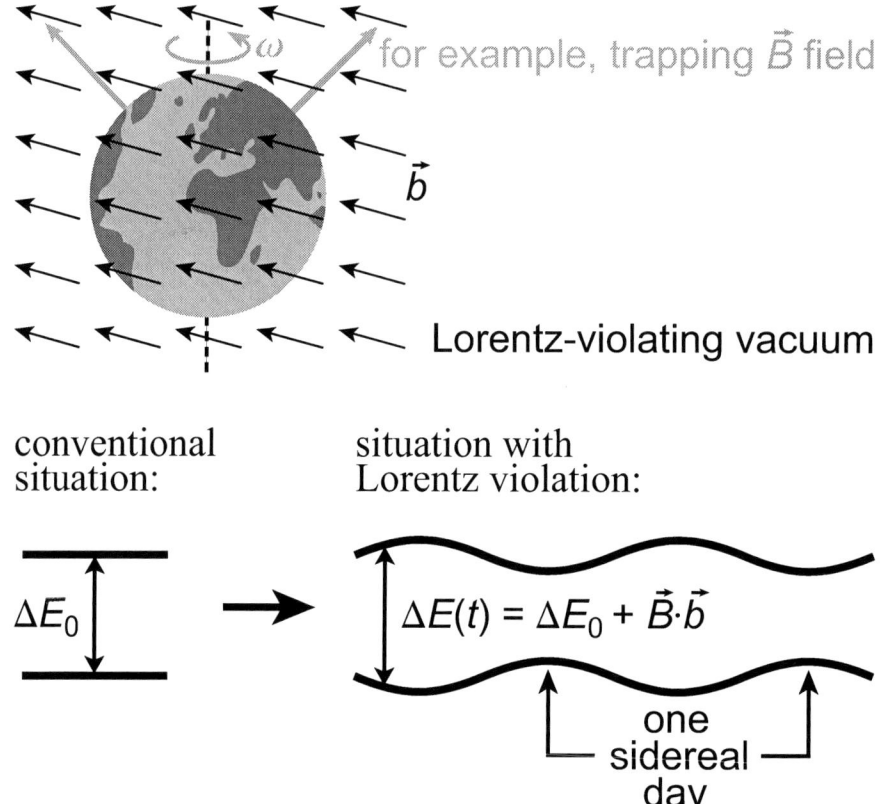

**FIGURE 6.** Sidereal variations. Experiments are typically associated with an intrinsic direction. For instance, particle traps usually involve a magnetic field. As the Earth rotates, this direction will change if the experiment is attached to the Earth. In the above figure, a trapping field $\vec{B}$ pointing vertically upward is shown at two times separated by approximately 12 hours (gray arrows). The angle between the Lorentz-violating background (black $\vec{b}$ arrows) and the orientation of the experiment is clearly different at these two times. An observable, such as an atomic transition, may for example acquire a correction $\sim \vec{B} \cdot \vec{b}$ that leads to the shown sidereal modulation.

transition, so that the experimental resolution is limited by the size of the inhomogeneity in the trapping magnetic field. The development of novel experimental techniques would appear necessary to achieve resolutions close to the natural linewidth.

**Hyperfine Zeeman transitions within the 1S state.** An alternative Lorentz and CPT test could measure the transition frequency between the Zeeman-split states within the 1S level itself. Even in the zero-magnetic-field limit, the SME predicts first-order effects for two of these transitions. Other transitions of this type, such as the conventional Hydrogen-maser line, can be well resolved in experiments.

**Tests in Penning traps.** The SME predicts that not only atomic energy levels can be shifted in the presence of Lorentz violation, but also proton and antiproton levels in

Penning traps. A calculation shows that only one SME coefficient ($b^\mu$ in the standard notation) leads to transition-frequency differences between the proton and its antiparticle. More specifically, the anomaly frequencies are changed in opposite directions for protons and antiprotons. This effect permits a clean observational bound on $b^\mu$ for the proton.

**Searches for sidereal variations.** Another general signature for Lorentz and CPT breakdown is the variation of measured quantities with the sidereal day. The anti-CPT theorem implies that CPT breakdown always comes with Lorentz violation, which in turn is typically accompanied by the loss of isotropy. Thus, experimental effects will generally depend on the direction. As the laboratory is attached to the rotating Earth, its orientation will change continually leading to sidereal modulations of signals. This situation is schematically depicted in Fig. 6. Note that sidereal-variation tests are not confined exclusively to H–H̄ spectroscopy, but they can also be performed in the context of other rotation-violation searches. Recent experiments with H-masers employing ingenious experimental techniques are based on such modulations [37].

# SUMMARY

Although Lorentz and CPT symmetry are deeply ingrained in the currently accepted laws of physics, there are a variety of candidate underlying theories that could generate the breakdown of these symmetries. The sensitivity attainable in matter–antimatter comparisons offers the possibility for CPT-violation searches with Planck precision. Lorentz tests open an additional avenue for CPT measurements because CPT breakdown implies Lorentz violation.

A potential source of Lorentz and CPT violation is spontaneous symmetry breaking in string theory. Since this mechanism is theoretically very attractive and since strings show great potential as a candidate fundamental theory, this Lorentz-violation source is particularly promising. Lorentz and CPT violation can also originate from spacetime-dependent couplings: the gradient of such couplings selects a preferred direction in the effective vacuum. This mechanism for Lorentz breaking might be of interest in light of recent claims of a time-dependent fine-structure parameter and the presence of varying scalar fields in many cosmological models.

The leading-order Lorentz- and CPT-violating effects resulting from Lorentz-symmetry breakdown in approaches to fundamental physics are described by the SME. At the level of effective quantum field theory, the SME is the most general dynamical framework for Lorentz and CPT breaking that is compatible with the fundamental principle of coordinate independence. Experimental investigations are therefore best performed within the SME.

Cold antiprotons are excellent high-sensitivity tools in experimental searches for Planck-scale physics. Suppressed and unsuppressed effects exist for 1S–2S transitions in H and H̄. Leading-order shifts are also predicted in the 1S hyperfine Zeeman levels, which offers the possibility of alternative measurements. Further possibilities for Lorentz- and CPT-violation searches with antiprotons exist in Penning traps, where anomaly frequencies are affected. In general, tests with cold antiprotons probe parameter combinations inaccessible by other experiments.

# ACKNOWLEDGMENTS

The author would like to thank Yasunori Yamazaki for organizing this stimulating meeting, for the invitation to attend, and for partial financial support.

# REFERENCES

1. See, e.g., *The Hydrogen Atom*, edited by G.F. Bassani, M. Inguscio, and T.W. Hänsch, Springer-Verlag, Berlin, 1989.
2. For a nontechnical overview, see, e.g., V.A. Kostelecký, Scientific American **291**, 92 (September 2004).
3. G. Baur *et al.*, Phys. Lett. B **368**, 251 (1996); G. Blanford *et al.*, Phys. Rev. Lett. **80**, 3037 (1998).
4. M. Amoretti *et al.*, Nature **419**, 456 (2002); G. Gabriels *et al.*, Phys. Rev. Lett. **89**, 213401 (2002).
5. E. Widmann *et al.*, Phys. Rev. Lett. **89**, 243402 (2002).
6. See, e.g., R.G. Sachs, *The Physics of Time Reversal*, University of Chicago Press, Chicago, 1987.
7. V.A. Kostelecký and S. Samuel, Phys. Rev. D **39**, 683 (1989); Phys. Rev. Lett. **63**, 224 (1989); **66**, 1811 (1991); V.A. Kostelecký and R. Potting, Nucl. Phys. B **359**, 545 (1991); Phys. Lett. B **381**, 89 (1996); Phys. Rev. D **63**, 046007 (2001); V.A. Kostelecký *et al.*, Phys. Rev. Lett. **84**, 4541 (2000).
8. G. Amelino-Camelia *et al.*, Nature (London) **393**, 763 (1998); D. Sudarsky *et al.*, Phys. Rev. D **68**, 024010 (2003).
9. F.R. Klinkhamer, Nucl. Phys. B **578**, 277 (2000).
10. V.A. Kostelecký *et al.*, Phys. Rev. D **68**, 123511 (2003); O. Bertolami *et al.*, Phys. Rev. D **69**, 083513 (2004).
11. S. Eidelmann *et al.* [Particle Data Group], Phys. Lett. B **592**, 1 (2004).
12. O.W. Greenberg, Phys. Rev. Lett. **89**, 231602 (2002).
13. For a somewhat more pedestrian exposition, see O.W. Greenberg, hep-ph/0309309.
14. D. Colladay and V.A. Kostelecký, Phys. Rev. D **55**, 6760 (1997); **58**, 116002 (1998); V.A. Kostelecký and R. Lehnert, Phys. Rev. D **63**, 065008 (2001); V.A. Kostelecký, Phys. Rev. D **69**, 105009 (2004); R. Bluhm and V.A. Kostelecký, Phys. Rev. D **71**, 065008 (2005).
15. R. Lehnert, Phys. Rev. D **68**, 085003 (2003).
16. G. Barenboim *et al.*, JHEP **0210**, 001 (2002).
17. G. Barenboim and J. Lykken, Phys. Lett. B **554**, 73 (2003).
18. O.W. Greenberg, Phys. Lett. B **567**, 179 (2003).
19. M.S. Berger and V.A. Kostelecký, Phys. Rev. D **65**, 091701(R) (2002); H. Belich *et al.*, Phys. Rev. D **68**, 065030 (2003); M.S. Berger, Phys. Rev. D **68**, 115005 (2003).
20. V.A. Kostelecký and M. Mewes, Phys. Rev. D **66**, 056005 (2002).
21. S. Coleman and S.L. Glashow, Phys. Rev. D **59**, 116008 (1999).
22. OPAL Collaboration, R. Ackerstaff *et al.*, Z. Phys. C **76**, 401 (1997); DELPHI Collaboration, M. Feindt *et al.*, preprint DELPHI 97-98 CONF 80 (1997); BELLE Collaboration, K. Abe *et al.*, Phys. Rev. Lett. **86**, 3228 (2001); BaBar Collaboration, B. Aubert *et al.*, hep-ex/0303043. FOCUS Collaboration, J.M. Link *et al.*, Phys. Lett. B **556**, 7 (2003).
23. V.A. Kostelecký and R. Potting, Phys. Rev. D **51**, 3923 (1995).
24. D. Colladay and V.A. Kostelecký, Phys. Lett. B **344**, 259 (1995); Phys. Rev. D **52**, 6224 (1995); V.A. Kostelecký and R. Van Kooten, Phys. Rev. D **54**, 5585 (1996); O. Bertolami *et al.*, Phys. Lett. B **395**, 178 (1997); N. Isgur *et al.*, Phys. Lett. B **515**, 333 (2001).
25. V.A. Kostelecký, Phys. Rev. Lett. **80**, 1818 (1998); Phys. Rev. D **61**, 016002 (2000); Phys. Rev. D **64**, 076001 (2001).
26. D. Bear *et al.*, Phys. Rev. Lett. **85**, 5038 (2000); D.F. Phillips *et al.*, Phys. Rev. D **63**, 111101 (2001); Phys. Rev. A **62**, 063405 (2000); V.A. Kostelecký and C.D. Lane, Phys. Rev. D **60**, 116010 (1999); J. Math. Phys. **40**, 6245 (1999).
27. R. Bluhm *et al.*, Phys. Rev. Lett. **88**, 090801 (2002).
28. F. Canè *et al.*, Phys. Rev. Lett. **93**, 230801 (2004).
29. H. Dehmelt *et al.*, Phys. Rev. Lett. **83**, 4694 (1999); R. Mittleman *et al.*, Phys. Rev. Lett. **83**, 2116 (1999); G. Gabrielse *et al.*, Phys. Rev. Lett. **82**, 3198 (1999); R. Bluhm *et al.*, Phys. Rev. Lett. **82**,

2254 (1999); Phys. Rev. Lett. **79**, 1432 (1997); Phys. Rev. D **57**, 3932 (1998); C.D. Lane, hep-ph/0505130.

30. L.-S. Hou *et al.*, Phys. Rev. Lett. **90**, 201101 (2003); R. Bluhm and V.A. Kostelecký, Phys. Rev. Lett. **84**, 1381 (2000).
31. H. Müller *et al.*, Phys. Rev. D **68**, 116006 (2003); R. Lehnert, J. Math. Phys. **45**, 3399 (2004).
32. S.M. Carroll *et al.*, Phys. Rev. D **41**, 1231 (1990); V.A. Kostelecký and M. Mewes, Phys. Rev. Lett. **87**, 251304 (2001); J. Lipa *et al.*, Phys. Rev. Lett. **90**, 060403 (2003); Q. Bailey and V.A. Kostelecký, Phys. Rev. D **70**, 076006 (2004); R. Lehnert and R. Potting, Phys. Rev. Lett. **93**, 110402 (2004); Phys. Rev. D **70**, 125010 (2004).
33. V.W. Hughes *et al.*, Phys. Rev. Lett. **87**, 111804 (2001); R. Bluhm *et al.*, Phys. Rev. Lett. **84**, 1098 (2000); E.O. Iltan, JHEP **0306**, 016 (2003).
34. D.L. Anderson *et al.*, Phys. Rev. D **70**, 016001 (2004).
35. S. Coleman and S.L. Glashow, Phys. Rev. D **59**, 116008 (1999); V. Barger *et al.*, Phys. Rev. Lett. **85**, 5055 (2000); J.N. Bahcall *et al.*, Phys. Lett. B **534**, 114 (2002); V.A. Kostelecký and M. Mewes, Phys. Rev. D **70**, 031902 (2004).
36. V.A. Kostelecký and M. Mewes, Phys. Rev. D **69**, 016005 (2004).
37. M.A. Humphrey *et al.*, Phys. Rev. A **68**, 063807 (2003).

# The Deepest Symmetries of Nature: CPT and SUSY [1]

## Dezső Horváth[†]

*KFKI Research Institute for Particle and Nuclear Physics, H–1525 Budapest, Hungary*
*Institute of Nuclear Research (ATOMKI), Debrecen, Hungary*

**Abstract.** The structure of matter is related to symmetries on every level of study. CPT symmetry is one of the most important laws of field theory: it states the invariance of physical properties when one simultaneously changes the signs of the charge and of the spatial and time coordinates of particles. Although in general opinion CPT symmetry is not violated in Nature, there are theoretical attempts to develop CPT-violating models. The Antiproton Decelerator at CERN has been built to test CPT invariance.

Several observations imply that there might be another deep symmetry, supersymmetry (SUSY), between basic fermions and bosons. SUSY assumes that every fermion and boson observed so far has supersymmetric partners of the opposite nature. In addition to some theoretical problems of the Standard Model of elementary particles, supersymmetry may provide solution to the constituents of the mysterious dark matter of the Universe. However, as opposed to CPT, SUSY is necessarily violated at low energies as so far none of the predicted supersymmetric partners of existing particles was observed experimentally. The LHC experiments at CERN aim to search for these particles.

**Keywords:** CPT invariance, supersymmetry, Higgs mechanism, symmetry breaking

## SYMMETRIES IN PARTICLE PHYSICS

Symmetries in particle physics are even more important than in chemistry or solid state physics. Just like in any theory of matter, the inner structure of the composite particles are described by symmetries, but in particle physics everything is deduced from the symmetries (or invariance properties) of the physical phenomena or from their violation: the conservation laws, the interactions and even the masses of the particles.

The conservation laws are related to symmetries: the Noether theorem states that a global symmetry leads to a conserving quantity. The conservation of momentum and energy are deduced from the translational invariance of space-time: the physical laws do not depend upon where we place the zero point of our coordinate system or time measurement; and the fact that we are free to rotate the coordinate axes at any angle is the origin of angular momentum conservation.

The symmetry properties of particles with half–integer spin (fermions) differ from those with integer spin (bosons). The wave function describing a system of fermions changes sign when two fermions switch quantum states whereas in the case of bosons there is no change; all other differences can be deduced from this property.

[1] Invited paper presented at *Workshop on Physics with Ultra Slow Antiproton Beams, RIKEN, Wako, Japan, 14-16 March 2005*

**TABLE 1.** The elementary fermions of the Standard Model. L stands for left: it symbolizes that in the weak isospin doublets left-polarized particles and right-polarized antiparticles appear, their counterparts constitute iso-singlet states. The apostrophes of down-type quarks denote their mixed states for the weak interaction.

| | fermion doublets $(S = 1/2)$ | charge $Q$ | isospin $I$ |
|---|---|---|---|
| leptons | $\begin{pmatrix} \nu_e \\ e \end{pmatrix}_L \begin{pmatrix} \nu_\mu \\ \mu \end{pmatrix}_L \begin{pmatrix} \nu_\tau \\ \tau \end{pmatrix}_L$ | $0$<br>$-1$ | $+1/2$<br>$-1/2$ |
| quarks | $\begin{pmatrix} u \\ d' \end{pmatrix}_L \begin{pmatrix} c \\ s' \end{pmatrix}_L \begin{pmatrix} t \\ b' \end{pmatrix}_L$ | $+2/3$<br>$-1/3$ | $+1/2$<br>$-1/2$ |

Being an angular momentum the spin is associated with the symmetry of the rotation group and it can be described by the $SU(2)$ group of the Special (their *determinant* is 1) Unitary $2 \times 2$ matrices. When we increase the degrees of freedom, we get higher symmetry groups of similar properties. The next one, $SU(3)$, which we shall use later, is the symmetry group of Special Unitary $3 \times 3$ matrices. It can be visualized the following way. A half–spin particle has two possible fundamental states (two *eigenstates*), spin up and spin down. In the case of the $SU(3)$ symmetry group there are three eigenstates with an $SU(2)$ symmetry between any two of them.

We can also *decrease* the degrees of freedom and we get the $U(1)$ group of unitary $1 \times 1$ matrices which are simply complex numbers of unit absolute value. This is the symmetry group of the *gauge transformations* of electromagnetism. This *gauge symmetry* means, e.g., in the case of electricity a free choice of potential zero: as shown by the sparrows sitting on electric wires the *potential difference* is the meaningful physical quantity, not the potential itself. The $U(1)$ symmetry of Maxwell's equations leads to the conservation of the electric charge, and, in the more general case, the $U(1)$ symmetry of the Dirac equation, the general equation describing the movement of a fermion, causes the conservation of the number of fermions [1].

The important role of symmetries in particle physics is well expressed by the title of the popular scientific journal of SLAC and FERMILAB: *symmetry — dimensions of particle physics*.

## SYMMETRIES IN THE STANDARD MODEL

According to the Standard Model of elementary particles the visible matter of our world consists of a few point-like elementary particles: fermions, quarks and leptons, and bosons (see Table 1).

The hadrons, the mesons and the baryons, are composed of quarks, the mesons are quark-antiquark, the baryons three-quark states.

The three basic interactions are deduced from local gauge symmetries. By requiring that the Dirac equation of a free fermion were invariant under local (i.e. space-time

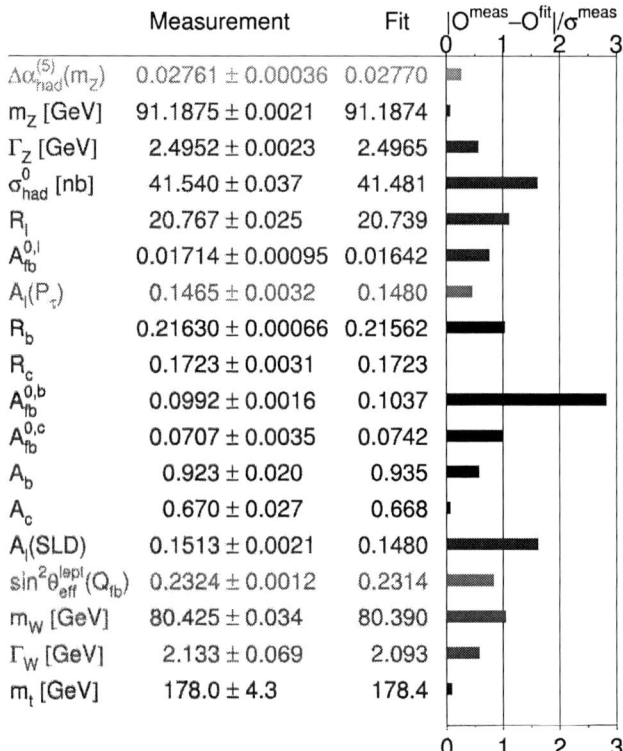

| | Measurement | Fit | $\|O^{meas}-O^{fit}\|/\sigma^{meas}$ 0 1 2 3 |
|---|---|---|---|
| $\Delta\alpha_{had}^{(5)}(m_Z)$ | $0.02761 \pm 0.00036$ | 0.02770 | |
| $m_Z$ [GeV] | $91.1875 \pm 0.0021$ | 91.1874 | |
| $\Gamma_Z$ [GeV] | $2.4952 \pm 0.0023$ | 2.4965 | |
| $\sigma_{had}^0$ [nb] | $41.540 \pm 0.037$ | 41.481 | |
| $R_l$ | $20.767 \pm 0.025$ | 20.739 | |
| $A_{fb}^{0,l}$ | $0.01714 \pm 0.00095$ | 0.01642 | |
| $A_l(P_\tau)$ | $0.1465 \pm 0.0032$ | 0.1480 | |
| $R_b$ | $0.21630 \pm 0.00066$ | 0.21562 | |
| $R_c$ | $0.1723 \pm 0.0031$ | 0.1723 | |
| $A_{fb}^{0,b}$ | $0.0992 \pm 0.0016$ | 0.1037 | |
| $A_{fb}^{0,c}$ | $0.0707 \pm 0.0035$ | 0.0742 | |
| $A_b$ | $0.923 \pm 0.020$ | 0.935 | |
| $A_c$ | $0.670 \pm 0.027$ | 0.668 | |
| $A_l(SLD)$ | $0.1513 \pm 0.0021$ | 0.1480 | |
| $\sin^2\theta_{eff}^{lept}(Q_{fb})$ | $0.2324 \pm 0.0012$ | 0.2314 | |
| $m_W$ [GeV] | $80.425 \pm 0.034$ | 80.390 | |
| $\Gamma_W$ [GeV] | $2.133 \pm 0.069$ | 2.093 | |
| $m_t$ [GeV] | $178.0 \pm 4.3$ | 178.4 | |

**FIGURE 1.** The glory road of SM at LEP: the relative deviation of the measured quantities from the predictions of the Standard Model [3] (status of Winter 2005). At present the most deviating quantity is the forward-backward asymmetry at the decay of Z bosons to b hadrons. For the definitions see [4].

dependent) $U(1) \otimes SU(2)$ transformations one gets the Lagrange function of the electroweak interaction with massless mediating gauge bosons (photon γ and weak bosons Z and W±). Adding local $SU(3)$ results in the strong interaction (quantum chromodynamics) with 8 massless gluons as mediating particles. And, finally, adding a two-component complex Higgs-field with its 4 degrees of freedom, which breaks the $SU(2)$ symmetry will put everything in place: produces masses for the weak bosons (and for the fermions as well) and creates the Higgs boson [5], the scalar particle badly needed to make the theory renormalizable (to remove divergences).

The Standard Model is an incredible success: its predictions are not contradicted by experiment, any deviation encountered in the last 30 years disappeared with the increasing precision of theory and experiment. For a complete comparison one should consult the tables and reviews of the Particle Data Group [4], Fig 1 presents a brief view. The only missing piece is the Higgs boson; however, it is a strong indirect evidence for its existence that the goodness-of-fitting of the electroweak parameters shows a

56

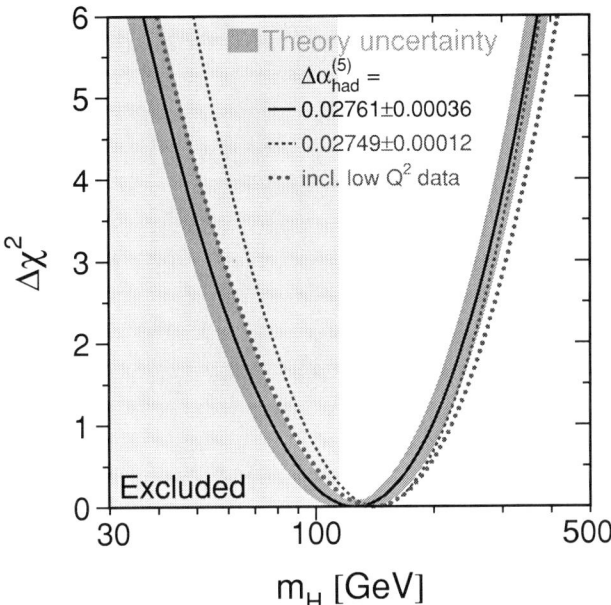

**FIGURE 2.** The goodness-of-fitting of the electroweak parameters as a function of the mass of the Standard Model Higgs boson [3] (status of Winter 2005). The best fit indicates a light Higgs boson and the LEP searches excluded Higgs masses up to 114.4 GeV [6].

definite minimum at light Higgs masses (Fig. 2). The direct searches at LEP excluded the Standard Model Higgs boson up to masses of 114.4 GeV (with a confidence limit of 95%), whereas the fitting seems to limit it from above as well. Thus within the framework of the Standard Model the mass of the Higgs boson should be in the interval $114.4 < M_H < 260$ GeV (with a 95 % confidence).

## ANTIPARTICLES AND *CPT* INVARIANCE

All fermions have *antiparticles*, anti-fermions which have identical properties but with opposite charges. The different abundance of particles and antiparticles in our Universe is one of the mysteries of astrophysics: apparently there is no antimatter in the Universe in significant quantities, see, e.g., [2]. If there were antimatter galaxies they would radiate antiparticles and we would see zones of strong radiation at their borders with matter galaxies, but the astronomers do not see such a phenomenon anywhere.

An extremely interesting property of *antiparticles* is that they can be treated mathematically as if they were *particles* of the same mass and of oppositely signed charge of the same absolute value *going backward in space and time*. This is the consequence of one of the most important symmetries of Nature: $CPT$ invariance [1, 4]. It states that the following operations:

- charge conjugation (i.e. changing particles into antiparticles), $C\psi(r,t) = \overline{\psi}(r,t)$;
- parity change (i.e. mirror reflection), $P\psi(r,t) = \psi(-r,t)$, and
- time reversal, $T\psi(r,t) = \psi(r,-t)$

when done together do not change the physical properties (i.e., the wave function or in the language of field theory the *field function* $\psi(r,t)$) of the system:

$$CPT\psi(r,t) = \overline{\psi}(-r,-t) \sim \psi(r,t). \tag{1}$$

This means that, e.g., the annihilation of a positron with an electron can be described as if an electron came to the point of collision, irradiated two or three photons and then went out backwards in space-time.

If we build a clock looking at its design in a mirror, it should work properly except that its hands will rotate the opposite way and the lettering will be inverted. The laws governing the work of the clock are invariant under space inversion, i.e. conserve parity. As we know, the weak interaction violates parity conservation, unlike the other interactions. The weak forces violate the conservation of $CP$ as well. $CPT$ invariance, however, is still assumed to be absolute. Returning to the example of the clock, a $P$ reflection means switching left to right, a $C$ transformation means changing the matter of the clock to antimatter, and the time reversal $T$ means that we play the video recording of the movement of the clock backward.

## TESTING *CPT*

$CPT$ invariance is so deeply embedded in field theory that many theorists claim it is impossible to test within the framework of present-day physics. Indeed, in order to develop $CPT$-violating models one has to reject quite fundamental axioms as Lorentz invariance or the locality of interactions [7, 8, 9, 10].

As far as we know, the Standard Model is valid up to the Planck scale, $\sim 10^{19}$ GeV. Above this energy scale we expect to have new physical laws which may allow for Lorentz and $CPT$ violation as well [7]. Quantum gravity [8, 9] could cause fluctuations leading to Lorentz violation, or loss of information in black holes which would mean unitarity violation. Also, a quantitative expression of Lorentz and $CPT$ invariance needs a Lorentz and $CPT$ violating theory [7]. On the other hand, testing $CPT$ invariance at low energy should be able to limit possible high energy violation.

$CPT$ invariance is so far fully supported by the available experimental evidence and it is absolutely fundamental in field theory. Nevertheless, there are many experiments trying to test it. The simplest way to do that is to compare the mass or charge of particles and antiparticles. The most precise such measurement is that of the relative mass difference of the neutral K meson and its antiparticle which has so far been found to be less than $10^{-18}$ [4]. CERN has constructed its *Antiproton Decelerator* facility [12] in 1999 in order to test the $CPT$ invariance by comparing the properties of proton

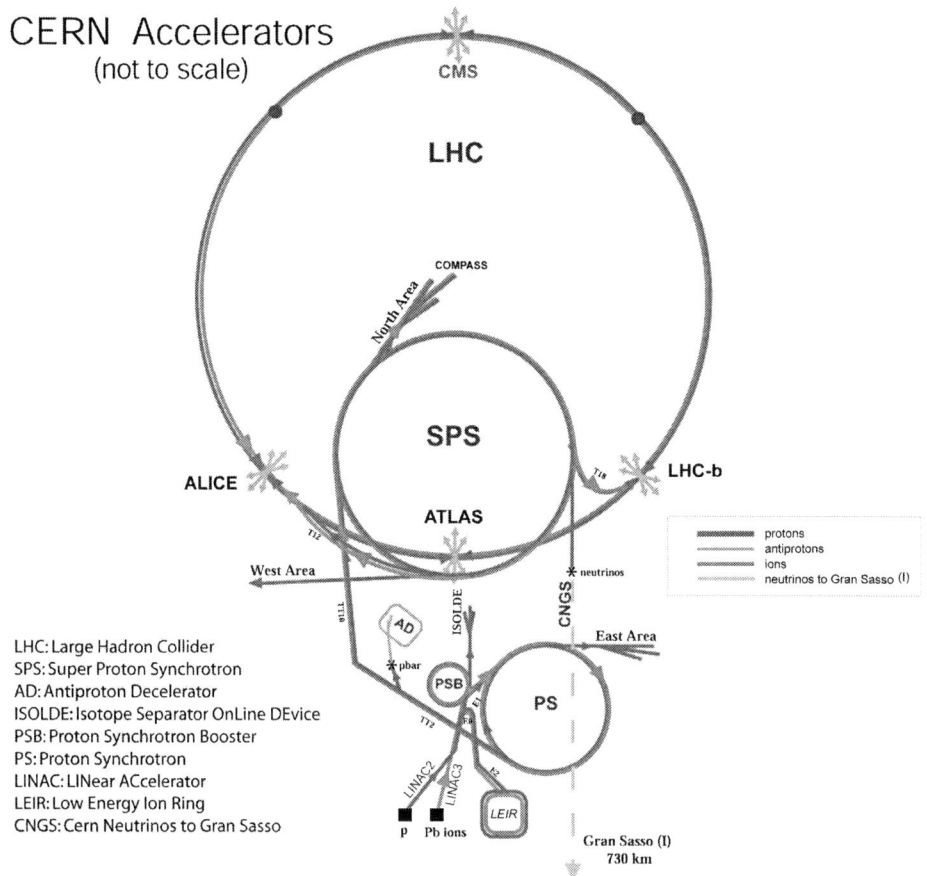

**FIGURE 3.** The accelerator complex of CERN. The LINAC2 linear accelerator and the PSB booster feed protons into the PS proton synchrotron, which accelerates them to 25 GeV/$c$ and passes them to the experiments in the East Area or to the SPS super proton synchrotron for further acceleration and once every 100 seconds into an iridium target to produce antiprotons. The antiprotons are collected at 3.5 GeV/$c$ by the AD where they are decelerated in three steps to 100 MeV/$c$. The PS also accelerates heavy ions for the SPS North Area experiments and until 2000 it did accelerate electrons and positrons for the LEP Large Electron Positron collider which was dismounted to be replaced by the LHC Large Hadron Collider in 2007.

and antiproton and those of hydrogen and antihydrogen (Fig. 3). The results of the AD experiments, ATHENA, ATRAP and ASACUSA are well summarized by their speakers at this conference.

Note that antiproton gravity is not of this category. The *CPT* theorem only says that an apple should fall towards Earth the same way as an anti-apple to anti-Earth, it is the weak equivalence principle which should make an anti-apple fall to Earth the same way.

# LOST SYMMETRIES?

I should like to start this section with a quotation of the great paper of Frank Wilczek [11] when speaking about the spontaneously broken gauge symmetries: *"According to this concept, the fundamental equations of physics have more symmetry than the actual physical world does"*. We believe the *CPT symmetry* being fundamental and absolute, with no violation (at least below the Planck scale). The $SU(3)$ global gauge invariance has no violation and conserves the color charge; as a local gauge invariance it gives rise to the strong (color) interaction.

The $U(1) \times SU(2)$ gauge invariance is, however, spontaneously broken by the Higgs field, and that breaking is needed to give rise to the electroweak interaction, to give masses to the weak bosons (and generally to all particles) and to produce the Higgs boson which helps to regularize the theory. Thus one can be unhappy with the Higgs mechanism that it breaks a nice symmetry of the Dirac equation, but it is needed to make the Standard Model work.

In spite of its great success in interpreting all the experimental data there are several problems with the Standard Model.

- The calculated mass of the Higgs boson quadratically diverges due to radiative corrections (*naturalness* or *hierarchy* problem). These divergences should be cancelled if fermions and bosons existed in pairs as their contribution would have the same order with opposite signs.
- Dark matter and dark energy seems to give the dominant mass of the Universe. What is it that we observe its gravity only?
- Gravity does not fit in the system of gauge interactions (strong, electromagnetic, weak).
- In the Standard Model the three gauge couplings belonging to the three local gauge symmetries, $U(1)$, $SU(2)$ and $SU(3)$ seem to converge at $\sim 10^{16}$ GeV but do not quite meet.

# SUPERSYMMETRY (SUSY)

All these problems of the Standard Model would be solved if the fermions and bosons existed in exact symmetry, i.e. every fermion had a corresponding boson partner and vice versa. This fermion–boson symmetry is called supersymmetry or SUSY [11, 13]. The basic properties of the hypothetical partner particles are listed in Table 2.

As from the point of view of weak interactions each fermion has two different states, the left-polarized fermions (and right-handed anti-fermions) are in the weak doublets shown in Table 1 whereas the right-handed fermions and their left-handed anti-particles are weak singlet states. Correspondingly, they must have different partners in the supersymmetric world as well. However, although an electron's mass does not depend, of course, its polarization, the left-handed and right-handed scalar electron (selectron) are predicted to be indeed different particles with different masses.

For characterizing the SUSY particles a clever quantum number is introduced, the $R$ parity: $R = (-1)^{2S+3B+L}$ where $S$, $B$ and $L$ are the spin, baryon number and lepton

**TABLE 2.** Partner particles in a supersymmetric world. They have identical charges (electric, color, fermion) but different spins (less by $\frac{1}{2}$). R parity is defined as $R = (-1)^{2S+3B+L}$ where $S$ is the spin, $B$ is the baryon number and $L$ is the lepton number.

| Property | particle | ordinary | SUSY |
|---|---|---|---|
| R parity | | +1 | -1 |
| Spin | fermion | $S = \frac{1}{2}$ | $S = 0$ |
| | gauge boson | $S = 1$ | $S = \frac{1}{2}$ |
| | Higgs boson | $S = 0$ | $S = \frac{1}{2}$ |
| | graviton | $S = 2$ | $S = \frac{3}{2}$ |
| Chirality | fermion | $X_L, X_R$ | $\tilde{X}_1, \tilde{X}_2$ |
| Mass | fermion | $M(X_L = X_R)$ | $M(\tilde{X}_1) \neq M(\tilde{X}_2)$ |

**TABLE 3.** Elementary fermions with their assumed SUSY partners, called scalar fermions or sfermions.

| Leptons $(S = \frac{1}{2})$ | scalar leptons $(S = 0)$ |
|---|---|
| $e, \mu, \tau$ | $\tilde{e}, \tilde{\mu}, \tilde{\tau}$ |
| $\nu_e, \nu_\mu, \nu_\tau$ | $\tilde{\nu}_e, \tilde{\nu}_\mu, \tilde{\nu}_\tau$ |
| Quarks $(S = \frac{1}{2})$ | scalar quarks $(S = 0)$ |
| u, d, c, s, t, b | $\tilde{u}, \tilde{d}, \tilde{c}, \tilde{s}, \tilde{t}, \tilde{b}$ |

number. For the leptons $B = 0$ and $L = 1$, for the quarks $B = 1/3$ and $L = 0$ with $S = 1/2$, for the gauge bosons $B = L = 0$ and $S = 1$ and for the Higgs boson $S = B = L = 0$, they all have $R = 1$, whereas for their SUSY partners $R = -1$.

Table 3 lists the elementary fermions with their assumed SUSY partners. The antiparticles and their anti-partners are not listed. The SUSY partner of a particle is denoted by a tilde above the particle symbol, thus the symbol of a scalar quark or squark is q̃, that of the stau is τ̃.

The SUSY partners of the gauge and Higgs bosons are listed in Table 4. The supersymmetric extensions of the Standard Model need two Higgs doublets, separately for up-type and down-type fermions of the weak doublets, and that results in 4 complex Higgs fields (with 8 degrees of freedom), two neutral and two charged ones, with corresponding partners on the SUSY side. The spontaneous symmetry breaking (Higgs mechanism) takes 3 degrees of freedom away to create masses (longitudinal polarizations) for the three weak bosons, $W^\pm$ and Z, and five Higgs bosons are left, h, H, A, $H^+$ and $H^-$. The degrees of freedom are equal on each side as the four higgsinos are fermions with two polarizations each.

Supersymmetry is obviously broken in Nature as we cannot see such particles: if they exist they must have much larger masses then their ordinary partners. One can ask: why should we need a broken symmetry, what is it good for? In the case of the Higgs mechanism we started with a Dirac equation of a point-like fermion and added a Higgs

**TABLE 4.** The SUSY partners of the elementary (gauge and Higgs) bosons.

| Elementary boson | spin | SUSY partner | spin |
|---|---|---|---|
| photon: $\gamma$ | 1 | photino: $\tilde{\gamma}$ | $\frac{1}{2}$ |
| weak bosons: | 1 | zino: $\tilde{Z}$ | $\frac{1}{2}$ |
| $Z, W^+, W^-$ | 1 | wino: $\tilde{W}^+, \tilde{W}^-$ | $\frac{1}{2}$ |
| gluons: $g_1, ... g_8$ | 1 | 8 gluinos: $\tilde{g}_1, ... \tilde{g}_8$ | $\frac{1}{2}$ |
| Higgs fields $H_1^0, H_2^0, H_1^+, H_2^-$ | 0 | higgsinos $\tilde{H}_1^0, \tilde{H}_2^0, \tilde{H}_1^+, \tilde{H}_2^-$ | $\frac{1}{2}$ |
| graviton | 2 | gravitino | $\frac{3}{2}$ |

field which breaks that symmetry. The Higgs mechanism breaks an existing symmetry whereas SUSY introduces a non-existing one, both serve to make a theory more rational and consistent. The advantage of SUSY is demonstrated in Fig. 4 for the unification of gauge interactions: in the Standard Model the three gauge couplings get close, but do not converge at high energies, whereas in supersymmetric models there is a perfect convergence at $\sim 10^{16}$ GeV, the grand unification energy. The difference is due to the presence of extra particles in the case of SUSY which provides more loop corrections.

Introducing supersymmetry brings both positive and negative consequences. The advantages are the following:

- It brings back the naturalness of theory by eliminating the hierarchy problem: the appearance of the SUSY partners cancels those enormous corrections which caused, e.g., the mass of the Higgs boson to be calculated by the difference of twelve orders of magnitude larger quantities.
- There is a nice SUSY candidate for the cold dark matter of the Universe which should constitute about 23 % of its mass: the lightest supersymmetric particle which cannot decay to anything else and cannot interact with ordinary matter.
- It helps the unification of gauge interactions (Fig. 4, even including gravitation as well.

However, SUSY also has weak points, raises news questions and leaves certain problems unsolved:

- It is not clear at all what mechanism causes the apparent breaking of supersymmetry. Note that this violation cannot necessarily be considered to be very strong if one compares the presently accessible laboratory energies with those of the grand unification or Planck scale.
- There are many possible ways to include SUSY in the Standard Model and as a result there are many-many different SUSY models.
- Supersymmetry introduces many (more than a hundred) new parameters in the Standard Model which had originally only 19 ones (if one neglects the neutrino masses). Of course, the parameter sets have to be reduced with more-or-less reasonable assumptions and simplifications: the convergence of the gauge interactions

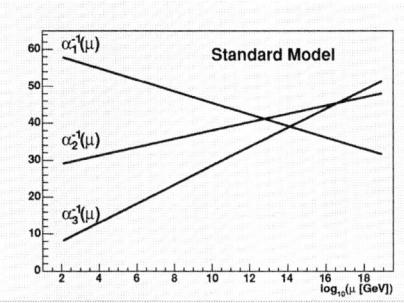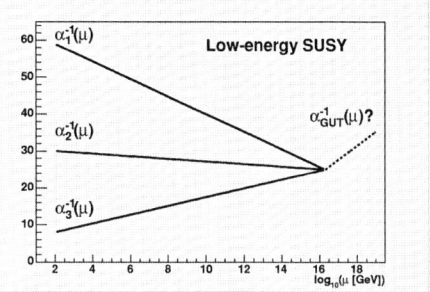

**FIGURE 4.** The unification of gauge interactions [11]. In the Standard Model the three gauge couplings get close, but do not converge at high energies, whereas in supersymmetric models there is a perfect convergence at $\sim 10^{16}$ GeV.

helps a lot, and the masses of the particles are usually assumed to converge as well.

- No SUSY particle has been seen below $\tilde{m} \sim 100$ GeV, although all experiments were searching for them.

## MINIMAL SUPERSYMMETRIC STANDARD MODEL

An experimental search for new particles needs precise predictions about its properties, and for that one has to drastically reduce the number of parameters. At present there are quite a few such supersymmetric extensions of the Standard Model, the most popular one being the Minimal Supersymmetric Standard Model (MSSM). It simplifies the general approach with reasonable boundary conditions, assuming a general convergence of masses and couplings at the Grand Unification energy (GUE $\sim 10^{14} - 10^{16}$ GeV) and adds just the following six new parameters to the Standard Model:

- $m_{1/2}$: fermion masses at GUE;
- $m_0$: boson masses at GUE;
- $A_0$: SUSY breaking (X–Y–Higgs) coupling constants at GUE;
- $\tan\beta = v_1/v_2$: ratio of the vacuum expectation values of the upper and lower Higgs fields;
- $m_A$: mass of a Higgs boson;
- $\mu$: mixing parameter of the higgsinos.

There are several versions of this model, some has less parameters, e.g. omit $m_A$ and keep the sign of $\mu$ only (those are called $4\frac{1}{2}$ parameter models).

## SEARCH FOR SUSY PHENOMENA

The first problem with these searches is the fact that if particle states can mix, i.e. the mixing is not prohibited by conservation laws, then they will. As an experiment usually looks for eigenstates one has to calculate the cross sections for those. The fermionic SUSY partners of the SM bosons mix into *charginos*,

$$\{\tilde{W}^+, \tilde{W}^-, \tilde{H}_1^+, \tilde{H}_2^-\} \Rightarrow \{\tilde{\chi}_1^\pm, \tilde{\chi}_2^\pm\} \qquad (2)$$

and neutralinos:

$$\{\tilde{\gamma}, \tilde{Z}, \tilde{H}_1^0, \tilde{H}_2^0\} \Rightarrow \{\tilde{\chi}_1^0, \tilde{\chi}_2^0, \tilde{\chi}_3^0, \tilde{\chi}_4^0\} \qquad (3)$$

in order of increasing mass.

In order to search for those new particles one needs observable properties, i.e. mass and cross-section predictions. Generally one assumes that the SUSY particles are created in pairs, and decay to ordinary and SUSY particles. The end of the decay chain in the SUSY sector (assuming no R-parity violation) is the lightest SUSY particle (LSP) which has nowhere to decay. In the experimentlists' favorite MSSM models usually the $\tilde{\chi}_1^0$ neutralino is assumed to be the LSP. Another popular group is that of the *gauge mediated SUSY breaking, GMSB* models whose LSP is the $\tilde{G}$ gravitino.

SUSY particles are continuously searched for at every particle physics facility, the largest ones, CERN's Large Electron Positron (LEP) collider (Fig. 3) and Fermilab's Tevatron, devoted great efforts to such searches, so far with no success.

The main problem is how to distinguish SUSY reactions from ordinary events allowed by the Standard Model. For instance, when one looks for scalar lepton formation in electron-positron collisions, they are expected to be created in pairs,

$$e^+e^- \rightarrow \tilde{\ell}^+\tilde{\ell}^- \qquad (4)$$

and decay, e.g., to ordinary leptons like

$$\tilde{\ell}^\pm \rightarrow \tilde{\chi}_1^0 \ell^\pm \qquad (5)$$

with model-dependent cross-sections. Thus one should look for

$$e^+e^- \rightarrow \ell^+\ell^- + \text{missing energy.} \qquad (6)$$

However, the pair production of W bosons can give a very similar reaction,

$$e^+e^- \rightarrow W^+W^- \rightarrow \ell^+\nu\ell^-\bar{\nu} \qquad (7)$$

producing a substantial and almost irreducible background The only hold is the spin difference leading to slightly different angular distributions. No having seen signs of SUSY particles the experiments use statistical methods to limit the parameter space of the various models. The searches of the four LEP experiments are summed statistically up and gave the result that no SUSY particle is seen with masses below 90-100 GeV/$c^2$, close to the kinematic limit of LEP.

Fig. 3 presents the LHC, the Large Hadron Collider, as it is scheduled to operate from 2007 on. Two general-purpose detectors are being built for it, ATLAS [14] and CMS [15], each representing international collaborations with more than 2000 scientists. The other two, LHCb [16] and ALICE [17] are more specialized: as their names suggest

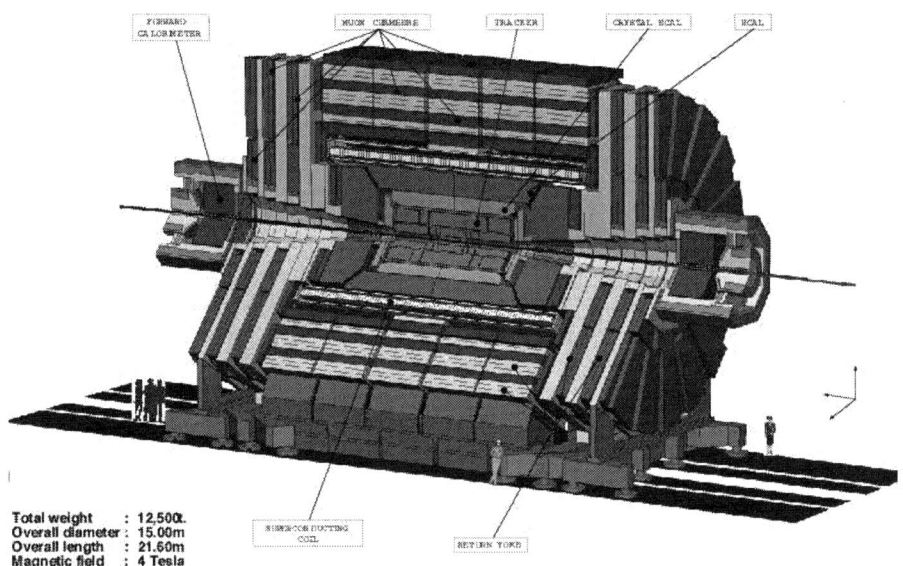

## CMS
## A Compact Solenoidal Detector for LHC

Total weight : 12,500t.
Overall diameter : 15.00m
Overall length : 21.60m
Magnetic field : 4 Tesla

**FIGURE 5.** The Compact Muon Solenoid (CMS) calorimeter of the Large Hadron Collider (LHC). [15].

ALICE is oriented towards heavy ion physics whereas LHCb towards the physics of the b quark. The main physics aim of ATLAS and CMS is the discovery and thorough study of the Higgs boson(s), but they are also developing means to observe SUSY particles if they exist. The author is a member of the CMS Collaboration, so the CMS detector (Fig. 5) will be used as an example how this work will proceed.

The ATLAS and CMS detectors are at the moment the largest detectors on Earth. CMS is somewhat smaller but much heavier, it weighs 12500 tons and contains more iron than the Eiffel tower in Paris. It has the largest superconducting solenoid: it keeps a $B = 4$ Tesla magnetic field in its 6 m diameter, 12.5 m long cylindrical volume. The proton bunches of the LHC will collide at 40 MHz frequency, and when the LHC achieves its design luminosity, 10-20 p-p interactions are expected to happen at every bunch crossing, i.e. at every 25 ns. Moreover, the proton is a composite particle consisting of three valence quarks and a lot of gluons, thus a high-energy p-p collision means a spray of jets, mostly along the beam direction. It needs an extremely intelligent trigger to pick and store those events only where we expect to see something interesting. The event filter will be done at the data rate of 500 GB/sec, using about 4000 computers. We expect to store about 10 petabyte of data per year and to generate the same amount of Monte Carlo simulations. Such an amount of data cannot be processed by a single site as was done earlier at CERN, that will be done by the LHC Computing Grid system which includes more than 80 computer centers all over the world.

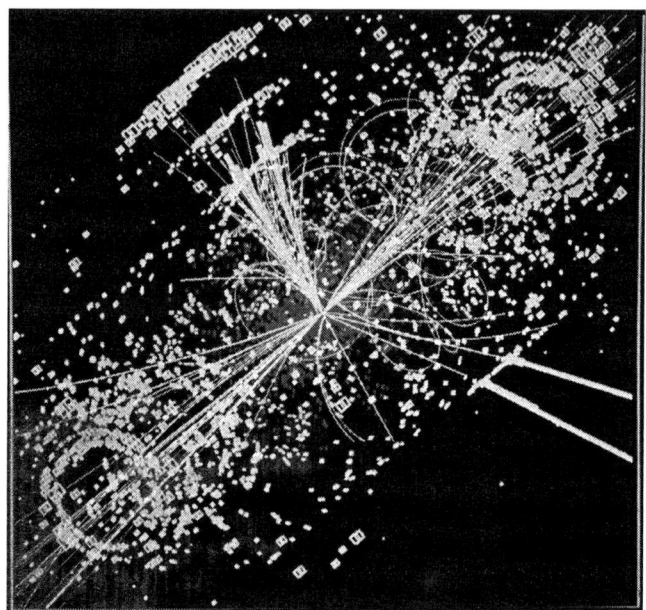

**FIGURE 6.** A simulated H → ZZ → eeqq event. A Higgs boson produced in proton–proton collision decays into two Z bosons; one Z decays into an electron–positron pair, the other one into a quark pair and the quarks produce hadron jets [15].

Fig. 6 shows a simulated CMS event: a Higgs boson produced in proton–proton collision decays into two Z bosons; one Z decays into an electron–positron pair, the other one into a quark pair and the quarks produce hadron jets [15]. It is clear from the picture that one can hope to analyze those events only where a substantial amount of energy is flowing orthogonally to the beam direction (*transverse* momentum or energy).

Because of the appearance of a non-interacting particle, the LSP, SUSY events should have another characteristic feature: *missing transverse momentum*, i.e. an unbalanced transverse momentum distribution. As the p–p collision produces mostly hadrons, the easiest way to identify nice new events is by looking for leptons with high transverse momentum. For instance, a gluino decay can produce a lepton cascade:

$$\tilde{g} \rightarrow \tilde{b}\,\bar{b} \rightarrow \tilde{\chi}_2^0\,\bar{b}\,b \rightarrow \tilde{\ell}^+\,\ell^-\,\bar{b}\,b \rightarrow \tilde{\chi}_1^0\,\ell^+\,\ell^-\,\bar{b}\,b \tag{8}$$

A new particle can be discovered by observing a kinematic cutoff in the invariant mass spectra of certain sets of detected particles, lepton or jet pairs or triplets, and the mass of the new particle will be deduced from the cutoff energy (Fig. 7).

Of course, it is impossible make measurements for all parameter values of all models. Close collaboration between theoreticians and experimentalists produced a set of *benchmark* points in the parameter space of the constrained MSSM and other models with properly predicted SUSY properties and reaction probabilities. Those will be thoroughly investigated using the collected data.

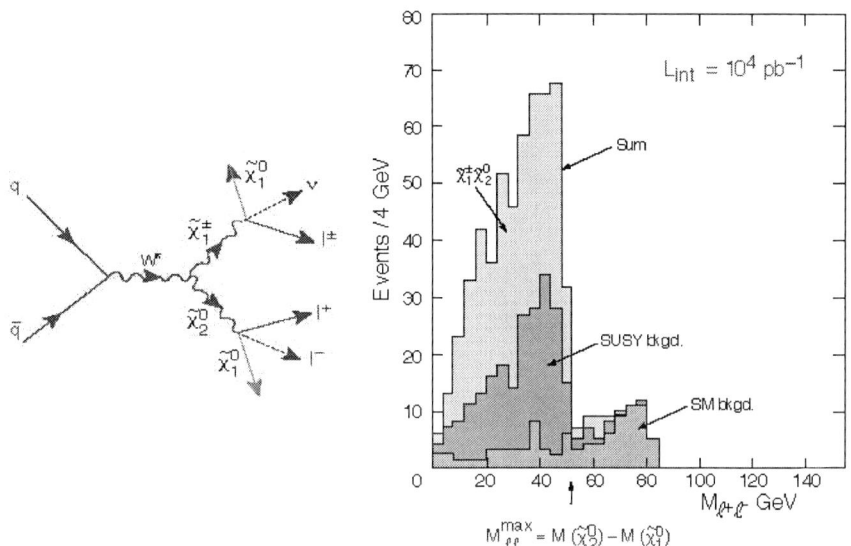

**FIGURE 7.** A hypothetical SUSY event and its appearance in the di-lepton invariant mass spectrum [15].

# CONCLUSION

There is no conclusion yet: the Standard Model stands as it is in spite of its theoretical difficulties. We could not find any Higgs boson yet, but we think it will be there. We also hope for supersymmetry: it is very nice even though broken.

# ACKNOWLEDGMENTS

The present work was supported by the Hungarian National Research Foundation (Contracts OTKA T042864 and T046095) and the Marie Curie Project TOK509252. My participation at the conference was made possible by the financial help of the organizers.

# REFERENCES

1. F. Halzen, A.D. Martin, *Quarks and Leptons: An Introductory Course in Modern Particle Physics*, Wiley, New York, 1984.
2. Cohen, A.G., De Rujula, A., Glashow, S.L., *A Matter - Antimatter Universe?*, Astrophys. J., **495**, 539 (1998).
3. The LEP Electroweak Working Group, home page, http://lepewwg.web.cern.ch/LEPEWWG/
4. Particle Data Group, S. Eidelman *et al.*, *Review of Particle Physics*, Physics Letters B **592**, 1 (2004). (URL: http://pdg.lbl.gov).

5.  P. Higgs: *My Life as a Boson: The Story of 'The Higgs'*, Int. J. Mod. Phys. A, Proc. Suppl. **17**, 86 (2002).
6.  The LEP Collaborations, A. Heister *et al.*: *Search for the Standard Model Higgs Boson at LEP*, Physics Letters B **565**, 61 (2003).
7.  V. A. Kostelecký: *Gravity, Lorentz Violation, and the Standard Model*, Phys.Rev.D **69**, 105009 (2004).
8.  N. E. Mavromatos: *CPT Violation and Decoherence in Quantum Gravity*, Lecture Notes in Physics (in press), E-print gr-qc/0407005.
9.  F. R. Klinkhamer, Ch. Rupp: *Space-Time Foam, CPT Anomaly, and Photon Propagation*, Phys.Rev.D **70**, 045020 (2004).
10. R. Lehnert: *Lorentz and CPT Tests Involving Antiprotons*, Talk at the present conference.
11. F. Wilczek: *In search of symmetry lost*, Nature **433**, 239 (2005).
12. The Antiproton Decelerator at CERN, `http://psdoc.web.cern.ch/PSdoc/acc/ad/index.html`
13. S.P. Martin: *A Supersymmetry Primer*, In: G.L. Kane, ed., *Perspectives of Supersymmetry*, World Scientific, Singapore, 1997, pp. 1-98 (E-print `hep-ph/9709356`).
14. ATLAS (*A Toroidal LHC ApparatuS*): *http://atlas.web.cern.ch/Atlas/*
15. *CMS* (Compact Muon Solenoid): http://cmsinfo.cern.ch/
16. LHCb (*The Large Hadron Collider beauty experiment*): *http://lhcb.web.cern.ch/lhcb/*
17. *ALICE* (A Large Ion Collider Experiment): http://pcaliweb02.cern.ch/NewAlicePortal/en/index.html

# ATOMIC PHYSICS

# Sub-Femtosecond Correlated Dynamics Explored with Antiprotons

J. Ullrich[1], A. Dorn[1], D. Fischer[1], R. Moshammer[1], B. Najjari[1], M. Schulz[3,2,1], A. Voitkiv[1], C.P. Welsch[1,4]

[1]Max-Planck-Insitut für Kernphysik, Saupfercheckweg 1, D-69117 Heidelberg, Germany
[2]University of Missouri-Rolla, Rolla, Missouri 65409-0640, USA
[3]Institut für Kernphysik, August-Euler Str. 6, D-60486 Frankfurt, Germany
[4]CERN, Geneva, Switzerland

**Abstract:** Ionizing collisions of antiprotons with atoms or molecules at energies between a few keV up to about one MeV provide a unique tool to explore correlated dynamics of electrons at large perturbations on a time scale between several femtoseconds (1 fs = $10^{-15}$ s) down to some tens of attoseconds (1 as = $10^{-18}$ s). Exploiting and developing many-particle imaging methods – Reaction-Microscopes – integrated into a novel ultra-low energy storage ring (USR) for slow antiprotons will enable to access for the first time fully differential cross sections for single and multiple ionization in such collisions. Moreover, the formation of antiprotonic atoms, molecules or of protonium might be explored in kinematically complete experiments yielding unprecedented information on (n,l)-distributions of captured antiprotons as well as precise spectroscopic data of the respective energy levels.
In this contribution the present status on single and double ionization by antiproton and ion impact is highlighted pointing to the puzzling discrepancies between experiments and theoretical predictions. The design status of the USR as a central element of the proposed facility for low-energy antiproton and ion research (FLAIR) at GSI will be shortly presented.

## INTRODUCTION

It was in 1929 when Dirac wrote: „The general theory of quantum mechanics is now almost complete..." "The underlying physical laws necessary for the mathematical theory of a large part of physics and the whole of chemistry are thus completely known, and the difficulty is only that the exact application of these laws lead to equations much too complicated to be soluble" [1]. Single and double ionization of hydrogen or helium by charged particle impact, i.e. the dynamical three- or four-particle quantum problem, is until the present day one of the most striking, simple and thus fundamental examples, where a reliable prediction of the basic properties of the system under consideration remains to be challenging.

It was not before 1999 when it was claimed that the three-body Coulomb problem has been solved in a numerical consistent way [2] followed by a series of papers ([3,4] and references therein) culminating in the statement in 2003 that the ionization amplitude in general has been formulated mathematically correct [4] providing the theoretical explanation for the success of several state-of-the-art large-scale numerical approaches including the one of 1999 [2,3,5,6]. In practice, however, this class of "exact" theories has only been applied to calculate excitation, single ionization or ionization-excitation [7] cross sections for hydrogen or helium targets in collisions with electrons and for photo double ionization of helium. Double ionization in any other situation, single ionization induced by intense lasers, or in collisions with antiprotons and ions from keV protons to swift (relativistic) highly-charged ions, not to speak about multiple ionization of many-electron atoms, molecules or clusters require to introduce sometimes severe theoretical approximation methods.

Puzzling discrepancies observed for reactions as simple as single ionization of helium on the level of total cross sections for impact of antiprotons (for a recent paper see [8]), on the level of differential cross sections for collisions with slow protons (see e.g. [9]) or for laser-pulse impact (see e.g. [10]), on the level of differential as well as fully differential cross sections (FDCS) for fast ion and proton encounters (see e.g. [11-15]) strikingly uncover our fundamental difficulties to describe the dynamics of correlated quantum systems – only four active particles in the reactions mentioned above – on the time scale of femto- to attoseconds, i.e. in a regime that is comparable to the classical bound-state revolution time of active electrons in light systems.

## ANTIPROTONS: The ULTIMATE TOOL TO EXPLORE SUB-FEMTOSECOND CORRELATED DYNAMICS

Motivated by the need to probe and by the vision to possibly control correlated motion of bound electrons on exactly that time-scale, huge efforts are being undertaken in modern laser technology, in "Attosecond Science", to produce sub-femtosecond pulses with a record of few hundred attoseconds achieved recently [16]. However, clean electromagnetic and sufficiently strong half-cycle pulses of few femtoseconds down to a few tens of attoseconds could not be produced with state-of-the-art lasers and will hardly be available in the foreseeable future.

As visualized in Figure 1 and depicted in Figure 2, antiprotons passing atoms or molecules at energies below about 1 MeV down to 1 keV just generate such pulses. Here, a characteristic interaction time $\tau \sim b/v$ between projectile and target may be defined for a given projectile velocity v and impact parameter b. In ref [17] it was demonstrated that for ionizing keV proton on helium collisions, the measured transverse momentum transfer by the projectile $P_{P\perp}$ is reasonably well related to an impact parameter (on the basis of an assumed scattering potential). Thus, for fixed $P_{P\perp}$ and given projectile velocity, the collision time i.e. the duration of the half-cycle pulse as illustrated in Figure 1 can be well estimated within about a factor of two [17]. Using Classical Trajectory Monte Carlo Calculations (CTMC) in [17] typical impact parameters between 1 a.u. and 10 a.u. have been obtained for the majority of singly ionizing collisions at 15 keV proton energy (1 a.u. = $0.52 \cdot 10^{-10}$ m) corresponding to collision times $\tau \sim 30 - 300$ as. Assuming similar regimes of relevant impact parameters for collision energies between 1 keV and 1 MeV antiproton energies (which is certainly not completely true but nevertheless a good estimate) accessible interaction times range from a few attoseconds (b = 1 a.u. at 1 MeV) to more than one femtosecond (b = 10 a.u. at 1 keV) as roughly indicated in Figure 2. Since the classical orbiting time of an electron bound with 13.6 eV in the ground state of atomic hydrogen is around 150 as, this is exactly the relevant and interesting time scale that is hardly accessible by any other means. (It should be noted that shorter times on the order of few attoseconds even approaching the zeptosecond regime (1 zs = $10^{-21}$ s) are easily achievable in fast to relativistic ion- or electron-atom collisions [18] and have been extensively explored. Here, however, for outer-shell processes, the momentum transfer to the projectile becomes very small, comparable to the momenta of the ionized electrons, such that a good experimental estimate of the collision time is difficult and only averaged values might be assumed from theoretical impact parameter ranges.)

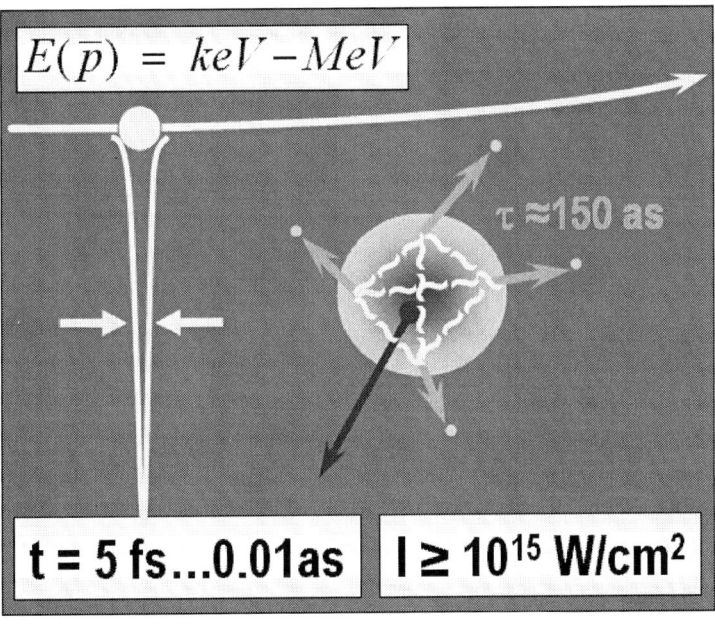

**FIGURE 1.** Illustration of ultra-short, intense half-cycle pulses generated by fast antiprotons in collisions with atoms, molecules or clusters.

Certainly, different from a laser, the interaction time during a collision is not an observable and can only be estimated on the basis of an assumed or calculated interaction potential. This has been done here just for the visualisation and illustration of relevant time scales and to provide insight in comparison with short-pulse laser research. It should be kept in mind, however, that the relevant quantity in any interaction is the total action, i.e. the energy integrated over time which is well defined for a given momentum transfer and which is reflected in the perturbation expansion parameter Z/v. Thus, fixing the momentum transfer, as it is standard in state-of-the-art collision experiments, provides the ultimate test for theory under unsurpassed clean conditions. These conditions are usually much better defined than in present measurements with intense short-pulse lasers where one has to average over all intensities in the focus of the beam (see e.g. [19] for single ionization).

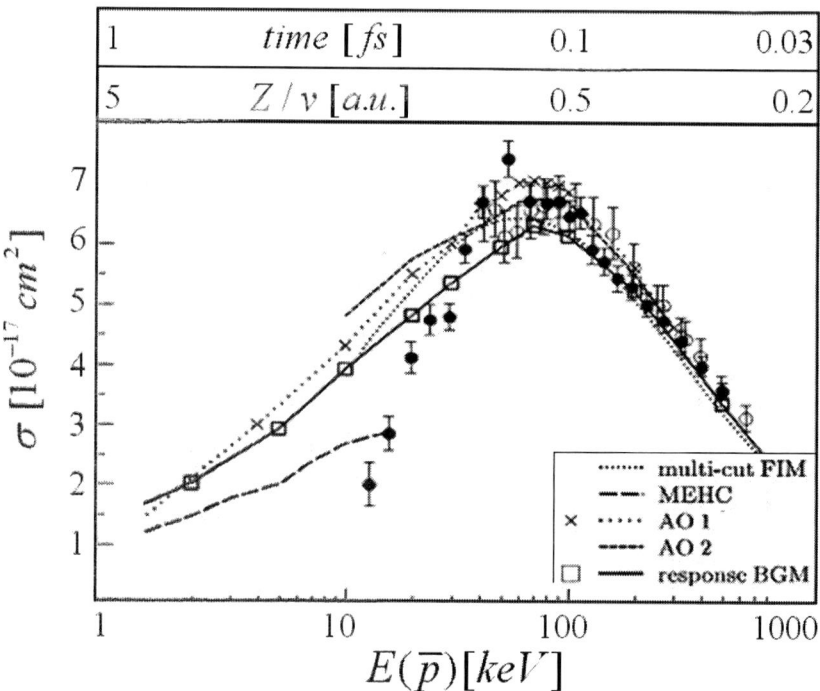

**FIGURE 2.** Cross section for single ionization of He by antiproton impact. Full and open circles: experiment [20, 21]. Various lines: theoretical results: multi-cut FIM (Forced Impulse Method) [22]; MEHC: Multi-Electron Hidden Crossings model (for reference see [8]); AO1,2: Atomic Orbital coupled channel calculations (for reference see [8]); response BGM (Basis Generator Method) [8]. The picture is taken from ref. [8].

As compared to proton or electron projectiles, slow antiprotons offer new and unique possibilities. In slow proton encounters, for example, electron capture by the projectile strongly dominates complicating the theoretical analysis considerably, whereas ionization or excitation are the only relevant reaction channels that occur for antiproton impact at keV energies (antiproton capture to form antiprotonic atoms is quite small at energies above about 20 eV and can be safely neglected). Compared to electrons, even ultra-low velocities of 0.03 a.u. and effective collision times of several fs may be explored with a lower limit at an energy of 24.6 eV, the threshold for direct impact ionization of He (v = 1 a.u. = 1/137·c; c: speed of light). Due to their smaller mass, electrons require a minimum velocity of 1.3 a.u. to overcome this threshold and, thus, collision times longer than about 100 as are not accessible under well defined conditions in ionization experiments. (Although very close to threshold interaction times approaching the orbiting time of the atomic electron are certainly possible for electron impact as well, the situation becomes much more complicated for theory since, for example, exchange processes, negative ion resonances etc. have to be considered.)

Hence, slow antiprotons provide an unsurpassed, precise and presumably the simplest tool for theory to study many-electron dynamics in the few atto- to femtosecond time-regime, comparable to the revolution time of bound electrons. Here, the interaction strength maximizes, which is important for the future control of this motion, and consequently, the experimental [20, 21] as well as theoretical ionization cross sections for helium (see [8] and references therein) peak as illustrated in Figure 2.

At the same time, even without the above-mentioned complications that one faces for proton or electron impact, this regime poses utmost challenges to theory. Due to the strong coupling any perturbative Ansatz (valid for $Z/v << 1$) in powers of the interaction strength $Z/v$ must fail ($Z$, $v$: projectile charge and velocity in atomic units, a.u., respectively). By reducing the velocity, huge values up to $Z/v = 5$ for the perturbation can be reached as indicated in the figure (until now, similar perturbations have only been achieved in MeV highly-charged heavy-ion collisions corresponding to much shorter interaction times in the regime of few attoseconds). Moreover, as demonstrated in numerous approximation calculations (various curves in the figure, for details see [8]) the correlation between both electrons or at least the time-dependent response of the second electron (not being active in the sense that it remains in a bound state) has to be taken into account. Even on the level of total cross sections pronounced differences between the predictions of different theories arise and all of them are in poor agreement with available experimental data in the low-energy regime. Moreover, even if theoretical predictions are quite close to each other, some of them come to drastically different, even contradictory interpretations. For example, on the basis of the Forced Impulse Method (FIM) results [22] time-dependent electronic correlation, beyond a Hartree-Fock treatment, is claimed to be essential, whereas the equally good predictions within the Basis Generator Model (BGM) [8] including microscopic response lead to the result that an effective one-particle picture is sufficient if the response of the second electron to the time-dependent field is properly taken into account. In the strict sense, this means that dynamical electron-electron correlation is not needed at all!

To conclude, low-energy antiproton collisions provide the ultimate benchmark for the development of strong-field non-perturbative theories in the presence of correlation. This is especially true if fully differential cross sections are explored for single, double and multiple ionization, as will be demonstrated below.

## REACTION MICROSCOPES

Experimentally, a break-through has been achieved with the recent development of "Reaction-Microscopes" [23], i.e. many-particle imaging techniques that allow one to measure the momentum vectors of several outgoing electrons or ions simultaneously with large solid angle and excellent resolution. For the first time, kinematically complete experiments that cover a large part of the final state-momentum space became feasible for charged particles or laser pulse impact single ionization of helium, for double ionization of helium in any situation, for ionization-excitation [7] or even for laser assisted single-ionization reactions in collisions with electrons [24]. Due to space limitations for the present article the reader is referred to the above-cited literature as well as to extended recent reviews where the technique as well as plenty of results are discussed in detail [25-27]).

# FDCS FOR SINGLE IONIZATION OF HELIUM BY FAST ION IMPACT: A BRIEF SUMMARY

Numerous fully differential cross section measurements, so-called (e,2e) experiments have been performed for single ionization by electron impact since the pioneering work of Ehrhardt et al [28] (for a review see e.g. [29]), most of them under certain, restricted geometrical conditions in so-called coplanar geometry (see below). Due to immeasurably tiny scattering angles of heavy ionic projectiles, kinematically complete data for ion impact have only become available with the advent of Reaction-Microscopes. Whereas pioneering results were reported a decade ago [30], first FDCS were not published until 2001 [31]. Covering a large part of the final-state phase space they provided, at the same time, first information for electron emission into all three spatial directions, i.e. compared to electron impact for hitherto unexplored geometries.

Inspecting Figure 2, one observes that all theoretical predictions merge at high energies on the level of total cross sections. Here, at small perturbations, the first Born approximation (FBA) is fully applicable, cross sections scale with $(Z/v)^2$ and thus, become independent of the sign of the projectile charge such that electron and ion impact should yield identical results. In view of numerous (e,2e) measurements in this regime, it came as a big surprise that three-dimensional experimental FDCS for ion impact were found to be in significant disagreement with all state-of-the-art theoretical predictions for single ionization at $Z/v = 0.1$, well within the perturbative regime ($Z/v \ll 1$) [13]. In Figure 3 (top left), a 100 MeV/u $C^{6+}$ projectile ion impinges with momentum $\vec{P}_0$ along the vertical arrow onto a helium atom. Being deflected and emerging along the upwards directed arrow, it ionizes the atom, imposing a momentum transfer $\vec{q}$ to the system as illustrated by the arrow pointing to the right. Electrons of 6.5 eV are emitted into all spatial directions with an intensity represented by the distance of each point on the three-dimensional surface from the origin.

Theoretically, as can be seen from the result of advanced calculations (Figure 3, bottom left), an emission minimum is predicted in the plane perpendicular to the "scattering plane", i.e. the plane surrounded by the dotted line containing the momentum transfer vector $\vec{q}$ (also called coplanar geometry). Whereas this minimum has been consistently observed within the scattering plane over more than 30 years in numerous previous ionization experiments using electron projectiles in agreement with the present results for ion impact, a severe disagreement between our data and all theoretical predictions is found in the perpendicular plane, never explored in any experiment before. Despite considerable efforts from various sides, and quite some explanation attempts (see e.g. [13, 32, 33], the reason for this discrepancy has not yet been conclusively determined.

By increasing the projectile charge state while staying at relatively high velocities in order to minimize capture channels (scaling with $\sim v^{-11}$) the regime of strong perturbations up to $Z/v = 4.4$ has been explored using 3.6 MeV/u $Au^{53+}$ projectiles (Figure 3, upper right). As might have been expected, the experimental results deviate dramatically from the theoretical prediction (Figure 3, bottom right) not only in shape, but more severe in absolute magnitude (not shown here). It should be mentioned that

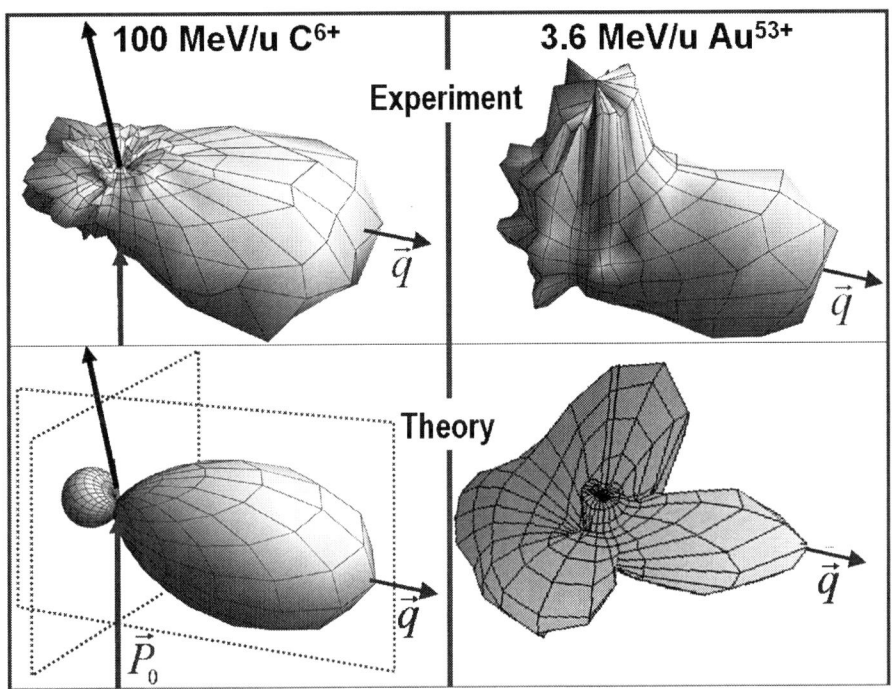

**FIGURE 3.** Measured (upper row) and calculated (lower row) three-dimensional intensity distributions of low-energy electrons (6.5 eV left and 20 eV right) emitted in singly ionizing 100 MeV/u $C^{6+}$ and 3.6 MeV/u $Au^{53+}$ on helium collisions for fixed momentum transfer $\vec{q}$ along the arrows pointing to the right (q = 0.75 a.u. left and 1.0 a.u. right).

even in the strongly non-perturbative regime, total as well as differential electron emission cross sections are in excellent agreement with the theoretical predictions (see e.g. [34-36]). Discrepancies only arise as soon as cross sections differential in any projectile parameter are investigated and are usually severe in FDCS, i.e. when the quantum four-particle problem is explicitly addressed.

In order to investigate contributions of higher-order terms in the perturbation expansion on the level of FDCS at energies, where the proton-antiproton difference, i.e. deviations from a first order theory are negligibly small for total cross sections, we have performed a series of systematic calculations for 200 keV, 500 keV and 1 MeV [37]. In Figure 4 we present one example, a FDCS for 500 keV proton or antiproton impact. Whereas only tiny differences can be observed between proton and antiproton impact in total or singly differential cross sections as a function of the electron energy shown in Figure 5 and the FBA does an excellent job (dotted line), the FDCS for $p, \overline{p}$ exhibit large differences between each other and both deviate strongly from the FBA result (dotted line). Clearly, the second order calculations have not converged (dash-dotted and dash-dot-dot line for $p, \overline{p}$, respectively), such that all orders have to be taken into account (thick full and dashed lines, respectively) even at this relatively high impact velocity at a still small perturbation of $Z/v \approx 0.2$.

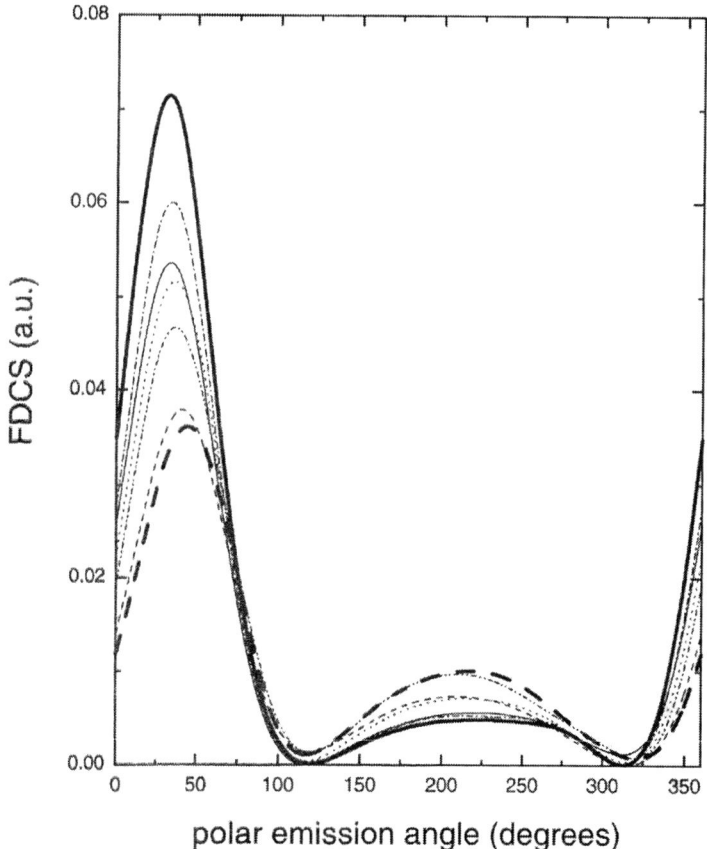

**FIGURE 4.** Theoretical FDCS for 500 keV proton (thick full line) and antiproton (thick dashed line) impact ionization of hydrogen in all orders of the perturbation (see [37]). Dotted line: First Born approximation. Dash-dot (dot) lines: Second Born approximation for proton (antiproton) impact. The momentum transfer is $|\vec{q}| = 0.2$ a.u. and the energy of the emitted electron is $E_e = 20$ eV.

In summary, first measurements for ion impact at various perturbations as well as theoretical studies clearly demonstrate that only FDCS provide a reliable test ground for theory. From the experience gained with positive ions, it is expected that even in situations where theoretical predictions agree with each other on the level of total cross sections for single ionization of helium by antiproton impact, e.g. above the cross section maximum towards higher impact energies as shown in Figure 2, significant differences might arise between different theoretical results and experimental data when FDCS are explored. At lower energies, where the theoretical predictions deviate from each other as well as from available data even for total cross sections, FDCS for antiproton impact will provide the ultimate benchmark for theoretical models to describe sub-femtosecond correlated dynamics.

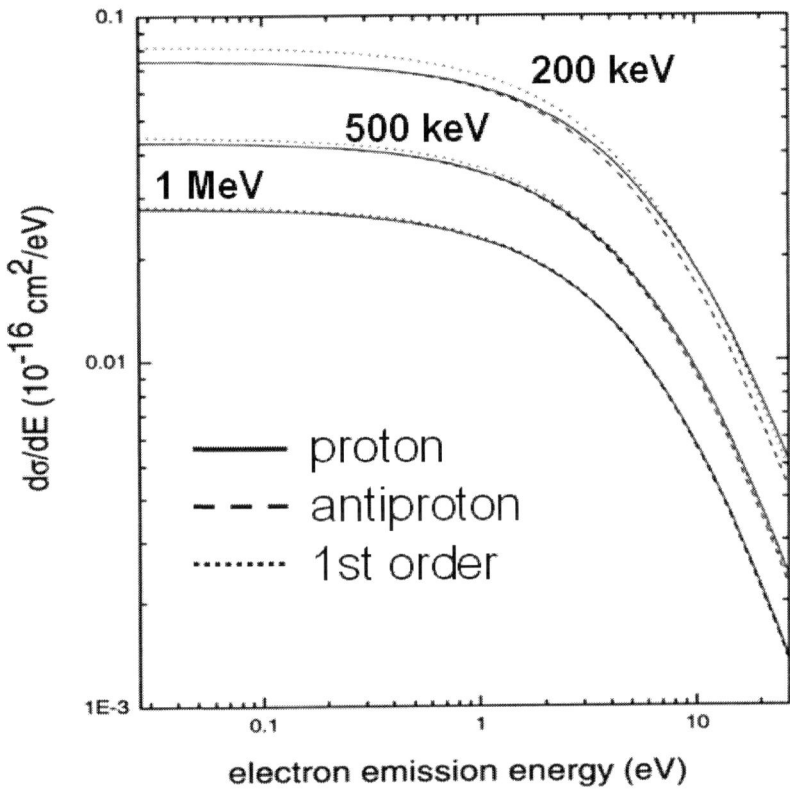

**FIGURE 5.** Single differential cross sections as a function of the ejected electron energy for single ionization of hydrogen by proton (full lines) and antiproton impact (dashed lines) as well as FBA results (dotted line) at various collision energies as indicated in the Figure (from [37]).

# ON THE CHARGE SIGN DEPENDENCE OF DOUBLE IONIZATION CROSS SECTIONS

A huge amount of data has been assembled for the ratio R of double to single ionization of helium for various projectiles, ranging from protons to bare uranium at velocities of a few hundred keV up to 1.5 GeV/u, with the goal to disentangle the role of electron-electron correlations. As convincingly shown in Figure 6, all of these data closely fall on one common curve if plotted versus the squared inverse perturbation (see e.g. [38] and references therein). It has been discussed in many papers that the linear behaviour of the ratio as a function of $(Z/v)^{-2}$ at larger perturbations is due to the second order amplitude dominating double ionization, i.e. each of the two electrons independently interacts with the projectile. At small perturbations instead, the ratio becomes independent of Z/v, only one virtual photon is exchanged in the FBA and double ionization is purely a result of correlations.

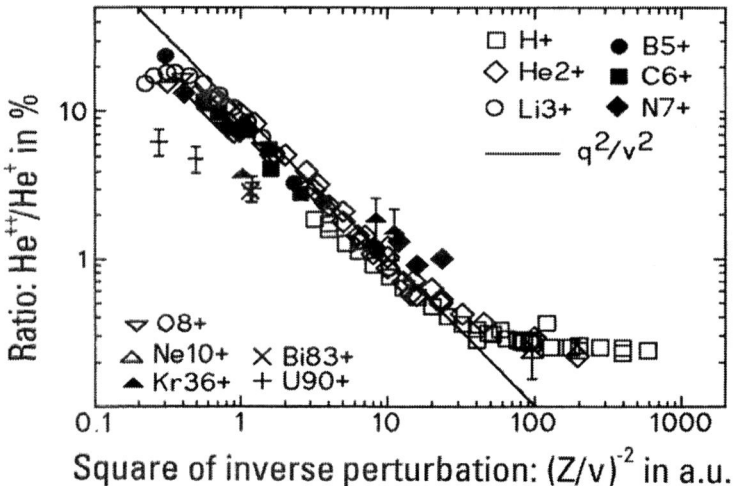

**FIGURE 6.** Ratio of double to single ionization of helium in collisions with various projectiles as indicated in the Figure as a function of the squared inverse perturbation (full line). For references see [38].

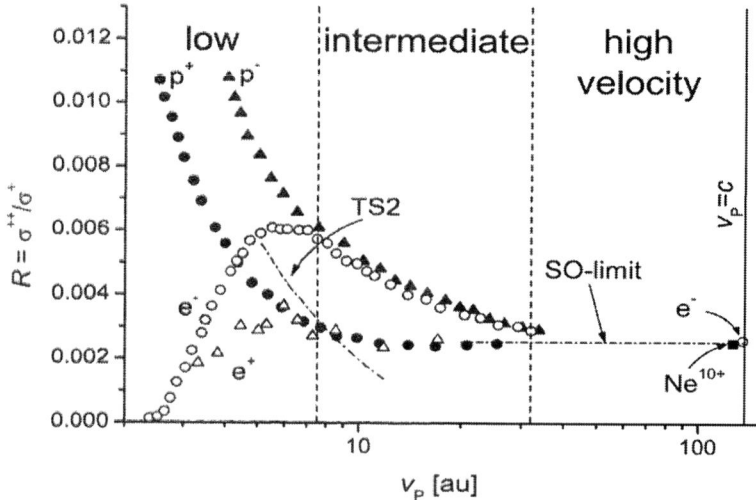

**FIGURE 7.** The ratio R of the total double-to-single ionization cross section for helium as a function of the projectile velocity for protons (full circles), antiprotons (full triangles), electrons (open circles), positrons (open triangles) and $Ne^{10+}$ (full square) from ref. [40] (for references see [40]). The dash-dotted lines denote the velocity dependence of the "pure" second-order contribution (two-step 2: TS2) without any electron-electron correlation and of the high-velocity (low perturbation) limit where the emission of the second electron is a result of correlations only, estimated here within the shake-off approximation (SO). Since all of the projectiles apart from $Ne^{10+}$ are singly charged, the velocity scale is just in inverse perturbation scale as in Figure 6.

As pointed out by McGuire [39], interferences and thus, a projectile charge sign dependence of the ratio is expected in regimes, where first ("SO" in Figure 7) and second order ("TS2" in Figure 7) amplitudes are of about the same magnitude. This has been impressively demonstrated in experiments with antiprotons and positrons and is illustrated in Figure 7 [40]. Several theoretical models can describe this behaviour but nevertheless, similar to the situation for single ionization, there is an ongoing discussion on the importance of dynamic correlation during the collision on a sub-femtosecond time scale (see e.g. [41] and references therein).

Again, the ultimate test of theoretical models to describe the time-dependent motion of two electrons in the simplest two electron atom will be fully differential cross sections, which are however, much more difficult to obtain for double ionization. Only recently such experiments have become feasible for electron impact [42-44] and just one single experiment has been performed for ion impact [40] until now, still suffering from limited statistical significance. In Figure 8 we present experimental (left column) and theoretical (right column) FDCS in coplanar geometry, where two electrons of identical energy and the recoil ion are emitted within the scattering plane for electron impact at v = 12.2 a.u. (upper row) and proton impact v = 15.5 a.u. (lower row). At these velocities the mass effect can be considered small as can be seen from Figure 7 and, hence, proton and electron data can be compared to explore the charge sign dependence of double ionization. Momentum transfers and energies are given in the Figure, the full bar in the middle of the picture denotes the accepted momentum transfers between 0.8 and 1.4 a.u.. The full lines indicate the symmetry axis obtained in any first order theory. Dashed lines denote angular correlations that are forbidden in the dipole approximation and dotted lines encircle areas where full experimental acceptance is guaranteed.

Two features are evident for proton as well as for electron impact. First, the final state angular distributions are strongly structured on the one hand by the dipole selection rules showing clear remnants of nodes at positions of the dashed lines and on the other hand by the Coulomb repulsion in the continuum forbidding the emission of equal energy electrons under identical angles. Second, two more or less pronounced maxima are visible, that can be interpreted as the binary (the two innermost maxima) and the recoil peaks (the two outer maxima), respectively where the sum momentum of the two electrons is along or opposite to the momentum transfer. Clear differences between electron and proton impact (electron and positron impact for theory) are discernible. First, very similar to single ionization, the binary peak is more pronounced for positive projectile charge with the recoil peak barely visible at all in the experiment whereas the recoil peak is clearly seen for negative charge state sign of the projectile. Second, the proton data seem to be quite symmetric with respect to the full line whereas the recoil-peak in the electron impact data is significantly shifted pointing to the importance of higher order contributions.

In ref [40] these differences, with a similar systematic behaviour for single ionization seen in Figure 4, have been interpreted in a classical picture. At medium impact parameters of about one atomic unit, contributing most to double ionization at the given velocity according to calculations, negative projectiles tend to push electrons "into their parent helium atom", thus enhancing their probability for being deflected

81

off the nucleus giving rise to an enhanced "recoil-peak". Positively charged particles instead, more likely deflect electrons "away from the nucleus" thus emphasizing clean binary situations. Clearly, this systematic trend is observed in the coupled channel calculations as well [45]. Moreover, slight deviations from a perfect symmetry along the given axis are found in principal agreement with the experimental findings, although they are not as pronounced as in the measurement.

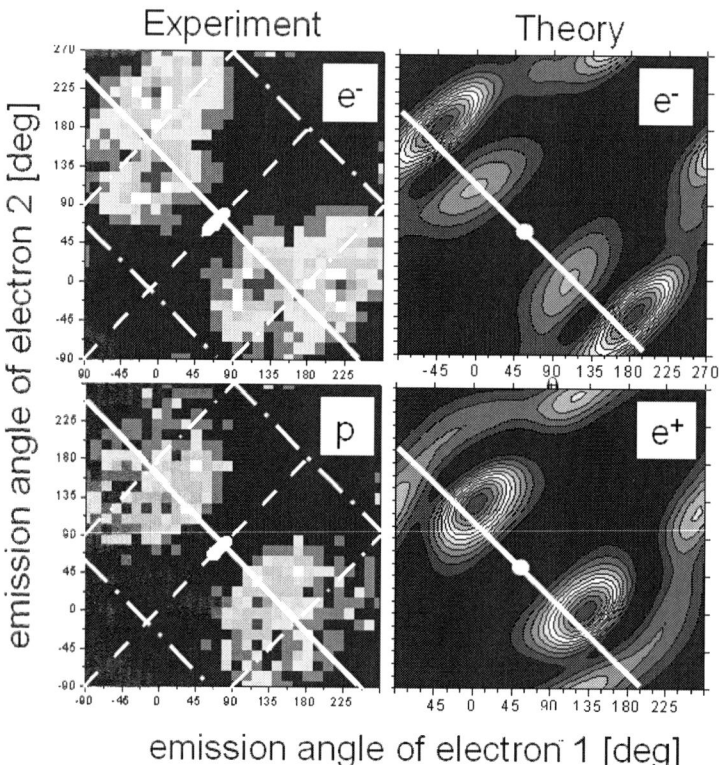

**FIGURE 8.** Fully differential cross sections for double ionization of helium in coplanar geometry as a function of the emission angles of the two low-energy (target) electrons. Left column: Experiment for electron (upper panel) and proton (lower panel) impact from [40]. Momentum transfer $0.8 \leq q \leq 1.4 a.u.$ indicated by the white bar. Electrons of equal energy are considered for $E_{e1} = E_{e2} < 25 \ eV$ (see text). Right column: Convergent Close Coupling calculations (CCC) for electron (upper panel) and positron (lower panel) impact for $E_{e1} = E_{e2} = 5 \ eV$ and $\bar{q} = 0.6 \ a.u.$. [45]

Due to limited statistical significance of the experimental data especially for proton impact, enforcing large acceptance angles and broad energy windows to be used for both electrons and the recoiling ion, and due to the lack of theoretical models for ion impact that include higher order contributions, a more detailed quantitative comparison is not yet possible. In addition, similar experimental data at lower energies where the proton-antiproton difference in the ratio becomes larger can be accessed experimentally with electrons but cannot straight forward be compared to proton results due to the increasing influence of the threshold mass effect. Thus, on the experimental side, rigorously novel techniques and approaches are required in order to deliver high-quality benchmark data for the investigation of the correlated two-electron quantum problem in double ionization of helium.

## The USR WITH IN-RING REACTION-MICROSCOPE AT FLAIR

A Reaction-Microscope that can be integrated into a new-designed electrostatic storage ring for slow antiprotons is envisioned to represent the experimental break-through that should allow performing, for the first time, differential single and multiple ionization cross section measurements for antiprotons colliding with atoms, molecules and clusters. The luminosity for in-ring experiments will be increased by at least five orders of magnitude compared to a single pass situation and cross sections as small as $10^{-22}$ cm$^2$ will become accessible under single collision conditions. Total, as well as any differential cross sections up to FDCS including ionization-excitation reactions will become measurable serving as benchmark data for theory. Several theory groups world wide now concentrate efforts to solve the fundamental few-body Coulomb problem for half-cycle pulses in the atto-to femtosecond time-regime as generated in antiproton collisions after the success for electron projectiles, as has been mentioned in the introductory part of the paper. Moreover, the formation of antiprotonic atoms, molecules or of protonium might be explored in kinematically complete experiments if ultra-low energies can be realized in the ring, yielding unprecedented and important information on (n, l)-distributions of captured antiprotons [46] as well as precise spectroscopic data of the respective energy levels [47].

In order to achieve these ambitious goals, challenging developments in both, storing and imaging techniques have to be achieved. To mention some of them, electron cooling must be demonstrated at electron energies as low as 10 eV, deceleration of antiprotons must be realized in an electro-"static" storage ring for the first time, a Reaction-Microscope has to be implemented in such a way that the electric and magnetic projection fields don't disturb the circulating beam and, finally, the stored beam has to be bunched into a single circulating ion packet of not more than 2 ns length in order to provide a timing signal for the Reaction Microscope. Electron cooling at low energies seems to be feasible on the basis of excellent results achieved with the new electron cooler at the TSR in Heidelberg that is based on ultra-cold, laser induced electron emission from a pre-cooled GaAs surface [48]. Moreover, a Reaction Microscope has been designed to be integrated into a storage ring and is anticipated to be implemented into the ESR (experimental storage ring) of GSI at the end of 2005, such that experimental experience with such machines will be available soon.

In order to access the remaining challenges – deceleration as well as beam bunching at low energies – and to enable at the same time novel experiments with stored and cooled molecular ions as well as highly charged ions from the Heidelberg EBIT (Electron Beam Ion Trap) the development of a test-ring has been successfully launched at the Max-Planck-Institute for Nuclear Physics (MPI-K) in Heidelberg [49].

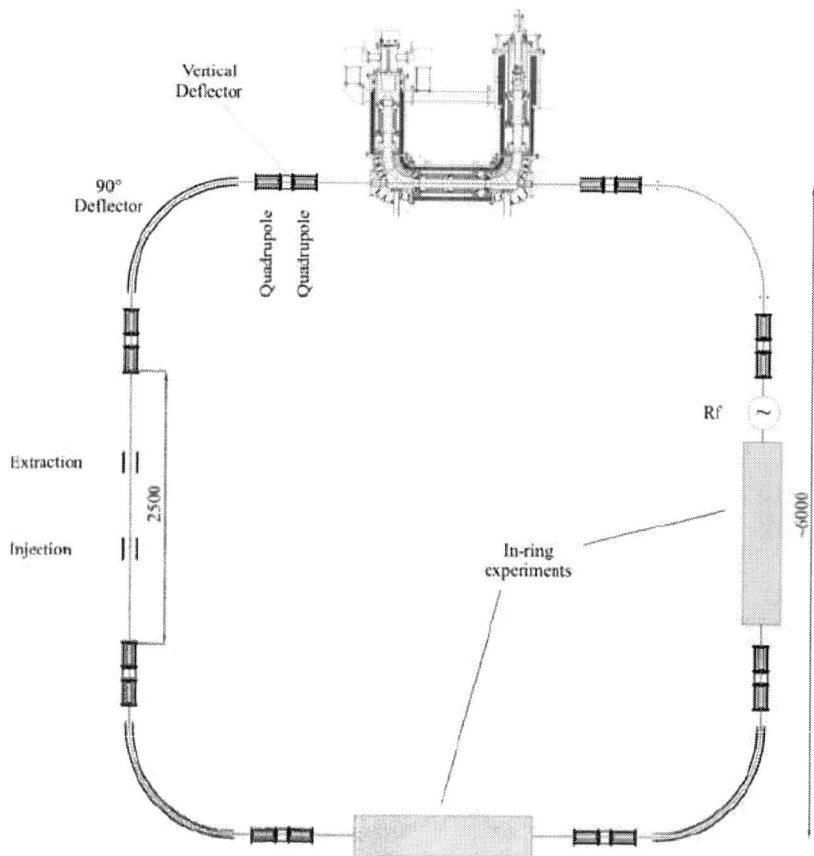

**FIGURE 9.** Outline of the new electro-static ultra-low energy storage ring (USR) for antiprotons at the new Facility for Low-energy Antiproton and Ion Research (FLAIR).

This Cryogenic Storage Ring (CSR) will be operated at cryogenic wall temperatures of below 10 K in order to achieve supreme vacuum conditions guaranteeing long storage times for molecular ions such that they can vibrationally and even rotationally relax into their ground-state [50].

In Figure 9 a schematic sketch of the anticipated ultra-low energy storage ring (USR) to be integrated into the proposed facility for low-energy antiproton and ion research (FLAIR), as described in detail in the technical design report [51], is shown. In short, the symmetric four-sided machine with a circumference of about 30 m shall store antiprotons at energies between 20 and 300 keV. Cylindrical $90^0$-deflectors along with quadrupole doublets with integrated steerers will keep the beam on stable orbits and several stable working points have been found in numerical calculations. In the straight sections of the ring, the electron cooler and the Reaction Microscope along with a supersonic gas jet will be placed, single turn injection as well as fast and slow extraction by pulsed deflectors will be realized and the remaining open side is available to implement further in-ring experiments. Here, a merged positron beam is presently envisaged in order to form anti-hydrogen in flight. In addition to the in-ring experiments to be performed, and very importantly, the ring will serve as an efficient deceleration and cooling device to deliver a low emittance ($\varepsilon \sim 10$ mm·mrad), mono-energetic ($\Delta p/p \sim 10^{-3}$) 20 keV antiproton beam to various trap experiments.

# SUMMARY

The understanding and the envisioned future control of the correlated quantum-motion of few-electron systems on ultra-short femto- to attosecond time scales are among the most attractive and challenging problems in contemporary atomic, molecular and laser physics. It has been shown in this contribution that slow antiprotons provide a unique tool to experimentally access this time regime, being typical for the dynamics of outer-shell electrons in atoms and molecules which, in turn, form chemical bonds and, thus, determine our every-day life.

At the same time we have provided striking evidence, that this motion is only poorly understood. Whereas tremendous progress has been achieved in the theoretical description of electron impact induced correlated dynamics, troubling and unresolved discrepancies between experiment and theory are found for the simplest and fundamental ion- or antiproton-impact induced reactions, like single or double ionisation of helium. Recent experiments have brought to light, that only fully differential cross sections, i.e. the complete determination of the final state momenta in such reactions can serve as the ultimate benchmark for theory. Certainly, agreement on the level of total or even single differential cross sections does not imply understanding of correlated quantum dynamics!

Finally, planned next-generation experiments with stored and cooled antiprotons at the new FLAIR facility at GSI, using advanced in-ring imaging techniques have been shortly described.

# REFERENCES

1. P.A.M. Dirac Proc. Roy. Soc. London A123 (1929) 714

2. T.N. Rescigno, M. Baertschy, W.A. Isaacs, and C.W. McCurdy, Science 286 (1999) 2474; M. Baertschy, T.N. Rescigno, C.W. McCurdy, Phys. Rev. A 64 (2001) 022709

3. I. Bray, Phys. Rev. Lett. 89 (2002) 273201-1

4. A.S. Kadyrov, A.M. Mukhamedzhanov, A.T. Stelbovics, and I Bray, Phys. Rev. Lett. 91 (2003) 253202-1

5. K. Bartschat, E.T. Hudson, M.P, Scott, P.G. Burke, and V.M. Burke, J. Phys. B 29 (1996) 115

6. M.S. Pindzola, D. Mitnik, and F. Robicheaux, Phys. Rev. A 62 (2000) 062718

7. G. Sakhelashvili, A. Dorn, C. Höhr, J. Ullrich, Phys. Rev. Lett., in print

8. M. Keim, A. Achenbach, H.J. Lüdde and T. Kirchner, Phys. Rev. A 67 (2003) 062711-1

9. E.Y. Sidky, C. Illescas and C.D. Lin, Phys. Rev. Lett. 85 (2000) 1634

10. A. Rudenko, K. Zrost, C.D. Schröter, V.L.B. de Jesus, B. Feuerstein, R. Moshammer and J. Ullrich, J. Phys. B 37 (2004) L407

11. A. Hasan, N.V. Meydanyuk, B.J. Fendler, A. Voitkiv, B. Najjari, and M. Schulz, J. Phys. B. 37 (2004) 1923

12. N.V. Meydanyuk, A. Hasan, M. Foster, B. Tooke, E. Nanni, D.H. Madison, and M. Schulz, PRL submitted

13. M. Schulz, R. Moshammer, D. Fischer, H. Kollmus, D.H. Madison, S. Jones and J. Ullrich, Nature, 422 (2003) 48

14. R. Moshammer, A. Perumal, M. Schulz, V.D. Rodríguez, H. Kollmus, R. Mann, S. Hagmann, and J. Ullrich, Phys. Rev. Lett. 87 (2001) 223201

15. M. Schulz, R. Moshammer, A.N. Perumal, and J. Ullrich, J. Phys. B 35 (2002) L161

16. M. Hentschel, R. Kienberger, Ch. Spielmann, G.A. Reider, N. Milosevic, T. Brabec, P. Corkum, U. Heinzmann, M. Drescher, F. Krausz, Nature, 414 (2001) 509

17. R. Dörner, H. Khemliche, M.H. Prior, C.L. Cocke, J.A. Gary, R.E. Olson, V. Mergel, J. Ullrich, H. Schmidt-Böcking; Phys. Rev. Lett. 77 (1996) 4520

18. R. Moshammer et al., Phys. Rev. Lett. 79 (1997) 3621

19. A. Rudenko, K. Zrost Th. Ergler, A.B. Voitkiv, B. Najjari, V.L.B. Jesus, B. Feuerstein, C.D. Schröter, R. Moshammer, and J. Ullrich J. Phys. B, in print

20. L.H. Anderson, P. Hvelplund, H. Knudsen, and S.P. Møller, Phys. Rev. Lett. 57 (1986) 2147

21. P Hvelplund, H. Knudsen, U. Mikkelsen, E. Morenzoni, S.P. Møller, E. Uggerhøj, and T. Worm, J. Phys. B 27 (1994) 925

22. T. Bronk, J.F. Reading, and A.L. Ford, J. Phys. B 31 (1998) 2477

23. R. Moshammer, M. Unverzagt, W. Schmitt, J. Ullrich and H. Schmidt-Böcking, Nucl. Instr. Meth. B108 (1996) 425

24. C. Höhr, A. Dorn, B. Najjari, D. Fischer, C.D. Schröter, and J. Ullrich, Phys. Rev. Lett. (2005) accepted

25. R. Dörner, V. Mergel, O. Jagutzki, L. Spielberger, J. Ullrich, R. Moshammer, H. Schmidt-Böcking, Physics Reports 330 (2000) 95-192

26. J. Ullrich, R. Moshammer, A. Dorn, R. Dörner, L.Ph. H. Schmidt, H. Schmidt-Böcking, Reports on Progress in Physics 66 (2003) 1463

27. *Many-Particle Quantum Dynamics in Atomic and Molecular Fragmentation* edited by J. Ullrich and V.P. Shevelko, Springer series on atomic, optical, and plasma physics 35 Springer Berlin (2003)

28. H. Ehrhardt, M. Schulz, T. Tekaat, and K. Willmann, Phys. Rev. Lett. 22 (1969) 89

29. A. Lahmam-Bennani, J. Phys. B 24 (1991) 2401

30. R. Moshammer, J. Ullrich, M. Unverzagt, W. Schmidt, P. Jardin, R.E. Olson, R. Mann, R. Dörner, V. Mergel, U. Buck and H. Schmidt-Böcking, Phys. Rev. Lett. 73 (1994) 3371-3374

31. M. Schulz, R. Moshammer, D.H. Madison, R.E. Olson, P. Marchalant, C.T. Whelan, H.R.J. Walters, S. Jones, M. Foster, H. Kollmus, W. Schmitt, A. Cassimi, and J. Ullrich, J. Phys. B 34 (2001) L305

32. D. H. Madison, D. Fischer, M. Foster, M. Schulz, R. Moshammer S. Jones, and J. Ullrich, Phys. Rev. Lett. 91 (2003) 253201-1

33. J. Fiol and R.E. Olson, J. Phys. B 37 (2004) 3947

34. P.D. Fainstein, L. Gulyás, F. Martín, and A. Salin, Phys. Rev. A 53 (1996) 3243

35. W. Schmitt, J. Ullrich, R. Moshammer, F. S. C. O'Rourke, H. Kollmus, L. Sarkadi, R. Mann, R.E. Olson, Phys. Rev. Lett. 16 (1998) 4337

36. T. Kirchner, L. Gulyás, M. Schulz, R. Moshammer, and J. Ullrich, Phys. Rev. A. 65 (2002) 042727

37. A. Voitkiv and J. Ullrich, Phys. Rev. A 67 (2003) 062703

38. J. Ullrich, R. Dörner, H. Berg, C.L. Cocke, J. Euler, K. Froschauer, S. Hagmann, O. Jagutzki, S. Lencinas, R. Mann, V. Mergel, R. Moshammer, H. Schmidt-Böcking, H. Tawara and M. Unverzagt, Nucl. Instr. Meth. B87 (1994) 70

39. J.H. McGuire, Phys. Rev. Lett. 49 (1982) 1153

40. D. Fischer, R. Moshammer, B. Feuerstein, A. Dorn, C.D. Schröter, J.R. Crespo Lopez-Urrutia, C. Höhr, H. Kollmus, S. Hagmann, and J. Ullrich Phys. Rev. Lett. 90 (2003) 243201-1

41. C. Díaz, F. Martín, and A. Salin, J. Phys. B 33 (2000) 4373

42. A. Taouil, A. Lahmam-Bennani, A. Duguet, and L. Avaldi, Phys. Rev. Lett. 81 (1998) 3548

43. A. Dorn, R. Moshammer, C. D. Schröter, T. Zouros, W. Schmitt, H. Kollmus, R. Mann, J. Ullrich, Phys. Rev. Lett. 82 (1999) 2496

44. A. Dorn, A. Kheifets, C.D. Schröter, B. Najjari, C. Höhr, R. Moshammer, and J. Ullrich, Phys. Rev. Lett. 86 (2001) 3755

45. A.S. Kheifets, Phys. Rev. A 69 (2004) 032712

46. J.S. Cohen, Phys. Rev. A 62 (2000) 022512-1

47. D. Fischer, B. Feuerstein, R. D. DuBois, R. Moshammer, J. R. Crespo Lopez-Urrutia, I. Draganic, H. Lörch, A. N. Perumal, and J. Ullrich, J. Phys. B **35** (2002) 1369

48. Progress Report 2003/04, Max-Planck-Institute for Nuclear Physics (2005), in print

49. J. Ullrich, A. Wolf, R. von Hahn, M. Grieser, J.R. Crespo López-Urrutia, D.A. Orlov, C.D. Schröter, D. Schwalm, C.P. Welsch, D. Zajfman, and X. Urbain *Next Generation Low-Energy Storage Rings for Antiprotons, Molecules, and Atomic Ions in Extreme Charge States.* Proposal to the Max-Planck Society (2004)

50. D. Zajfman, A. Wolf, D. Schwalm, D.A. Orlov, M. Grieser, R. von Hahn, C.P. Welsch, J.R. Crespo López-Urrutia, C.D. Schröter, X. Urbain, and J. Ullrich, submitted

51. FLAIR Technical Design Report  http://www-linux.gsi.de/~flair/

# Ionization Dynamics in p and p̄ on Helium collisions

M. Schöffler[1], H. Schmidt-Böcking[1], L. Ph. H. Schmidt[1], O. Jagutzki[1], V. Mergel[1], R. Dörner[1], J. Titze[1], K. Khayyat[1], Th. Weber[1], A. L. Godunov[2], C. T. Whelan[2], J. Walters[3]

[1] *Institut für Kernphysik der J. W. Goethe-Universität, Max von Laue-Straße 1, 60438 Frankfurt am Main, Germany*

[2] *Department of Physics, Old Dominion University, Norfolk, Virginia 23529-0116, USA*

[3] *Department of Applied Mathematics and Theoretical Physics, Queen's University, Belfast, BT7 1NN, UK*

**Abstract.** Differential cross sections for He single ionization in fast p and p̄ impact are presented and compared to theory. To investigate possible correlation effects on the double ionization cross section in p̄ collisions the in momentum space complete differential cross sections for fast p on He transfer ionization processes has been investigated and the influence of correlation effects, i.e. the influence of so called "off-shell" contributions in the He ground state wave function, has been measured.

**Keywords:** proton, antiproton
**PACS numbers: 03.65.Nk, 34.50.FA, 34.80.Pa**

## INTRODUCTION

It is one of the most interesting puzzles in atomic collision physics that double ionization cross sections of He in p and collisions differ by about a factor two, whereas the single ionization cross sections are nearly identical [1]. Many experimental and theoretical investigations were performed to solve this puzzle [1-6]. But so far no satisfactory explanation has been given to identify the origin of these differences. Recent theoretical work by Diaz et al. [7,8] and Morishita et al. [9,10] explores these effects with respect to electron-electron correlations in the He ground state wave function. They conclude that these correlation effects, i.e. "off-shell" contributions in the He ground state, lead to higher double ionization cross sections in p̄ impact. In this paper we will present the comparison of differential cross sections in p and p̄ on He collisions for single ionization and discuss the correlated double

CP793 *Physics with Ultra Slow Antiproton Beams*
edited by Yasunori Yamazaki and Michiharu Wada
© 2005 American Institute of Physics 0-7354-0282-5/05/$22.50

ionization from the highly correlated "off-shell" contributions in fast p on He transfer ionization processes.

As more differential the measured cross sections are as more insight they provide on the many-particle ionization mechanism. Measuring in momentum space complete differential ionization cross sections the ultimate stage of differential cross section is revealed. For such highly differential cross sections the final-state momentum pattern differ strongly for different projectiles like photons, ions or electrons, particularly if double and multiple ionization processes are considered. Even when the total momentum transfer from projectile to the target object is negligibly small compared to the initial-state momenta the final-state momentum pattern show characteristic features for different momentum transfers and projectiles. The final-state momentum pattern for photons strongly depend on the photon polarization and on the ratio of photon energy compared to the electron binding energy. The photon absorption process (photo effect) creates a very different pattern than the photon scattering process (Compton effect). Going from slow to fast ions the ionization process varies dramatically due to different ionization processes like molecular promotion in slow collisions or ionization by virtual photon absorption processes in fast collisions. Dependent on the collision process the final-state momentum pattern clearly depend on initial- or final-state correlations. In spite of great progress in theory the details of multiple ionization dynamics are not sufficiently well understood.

To measure complete differential cross sections in momentum space, i.e. performing a multi-coincidence study between very slow recoil ions and low energetic electrons new state-of-the-art imaging techniques have to be applied. One one these successful techniques is "COLTRIMS" (COLd Target Recoil Ion Momentum Spectroscopy) [11,12].

## THE COLTRIMS IMAGING TECHNIQUE

The recent development of the COLTRIMS technique provides a coincident multi-fragment imaging technique for eV and sub-eV fragment detection. In its completeness it is as powerful as the bubble chamber in high energy physics. Based on state-of-the-art cooling techniques (super sonic jets, MOT etc.) and nuclear physics imaging methods fragmentation processes of atoms, molecules, clusters etc. induced by single photon absorption, electron or ion impact can completely be explored in momentum space with high resolution. In recent benchmark experiments quasi snapshots of the correlated dynamics between electrons and nuclei has been made for atomic and molecular objects [13,14].

So far it was generally assumed that the observation of correlated dynamics of electrons and the nuclei in atoms or molecules is beyond any technical realization, since it requires detection techniques, which must image the electronic motion in the low attosecond domain. Furthermore it requires the observation of several such particles at the same moment with high momentum resolution. With the recent development of the COLTRIMS imaging technique this observation window is now opened. Instead of measuring two transition energies in atoms or molecules at two different moments a few femtosec separated in time (e.g. with pump and probe laser technique) here the atom or molecule is very suddenly (a few attosec time duration)

90

fragmented into several free particles. Thus the time is too short that the momenta of these ejected particles can be changed by more than a few percent with respect to the initial momenta. The angular correlation and the momenta of these fragments contain the information on the correlated dynamics of the initial many-particle state. This requires a momentum resolution of at least a factor of ten better than the size of the initial momenta, i.e. typically 0.1 a.u.. These momenta correspond to electron energies of about 100 meV or ionic recoil energies of less than 100 μeV.

The principle of the COLTRIMS-method, namely measuring the momentum of the emitted charged particles from an atomic fragmentation process is as simple as determining the trajectory of a thrown stone. From knowing the position, from where the stone was slung and where it hits the target as well as measuring its time-of-flight, the trajectory of the stone and thus its initial velocity vector can precisely be determined. Furthermore, in order to achieve good precision we have to know whether the person, who throws the stone, was at rest in the frame of observation or with which relative velocity this person was moving. Thus to obtain optimal momentum resolution for the exploding fragments one has to bring the fragmenting object to a complete rest in the frame of measurement before the reaction occurs, i.e. if the object is a gas atom or molecule one has to cool it down to sub-milli Kelvin temperatures.

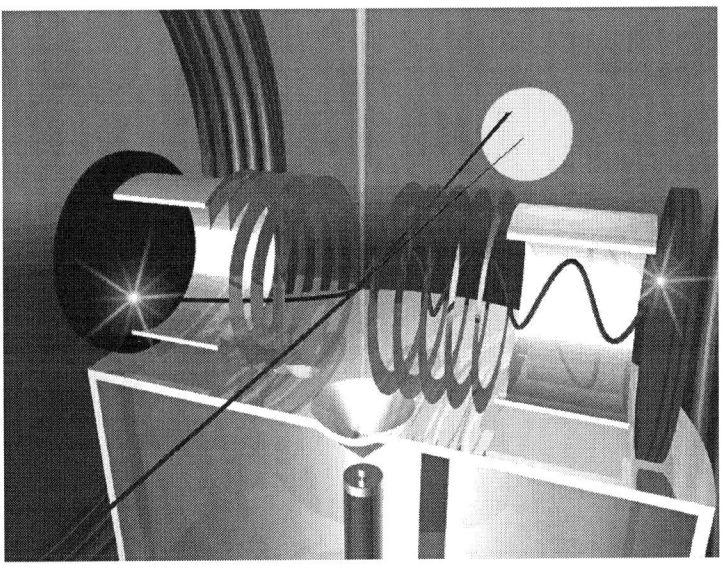

**FIGURE 1**: Artist view of the COLTRIMS imaging system [11,12].

In figure 1 the principle of the new reaction microscope (synonym: COLTRIMS) is presented. In a well designed electric field configuration (static or pulsed as well as with a superimposed magnetic field [11]) the positively as well as the negatively charged fragments are projected (typically with $4\pi$ solid angle) on two position-sensitive detectors. Measuring the impact position on the detector (typically < 0,1mm resolution) and the time-of-flight of the fragment (TOF) after the moment of fragmentation till hitting the detector, the particle trajectory, i.e. the particle momentum after fragmentation, can be determined. To improve its momentum resolution electrostatic lenses can be incorporated into the projection system, thus the influence of the unknown size of the target region, from where the fragments originate, can completely be eliminated [12]. To detect also the higher energetic fragments magnetic fields as well as pulsed electric fields can be used to improve multi-coincidence efficiency. If particle detectors based on fast delay-line position read-out are used multi-hit detection is possible. Even two particles hitting the detector at the "same" moment ($t_1 - t_2 < 1$ ns) can simultaneously be detected. The number of detected multi-hits is practically only limited by the electronics needed to store in event mode all information. In future even up to 100 particles per microsec might be detectable. Thus the COLTRIMS method is indeed powerful like an advanced bubble chamber system or even comparable with modern Time-Projection-Chamber systems used in high energy physics. Furthermore the rate of fragmentation processes can exceed several 100 kHz.

## DATA AND DISCUSSION

To reveal the mechanisms responsible for the observed [1,5] differences in total double ionization cross sections between p [15] and $\bar{p}$ impact Khayyat et al. [16] have measured double differential single ionization cross sections for p and $\bar{p}$ impact using the COLTRIMS imaging technique. Their experiment measured for 1 MeV ion impact the longitudinal momentum distribution of the emitted electron as well as of the emitted recoil ion. This distribution reflects directly the differential energy loss for the single ionization process. In figure 2 the electron momentum distributions for 1 MeV p and $\bar{p}$ impact are presented. The points represent the $\bar{p}$ and the solid line the p differential cross sections.

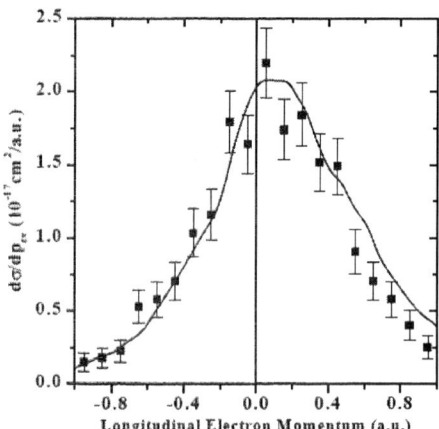

**FIGURE 2:** Electron longitudinal momentum distributions for 1 MeV p (solid line) and $\bar{p}$ impact (points) [16]

Within the experimental error bars the p and $\bar{p}$ data agree nearly perfectly in relative shape as well as in absolute height. Only in forward direction is a slight difference, the proton data are higher probably due to the attractive electron-p force, where the negative $\bar{p}$ pulls the electrons a little more backward. This effect is due to the final state interaction. Its influence on total cross sections is a few percent only.

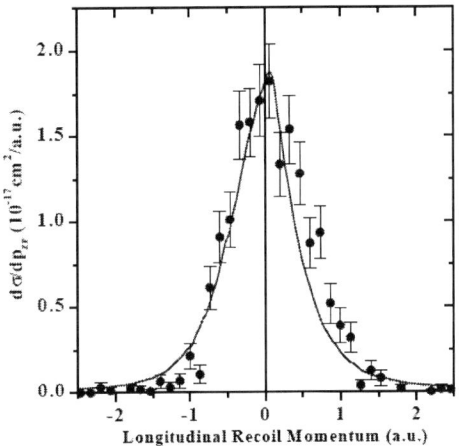

**FIGURE 3:** Recoil ion longitudinal momentum distributions for 1 MeV p (solid line) and $\bar{p}$ impact (points) [16]

Comparing the recoil data (figure 3) the final-state interaction (because of the positive recoil charge) favors forward emission for $\bar{p}$ compared the positive p. The absolute height is again in nice agreement yielding nearly identical total cross sections.

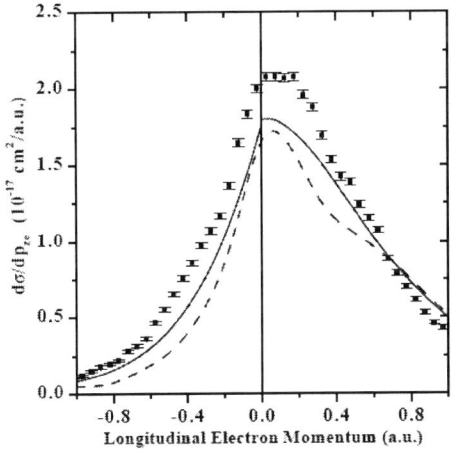

**FIGURE 4:** Electron longitudinal momentum distributions for 1 MeV p (points) in comparison with CDW (solid line) and CTMC (dashed line) calculations [3,16,17]

More severe is the observation that the best available ionization theories [3,16] do not perfectly describe the shape and absolute height of the differential cross sections even for protons (see figure 4). From these data we can conclude that the dominant single ionization mechanisms for p as well as p̄ impact obviously are not different.

Considering the reported differences in the total cross sections for double ionization for p and p̄ impact Morishita et al [9,10] explained such differences due to higher angular momentum components in the He ground state wave function. These higher angular momentum components are highly correlated electron pair states, so called "virtually excited" states, in nuclear physics called "off-shell" states. They are lying in the continuum. If the p̄ interact with only one of these virtually in the continuum excited electrons the other electron in this pair state is always emitted too to the continuum yielding double ionization. Very little is known on the dynamical correlation of these states. Thus p̄ must have a higher interaction probability via virtual photon exchange with these states.

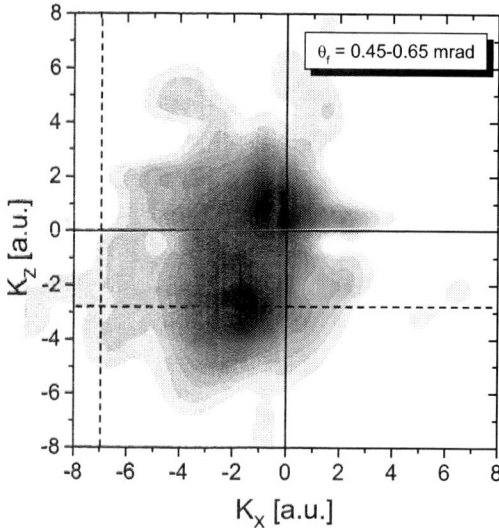

**FIGURE 5**: $He^{2+}$ recoil ion momentum distribution in the H° scattering plane for 1 MeV p on He, H° scattering angle of 0.55 mrad. The z-axis is given by the impacting p, the x-axis points transverse to the z-direction into the direction of the scattered H°. The recoil peak at $(K_z, K_x) = (0.8, -0.6)$ represents the EETI channel, the recoil peak at $(K_z, K_x) \approx (-2.8, -1.7)$ the TTI channel. The NTI would be located at the crossing of the dashed lines $(K_z, K_x) = (-2.8, -7.0)$ [18,19].

Since presently no p̄ beam with enough intensity is available we have investigated these virtually excited continuum states of the He ground state with fast proton beams. The p + He => H° + $He^{2+}$ + e transfer ionization channel allows us to study selectively these tiny contributions. Different mechanisms like the nucleus-electron Thomas process accompanied by not correlated ionization of the second electron (NTI) or the electron – electron Thomas process (EETI) contribute to transfer ionization. At large impact parameter, i.e. very small transverse momentum of the scattered H°, the kinematical capture channel dominates, where one electron with the resonant speed of the projectile can be captured by the proton. Here the electron tunnels through the p-He Coulomb barrier, thus we call this channel Tunneling Transfer Ionization (TTI). Using the high resolution COLTRIMS technique all channels can be clearly separated. In figure 5 the recoil momenta are shown for 1 MeV p impact and for H° scattering angles of 0.55 mrad. We can see that the TTI channel dominates and the nucleus-electron Thomas NTI channel is even not visible [18,19].

For the TTI process at large impact parameter the second electron is always backward emitted and the backward momentum increases with increasing proton velocity. The momentum vectors of the recoil ion and of the emitted electron are all pointing into the scattering plane indicating the non isotropic emission due to the non-$s^2$ contributions in the He ground state wave function. Thus the observed final state momentum pattern of the TTI process clearly indicates that the TTI process proceeds only via the non-$s^2$ contributions, i.e. the highly correlated angular momentum part of the He ground state wave function [19].

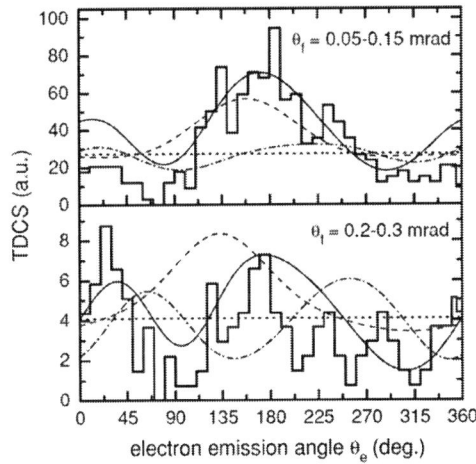

FIGURE 6: The triple differential cross section of transfer-ionization of helium by 630 keV proton impact as a function of the electron emission angle for several H° scattering angles. The cross sections were averaged over the electron emission energies ($Ee$ = 2.5–7.5 eV). Theory: solid line—calculations include both 'transfer first' and 'ionization first' mechanisms with angular and radial correlation in the initial state; chain line—the initial state includes radial correlation only; dashed line—calculations with 'transfer first' only that includes both angular and radial correlation; dotted line—'transfer first' calculations with radial correlation only. Experiment: COLTRIMS measurements normalized to theoretical cross sections around the peak value [20,21].

To present more evidence for the above outlined conclusions in figure 6 the calculations of Godunov et al. [20] together with the measurement of Schöffler et al. [21] are shown for 630keV proton impact. The theory is a first order approximation based on a very accurate multi configuration ground state wave function. It predicts for the non-$s^2$ contributions a very strong angular correlation of the emitted electron with respect to the other electron and a strong variation with impact parameter. The experimental data and the calculations (solid line in figure 6) are in very good agreement supporting fully the assumption that TTI reveals the non-$s^2$ contributions in the He ground state. The dotted line is $s^2$ only in clear disagreement with the data.

The TTI transfer ionization channel in fast p on He collisions provides the tools to reveal the absolute fractions of the highly correlated non-$s^2$ contributions in the He ground state and to test in a so far unprecedented manner the long range correlations in the He ground state. Thus very accurate correlated He ground state wave functions are now available which can be the basis of new calculations for the double ionization in $\bar{p}$ on He collisions. All these double ionization shake-off processes proceed via virtual photon exchange of the projectile with the He target. Why $\bar{p}$ and p should interact via virtual photons with those non-$s^2$ contributions in different ways remains somewhat unclear.

The new $\bar{p}$ facility with electrostatic storage rings for slow $\bar{p}$ at GSI will provide intense $\bar{p}$ beams to study the $\bar{p}$ ionization mechanisms in great detail. The here listed questions may then be answered by new fully differential data on He double ionization.

## ACKNOWLEDGEMENT

We acknowledge the great help of C. L. Cocke, E. Weigold, H. J. Lüdde, A. Kheifets and others for the numerous discussions in understanding the transfer ionization data. This work was supported by DFG, BMBF, GSI, and Roentdek GmbH.

## REFERENCES

1. L. H. Andersen, P. Hvelplund, H. Knudsen, S. P. Möller, K. Elsener, K.-G. Rensfelt, E. Uggerhoj, *Phys. Rev. Lett.* **57**, 2147 (1986)
2. J. F. Reading, L. Ford, *Phys. Rev. Lett.* **58**, 543 (1987).
3. R. E. Olson, *Phys. Rev. A* **36**, 1519 (1987).
4. L. H. Andersen, P. Hyelplund, H. Knudsen, S. P. Moller, A. H. Sorensen, K. Elsener, K.-.Rensfelt, E. Uggerhoj, *Phys. Rev. A* **36**, 3612 (1987).
5. L. H. Andersen, P. Hvelplund, H. Knudsen, S. P. Möller, J. O. P. Pedersen, S. Tang-Petersen, E. Uggerhoj, K. Elsener, and E. Morezoni, *Phys. Rev. A* **40**, 7366 (1989).
6. L. Ford, J. F. Reading, *J. Phys. B* **27**, 4215-4227 (1994).
7. C. Diaz et al., *J. Phys. B* **33**, 4373 (2000).
8. C. Díaz, F. Martín, S. Salin, *J. Phys. B* **35**, 2555 (2002).
9. T. Morishita, K. Hino, T. Edamura, D. Kato, S. Watanabe, and M. Matsuzawa, *J. Phys. B* **34**, L475-484 (2001).
10. T. Morishita, T. Sasajima, S. Watanabe, and M. Matsuzawa, *NIMB* **214**, 144 (2004).
11. J. Ullrich, R. Moshammer, R. Dörner, O. Jagutzki, V. Mergel, H. Schmidt-Böcking, L. Spielberger, *J. Phys. B* **30**, 2917, „Topical Review" (1997)
12. R. Dörner, V. Mergel, O. Jagutzki, L. Spielberger, J. Ullrich, R. Moshammer and H. Schmidt-Böcking, *Phys. Rep.* **330**, 96 (2000)

13. A. Becker, R. Dörner and R. Moshammer, *J. Phys. B* **38** S753-S772 (2005)
14. J. Ullrich, R. Moshammer, A. Dorn, R.Dörner, L. Ph. H. Schmidt and H. Schmidt-Böcking, *Rep. Prog. Phys.* **66,** 1463-1545 (2003)
15. Th. Weber, Kh. Khayyat, R. Dörner, V. Mergel, O. Jagutzki, L. Schmitt, F. Afaneh, A. Gonzales, C. L. Cocke, A. L. Landers, H. Schmidt-Böcking, *J. Phys. B* **33,** 3331 (2000).
16. Kh. Khayyat, T. Weber, R. Dörner, M. Achler, V. Mergel, L. Spielberger, O. Jagutzki, U. Meyer, J. Ullrich, R. Moshammer, W. Schmitt, H. Knudsen, U. Mikkelsen, P. Aggerholm, E. Uggerhoej, S.P. Moeller, J. H. Rodriguez, S. F. C. O'Rourke, R. E. Olson, P. D. Fainstein, J. H. McGuire, H. Schmidt-Böcking, *J. Phys. B* **32,** L73 (1999).
17. P.D. Fainstein, J. Phys. B, **29**, L763 (1996)
18. V. Mergel, R. Dörner, Kh. Khayyat, M. Achler, T. Weber, H. Schmidt-Böcking, H.J. Lüdde, *Phys. Rev. Lett.* **86**, 2257 (2001)
19. H. Schmidt-Böcking, V. Mergel, R. Dörner, C. L. Cocke, O. Jagutzki, L. Schmidt, Th. Weber, H. J. Lüdde, E. Weigold, J. Berakdar, H. Cederquist, H. T. Schmidt, R. Schuch and A. S. Kheifets, *Europhys. Lett.* **62** (4), 477-483 (2003)
20. A. Godunov, C. T. Whelan, H. R. J. Walters, *J. Phys. B* **37**, L201-L208 (2004)
21. M. Schöffler, A. L. Godunov, C. T. Whelan, H. R. J. Walters, V. S. Schipakov, V. Mergel, R. Dörner, O. Jagutzki, L. Ph. H. Schmidt, J. Titze, E. Weigold and H. Schmidt-Böcking, *J. Phys. B* **38**, L123-L128 (2005)

# Energy and Angular Distributions of Electrons Emitted in Capture of Antiprotons by Helium and Neon

James S. Cohen

*Theoretical Division, Los Alamos National Laboratory,*
*Los Alamos, New Mexico 87545*

**Abstract.** Using results of a previous theoretical treatment of antiproton capture by helium and neon atoms, the energies and angular distributions of the electrons emitted in the capture process are analyzed for various incident antiproton energies. The emitted electron energies are considerably higher for the neon target than for the helium target. The electron energies increase with increasing incident antiproton energies, but this dependence is fairly weak. The angular distributions of the emitted electrons are approximately isotropic. They are similar for helium and neon and depend only weakly on the antiproton energy. The angular distributions of uncaptured antiprotons at similar and somewhat higher collision energies are also presented.

**Keywords:** antiproton, capture, ionization, angular distribution, helium, neon
**PACS:** 36.10.-k, 34.10.+x, 25.43.+t, 03.65.Sq

## INTRODUCTION

In capture of antiprotons ($\bar{p}$), one or more electrons are ionized and carry off the binding energy. In this process, multiple electronic continua come into play and electron–electron as well as antiproton–electron correlation may be important. Many $\bar{p}$-capture experiments have been done with the noble-gas atoms, as well as with simple and complex molecules. However, these experiments were done in media where multiple collisions occurred and only the resulting x rays were detected. It has previously not been possible to measure the cross section for capture at a specific incident energy or to detect the emitted particle energies and angles resulting from a single collision. The recently developed source of slow antiprotons trapped in near vacuum (antiproton decelerator, AD, at CERN), has changed this picture. The anticipated experiments [1, 2] will provide a rigorous test on our understanding [3]. The purpose of the present calculations is to provide some guidance for the requirements of experimental detectors, including the energies and angles of the emitted particles, as well as the magnitudes of the cross sections.

Most of the detailed theory has been done for the hydrogen atom [4, 5], while capture by this simplest target is yet to be examined experimentally. Several theoretical treatments of the hydrogen atom have shown that the capture energy primarily goes to target ionization with the ionized electron carrying off relatively little kinetic energy, as predicted by the adiabatic-ionization model [6]. Capture by multielectron atomic targets involves additional questions: (i) are more than one electron ionized, (ii) is the target atom left in its ground electronic state immediately after capture, and (iii) are the

*CP793 Physics with Ultra Slow Antiproton Beams*
edited by Yasunori Yamazaki and Michiharu Wada
© 2005 American Institute of Physics 0-7354-0282-5/05/$22.50

electrons ionized with very low kinetic energy? The analog to the adiabatic-ionization model would suggest the answer to (i) is the minimum number consistent with a positive answer to (ii) and (iii).

Treating multi-electron targets requires a theoretical method capable of treating all electron dynamics, including multiple ionization and correlation. Previous work showed that muon-electron correlation in capture by the hydrogen atom is essential [7]. Except for the one-electron atomic target, these demands exceed the capability of any existing fully quantum-mechanical method, but are treatable by the quasiclassical method known as fermion molecular dynamics (FMD) [8]. The FMD method uses pseudopotentials to approximate quantum-mechanical behavior and formulate the problem within Hamilton's equations of motion. It accounts for the three-dimensional, correlated motion of all electrons. It may be noted that obtaining detailed differential results, like the energies and angles of the particles in the final state, in addition to the total capture cross section, is much easier with a quasiclassical method than with a completely quantum-mechanical method. The capture cross sections for helium and neon have been previously published [9]. In the present work, these results are further analyzed to obtain the differential quantities.

## THE FERMION MOLECULAR DYNAMICS (FMD) METHOD

The FMD method [8] utilizes the Kirschbaum-Wilets ansatz for atomic structure [10]. In this model, pseudopotentials $V_H$ and $V_P$ constrain the quasiclassical dynamics to satisfy the Heisenberg uncertainty and Pauli exclusion principles, respectively. The multielectron atom, which does not exist classically, is stabilized and possesses a shell structure [11]. Similar terms are included for the exotic atom structure, but have little effect since it is formed in highly excited states, which behave nearly classically according to the correspondence principle.

The FMD effective Hamiltonian for the system is written

$$H_{\mathrm{FMD}} = H_0 + V_H + V_P \tag{1}$$

where $H_0$ is the usual Hamiltonian of the system containing the kinetic energies of all particles and the Coulomb potentials for all pairs of particles. The extra terms are of the form

$$V_H = \sum_{i=1}^{N_e} \frac{\hbar^2}{\mu_{ni} r_{ni}^2} f(r_{ni} p_{ni}; \xi_H, \alpha_H) + \frac{\hbar^2}{\mu_{n\bar{p}} r_{n\bar{p}}^2} f(r_{n\bar{p}} p_{n\bar{p}}; \xi_H, \alpha_H) \tag{2}$$

and

$$V_P = \sum_{i=1}^{N_e} \sum_{j=i+1}^{N_e} \frac{\hbar^2}{\mu_{ij} r_{ij}^2} f(r_{ij} p_{ij}; \xi_P, \alpha_P) \delta_{s_i, s_j} \tag{3}$$

where the sums are over the $N_e$ electrons, $r_{ni}$ ($p_{ni}$) is the relative distance (momentum) of electron $i$ with respect to the nucleus $n$, $r_{n\bar{p}}$ ($p_{n\bar{p}}$) is the relative distance (momentum) of the antiproton with respect to the nucleus, $r_{ij}$ ($p_{ij}$) is the relative distance (momentum) of electron $j$ with respect to electron $i$, $s_i$ ($s_j$) is the spin of electron $i$ ($j$), and $\delta$ is the

Kronecker delta function. Hamilton's classical equations of motion are solved with the Hamiltonian $H_{FMD}$.

The constraining potentials are embodied by the function [10]

$$f(rp; \xi, \alpha) \equiv \frac{\xi^2}{4\alpha} \exp\left\{ \alpha \left[ 1 - \left(\frac{rp}{\xi\hbar}\right)^4 \right] \right\}. \tag{4}$$

where the parameter $\xi$ reflects the size of the core (Heisenberg or Pauli) while $\alpha$ is a hardness parameter. The only way in which we deviate from the original prescription of Kirschbaum and Wilets is to use values of $\xi_H$ and $\xi_P$ optimized for the target atom, rather than the universal values [11], and softer values of $\alpha_H$ and $\alpha_P$ as recommended by Beck and Wilets [12], $\alpha_H = 2.0$ and $\alpha_P = 1.0$.

The values of $\xi_H$ and $\xi_P$ were determined by nonlinear least-square fits of $H_{FMD}$ to match the experimental first ionization potential (IP$_1$) and nonrelativistic Hartree-Fock total binding energy ($E_{tot}$) of the target atom: $\xi_H = 0.9343$ for He; $\xi_H = 1.1311$ and $\xi_P = 1.511$ for Ne. This procedure is appropriate since the first ionization potential is expected to be most important for the capture dynamics and the higher ionization potentials will be correct on the average if the total energy is accurate.

The present calculations treat all electrons explicitly. The $6(N_e + 2)$ equations of motion, for the antiproton, the nucleus, and the target electrons, are solved in the laboratory frame. The target nucleus is initially stationary at the origin, and the projectile is started at $x = -10a_0$ with impact parameter $y = b$ chosen by uniform sampling of ranges of $b^2$ until converged. The target is initially orientated by a random Euler rotation.

Some indication of the accuracy of the FMD method can be gained from the FMD calculation of capture by the hydrogen atom [5], which is in excellent agreement with an accurate quantum-mechanical calculation [13]. Quantum-mechanical all-electron calculations for multielectron targets are beyond present theoretical capabilities. However, the FMD calculations, as well as calculations using the Fermi-Teller method (based on the Thomas-Fermi model of the atom), indicate that electrons inside the valence shell cannot be ignored. An effective one-electron method used for capture of antiprotons by the noble-gas atoms gave cross sections much too small for all the targets except helium [14].

## RESULTS

### Capture cross sections and electron ionization

Antiproton capture cross sections have been calculated for all the noble-gas atoms using the FMD method [15]. The results are given in Figure 1 as a function of the c.m. system energy. The capture cross sections extend to collision energies well over 100 eV for all the atoms except He. For helium, the capture cross section decreases rapidly at collision energies exceeding its first ionization potential [16, 9]; this behavior is similar to that of the hydrogen atom and can be interpreted as quasi-adiabatic ionization (strictly adiabatic ionization of helium cannot occur since the united-atom limit of $\bar{p}$+He is bound).

100

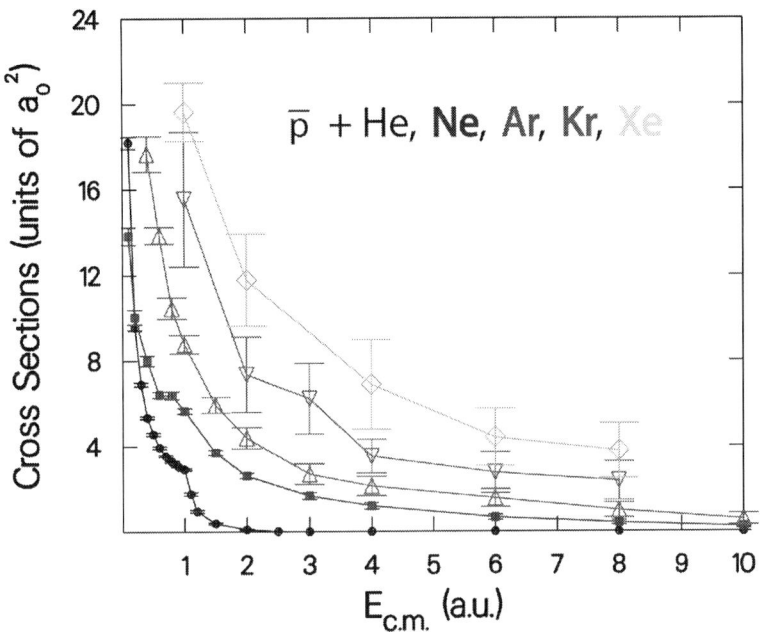

**FIGURE 1.** Cross sections for antiproton capture by He (circle), Ne (square), Ar (up triangle), Kr (down triangle), and Xe (diamond) as a function of incident antiproton energy, with Monte Carlo error bars of the FMD calculation [15].

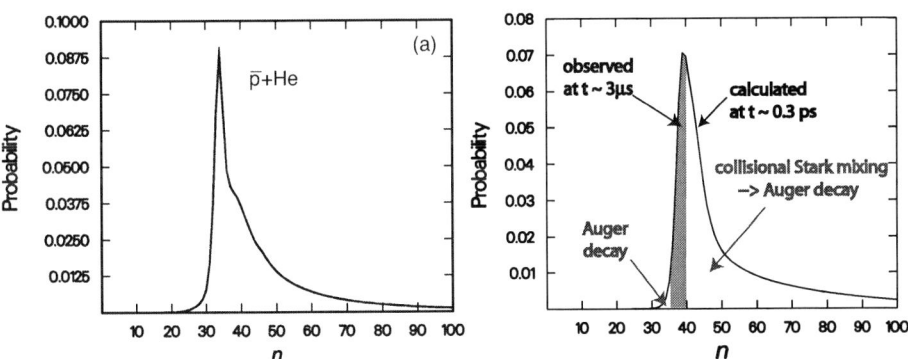

**FIGURE 2.** (a) Principal quantum number $n$ distributions (integrated over capture collision energies). These distributions were obtained at a time of $\sim 0.3$ ps after the arrival of the $\bar{p}$ in a thick target. At the peak both electrons are ionized, but in the region of the shoulder and higher some atoms retain an electron; (b) Principal quantum number distributions for the atoms with one electron remaining at time $\sim 0.3$ ps, with comparison to the experimental observation at a much longer time of $\sim 3$ $\mu$s [17].

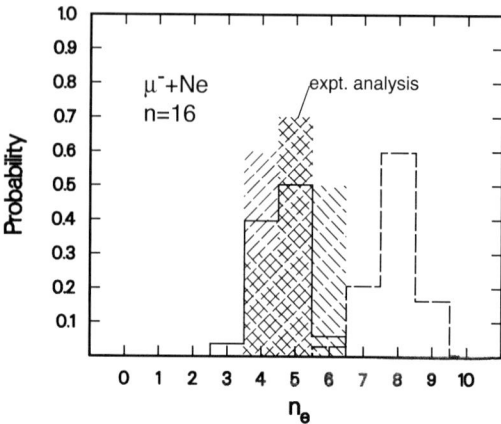

**FIGURE 3.** Number of residual bound electrons [9] when $\mu^-$Ne reaches $n = 16$. The dashed histogram includes weakly bound electrons (case i); the solid histogram counts only electrons that would still be bound after electron–electron Auger relaxation (case ii). The hatched area shows the results of the experimental analysis [22]; the cross-hatched area corresponds to $n_e = 4.7$.

The distribution of initial capture $n$ quantum numbers are shown in Figure 2 for $\bar{p}$ capture by He. At the time analyzed ($\sim 0.3$ ps) 20% of the atoms still retain one electron (another FMD calculation [16] found this fraction to be 22%). A similar conclusion has been reached by comparing the Auger and radiative lifetimes of the states [18]. Experiments have shown that only $\sim 3\%$ of the atoms retain an electron until much longer $\mu$s times, but the quasiclassical treatment is not valid for such long times. The electrons in the long-lived atoms reside in near-circular orbits, $l = n - 1$. Experimentally, long-lived states with $n > 40$ are absent [17]. The absence of states with $n > 50$ at long times is easily explained since initial states with $n > 50$, but $l < 49$ (approximately the largest $l$ populated [3]), will not be metastable. Theory indicates that states with $40 < n < 50$ do exist at earlier times but are destroyed by collisions before the experimental observations at $\mu$s times [18, 19, 20]. There is also some experimental evidence for such quenching [21].

Unlike helium, the capture cross sections for the heavier atoms display no obvious alteration in behavior at collision energies near their ionization energies, which is the maximum collision energy at which ionization of a single electron, carrying off zero kinetic energy, suffices to capture the incident exotic particle. We interpret this smoothness as due to the fact that electron correlation in the heavier atoms enables excitation or ionization of additional electrons. Generally, the capture cross sections at a given energy increase monotonically with the $Z$ of the noble gas.

The increase in the cross sections, especially at higher $E$, with increasing $Z$ suggests that electron correlation is important, so elimination of electrons in favor of a core potential would have to be done with great care. Although the quasiclassical method may overestimate correlation, it seems clear that multiple electrons participate in the capture and any one-electron method is destined to fail.

Since it is found that the kinetic energies of the ionized electrons are relatively small

[9], it is evident that captures at high energies require ionization of multiple electrons. In subsequent internal Auger processes, more electrons may be ionized if not refilled in collisions. We certainly want to characterize the state before this time, even though the "instant" of capture is not a well-defined quantity. To model an existing experimental analysis for $\mu^-$ capture in neon [22], calculations have been done in which each $\mu^-$+Ne trajectory is followed until the principal quantum number $n = 16$ is reached. The number of electrons remaining at this point is important to the subsequent cascade. The ion is actually in a shake-up state and some electrons will be spontaneously removed by purely electron-electron interactions.

We attempt to deal with this situation by utilizing two limiting measures of the number of bound electrons: (i) counting all remaining electrons, no matter how weakly bound and (ii) counting only the electrons that would remain bound if electronic Auger processes proceeded to completion while the exotic particle remained frozen in the level $n = 16$. The latter measure is approximated by comparing the energy of the electronic system with the *ground-state* energy of the ion having just lower energy. The minimum number of residual electrons is obtained by supposing that each ejected electron carries off negligible kinetic energy. The *effective* number of electrons should lie in between these two measures. The numbers of bound electrons, determined by these two measures, are shown in Figure 3. The experimental analysis [22] found a best fit for $\mu^-$Ne at $n = 16$ with $n_e = 4.7^{+0.8}_{-0.3}$. The fractional value of $n_e$ was interpreted as a mixture of the next lower and next higher integer charges, though higher and lower charge states cannot really be precluded. This experimental distribution agrees well with the measure (ii). This finding is consistent with both the fast reaction times for the light electrons and the weak interaction between the muon and very diffuse electrons. A similar situation can be expected to hold for $\bar{p}$Ne.

## Electron energy and angular distributions

As discussed above, the number of electrons ionized in capture of an antiproton (or negative muon) is not a well-defined quantity unless a specific excited state of the exotic atom is specified. Additional electrons may be ionized in the Auger-radiative-collisional cascade toward the ground state, which will ultimately be reached unless particle decay (in the case of $\mu^-$) or nuclear annihilation (in the case of $\bar{p}$) occurs first. The energy, and possibly angle, of the electron can be expected to depend on the stage at which emission occurs. All are emitted by Auger processes, but an Auger process due to the $\bar{p}$-$e^-$ interaction is quite different from one due to a $e^-$-$e^-$ interaction. We, somewhat arbitrarily, consider only the electrons that are emitted before the $\bar{p}$ is definitely captured (determined on the basis that the $\bar{p}$ no longer has sufficient energy available to escape).

The kinetic energies and the angular distributions of these electrons are shown in Figures 4 and 5 for helium and neon. The jaggedness of the curves is thought to be mostly due to the Monte Carlo statistics. Electrons at distances greater than $3a_0$, at the end of the trajectory integration, are assumed to be ionized. This choice of distance is rather arbitrary, but is mandated, in part, by the criteria previously used to determine the final state, which was concerned only with converging the antiproton capture. Anyway,

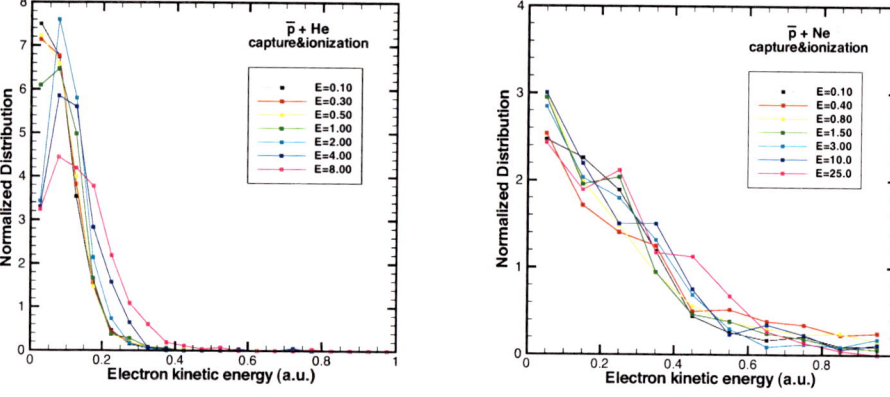

**FIGURE 4.** Energy distributions of electrons emitted in collisions of $\bar{p}$ with the helium and neon atoms at various collision energies (a.u. in the c.m. system). Each data point shown lies at the center of a histogram bin; the points are connected by straight line segments. At collision energies below the target ionization potential (0.90 a.u. for He and 0.79 a.u. for Ne), all $\bar{p}$ are captured; at higher energies a decreasing fraction of the $\bar{p}$ are captured.

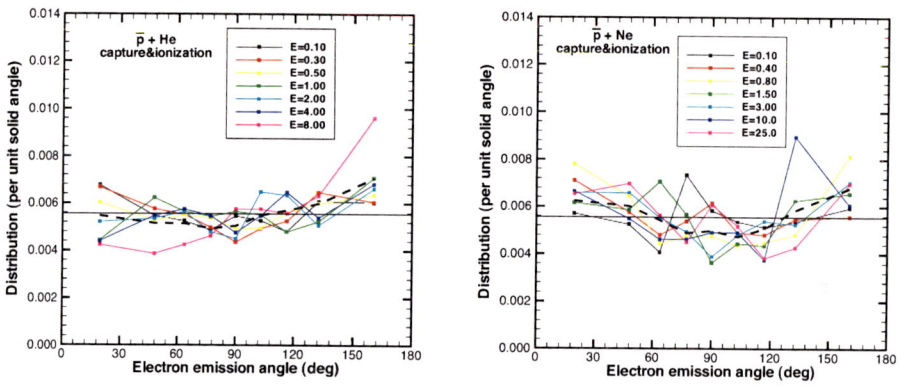

**FIGURE 5.** Angular distributions (in the lab frame) of electrons emitted in collisions of $\bar{p}$ with helium and neon atoms at various collision energies (a.u. in the c.m. system). The centers of the histograms are marked, with each bin containing a equal solid angle measure; these data points are connected by straight line segments. The jaggedness is within the Monte Carlo statistics. The average distribution at all calculated collision energies (including some not shown) is shown by a dashed curve. The horizontal line shows an isotropic distribution for comparison.

the present results are not found to be too sensitive to the choice of ionization distance, although the ending time of the trajectory integrations does cause subsequent ionizations due to internal Auger ionization to be missed.

The average numbers of electrons ionized, at this point in time, is 1.7 for helium and 3.9 for neon. These numbers should not be taken too seriously for three reasons: (i) the calculation has been stopped at relatively short times, (ii) the quasiclassical treatment is not expected to be accurate for $e^-$-$e^-$ Auger processes, and (iii) radiation is not taken into account. Such a short time is probably not resolvable in experiments. Radiation does not play a significant role in the initial capture, but is important in the subsequent cascade. It is evident in Figure 3 that more electrons are ionized at later times due to the relaxation of the shake-up state.

The kinetic energies of the emitted electrons are shown in Figure 4. They peak at very low, but nonzero, energies. The energies increase somewhat as the incident antiproton energy increases. This increase, as well as a peak at nonzero energy, is consistent with a diabatic interpretation, in which the ionization width is zero at the crossing into the continuum and higher collision energies enable deeper penetration before the occurrence of ionization. It is found that the electron energies are similar whether or not capture occurs, so only the totals are shown in Figure 4. Capture occurs if the sum of the potential and kinetic energies of the ejected electrons exceeds the incident antiproton energy (in the c.m. system). Thus capture always accompanies ionization if the collision energy is lower than the first ionization potential. Likewise, capture requires ionization of two or more electrons if the incident energy exceeds the sum of the first ionization potential and the maximum probable electron energy. It is evident from Figure 1, where the capture cross section for helium falls off rapidly at energies exceeding the first ionization potential, that capture by helium normally entails only single ionization. The second electron is usually ejected fairly quickly thereafter as the antiproton cascades to a lower level, but, as seen in Figure 2, the electron sometimes stays bound in a metastable state until much longer times.

For neon, it can be seen in Figure 1 that capture is still significant at collision energies much higher than the sum of the first ionization potential (0.79 a.u.) and the maximum probable electron energy ($\lesssim 1.0$ a.u.). This demonstrates the essential role of multiple ionization in capture by the higher-$Z$ elements. The distributions shown include all ejected electrons (up to the time when capture is deemed irreversible). The electrons subsequently ejected in the cascade of the many-electron target may be expected to have larger energies due to the larger quantum transition energies and to be distributed even more isotropically.

The angular distributions of these ionized electrons are shown in Figure 5. With the number of trajectories run, there does not appear to be any statistically significant difference between the angular distributions for different energies. The distributions appear to be approximately isotropic, with possibly a small deficiency in the orthogonal direction. The angular distributions for helium and neon are similar.

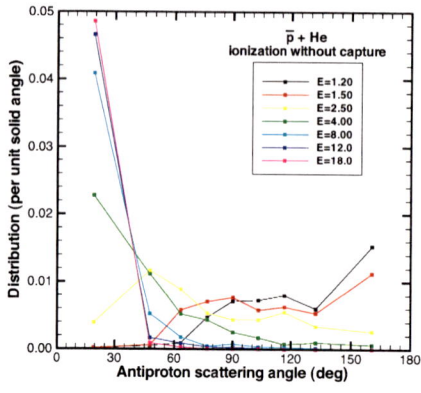
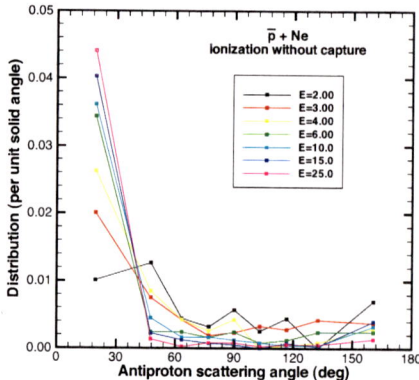

**FIGURE 6.** Angular distributions (in the lab frame) of antiprotons scattered in collisions of $\bar{p}$ with the helium and neon atoms where ionization occurs but the $\bar{p}$ is not captured. Each data point shown designates the center of a histogram bin.

## Angular distributions of scattered antiprotons at low energies

It may also be informative to detect the antiprotons inelastically scattered by the targets but not captured. These particles can be distinguished from the elastically scattered particles by their reduced energies. Their angular distributions are shown in Figure 6. At collision energies where some capture also occurs, the scattering angles can be seen to be quite large and peak in the backward direction. As the collision energy increases, the scattering moves increasingly toward the forward direction.

## ACKNOWLEDGMENTS

This work was performed under the auspices of the U.S. Department of Energy. I would like to thank Professor Y. Yamazaki for suggesting this calculation.

## REFERENCES

1. Y. Yamazaki, Nucl. Instrum. Methods B **154**, 174 (1999).
2. Y. Yamazaki, Hyperfine Interact. **138**, 141 (2001).
3. For a review of all theoretical methods used to treat capture of antiprotons, see J. S. Cohen, Rep. Prog. Phys. **67** 1769 (2004).
4. J. S. Cohen, in *Electromagnetic Cascade and Chemistry of Exotic Atoms*, edited by L. M. Simons, D. Horváth, and G. Torelli (Plenum Press, New York, 1990), Vol. 52 of the Ettore Majorana International Science Series, p.1.
5. J. S. Cohen, J. Phys. B **31**, L833 (1998).

6.  A. S. Wightman, Phys. Rev. **77**, 521 (1950).
7.  N. H. Kwong, J. D. Garcia, and J. S. Cohen, J. Phys. B **22**, L633 (1989).
8.  L. Wilets and J. S. Cohen, Contemp. Phys. **39**, 163 (1998).
9.  J. S. Cohen, Phys. Rev. A **62**, 022512 (2000).
10. C. L. Kirschbaum and L. Wilets, Phys. Rev. A **21**, 834 (1980).
11. J. S. Cohen, Phys. Rev. A **51**, 266 (1995).
12. W. A. Beck and L. Wilets, Phys. Rev. A **55**, 2821 (1997).
13. K. Sakimoto, J. Phys. B **35**, 997 (2002).
14. J. S. Briggs, P. T. Greenland, and E. A. Solov'ev, J. Phys. B **32**, 197 (1999).
15. J. S. Cohen, Phys. Rev. A **65**, 052714 (2002).
16. W. A. Beck, L. Wilets, and M. A. Alberg, Phys. Rev. A **48**, 2779 (1993).
17. M. Hori *et al.*, Phys. Rev. Lett. **89**, 093401 (2002).
18. G. Ya. Korenman, Hyperfine Interact. **101/102**, 463 (1996).
19. S. Sauge and P. Valiron, Chem. Phys. **265**, 47 (2001).
20. J. E. Russell, Phys. Rev. A **65**, 032509 (2002).
21. G. Bonomi, Nucl. Phys. **A692**, 137c (2001).
22. K. Kirch *et al.*, Phys. Rev. A **59**, 3375 (1999).

**ANTIHYDROGEN**

# Detecting Antihydrogen:
# The Challenges and the Applications

Makoto C. Fujiwara

*TRIUMF, Canadian National Laboratory for Particle and Nuclear Physics*
*4004 Wesbrook Mall, Vancouver, British Columbia, V6T 2A3, Canada*

**Abstract.** ATHENA's first detection of cold antihydrogen atoms relied on their annihilation signatures in a sophisticated particle detector. We will review the features of the ATHENA detector and its applications in trap physics. The detector for a new experiment ALPHA will have considerable challenges due to increased material thickness in the trap apparatus as well as field non-uniformity. Our studies indicate that annihilation vertex imaging should be still possible despite these challenges. An alternative method for trapped antihydrogen, via electron impact ionization, will be also discussed.

**Keywords:** Antihydrogen, Fundamental Symmetries, Si vertex detectors.
**PACS:** 52.27.Jt, 36.10.-k, 39.10.+j

## INTRODUCTION

A long term goal of antihydrogen research is precision tests of CPT and other fundamental symmetries via spectroscopy comparisons of hydrogen and its antimatter counter part, antihydrogen. The ATHENA experiment, located at CERN, achieved a major milestone by demonstrating production of cold antihydrogen atoms [1]. The ATHENA data taking is now completed, and a new experiment is proposed to continue its effort. ALPHA, Antihydrogen Laser Physics Apparatus, will aim to trap cold antihydrogen in a magnetic trap [2]. Stable trapping of antihydrogen will likely open up various possibilities in fundamental physics with cold anti-atoms. This article reviews the design issues of the ALPHA antihydrogen vertex detector.

## ANTIHYDROGEN DETECTOR

The capability to detect the vertex position of antihydrogen annihilations was an important feature of ATHENA not only for the identification of cold antihydrogen production [1], but also in the developments leading to the first production [3, 4]. Having real time images of antiproton losses was essential as a diagnostic when optimizing the trapped particle manipulation in a nested Penning trap.

In order to trap neutral antihydrogen atoms in ALPHA, we plan use an octupole magnetic field configuration for radial confinement, and a mirror field configuration for axial confinement. As we enter the unexplored regime of neutral anti-atom trapping, we believe it is crucial to retain the vertex imaging as a diagnosis tool. There

*CP793 Physics with Ultra Slow Antiproton Beams*
edited by Yasunori Yamazaki and Michiharu Wada
© 2005 American Institute of Physics 0-7354-0282-5/05/$22.50

exists very little experimental information on the behaviour of trapped particles in multipolar, and mirror magnetic fields. We will likely encounter unexpected new effects. Given the situation, the importance of real time imaging, together with other plasma diagnosis techniques we developed for ATHENA [5, 6], cannot be overstated.

We note further that vertex detection capability will play an important role in the future phases of ALPHA. Even after stable trapping of neutral antihydrogen atoms is achieved, performing spectroscopy measurements with few atoms will be extremely challenging, given the expected low signal rates. A clean vertex determination will help demonstrate laser transitions of anti-atoms, e.g. via resonant photo-ionization, and discriminate against possible sources of backgrounds.

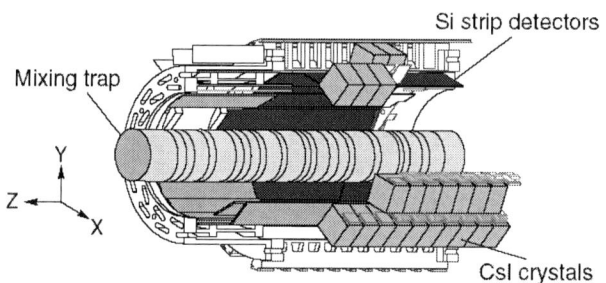

**FIGURE 1.** Schematic view of ATHENA antihydrogen detector [3, 4]. Two layers of double-sided Si strip detector and 192 CsI crystals surround the trap region. A 1.65 mm thick copper vacuum wall, between the trap and the detector, is not shown in the figure.

## THE CASE FOR VERTEX DETECTION IN ALPHA

In order to illustrate the power of the vertex detection, we will give in this section some examples from our experiences in ATHENA. Figure 2, taken from our first report [1], shows the difference in the vertex distributions between production data (cold mixing) and the control data (hot mixing). This provided an important piece of evidence for establishing the first production of cold antihydrogen atoms. Subsequent analyses using simulated vertex distributions showed that about 65% of annihilation events in Fig 2 (a) are due to antihydrogen and the rest due to background [7].

It should be stressed that detecting only antiproton annihilations, for example by external scintillators, is not sufficient evidence on its own for antihydrogen production, because of the existence of particle loss processes in the trap. Figure 3 further illustrates the importance of spatially sensitive detection. The axial vertex distributions are compared between the standard cold mixing and the "mixing" without positrons. A clear difference is observed. Suppression or inhibition of antihydrogen production due to causes such as plasma instabilities, vacuum deteriorations, or electrodes malfunctions, are sometimes otherwise difficult to identify, but they would show up immediately in the vertex distributions allowing rapid diagnosis of the system. An incomplete removal of the electrons from the trap (which is used to cool antiprotons in the first step of the mixing cycle) results in the vertex pattern similar to Fig. 3 (b) indicating suppressed antihydrogen production.

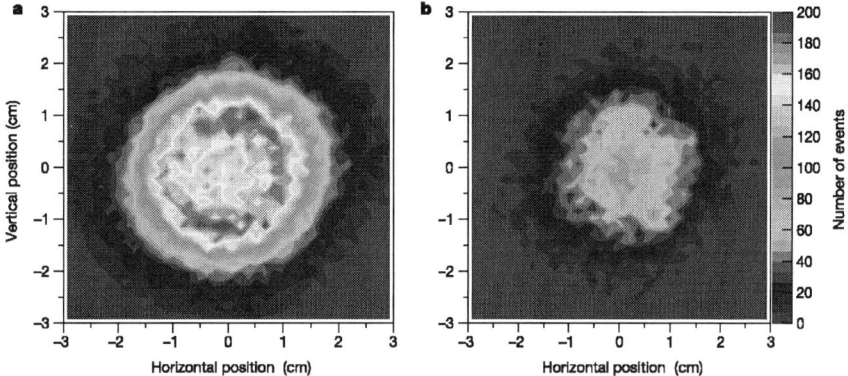

**FIGURE 2.** Schematic view of Fig. 2: X-Y distribution of antiproton annihilation vertices for (a) cold mixing and (b) hot mixing, obtained with the ATHENA detector. Both plots are normalized to the same number of mixing cycle. Enhanced annihilations on the trapped electrodes (inner radius 1.25 cm) in Fig (a) indicates antihydrogen production, whereas annihilations of the central part in Fig (b) represents antiproton annihilations on residual gas or ions.

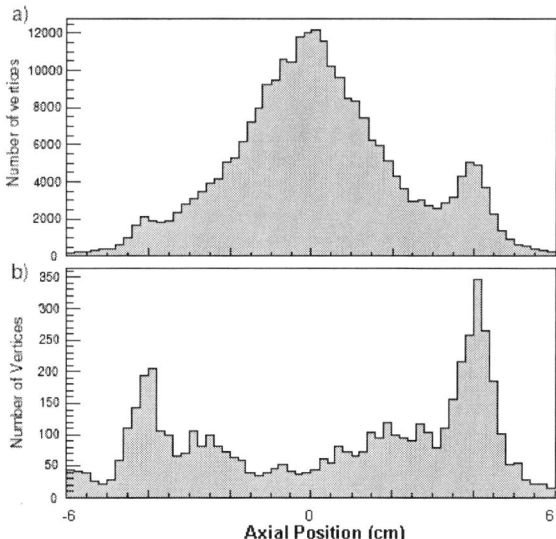

**FIGURE 3.** Axial (i.e. z) distribution of annihilation vertex for (a) standard cold mixing and (b) the data without positrons. Figure taken from [8].

While these observations attracted some interests on their own from the trap physics point of view, they have important implications for antihydrogen detection. Recall the radially symmetric distribution of annihilation at the trap wall for antihydrogen (Fig 2 a). The observation that *charged* particle loss at the wall results in hot spots, while the neutral antihydrogen atoms annihilates symmetrically, provides a new and effective signature of antihydrogen identification. An advantage is that we do not need to rely on the 511 keV detection, which are difficult due to its low efficiency and large background. In fact, in the recent runs of ATHENA, the presence or absence of the hot spots has been one of the most valuable diagnostic of our antihydrogen production processes.

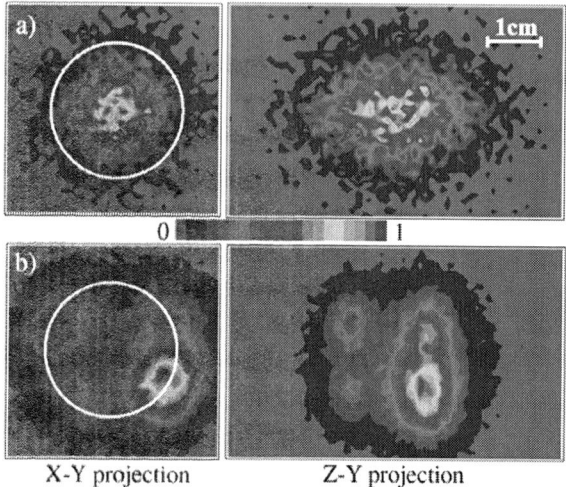

FIGURE 4. Annihilation images of trapped antiprotons in a harmonic Penning trap with (a) high residual gas density, and (b) low residual gas density. Both data are without positions. The electrode position is shown with a circle [4].

## DESIGN CONSIDERATIONS

The overall design of the ALPHA experiment will be driven by the requirements for neutral atom trapping. This implies, for the Si detector, that there will be a substantial amount of superconducting magnet materials between the trap and the detector. Our design goal is to retain the vertex position resolution similar to ATHENA of 4 to 5 mm ($\sigma$) as illustrated in the previous section.

### Modifications and Improvements over the ATHENA Detector

The ATHENA detector has worked well for the most part as discussed above, and we plant to adopt its basic design features. However, several modifications and improvements are foreseen.

The ATHENA detector did not allow the determination of the curvature of charged particle tracks in a 3T magnetic field, since it only had two double-sided layers. Hence the charged tracks were approximated with straight lines, and this was a dominating factor in the vertex resolution of 4-5 mm (see below). We plan to improve the tracking in the ALPHA vertex detector by having three or more Si layers. Gaseous detectors such as a multi-layered drift chamber or a time projection chamber would be an alternative possibility. The improved determination of the charged tracks will partly compensate the resolution loss due to the increased materials between the trap and the detector which are unavoidable.

511 keV gamma detection will be no longer essential in antihydrogen identification for the new generations of experiments such as ALPHA, as shown in Ref. [4, 7]. However, we will have a limited number of gamma detectors inside our magnet for the diagnosis of trapped positrons. Transmission of 511 keV gammas through a 1 cm thick Cu is of order of 25%, allowing sufficient count rates for the diagnosis of trapped cold positron plasma [9].

We envision operating the ALPHA detector at a temperature sufficiently higher than the ATHENA one, which were kept at 140K. Some of the problems we encountered in the ATHENA detector, in particular, deterioration of triggering capability and a steady increase in inactive modules, may be attributed to low temperature operations and repeated thermal cycling. By operating at a higher temperature, we hope to avoid these risks.

## Multiple Scattering

A very rough estimate of multiple scattering in the ALPHA neutral trap magnet can be obtained from the approximate Moliere formula [10]:

$$\theta_0 = \frac{13.6\,\mathrm{MeV}}{\beta c p} z \sqrt{x/X_0}\left[1 + 0.038\ln(x/X_0)\right] \qquad (1)$$

where $\theta_0$ is an rms scattering angle (projected on a plane), $p$, $\beta c$, and $z$ are the momentum, velocity, and charge number of incident particles, and $x/X_0$ is the thickness of the material in radiation length. In our case, $z=1$, $\beta 1$, and the average pion momentum $p \sim 300$ MeV/$c$. Our neutral trap magnet will be made of a superconducting alloy of NbTi/Cu. If we assume a Cu equivalent thickness of 1 cm for the scattering medium, we have $\theta_0 \sim 40$ mrad. For an averaged distance between the vertex and scattering material (i.e. the lever arm) of 4 cm, this corresponds to ~2 mm error in the measured track position at the vertex. This rough estimate indicates that the contribution to vertex resolution from the increased material thickness is comparable or smaller than the total resolution of the present ATHENA, the latter being dominated by the unmeasured track curvature. Therefore, if sufficiently accurate determination of the track curvature can be achieved by three or more layers of Si, the increased materials in ALPHA will be manageable in terms of vertex detection.

Detailed GEANT3 simulations have been performed to better estimate the effect of multiple scattering. Vertex resolutions, defined here as the rms distance between the true vertex and the reconstructed vertex, were calculated for different thicknesses of the material between the trap and the detector. Antiprotons are annihilated on the trap wall and pions are generated with realistic phase space and branching ratios. The straight line approximation was used for reconstruction routine. Table 1 summarizes the results of this simulation study.

**TABLE 1.** Simulated vertex resolutions ($\sigma$) for various material thicknesses

| Material Thickness Cu equivalent (mm) | Solenoid Filed | Resolution x (mm) | Resolution z (mm) | Comments |
|---|---|---|---|---|
| 1.65 | 3 T | 4.3 | 1.6 | ATHENA case |
| 1.65 | 0 T | 1.3 | 1.7 | ATHENA- without B field |
| 10.0 | 3 T | 5.0 | 3.0 | ALPHA - two Si layers |
| 10.0 | 0 T | 2.8 | 2.8 | ALPHA- known track curvature |

The first row is similar to the ATHENA case, where the Cu vacuum wall is 1.65 mm. An X resolution, much worse compared to Z, reflects the effect of magnetic field. This is confirmed in the second row, where the B field is turned off, and the X resolution is substantially improved. The remaining resolution in Row 2 is due primarily to multiple scattering. This comparison of the two cases indicates that ATHENA's 4 mm vertex resolution is dominated by the unmeasured track curvatures.

The third and forth rows are the simulations with increased scattering material. Instead of 1.65 mm, a 1 cm thick Cu, which is similar to the thickness of superconducting magnet material in ALPHA, is used in the calculations. With a 3 T solenoid field on (Row 3), the resolution is worsened compared to the standard ATHENA case (Row 1), due to the increased multiple scattering as expected. Recall that because of the straight line fits in the track reconstruction, the track curvature is unmeasured in this case. This level of vertex resolution is expected when only two Si layers are used in ALPHA. Row 4, with no B filed, isolates the effect of multiple scattering as in Column 2. The X plane resolution of 2.8 mm is in rough agreement with the estimate above using the Moliere formula. This value in Column 4 indicates that, even with a thick magnet, if the reconstruction error due to track curvature can be made negligible (e.g. by having three Si layers), the vertex resolution of the order of 3 mm can be achieved. This is well within our design goal.

The calculated distributions of the difference between true vertices and the reconstructed vertices, X(MC)-X(reconst), are graphically compared for the cases without magnetic field (Fig. 5, left), and with the 3T solenoid field (Fig 5, right).

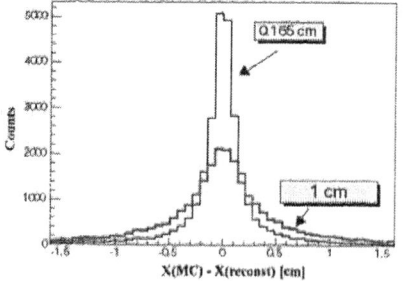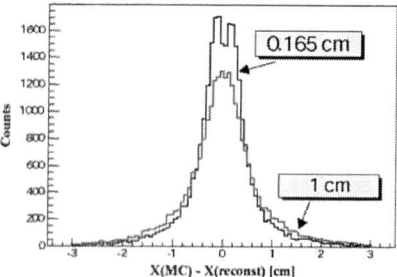

**FIGURE 5.** Calculated difference between the true vertices and reconstructed vertices for two different material thicknesses (as indicated in square boxes). Left: with out magnetic field, Right: with 3T solenoid field.

Note that magnitude of multiple scattering scales only as square root of the material thickness (see Equation 1), hence our conclusion here is not overly sensitive to the exact thickness of the magnet materials in the ALPHA apparatus.

## Other Effects

The effect of the multipolar field (as opposed to solenoid) on charged particle trajectories is expected to be small compared to the multiple scattering discussed above, and can in principle be corrected, given the charge and momentum of pions, and the magnetic field distribution. The axial field variation due to the mirror magnet could range from 1T to 3T, and its effect on the vertex resolution may be larger than the radial multipole field. A quantitative GEANT study is under way to study these effects.

Increased thickness of magnet material would result in a greater probability for the conversion of high energy gammas from neutral pion decays. The track pattern recognition routine in the off-line analysis software may need to be improved, should the increased multiplicity of charged tracks become problematic. Note that conversion events will likely have well-defined topology, since they mostly occurs at the neutral trap magnet at a fixed radius.

## TRIGGER AND DATA ACQUISITION

Improvements are foreseen for trigger and data acquisition system for the ALPHA detector, compared to ATHENA. A basic trigger scheme is illustrated in Figs. 6. A good feature of the IDEAS VA-TA readout chips used in ATHENA was their self-triggering capability. In our analysis [7], we showed that this trigger signal can be used as a proxy for antihydrogen signal in many cases, and some important physics results were obtained using this level 0 trigger signal [9,10]. We plan to retain this capability in the ALPHA. Higher level triggers (Fig. 6) can apply various cuts such as the multiplicity and the event topology as well as the trap conditions.

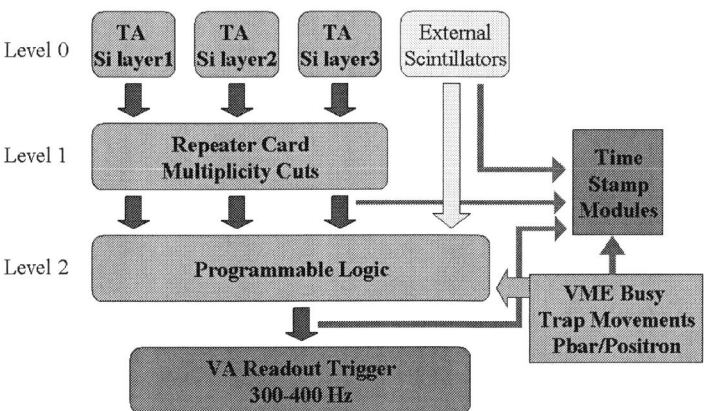

**FIGURE 6.** Vertex detector trigger scheme

117

Trigger events at each level, external annihilation scintillator signals, together with trap and other slow control activities, are time-stamped via multi-scalars which are synchronized to a 10 MHz atomic clock. Deadtime-less operation is possible for indefinite duration.

# ALIGNMENT

Relative mechanical alignment of each layer of Si detector with respect to one another will be determined using cosmic rays when the magnetic field is turned off. Alignment of the detector with respect the trap is somewhat more difficult, and we will use a method developed for the ATHENA detector [13]. Figures 7 and 8 show examples of such measurements. By moving the antiproton trap well, and measuring the annihilation positions, one can obtaine the correlations between trap well positions and measured annihilation positions. The detector z-position with respect to the trap is thus determined within 1 mm accuracy, a task which is otherwise difficult to achieve in our setup.

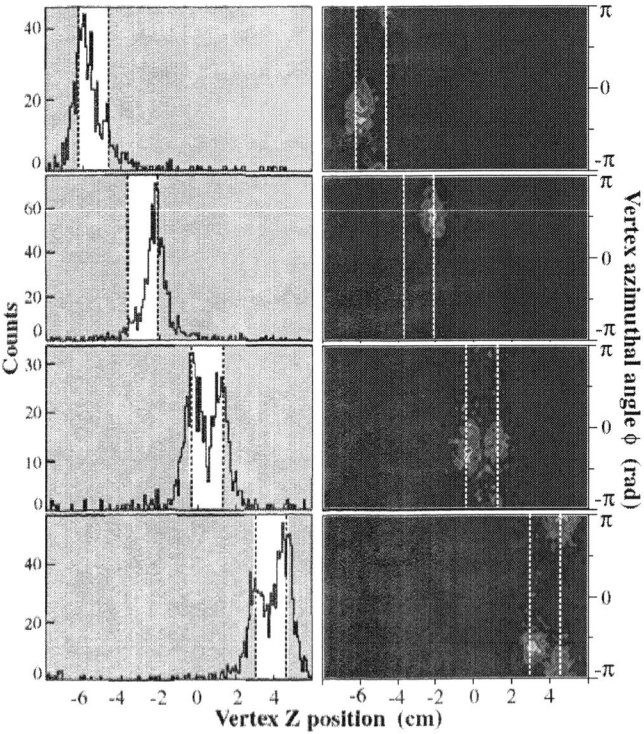

**FIGURE 7.** The projection of the antiproton annihilation distribution on the z axis (left column) and on the z-phi plane (right column) for four different confinement setup. The trap well positions are indicated by the unshaded regions, and the positions of the electrodes are depicted with dashed lines.

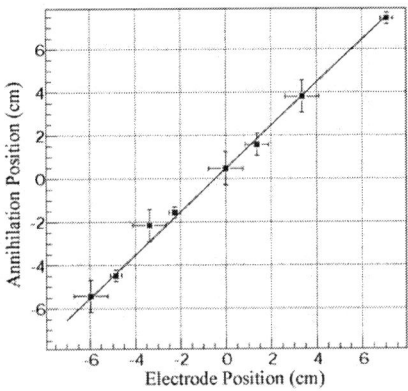

**FIGURE 8.** Correlation between the trap well position and the measured annihilation positions.

# ALTERNATIVE DETECTION SCHEME

A straightforward approach for demonstrating antihydrogen trapping would be to ramp down one of the trapping magnets, and observing the annihilations of the antihydrogen atoms leaking out of the trap. A technical challenge of this method may be rapid ramping of the superconducting magnet, in a time scale shorter than the trap lifetime. As well, if the ramping is slow, the annihilation signals have to compete with cosmic and other backgrounds. Here we consider another method using electron impact ionization.

## Electron Impact Ionization of Antihydrogen Atoms

By bombarding a beam of the electrons onto the trapped antihydrogen, the anti-atoms can be ionized. The antiprotons from the antihydrogen ionization can be collected in an electric potential well. If the electric trapping potential is larger than antiproton recoil energy (which should be less than few eV), antiprotons will be "born trapped". After accumulating a certain number of ionized antiprotons in the collecting well, they can be dumped onto the degrader in a usual manner [14], providing an efficient, nearly background free detection. See Fig. 9 for a schematic drawing.

## Detection Efficiency

The efficiency of this detection method can be estimated by assuming that the cross section is similar to that for electron impact ionization of hydrogen atoms. The latter can be found, e.g. in Ref. [15]. The rate for ionization (1/s) can be written: $R = \sigma_i \times n \times L \times flux$, where $\sigma_i$ (cm$^2$) is the electron impact ionization cross section, $n$ (cm$^{-3}$) the trapped antihydrogen number density, $L$ (cm) the length of antihydrogen cloud, and *flux* (1/s) is the flux of electron beam interacting with the antihydrogen cloud. Using the cross section for hydrogen of $\sim 10^{-16}$ cm$^2$ (at ~50 eV), and assuming 1000 antihydrogen is trapped in a 1 cm$^3$ volume (i.e. $nL \sim 1000$ cm$^{-2}$), a 1 mA electron beam interacting with the antihydrogen would give 1000 Hz ionization rate, i.e. all the antihydrogen are ionized in 1 sec. We note that we can vary the antiproton catching potential shape to explore the special distribution of trapped antihydrogen cloud. A potential difficulty is the transport of the electron beam through the multipole magnetic field. Detailed studies are in progress to determine the feasibility of this method. We note that realistic theoretical calculations for the cross section for electron impact antihydrogen ionization will be helpful.

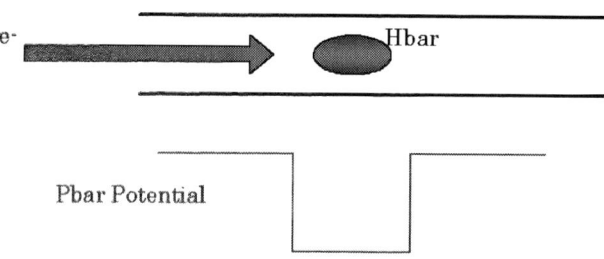

**FIGURE 9.** A schematic for electron impact ionization of trapped antihydrogen atoms.

## SUMMARY AND OUTLOOK

We have reviewed the features of the ATHENA vertex imaging detector, and the design challenges for a new detector for the ALPHA experiment. We have shown that the difficulties with unusually large amount of scattering materials could be overcome. Recently, this is confirmed via more detailed Monte Carlo simulations with three Si layers and helix fitting of the pion trajectories [16]. The final design of the detector is being worked out. An alternative method for trapped antihydrogen detection, based on electron impact ionization of antihydrogen, was also proposed in this article.

The ALPHA experiment aims to start its data taking at CERN AD in summer 2006. The vertex detector will be an unrivaled feature of ALPHA, which will allow detailed diagnostics of the trapped particles, as well as detection of cold antihydrogen atoms.

## ACKNOWLEDGMENTS

I would like to thank Professor Yasunori Yamazaki for the invitation to a very stimulating workshop. I also thank ATHENA and ALPHA collaborators for various discussions, and Professor Alberto Rotondi and Dr. Pablo Genova for valuable help with simulations. This work is supported in part by TRIUMF and by NSERC of Canada.

# REFERENCES

1. M. Amoretti et al., Nature (London), **419**, 456 (2002).
2. ALPHA Collaboration, Research Proposal to CERN (2005).
3. C. Regenfus, Nucl. Instrum. Methods A, **501**, 65 (2003).
4. M. C. Fujiwara et al., Phys. Rev. Lett. **92**, 065005 (2004).
5. M. Amoretti et al., Phys. Rev. Lett. **91**, 055001 (2003).
6. M. Amoretti et al., Phys. Plasma **10**, 3056 (2003).
7. M. Amoretti et al., Phys. Lett. B**578**, 23 (2004).
8. N. Madsen et al., Phys. Rev. Lett. **94**, 033403 (2005).
9. L. V. Jorgensen, accepted for Phys. Rev. Lett. (2005).
10. Particle Data Group, Review of Particle Physics, Phys. Lett. B**592**, 1 (2004).
11. M. Amoretti et al., Phys. Lett. B**583**, 59 (2004).
12. M. Amoretti et al., Phys. Lett. B**590**, 133 (2004).
13. M. C. Fujiwara et al., *Non-neutral Plasam Physics*, eds. M. Schauer et al., (AIP, New York, 2003) Vol. 5, p. 131.
14. M. C. Fujiwara et al., Hyperfine Interact **138**, 153 (2001).
15. NIST website, http://www.nist.gov.
16. P. Genova and M.C. Fujiwara, ALPHA technical communications (unpublished).

# ATRAP - on the way to trapped Antihydrogen

D. Grzonka*, D. Comeau†, G. Gabrielse**, F. Goldenbaum‡,
T. W. Hänsch§,¶, E. A. Hessels†, P. Larochelle**, D. Lesage**, B. Levitt**,
W. Oelert‡, H. Pittner§, T. Sefzick‡, A. Speck**, C. H. Storry†, J. Walz§ and
Z. Zhang‡

*Institut für Kernphysik, Forschungszentrum Jülich GmbH, 52425 Jülich, Germany
†York University, Department of Physics and Astronomy, Toronto, Ontario M3J 1P3, Canada
**Department of Physics, Harvard University, Cambridge, MA 02138
‡IKP, Forschungszentrum Jülich GmbH, 52425 Jülich, Germany
§Max-Planck-Institut für Quantenoptik, Hans-Kopfermann-Strasse 1, 85748 Garching, Germany
¶Ludwig-Maximilians-Universität München, Schellingstrasse 4/III, 80799 München, Germany

**Abstract.** The ATRAP experiment at the CERN antiproton decelerator AD aims for a test of the CPT invariance by a high precision comparison of the 1s-2s transition in the hydrogen and the antihydrogen atom.

Antihydrogen production is routinely operated at ATRAP [1],[2]. and detailed studies have been performed in order to optimize the production efficiency of useful antihydrogen. The shape parameters of the antiproton and positron clouds [3], the $n$-state distribution of the produced Rydberg antihydrogen atoms [2] and the antihydrogen velocity [4] have been studied. Furthermore an alternative method of laser controlled antihydrogen production was successfully applied [5].

For high precision measurements of atomic transitions cold antihydrogen in the ground state is required which must be trapped due to the low number of available antihydrogen atoms compared to the cold hydrogen beam used for hydrogen spectroscopy. To ensure a reasonable antihydrogen trapping efficiency a magnetic trap has to be superposed the nested Penning trap. First trapping tests of charged particles within a combined magnetic/Penning trap have started at ATRAP.

**Keywords:** Antihydrogen, Nested trap, Laser controlled recombination
**PACS:** 36.10-k

## INTRODUCTION

Antihydrogen was created the first time at relativistic velocities at CERN [6] and then at Fermilab [7] followed by the first production of antihydrogen within a Penning trap in 2002 by the ATHENA [8] and the ATRAP [1] experiments operated at the antiproton decelerator AD at CERN. The production in a nested Penning trap configuration is operated routinely whereas the trapping of antihydrogen has to be developed for precise physics studies of the antihydrogen system. For a recent review see [9].

The ATRAP experiment aims for a high precision test of the CPT invariance by a comparison of the spectral lines between the hydrogen and the antihydrogen atoms. Highest precision concerning the frequency measurement is expected from the two photon 1s-2s transition to the metastable 2s state with a natural line width of 1.3 Hz resulting in a possible precision of the transition frequency of $1 \cdot 10^{-15}$. For the hydrogen atom the 1s-2s transition was measured with a precision of $\Delta f/f = 1 \cdot 10^{-14}$ [10]. A similar precision is anticipated for the antihydrogen atoms. For the ratio of the

CP793 *Physics with Ultra Slow Antiproton Beams*
edited by Yasunori Yamazaki and Michiharu Wada
© 2005 American Institute of Physics 0-7354-0282-5/05/$22.50

frequencies (by a direct comparison of the hydrogen and antihydrogen atom) even higher accuracies are possible which could in principle be below the natural line width by comparing the line shapes.

Such high precisions call for high statistics of the detected transitions. A conventional scheme - as it was used for the hydrogen studies with a cold hydrogen beam - is not possible due to the limited amount of antihydrogen. Using hydrogen atoms a beam of these particles passed a region with an electric field resulting in Stark mixing of the 2s into the 2p states which decayed with a life time of 1.6 $ns$ to the ground state. The 2p-1s light resulting from a 1s-2s transition was measured as a function of the 1s-2s transition frequency. The technique proposed for the antihydrogen spectroscopy has to be different, a shelving scheme was proposed [11]. First, cold antihydrogen atoms in the ground state have to be trapped. Second, a continuous Lyman alpha laser will excite the antihydrogen sample to the 2p state which decays fast back to the ground state by emitting 2p-1s light observed by a photon detector. Third, the 1s-2s two photon transition is induced and if an antihydrogen atom is excited for the lifetime of the metastable 2s state there is no light from the 2p-1s transition. A scan of the 1s-2s frequency results in a reduction of the 2p-1s intensity at the correct 1s-2s frequency. This method would even work with a single atom.

To check the sensitivity of a CPT violation to the frequency shift of an atomic transition a suitable model like the standard model extension (SME) [12],[13] has to be consulted. In the SME CPT violating terms are introduced in the Lagrangian of the standard model where the strength of CPT violation is given by certain parameters on the energy scale which allows to better compare the sensitivity of different CPT tests. For a detailed discussion of the SME and the sensitivity of antihydrogen spectroscopy to CPT violation I refer to [14].

The trapping of antihydrogen can be done via the magnetic moment of the antihydrogen atom in a magnetic gradient field. The magnetic potential energy is given by $E_B \sim -\vec{\mu}\vec{B}$ with the magnetic moment $\mu$ and a magnetic field $B$, i.e. a state with a magnetic moment in field direction is a low field seeker, it is driven to the minimum of the B field. Unfortunately the potential energy is very low due to the small value of $\mu_B$ which corresponds to a temperature ($T = \mu_B B/k$, $\mu_B$ =5.788 $\cdot 10^{-5} eV/T$, $k$ = Boltzmann constant = $8.62 \cdot 10^{-5} eV/K$) of $0.67 K/T$. Therefore cold antihydrogen in the ground state is needed for an effective trapping of the produced antihydrogen. Assuming a Boltzmann distribution at a temperature of 4.2 $K$ (LHe) and a field gradient of 1T, achievable with superconducting magnets, only a few per cent of the antihydrogen atoms would be trapped which shows the importance to produce cold antihydrogen with a temperature of preferentially less than 4 $K$. Once the antihydrogen atoms are trapped laser cooling can be applied to further reduce the temperature necessary to achieve ultimate precision. Laser cooling with Lyman-$\alpha$ can reach a few $mK$ given by the Doppler limit and the photon recoil.

Especially for studies of the gravitational interaction between matter and antimatter which is another important topic in the antihydrogen experiments much lower temperatures are requested. Symphatetic cooling of positive antihydrogen atoms by laser cooled ions which can be cooled to temperatures below 100 $\mu K$ was proposed by Walz and Hänsch [15]. At such low temperatures a precise measurement of the acceleration of an

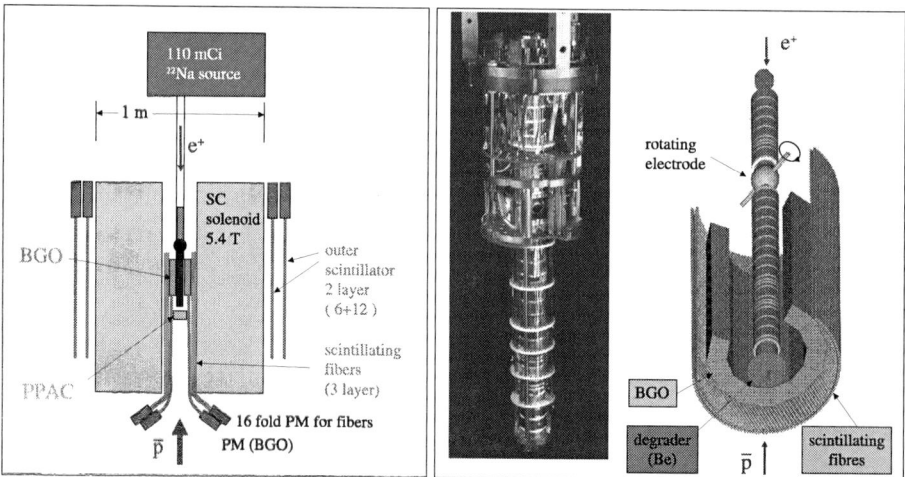

**FIGURE 1.** The experimental setup at ATRAP with a sketch of the whole system including superconducting solenoid and outer scintillator plates in the left part and details of the inner part on the right side. The stack of ring electrodes and the surrounding detector elements are shown.

antihydrogen atom in the gravitational field is possible which is up to now experimentally completely unknown.

The simplest configuration for a magnetic anithydrogen trap, a Ioffe-Pritchard design [16], [17] is a linear magnetic quadrupol for the radial, combined with two solenoids, called pinch coils, for the axial trapping. Such a magnetic trap has to be superposed the Penning trap used to confine the charged particles and where the antihydrogen is produced. Such a configuration of a combined Penning and Ioffe trap is not a straightforward setup because the trapping of charged particles requires a homogeneous magnetic solenoid field. A cloud of charged particles described by a charged plasma gets instable if the homogeneity of the magnetic field is disturbed [18] as it is largely the case with a magnetic quadrupole field. On the other hand for single charged particles stable orbits are possible at such conditions as shown by [19].

# EXPERIMENTS AT ATRAP

## The ATRAP experimental setup

The experimental studies at ATRAP were performed with the setup shown in fig. 1. A stack of ring electrodes within a superconducting solenoid with a field of 5.4 T builds the Penning traps for the charged particles. It is surrounded by detectors for the annihilation products which is rather limited due to the small magnet bore of only about 10 *cm* in diameter. BGO crystals were used for the $\gamma$ detection resulting from the positron annihilation. Scintillating fibers inside the magnet and scintillator plates outside the magnet detect charged annihilation products from the antiproton. The scintillating fiber detector consists of 3 layers, one with straight fibers and two helical fiber layers. This system detects annihilation events but the tracks of the charged particle can not be determined.

A tracking system is foreseen in a second setup with a magnet bore of about 50 *cm* in diameter. Here several layers of scintillating fibers will be installed which allow to reconstruct the annihilation vertex. Furthermore there is enough space for a superconducting Ioffe trap as well as comfortable laser access. This second setup will go into operation in 2006.

The Penning trap shown in fig. 1 includes a large number of electrodes which allow to generate very complex potential distributions needed for the various studies in the antihydrogen production. Cooled by liquid Helium to 4.2 *K* a vacuum of below $5 \cdot 10^{-17} mbar$ was achieved derived from the lifetime of trapped antiprotons [20].

## Trapping of antiprotons and positrons

The antiprotons are delivered by the AD at CERN [21]. Produced by a 26 $GeV/c$ proton beam of the proton synchrotron (PS) hitting a target, 3.5 $GeV/c$ antiprotons are collected and decelerated to 100 $MeV/c$. To reduce the losses stochastic (at 3.5 and at 2 $GeV/c$) and electron cooling (at 300 and 100 $MeV/c$) is applied. By fast extraction a bunch of about $3 \cdot 10^7$ antiprotons with a bunch length of about 90 *ns* is sent to the AD experiments every 86 *s*.

At ATRAP the antiprotons enter into the trap as indicated in fig. 1 from below after passing a thin foil, two parallel-plate avalanche counters (PPAC) as beam position monitor, a cell with a $He/SF_6$ gas mixture to adjust the energy loss to an optimum trapping efficiency and the Be degrader. At the electrode below the ball valve (see fig. 1) a -3 *kV* potential is applied which deflects antiprotons with sufficient low energies. When returning in the $\sim 15$ *cm* long trap also the degrader foil is switched to a potential of -3 *kV* to close the trap. Up to 25 000 antiprotons are trapped in this 3 *kV* potential well. While bouncing back and forth in the long well the antiprotons are cooled by electrons deposited before the trapping procedure in a number of short potential wells.

The time to the next AD bunch is sufficient to cool the antiprotons into the short wells by the electrons which themselves are cooled by synchrotron radiation to the 4.2 *K* surrounding. This allows a stacking of subsequent antiproton bunches from the AD in order to increase the number of trapped antiprotons for the antihydrogen experiments. For the next cycle the potential at the degrader is switched off to trap a second bunch of antiprotons which are again cooled into the short electron wells and a next bunch can be trapped.

This stacking method results in a linear increase of trapped antiprotons with the number of bunches experimentally verified up to 32 bunches. A more detailed description of the method is given in [22].

The positrons are delivered by a $^{22}Na$ source which is moved close to the trap entrance from above for the positron accumulation. The positrons pass a 2 $\mu m$ tungsten crystal, the transmission moderator, where a small part emerge from the crystal with eV energies by building highly excited positronium which is field ionized in the positron trap. At the

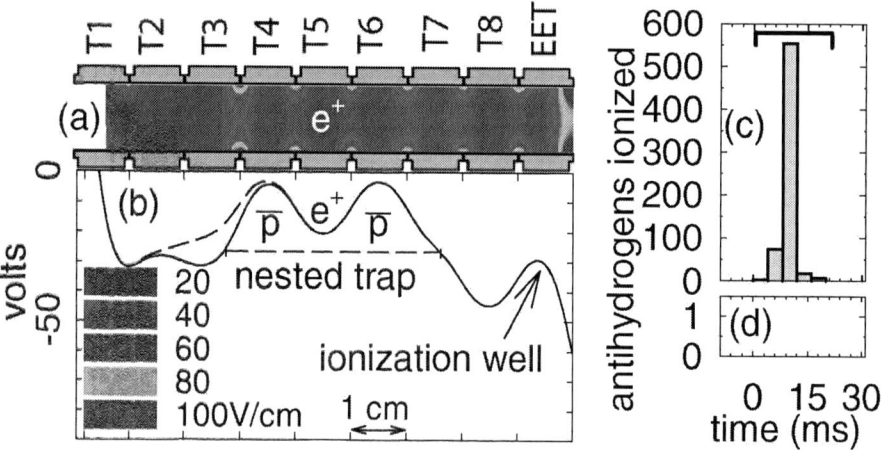

**FIGURE 2.** Antihydrogen production in a nested trap configuration. The upper part (a) shows the relevant trap electrodes with the electric field strength distribution. In the lower part (b) the potential on the center axis and on the right side a typical result of counting the number of antiprotons deposited in the ionization well with (c) and without (d) positrons in the trap is given.

ball valve another thick tungsten crystal is mounted, the reflection moderator, where also positronium is produced at the surface which moves to the trap region and is field ionized. For a detailed description of the positron loading technique see [23]. Up to $5 \cdot 10^6$ positrons with a loading rate of about $300 \, s^{-1}$ were accumulated for experiments during several hours before the antihydrogen experiments were performed.

## Antihydrogen production and detection

The formation of an antihydrogen atom out of a positron and an antiproton needs a third reaction partner to take up the binding energy which is a photon in the radiative recombination : $\bar{p} + e^+ \rightarrow \bar{H} + h\nu$ or a second positron in the three body recombination: $\bar{p} + e^+ + e^+ \rightarrow \bar{H} + e^+$. Other possible mechanisms are laser stimulated recombination ($\bar{p} + e^+ + h\nu \rightarrow \bar{H} + h\nu + h\nu$), field assisted recombination where the Coulomb potential of the antiproton is deformed by a pulsed field to catch the positron in analogy to the well known field ionization [24] or the use of positronium ($\bar{p} + (e^+ e^-) \rightarrow \bar{H} + e^-$) [25].

Most of the antihydrogen production studies at ATRAP were done with a nested well structure shown in fig. 2 [26]. A nested trap is generated by applying the required potential to the ring electrodes with a positron cloud trapped in the center. In the beginning a $\bar{p}$-cloud is located in the well at T2. By increasing the T3 potential for a short time the $\bar{p}$'s are launched towards the positron cloud bounce back and forth through the positrons and are cooled into the antiproton wells on both sides of the $e^+$ cloud. When the antiprotons pass through the positrons antihydrogen atoms can be build. The dominant production mechanism for the conditions at ATRAP (4.2 $K$ temperature, $n_{e^+} = 2 \cdot 10^7 \ cm^{-3}$) is the three body recombination with an expected event rate of $\Gamma = 6 \cdot 10^{-13}(4.2/T)^{9/2}n_{e^+}^2 n_{\bar{p}} s^{-1}$, ($\Gamma$ : recombination rate, $T$: temperature in $K$, $n_{e^+}$: number of positrons, $n_{\bar{p}}$: number of antiprotons) [26]. The expected radiative recombination rate of $\Gamma = 6 \cdot 10^{-11}(4.2/T)^{9/2}n_{e^+} n_{\bar{p}} s^{-1}$, is a factor of about 3000 smaller [26].

For the detection of antihydrogen a background free method via field ionization is used [1]. Neutral antihydrogen atoms are not reflected by the electric field in the nested trap region and may hit the trap wall and annihilate. Some of the antihydrogen atoms travel towards the electrode EET where a strong electric field exists, strong enough to ionize Rydberg antihydrogen atoms. Any antihydrogen atom ionized in this region deposits its antiproton in the potential well at EET. On the other hand no antiproton out of the nested trap is able to reach the EET well. After a production experiment is done all antiprotons in the nested trap are removed by opening the wells to minimize the background rate in the detectors. Now completely decoupled from the antihydrogen production experiment the number of antiprotons deposited into the EET well can be counted by slowly lowering the potential at T8. The advantage of this technique is an essentially background-free detection shown in fig. 2. In fig. 2 c the number of antiproton annihilations with time is given when the potential at T8 is linearly ramped up. In fig. 2d the same experiment is performed without positrons loaded into the trap resulting in zero background counts.

To increase the number of produced antihydrogen the driven production method was developed [2]. The antiprotons cooled into the wells beside the positron cloud were excited by radio frequency signals applied alternatively to the electrodes T4 or T6. By this method the antihydrogen production could be increased resulting typically in the order of 1000 trapped antiprotons in the ionization well which corresponds to about 100 000 produced antihydrogen atoms if an isotropic emission is assumed.

## Properties of the produced Antihydrogen atoms.

The high production rates achieved with the nested Penning trap configuration allowed detailed analysis. Important for the ongoing work towards trapped antihydrogen is the knowledge of the $n$-state distribution of the produced Rydberg atoms. With the potential structure shown in fig. 3 this distribution has been measured for high Rydberg states down to $n \sim 24$. In addition to the potential configuration in fig. 2 an analysis well is introduced at the electrodes ER and CS just in front of the ionization well.

Here an electric field is applied which ionizes Rydberg states down to a certain $n$ which then cannot reach the ionization well. Rydberg states at lower $n$ which survive the analysis well are ionized as usual at the ionization well and deposit their antiprotons. By scanning the potential of the analysis well the number of antihydrogen states produced in the different Rydberg states was determined. To normalize the individual experiments, on the other side of the production region a normalization well is arranged where field ionization with a fixed potential is performed. In fig. 4 the measured distribution is shown which follows a $F^{-2}$ relation up to a field strength $F$ of about 200 $V/cm$, the validity

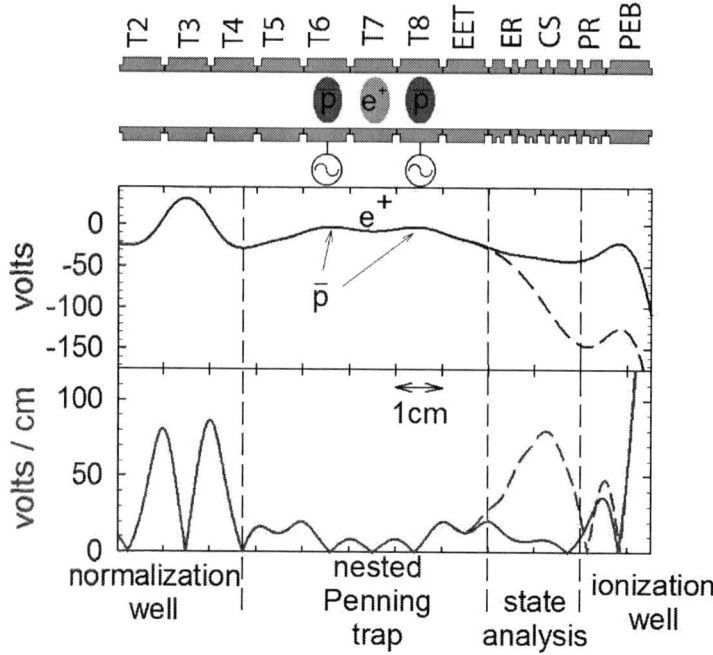

**FIGURE 3.** Potential and electric field structure applied for measurements of the *n*-state distribution. In the center is the nested Penning trap located, at the right side the state analysis region and the ionization well, and at the left side the normalization well.

range of a guiding center approximation (GCA) [27], [28], but at higher fields deviations from the simple power law start.

Another important information as described in the introduction is the velocity of the produced antihydrogen atoms. For the positron cloud as well as the cooled antiproton cloud a temperature of 4.2 $K$ can be assumed but for the antihydrogen production the antiprotons are launched trough the positron cloud resulting in some kinetic energy transferred to the antihydrogen.

To measure the velocity the potential structure shown in fig. 3 is used but now the analysis well potential is varying sinusoidal with a fixed frequency [4]. Depending on the frequency and on the velocity of the antihydrogen atoms the number of antihydrogen atoms reaching the ionization well varies. If the frequency is low the time window with a field strength below the ionization field for a certain $n$ state is rather long and even slow antihydrogen atoms have a chance to pass this region undisturbed. If the frequency is high only high velocity antihydrogen atoms can pass the region without being ionized. In fig. 5 the expected antihydrogen detection probability as a function of the frequency of the pre-stripping field is shown by solid lines assuming a Boltzmann distributed antihydrogen velocity together with the measured data. The measured velocity is clearly

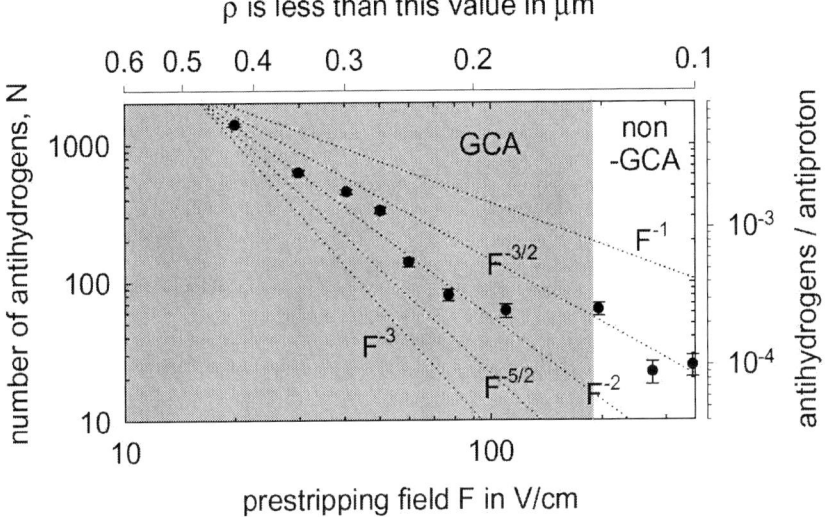

**FIGURE 4.** N-state population of the antihydrogen production in the nested Penning trap configuration.

above 4.2 $K$ corresponding to 0.36 meV. The data can be best described by a temperature of 2400 $K$ (200 $meV$) which would result in an extremely low trapping efficiency of antihydrogen.

Before trapping studies with antihydrogen are started the temperature has to be drastically reduced. Investigations of a possible reduction of the temperature of the produced antihydrogen by reducing the RF signals used for exciting the antiproton clouds have been started but up to now no improvement could be achieved. Studies in this direction will be continued with the upcoming beam time where also a launching of positrons through the antiproton cloud is foreseen.

## Laser controlled antihydrogen production

In view of the required low velocity antihydrogen atoms for trapping, another production technique has been applied [5]. It works via a double charge exchange process using highly excited Rydberg $Cs$ atoms. In fig. 6 a sketch of the process is shown. Cs atoms in a beam from a $Cs$ oven heated to 317 $K$ are excited to a well defined high Rydberg state by laser excitation. Then $Cs^*$ atoms pass through a positron cloud where positronium is created. The positronium drifts out of the positron cloud and may reach an antiproton cloud nearby where another charge exchange process can happen and antihydrogen is built. The $Cs$ excitation is done in a two step excitation scheme with a first $6S_{1/2} \to 6P_{1/2}$ induced by a diode laser at 852.2 $nm$ and a second $6P_{1/2} \to 37D_{5/2}$ transition with a copper vapor laser at 510.7 $nm$. For the chosen transition an excitation efficiency of about 24 % was calculated [29]. The strong magnetic field changes of course the level struc-

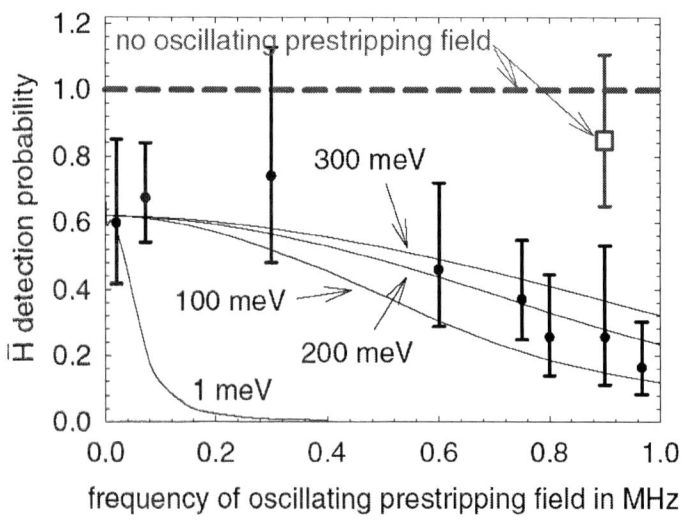

**FIGURE 5.** Probability for the detection of produced antihydrogen atoms as a function of the frequency of the oscillating pre-stripping field.

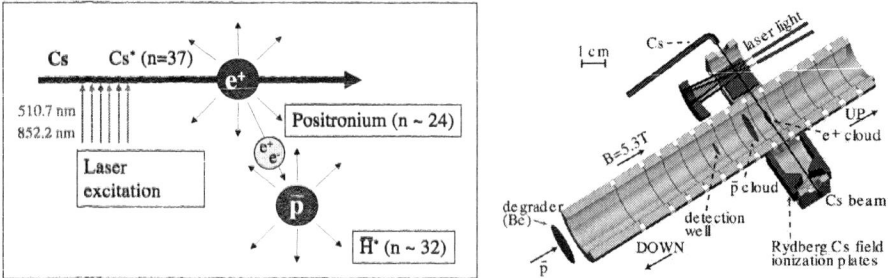

**FIGURE 6.** Sketch of the laser controlled antihydrogen production via double charge exchange using Rydberg *Cs* atoms.

ture which is not easily calculable but the actual state is not important for the charge exchange process, simply a high Rydberg state is requested. High Rydberg states result in very large cross sections for this charge exchange process which scales with $n^4$, the geometric area given by $n^4 \pi a_0^2$ with the radius of the Rydberg state given by $n^2 a_0$. Due to energy conservation Rydberg Positronium built in the charge exchange process is peaked at a $n$-state around 26. According to calculations under the applied conditions all positrons should be converted to Positronium [30]. In the second charge exchange process the positron is transferred to the antiproton resulting in Rydberg antihydrogen atoms with a $n$-state population expected to be peaked around $n=32$. The rate of produced antihydrogen atoms is apart from the annihilation losses determined by the solid

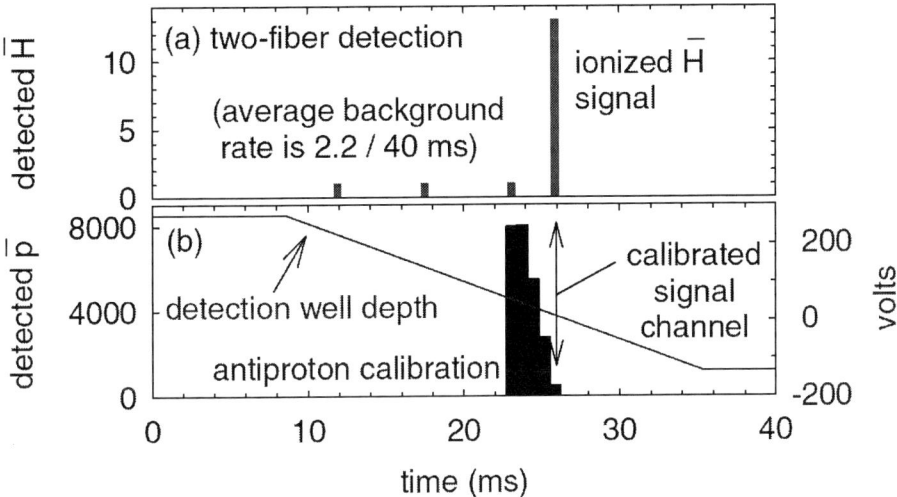

**FIGURE 7.** Result of a sequence of experiments to produce antihydrogen via a double charge exchange as described in the text.

angle and the charge exchange cross section. The lifetime of the Rydberg states is in the range of a few *ms*, sufficiently long to travel distances in the order of *m*.

In a typical experiment with $2 \cdot 10^6 e^+$ and $3 \cdot 10^5 \bar{p}$ about 100 antihydrogen atoms are expected assuming a cross section of $\sigma_{Ps\bar{p}} = 58 n_{Ps}^4 \pi a_0^2$ resulting from classical trajectory Monte Carlo (CTMC) calculations [30]. About 1 % of these antihydrogen reach the solid angle of the ionization well that on average 1 antihydrogen atom should be detected. The summed signals of a sequence of 6 experiments show a clear signal of these laser controlled antihydrogen production, see fig. 7.

The upper part shows the detector signals as a function of the ionization well depth which is ramped down with time. At about 25 *ms* when the detection well depth is around 0 Volt trapped antiprotons are expected to be released and annihilate as checked by the lower part of the figure where the detection well was filled by $\bar{p}$'s and emptied in the same way to calibrate the time scale. Compared to the nested well production technique running dominantly via TBR the production rate is rather low. An increase by a factor 10 to 100 seems to be possible by increasing the particle numbers and reducing the distance between positron and antiproton cloud.

A big advantage of this production technique is the expected low velocity of the produced antihydrogen which should be given by the velocity of the antiprotons cooled to 4.2 *K* and the limitation to a small range of populated *n*-states where the peak position is determined by the *Cs* Rydberg state chosen for the laser excitation. The low temperature at well defined Rydberg states is a good basis for the trapping of antihydrogen.

Of course the velocities of the produced antihydrogen atoms have to be checked. The technique used in the nested Penning trap studies is not suitable here due to the much lower production rates but the laser controlled production mechanism allows to measure the time between production and annihilation of each produced antihydrogen atom. The laser excitation is done by a pulsed copper vapor laser with a pulse width of about 20 $ns$. By adjusting the repetition rate with a mechanical chopper to a reasonable rate of 1 $kHz$ the annihilation time relative to the laser pulse can be measured for each antihydrogen annihilation. For the velocity also the track length is required which is not known without a tracking system in ATRAP-I. Therefore only time distributions for an event sample can be determined which are related to the mean velocities. Monte Carlo simulations of this process including Boltzmann distributions for the velocities in the particle clouds indicate a high sensitivity of the time distribution on the assumed temperature of the produced antihydrogen. The comparison of a measured time distribution with Monte Carlo generated ones will allow to determine the velocity with an accuracy of a few $K$ in the region of 4 $K$ with a reasonable statistics of a few thousand antihydrogen annihilations [31]. Very first experiments have been tried but the statistics was still much too low to draw any conclusions about the temperature.

## Stability of charged particles in a combined Penning/Ioffe trap

A basic question for the further studies in view of trapped antihydrogen is the stability of a charged particle cloud within a magnetic gradient field. The most efficient way of antihydrogen trapping within a Penning trap configuration is the superposition of a magnetic gradient, in the simplest case a magnetic quadrupole for the radial combined with two solenoids for the axial confinement. The trapped positrons or antiprotons can be considered as a charged plasma stabilized by the rotation in the magnetic field. The rotation is connected with an angular momentum which is conserved in a field with cylindrical symmetry providing the confinement of the plasma [18]. If the cylindrical symmetry is destroyed the confinement is no longer given and the particles are lost. A possible stability of a single charged particle in a combined Penning-Ioffe trap was investigated in Ref. [19] resulting in possible stable orbits which are connected with adiabatic invariants. As long as the particle density is low enough to limit collisons and if resonances are avoided a stable operation seems to be possible.

Experiments on stabilities of electron plasmas at high temperatures within a Penning trap with an additional magnetic quadrupole field [32] result in rapid losses even for weak quadrupole field strengths. Similar studies have been done for parameters more comparable to ATRAP conditions where the trapping time is rather limited strongly decreasing with increasing quadrupole field strength [33].

First tests of electron trapping in a Penning trap superposed by a magnetic quadrupole field have been performed at ATRAP where only a permanent quadrupole magnet has been used due to space limitations. If a permanent magnet is exposed to a solenoid field perpendicular to the magnetisation direction of the permanent magnet a reduction of the magnetisation is observed dependent on the solenoid field strength. A solenoid field of 3 T result in a reduction of the magnetisation by about 10 % which then stays constant as long as the field stays below 3 T. The quadrupole was installed in stead of the BGO detector. Within the available space a field gradient of only about 15 T/m has been achieved which is rather low. Electrons have been loaded into the trap by field emission off a wire. With about one million electrons and solenoid fields of 1, 2 and 3T the electron clouds were stable for more than one hour.

These first studies have to be confirmed by further measurements and extended to higher field gradients to check the suitable parameter range for the antihydrogen production and trapping within a combined Penning/Ioffe trap.

For further studies a superconducting magnet is designed where besides a quadrupole field configuration also higher multipoles are considered to reduce the disturbance of the axial symmetry. The advantage of a superconducting magnet is the flexibility in changing the field strength which is of course fixed with a permanent quadrupole.

# SUMMARY

At the ATRAP experiment operated at the antiproton decelerator AD at CERN antihydrogen is routinely produced within a nested Penning trap configuration. For the detection of antihydrogen a field ionization method was applied which is background-free and therefore gives a high sensitivity.

Precise spectroscopic studies aimed for at ATRAP, require trapped antihydrogen within a magnetic gradient field. Useful antihydrogen for trapping has to be cold enough and should be in the ground state. Before trapping studies can start it is therefore important to know the properties of the produced antihydrogen.

An essential parameter is the temperature of antihydrogen which has been measured at ATRAP to be at around $2400\ K$ in the nested trap experiments, much higher than the $4.2\ K$ LHe surrounding. Lower temperatures should be possible by an optimization of the working conditions. Studies in this direction have started and have to be continued in future experiments.

The other important feature is the $n$-state population in the production. Down to $n \sim 24$ the $n$-state distribution has been measured at ATRAP which is the basic for an efficient de-excitation. The optimization of the production towards more deeply bound states as well as the de-excitation needs the knowledge of the $n$-state distribution.

As a second approach a laser controlled production via a double charge exchange mechanism using highly excited Rydberg $Cs^*$ atoms has been demonstrated. The advantage of this method is the expected production of antihydrogen in well defined Rydberg states at the temperature of the antiproton cloud of $4.2\ K$. Further studies are needed to increase the production rate and to check the expected properties of the produced antihydrogen.

For an optimum trapping efficiency of antihydrogen production and trapping should be at the same place. A combined Penning/Ioffe trap is technologically feasible but the stability of charged particles in a Penning trap requires axial symmetry of the magnetic field which is destroyed if e.g. a quadrupole field is applied for the antihydrogen trapping. With higher multipoles a much lower field can be achieved in the trap center with high gradients at the edge which may be the solution of the problem.

First measurements on the stability of charged particle in a combined Penning/Ioffe trap have been performed which have to be continued and extended in order to develop the optimum trap configuration and working conditions for the trapping of antihydrogen.

## ACKNOWLEDGMENTS

The experimental studies at ATRAP were supported by the NSF and AFOSR of the US, the BMBF, MPG and FZ-Jülich of Germany, and the NSERC, CRC, CFI and OIT of Canada.

## REFERENCES

1. G. Gabrielse, N. S. Bowden, P. Oxley, A. Speck, C. H. Storry, J. N. Tan, M. Wessels, D. Grzonka, W. Oelert, G. Schepers, T. Sefzick, J. Walz, H. Pittner, T. W. Hänsch, and E. A. Hessels. *Phys. Rev. Lett.*, **89**, 213401 (2002).
2. G. Gabrielse, N. S. Bowden, P. Oxley, A. Speck, C. H. Storry, J. N. Tan, M. Wessels, D. Grzonka, W. Oelert, G. Schepers, T. Sefzick, J. Walz, H. Pittner, T. W. Hänsch, and E. A. Hessels. *Phys. Rev. Lett.*, **89**, 233401 (2002).
3. P. Oxley, N. S. Bowden, R. Parrott, A. Speck, C. H. Storry, J. N. Tan, M.Wessels, G. Gabrielse, D. Grzonka, W. Oelert, G. Schepers, T. Sefzick, J. Walz, H. Pittner, T. W. Hänsch, and E. A. Hessels. *Phys. Lett.*, **B595**, 60 (2004).
4. G. Gabrielse, A. Speck, C. H. Storry, D. Le Sage, N. Guise, D. Grzonka, W. Oelert, G. Schepers, T. Sefzick, H. Pittner, J. Walz, T. W. Hänsch, D. Comeau, and E. A. Hessels. *Phys. Rev. Lett.*, **93**, 073401 (2004).
5. C. H. Storry, A. Speck, D. Le Sage, N. Guise, G. Gabrielse, D. Grzonka, W. Oelert, G. Schepers, T. Sefzick, H. Pittner, M. Herrmann, J. Walz, T. W. Hänsch, D. Comeau, and E. A. Hessels. *Phys. Rev. Lett.*, **93**, 263401 (2004).
6. G. Baur, G. Boero, S. Brauksiepe, A. Buzzo, W. Eyrich, R. Geyer, D. Grzonka, J. Hauffe, K. Kilian, M. LoVetre, M. Macri, M. Moosburger, R. Nellen, W. Oelert, S. Passaggio, A. Pozzo, K. Röhrich, K. Sachs, G. Schepers, T. Sefzick, R. S. Simon, R. Stratmann, F. Stinzing, and M. Wolke. *Phys. Lett.*, **B368**, 251 (1996).
7. G. Blanford, D. C. Christian, K. Gollwitzer, M. Mandelkern, C. T. Munger, J. Schultz, and G. Zioulas. *Phys. Rev. Lett.*, **80**, 3037 (1998).
8. M. Amoretti, C. Amsler, G. Bonomi, A. Bouchta, P. Bowe, C. Carraro, C. L. Cesar, M. Charlton, M. J. T. Collier, M. Doser, V. Filippini, K. S. Fine, A. Fontana, M. C. Fujiwara, R. Funakoshi, P. Genova, J. S. Hangst, R. S. Hayano, M. H. Holzscheiter, L. J. Jorgensen, V. Lagomarsino, R. Landua, D. Lindelo, E. Lodi-Rizzini, M. Marcy, N. Madsen, G. Manuzio, M. Marchesoti, P. Montagna, H. Pruys, C. Regenfuß, P. Riedler, J. Rochet, A. Rotondi, G. Rouleau, G. Testera, A. Variola, T. L. Watson, and D. P. van der Werf. *Nature*, **419**, 456 (2002).
9. G. Gabrielse, *At. Mol. Opt. Phys.*, **50**, 1 (2004).
10. M. Niering, R. Holzwarth, J. Reichert, P. Pakasov, T. Udem, M. Weitz, T. W. Hänsch, P. Lemonde, G. Santarelli, M. Abgrall, P. Laurent, C. Salomon, and A. Clairon. *Phys. Rev. Lett.*, **84**, 5496 (2000).
11. T. W. Hänsch and C. Zimmermann. *Hyperfine Interact.*, **76**, 47 (1993).
12. D. Colladay and V. A. Kostelecký *Phys. Rev.*, **D55**, 6760 (1997), *Phys. Rev.*, **D58**, 116002 (1998).
13. V. A. Kostelecký and R. Lehnert, *Phys. Rev.*, **D63**, 065008 (2001).
14. R. Lehnert, *contribution to these proceedings*.
15. J. Walz and T. W. Hänsch. *General Relativity and Gravitation*, **36**, 361 (2004).
16. D. Pritchard, *Phys. Rev. Lett.*, **51**, 1336 (1986).
17. T. Bergemann, G. Erez and H. Metcalf, *Phys. Rev.*, **A35**, 1535 (1987).
18. T. M. O'Neil. *Phys. Fluids.*, **23**, 2216 (1980).
19. T. M. Squires, P. Yesley, and G. Gabrielse. *Phys. Rev. Lett.*, **86**, 5266 (2001).
20. G. Gabrielse, X. Fei, L. A. Orozco, R. L. Tjoelker, J. Haas, H. Kalinowsky, T. Trainor, and W. Kells. *Phys. Rev. Lett.*, **65**, 1317 (1990).
21. P. Belochitskii, T. Eriksson, and S. Maury. *Nucl. Instr. and Meth.*, **B214**, 176 (2004).

22. G. Gabrielse, N. S. Bowden, P. Oxley, C. H. Storry, J. N. Tan, M.Wessels, D. Grzonka, W. Oelert, G. Schepers, T. Sefzick, J. Walz, H. Pittner, T. W. Hänsch, and E. A. Hessels. *Phys. Lett.*, **B548**, 140 (2002).
23. J. Estrada, T. Roach, J. N. Tan, and G. Gabrielse. *Phys. Rev. Lett.*, **84**, 859 (2000).
24. C. Wesdorp, F. Robicheaux, L. D. Noordam, *Phys. Rev. Lett.*, **84**, 3799 (2000).
25. J. W. Humberston, M. Charlton, F. M. Jacobsen, and B. I. Deutch. *J. Phys.*, **B20**, L25 (1987).
26. G. Gabrielse, S. L. Rolston, L. Haarsma, and W. Kells. *Phys. Lett.*, **A129**, 38 (1988).
27. M. E. Glinsky and T. M. O'Neil. *Phys. Fluids*, **B3**, 1279 (1991).
28. G. Gabrielse, A. Speck, C. H. Storry, D. Le Sage, N. Guise, D. Grzonka, W. Oelert, G. Schepers, T. Sefzick, H. Pittner, J. Walz, T. W. Hänsch, D. Comeau, and E. A. Hessels. *Phys. Lett.*, **B**, to be published.
29. M. Herrmann *Master's thesis, TU München, Max Planck Institut für Quantenoptik, Garching*, (2003).
30. E. A. Hessels, D. M. Homan, and M. J. Cavagnero. *Phys. Rev.*, **A57**, 1668 (1998).
31. H. Pittner, *PhD thesis, Ludwig-Maximilians-Universität München* (2005).
32. E. P. Gilson, J. Fajans, *Phys. Rev. Lett.*, **90**, 015001 (2003).
33. P. Bowe et al., contribution to these proceedings, Proposal of the ALPHA collaboration, CERN-SPSC-2005-006.

# Prospects of CPT tests using antiprotonic helium and antihydrogen

Ryugo S. Hayano

*Department of Physics, University of Tokyo, 7-3-1 Hongo, Bunkyo-ku, Tokyo 113-0033, Japan*

**Abstract.** Testing CPT to the highest possible precision using the laser spectroscopy of antiprotonic helium atoms (a neutral three-body system consisting of an antiproton, a helium nucleus and an electron) is the current goal of ASACUSA collaboration at CERN AD. The present status and future prospects are discussed in the first half of the talk. Our program will be extended in the future to include the microwave spectroscopy of ground-state hyperfine splitting of antihydrogen. The physics motivations and possible measurement schemes are presented in the second half.

**Keywords:** CPT, antiproton, antiprotonic helium, antihydrogen
**PACS:** 36.10.-k, 25.43.+t, 34.90.+q

## INTRODUCTION

Testing CPT invariance to highest-possible precision is the common goal of the three experiments at the antiproton decelerator (AD) of CERN, ATHENA, ATRAP and ASACUSA. The first two collaborations concentrate on antihydrogen production and spectroscopy, while our collaboration, ASACUSA ( atomic spectroscopy and collisions using slow antiprotons. ), has so far used antiprotonic helium atoms, an exotic three-body metastable atom consisting of an antiproton, an electron and a helium nucleus.

Antihydrogens can now be routinely produced [1, 2], and the next step is to carry out spectroscopy. Although the highest CPT-test precision may be reached by comparing hydrogen and antihydrogen do to the long metastable lifetime of the $2s$ state, there are still hurdles to be cleared (*e.g.*, produce enough number of cold antihydrogen in the *ground* state, trap the cold antihydrogens, and let them interact with photons for a long-enough time), hence it may take some time before precision laser-spectroscopic techniques can be applied to antihydrogen.

Meanwhile, high-precision laser spectroscopy of antiprotonic helium[3], which measures $m_{\bar{p}}Q_{\bar{p}}^2$, has shown a steady progress over the last ten years[4, 5, 6, 7], and when combined with the (more precise) antiproton charge-to-mass ratio ($Q_{\bar{p}}/m_{\bar{p}}$) measured to $9 \times 10^{-11}$ by the TRAP group at CERN LEAR[8] provides the best baryonic CPT limits of $10^{-8}$ both in terms of mass ($|m_p - m_{\bar{p}}|/m_p$) and charge ($|Q_p + Q_{\bar{p}}|/Q_p$)[9].

ASACUSA will continue to improve the precision of antiprotonic helium spectroscopy, to better than $10^{-9}$, as described in the next section. At the same time, we are working towards future experiments on the ground-state hyperfine splitting (GS-HFS) measurement of antihydrogen. The physics motivations and possible measurement schemes are discussed towards the end.

CP793 *Physics with Ultra Slow Antiproton Beams*
edited by Yasunori Yamazaki and Michiharu Wada
© 2005 American Institute of Physics 0-7354-0282-5/05/$22.50

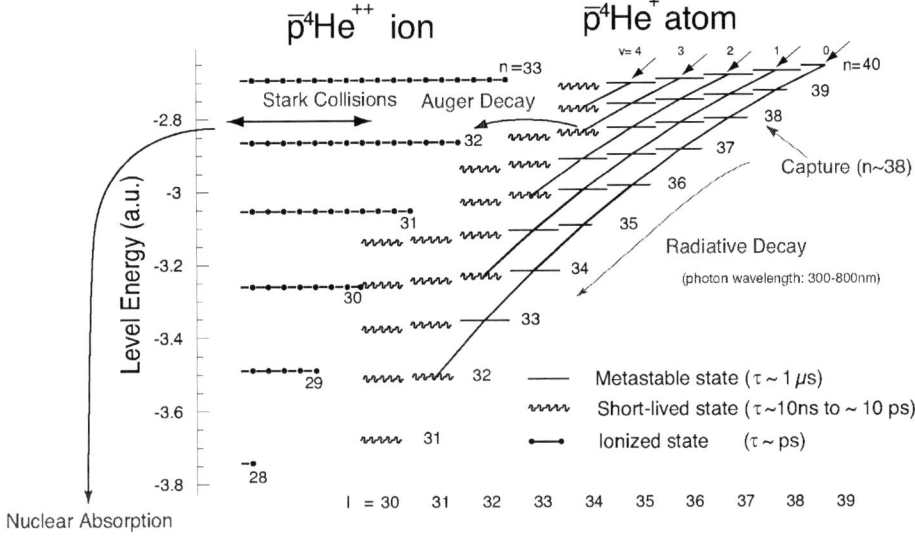

**FIGURE 1.** Level diagram of $\bar{p}\mathrm{He}^+$ in relation to that of $\bar{p}\mathrm{He}^{++}$. The solid and wavy bars stand for metastable and short-lived states, respectively, and the dotted lines are for $l$-degenerate ionized states.

## ANTIPROTONIC HELIUM ATOMS

### Naturally occurring antiproton traps

Antiprotonic helium atom (hereafter denoted $\bar{p}\mathrm{He}^+$) is a naturally-occurring antiproton trap that has the following remarkable features[3]: The atom can 'store' an antiproton for more than a microsecond. This 'longevity' occurs when the antiproton occupies a near-circular orbit having a large $n(\sim 38)$ and also large $\ell(\gtrsim 35)$. Unlike antihydrogen, it is not at all difficult to make $\bar{p}\mathrm{He}^+$. Just stop antiprotons in a helium target. Then, about 3% of the stopped antiprotons automatically become trapped in the metastable states. We usually use low-temperature ($T \sim 10$ K) helium gas as the target. The $\bar{p}\mathrm{He}^+$ atoms that are produced collide with the surrounding helium atoms, and are thermalized. Therefore, the antiprotonic helium atoms are already cold. The Doppler width for a typical single-photon transition is about 500 MHz.

Fig. 1 shows an energy level diagram of $\bar{p}^4\mathrm{He}^+$. The levels indicated by the continuous lines have metastable ($\gtrsim 1\mu$s) lifetimes and deexcite radiatively, while the levels shown by wavy lines are short lived ($\lesssim 10$ ns) and deexcite by Auger transitions to antiprotonic helium ion states (shown by dotted lines). Since the ionic states are hydrogenic, Stark collisions quickly induce antiproton annihilation on the helium nucleus, as indicated in Fig. 1.

# Progress of the antiprotonic helium laser spectroscopy

Laser spectroscopy of $\overline{p}\text{He}^+$ works as follows: As shown in Fig. 1, there is a boundary between metastable states and short-lived states. For example, $(n,\ell) = (39,35)$ is metastable, while $(n,\ell) = (38,34)$, which can be reached from (39,35) by an $E1$ transition, is short lived. Thus, if we use a laser ($\lambda = 597$ nm in this particular case) to induce a transition from (39,35) to (38,34), (and of course if an antiproton happens to be occupying the (39,35) level at the time of laser ignition), the antiproton is deexcited to the short-lived state, which then Auger-decays to an ionic $(n_i, \ell_i) = (32,31)$ state within $\lesssim 10$ ns. The ionic state is then quickly (usually within $\sim$ ps) destroyed by Stark collisions, leading to the nuclear absorption/annihilation of the antiproton. Adding all these together, we expect to see a sharp increase in the $\overline{p}$ annihilation rate in coincidence with the laser pulse. We measure the intensity of the laser-induced annihilation spike as a function of laser detuning, and compare the resonance-peak centroid $\nu_{\text{exp}}$ with the results of three-body QED calculations $\nu_{\text{th}}$ (calculated assuming $m_p = m_{\overline{p}}$ and $Q_p = -Q_{\overline{p}}$). See Fig. 2. So far, no statistically-significant deviation has been found. From this fact, we have concluded that 1) the three-body QED calculations have been performed with sufficient accuracy, and that 2) CPT holds within the experimental (and theoretical) error bars.

Our first experiment carried out in 1993[4] at LEAR had a precision of about 50 ppm ($5 \times 10^{-5}$). Theoretical predictions on the other hand scattered within about 1000 ppm. A breakthrough was made in 1995 by Korobov's first non-relativistic calculation, which improved the accuracy to some 50 ppm[10]. This was then improved to some 0.5 ppm by including relativistic corrections[11]. The precision of theoretical calculations continued to improve, and they have now reached $\lesssim 10^{-8}$[12, 13].

In competition, experimental error bars also continued to decrease. After the first success, we soon found out that although $\overline{p}\text{He}^+$ atoms are fairly stable against frequent collisions with helium atoms, the collisions induce frequency shift and broadening of the resonance lines[5]. By measuring the resonance centroids at different helium densities and by extrapolating to zero density, we reached a precision of 60 ppb in 2001[6]. This was the first physics result from the whole AD community.

Recently, we constructed a radio-frequency quadrupole decelerator (RFQD)[14] with which we can now decelerate the 5.3 MeV antiprotons extracted from AD to some 100 keV. This makes it possible to stop antiprotons in a very low density gas target ($\sim 10^{16-18}$ atoms/cm$^3$), eliminating the need for the zero-density extrapolation. Figure 2 shows the present status of the experiment-theory comparison for seven transitions in $\overline{p}^4\text{He}^+$ (left) and six transitions in $\overline{p}^3\text{He}^+$ (right)[7]. Experimental errors include the absolute-frequency calibration uncertainties, and are typically $\sim \pm 100$ ppb. The theoretical predictions of Korobov (squares) [12] and Kino (triangles)[13] are mostly in the experimental error bars, but for some transitions there are discrepancies of $\sim 100$ ppb between the two. These differences do not yet affect the final CPT limits, but as the experimental precisions are improved, they may eventually become the dominant error source in the CPT limits deduced from $\overline{p}\text{He}^+$ spectroscopy.

Our measurements, which yield $\sim m_{\overline{p}}Q_{\overline{p}}^2$, are then combined with the (more precise) antiproton vs proton cyclotron frequency comparison done by the TRAP group

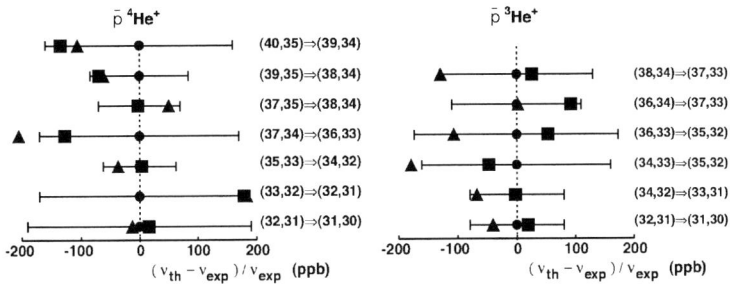

**FIGURE 2.** Comparisons between experimental result $v_{exp}$ (filled circles with errors) and theoretical predictions $v_{th}$ obtained assuming $m_p = m_{\bar{p}}$ and $Q_p = -Q_{\bar{p}}$ (squares [12] and triangles [13]). Taken from [7].

at LEAR[8]. This now provides the most stringent mass ($|m_{\bar{p}} - m_p|/m_p$) and charge ($|Q_{\bar{p}} + Q_p|/Q_p$) comparison between antiproton and proton of $10^{-8}$, as listed in the Review of Particle Physics[9].

## Discovery of metastable antiprotonic helium ions

Let us go back again to Fig. 1 and consider the fate of $\bar{p}He^{++}$ ions at very low target densities. The destruction of the $\bar{p}He^{++}$ states usually takes place in a matter of picoseconds due to the Stark collisions. If the $\bar{p}He^{++}$ ion is isolated in a vacuum, there are no collisions, and hence the $\bar{p}He^{++}$ states should become metastable (the radiative lifetimes of the circular states around $n_i \sim 30$ is several hundred ns). We, therefore, expect the prolongation of $\bar{p}He^{++}$ lifetimes at very low target densities.

This is exactly what we recently observed. In the left panel of Fig. 3, we show the annihilation spike produced by inducing the $\bar{p}^4He^+$ transition $(n, \ell) = (39, 35) \rightarrow (38, 34)$ measured by using the RFQD-decelerated beam at a low target density of $2 \times 10^{18}$ atom/cm$^3$. At this density, the decay time constant of the laser spike is still consistent with the Auger lifetime of the $(38, 34)$ level. However, as shown in the right panel, the shape of the laser-induced spike changes drastically in an ultra-low target density of $3 \times 10^{16}$ atoms/cm$^3$. Lifetime prolongation was also observed in the case of antiprotonic helium 3 ions. Systematic measurements of the ion lifetimes at various target densities have been carried out, which showed that the ionic-level lifetimes get shorter for larger principal quantum numbers.[15]

The long-lived antiprotonic helium ions $\bar{p}^4He^{++}$ and $\bar{p}^3He^{++}$ are quite interesting, since these are two-body systems and hence are practically free from theoretical errors. This motivates us to perform the laser spectroscopy of $\bar{p}He^{++}$. In principle, this appears possible, since we found that there is up to $\sim 50\%$ lifetime difference between the $\bar{p}He^{++}$ levels when the principle quantum number of the ion $n_i$ is changed by one unit[15].

Hence, if we use a laser to produce an ionic state, and then use *another* laser (in the UV region) to induce transitions between $n_i$ and $n_{i\pm 1}$, we should be able to observe a

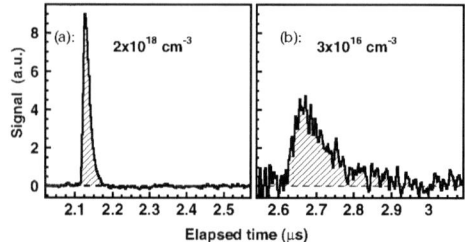

**FIGURE 3.** Annihilation spike produced by inducing the $\bar{p}\text{He}^+$ transition $(n, \ell) = (39, 35) \rightarrow (38, 34)$, measured at a high target density (a). A prolongation of the tail is observed at ultra-low densities (b), indicating the formation of long-lived $\bar{p}\text{He}^{++}$ ions.

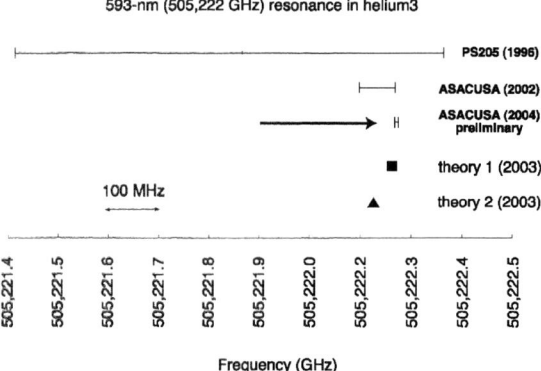

**FIGURE 4.** A preliminary result for one of the transitions measured in 2004. Note that the experimental error has become much smaller as compared with previous measurements, and hence the difference between the two theory calculations can no longer be ignored.

slight change in the decay time constant of the laser spike (such as in Fig. 3(b)). This is by no means an easy measurement, but is nevertheless an important one.

## Testing CPT or measuring a fundamental constant?

In 2004, we started a new series of measurements with a pulse-amplified continuous-wave (CW) laser system, in which the CW laser is stabilized and locked to an optical frequency comb[16]. Preliminary analyses look very promising (see Fig. 4, suggesting that we are likely to reach $\sim$ppb precision soon (in order to achieve this, we need continuing efforts of the theory community, so that the existing differences of some 100 ppb in the transition-frequency calculations are diminished).

In this way, the spectroscopy of $\bar{p}\text{He}^+$ is likely to play a leading role in baryonic CPT tests for some more time, but we will not be able to use $\bar{p}\text{He}^+$ to test CPT symmetry when we reach sub-ppb precision (the proton mass precision, measured in

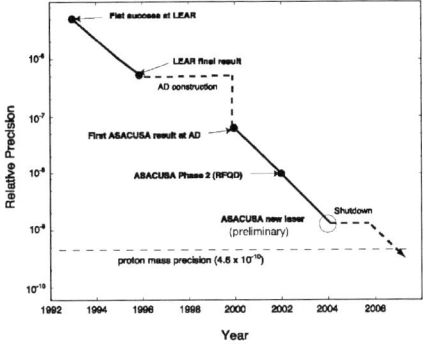

**FIGURE 5.** The progress of antiprotonic helium spectroscopy since the first success to the most recent measurement.

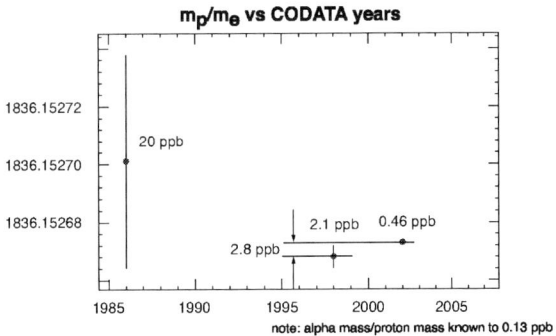

**FIGURE 6.** CODATA values for $m_p/m_e$ for the past years.

atomic unit, being $4.6 \times 10^{-10}$[17]). See Fig. 5. It is nevertheless important to measure $\overline{p}\mathrm{He}^+$ transition energies to the best of our ability (and also urge theorists to improve their calculations) since with the $\overline{p}\mathrm{He}^+$ spectroscopy we may be able to determine $m_{\overline{p}}/m_e$ as good as or better than $m_p/m_e$. Note that the CODATA value of $m_p/m_e$ moved as much as 2.8 ppb between 1998 and 2002 (see Fig. 6). It is hence important to compare the $m_{\overline{p}}/m_e$ value with the CODATA value as an independent check. If we believe that CPT is not violated at the level of ppb, the $\overline{p}\mathrm{He}^+$ spectroscopy may eventually contribute to the determination of a fundamental constant $m_p/m_e$.

# GROUND-STATE HYPERFINE SPLITTING OF ANTIHYDROGEN

## Motivation

ASACUSA is now investing heavily in the future ground-state hyperfine splitting (GS-HFS) measurement of antihydrogen. Since the ground state of antihydrogen has infinite lifetime, its high precision spectroscopy will give unprecedented accuracies in terms of CPT symmetry tests. In the case of hydrogen, the ground-state hyperfine splitting (GS-HFS) frequency $\nu_{HF}$ has been measured in a classic series of experiments which began in the 1930's with relatively simple atomic beam experiments, and culminated with maser experiments in the early 1970s which ultimately achieved a relative precision of order $10^{-12}$ (the latter technique is unfortunately not applicable to antimatter).

The 1S ground state of hydrogen is split due to the interaction of electron spin $\vec{S}_e$ and proton spin $\vec{S}_p$ according to $\vec{F} = \vec{S}_e + \vec{S}_p$ with quantum numbers $F = 0, 1$ (total spin) and $M = -1, 0, 1$ (projection of $F$ onto the magnetic field axis). The hyperfine splitting frequency between the $F = 0$ and $F = 1$ states $\nu_{HF}$ is given by the Fermi contact interaction, yielding

$$\nu_{HF} = \frac{16}{3} \left(\frac{M_p}{M_p + m_e}\right)^3 \frac{m_e}{M_p} \frac{\mu_p}{\mu_N} \alpha^2 c\, Ry(1 + \delta_{QED} + \delta_{Zemach}),$$

which is a direct product of the electron magnetic moment and the proton magnetic moment ($M_p$, $m_e$ denote proton and electron mass, $c$ the speed of light, $\alpha$ the fine structure constant, and $Ry$ the Rydberg constant). The QED correction $\delta_{QED}$ is due mostly to the anomalous electron g-factor and is of the order of 1000 ppm, while $\delta_{Zemach}$ is so-called the Zemach correction[18] which can be written as

$$\delta_{Zemach} = \frac{2Z\alpha m_e}{\pi^2} \int \frac{d^3 q}{q^4} \left[\frac{G_E(-q^2) G_M(-q^2)}{1 + \kappa} - 1\right],$$

where $G_E(-q^2)$ and $G_M(-q^2)$ are the electric and magnetic form factors of the proton, and $\kappa$ its anomalous magnetic moment. The Zemach correction, which is of the order of 30 ppm, therefore contains both the magnetic and charge distribution of the proton.

The GS-HFS frequency of antihydrogen is hence proportional to the spin magnetic moment of the antiproton, $\vec{\mu}_{\bar{p}}$, which is experimentally known only at the level of 0.3%. Below the level of several ppm accuracy, $\nu_{HF}$ also depends on the electric and magnetic form factors of the antiproton. The measurements of $\nu_{HF}(\overline{H})$ to a relative accuracy of better than $10^{-6}$ as discussed here will therefore yield an improvement of the value of $\vec{\mu}_{\bar{p}}$ by three orders of magnitude, and give some insight into the structure of the antiproton.

Furthermore, the only existing phenomenological extension of the standard model that includes CPT violations (the standard model extension – SME – of Kostelecky's group[19, 20, 21, 22, 23]) predicts that CPT violation in the 1S–2S transition is cancelled in first order, while for the hyperfine structure it is a leading-order effect. Although this model does not directly predict any CPT violation nor Lorentz invariance violation (LIV), it can be used as basis to compare CPT tests in different sectors, and as a guide where to look for possible CPT violating effects.

The SME-extension of the Dirac equation (for a free particle having a mass $m$) can be written as:

$$(i\gamma^\mu D_\mu - m - a_\mu \gamma^\mu - b_\mu \gamma_5 \gamma^\mu - \frac{1}{2} H_{\mu\nu} \sigma^{\mu\nu} + i c_{\mu\nu} \gamma^\mu D^\nu + i d_{\mu\nu} \gamma_5 \gamma^\mu D^\nu)\psi = 0,$$

where $\gamma$s, $D$s, $\sigma$s and $\psi$ are as in the standard Dirac equations, $a_\mu$ and $b_\mu$ are the CPT (and Lorentz) violating parameters, while $H_{\mu\nu}$, $c_{\mu\nu}$ and $d_{\mu\nu}$ are Lorentz violating,

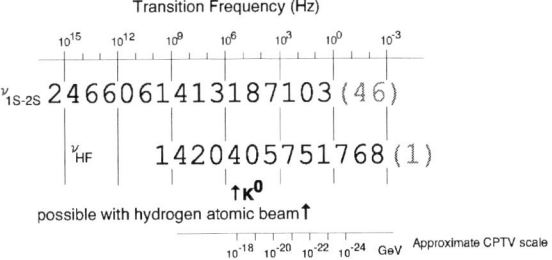

**FIGURE 7.** The current experimental value for the hydrogen 1S-2S frequency[24] (top) and the ground-state hyperfine frequency[25] (bottom), together with an approximate sensitivity to the CPT-violating parameter[23].

but CPT conserving parameters. The GS-HFS of antihydrogen is sensitive to the $b_\mu$ parameter[23] (also see Fig. 8).

Note that the CPT-violating parameters in SME have the dimension of energy (or frequency), hence it is advantageous to study CPT-violating effects in low-frequency transitions. Thus, within the SME framework, absolute energy (frequency) precision is important rather than relative precision $\delta\nu/\nu$. In other words, within the framework of SME, it is not necessary to measure the GS-HFS splitting to 18 digits (eventual goal of the 1S-2S laser spectroscopy) in order for the measurement to be competitive with the oft-quoted "most sensitive CPT bound" $|m_{K^0} - m_{\bar{K}^0}|/m_{\text{average}} < 10^{-18}$ [9]. This is because $|m_{K^0}c^2 - m_{\bar{K}^0}c^2|/h$ is constrained to $< 1.2 \times 10^5$ Hz in terms of frequency. The measurement of antihydrogen GS-HFS splitting to some $10^{-4}$ relative accuracy ($\Delta\nu \sim 100$ kHz) can thus already attain a sensitivity to the CPT-violating parameters as good as the $K^0 - \bar{K}^0$ comparison (see Fig. 7)[1].

## A possible method

We plan to apply a classical atomic beam method to the antihydrogen GS-HFS measurement[26]. The highest precision achieved for ordinary hydrogen using this method is $\delta\nu/\nu = 4 \times 10^{-8}$. If similar precision can be realized for antihydrogen, it will provide a more stringent test of the CPT symmetry than the $K^0 - \bar{K}^0$ mass difference. The atomic-beam measurement would involve:

1. An antihydrogen source - preferably a point-like source producing cold antihydrogens. The temperature should be as low as possible, but it is not necessary to produce $< 1$ K antihydrogen as would be required for antihydrogen trapping; a thermal source of some few tens of K would be sufficient.

---

[1] Note that the $K^0 - \bar{K}^0$ test (sensitive to the $a$ parameter) and the $H - \bar{H}$ test (sensitive to the $b$ parameter) cannot be directly compared; the purpose of the discussion here is to illustrate the order of magnitude of the achievable sensitivity to the CPT-violating parameters.

143

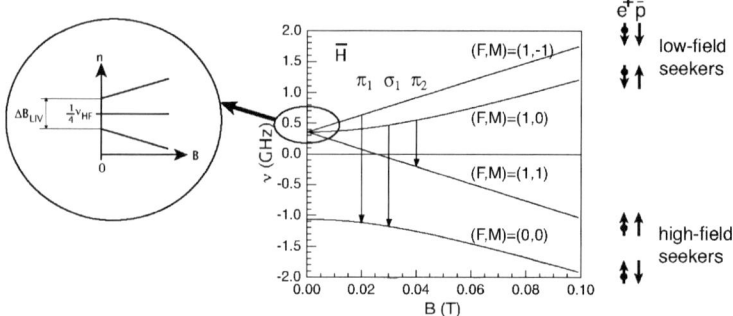

**FIGURE 8.** Right: Zeeman splitting of the ground state hyperfine levels of antihydrogen (Breit-Rabi diagram). The spin alignments of positron and antiproton in the high-field limit, when the spins are decoupled, is shown to the right. Left: Zero-field splitting of the $F = 1$ states in the presence of a CPT violating interaction as predicted by Bluhm et al. [23].

2. A sextupole magnet line to focus the antihydrogen atoms.
3. A microwave cavity (1.4 GHz) to resonantly induce spin flip.
4. Another sextupole magnet to focus the spin-flipped antihydrogen.
5. An antihydrogen detector.

The proposed strategy is hence quite different from that proposed for the 1S-2S laser spectroscopy.

The atomic-beam method works best if the source is point like. A simulation, assuming a point-like source with a $T = 50$ K thermal distribution, and a sextupole beam line optimized for $v = 350$m/s, shows that the beam line can transmit antihydrogen atoms within $\delta v = 25$m/s FWHM, which in turn corresponds to a line width $\delta v / v$ of $1.56 \times 10^{-6}$. The transmission (solid angle and the velocity acceptance included) is about $10^{-4}$ (see Fig. 9). With a colder source of $T = 4.2$ K, the transmission will be increased by a factor 10, while with a much hotter source of $T = 150$ K the transmission goes down only by a factor 2. This shows that while a cold source is desirable, a relatively hot source is still tolerated. A relatively large source size of some 1 cm$^3$ of nested Penning traps, and technical difficulties to interface a sextupole beam line to a superconducting solenoid with a large-enough solid angle, lead us to consider alternative antihydrogen production methods. Therefore, within ASACUSA, two R&D efforts are being conducted.

One is to use a two-frequency Paul trap to simultaneously confine positrons (with a few GHz microwave) and antiprotons (with a few MHz radiofrequency)[26]. Such a trap, if realized, will produce antihydrogen atoms within a very small sphere ($< 1$mm$^3$). The trap will have openings to extract antihydrogen atoms with a large-enough solid angle, and will also have openings for lasers to deexcite antihydrogen atoms to the ground state. In fact, due to the alternating electric field whose strength increase linearly towards outside, antihydrogen atoms in high-$n$ states are quickly dissociated and antiprotons and positrons are recycled. Only low-lying ($n \ll 10$) antihydrogen can leave the trap. This would ensure that only the ground-state antihydrogen atoms can reach the spin-

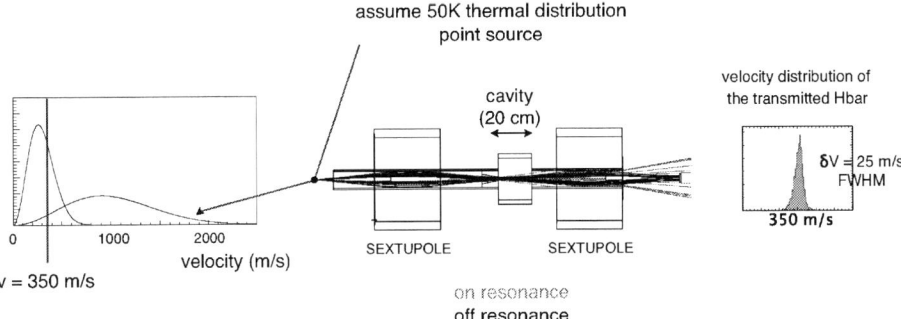

assume 50K thermal distribution
point source

cavity
(20 cm)

velocity distribution of
the transmitted Hbar

$\delta V = 25$ m/s
FWHM

350 m/s

velocity (m/s)

v = 350 m/s

SEXTUPOLE          SEXTUPOLE

on resonance
off resonance

**FIGURE 9.** Simulated trajectories of antihydrogen through a sextupole beam line. With an assumption of a point source having a $T = 50$ K thermal distribution, a narrow velocity band of $\delta v/v = 25$ m/s FWHM is accepted. The overall acceptance (including solid angle) is $10^{-4}$.

analyzing cavity.

Alternatively, we are developing a so-called cusp trap (a pair of Helmholtz-like coils energized in opposite directions, combined with an octupole electric field) to simultaneously trap antiprotons and positrons and to synthesize antihydrogen atoms[27]. The source size realizable with this method is larger than the Paul-trap case, but the strong magnetic-field gradient at the antihydrogen extraction port can in principle produce a spin-polarized antihydrogen beam. This would make it possible to eliminate the first sextupole magnet, thereby achieving a larger solid angle.

These are both in the R&D phase, still requiring tests with protons and electrons to demonstrate that the concepts work. However, in view of the fact that the antihydrogen atoms produced in nested traps are rather hot and not necessarily in the ground state, we believe it is important that we pursue alternative production methods as discussed above.

This work was supported by the Grant-in-Aid for Specially Promoted Research (15002005) of MEXT Japan, the Hungarian Scientific Research Fund (OTKA T033079 and TeT-Jap-4/98), and the Japan Society for the Promotion of Science.

## REFERENCES

1. M. Amoretti *et al.*, Nature **419**, 456 (2002).
2. G. Gabrielse *et al.*, Phys. Rev. Lett. **89**, 213401 (2002).
3. T. Yamazaki, N. Morita, R. S. Hayano, E. Widmann and J. Eades, Physics Reports **366**, 183 (2002).
4. N. Morita, M. Kumakura, T. Yamazaki, E. Widmann, H. Masuda, I. Sugai, R. S. Hayano, F. E. Maas, H. A. Torii, F. J. Hartmann, H. Daniel, T. von Egidy, B. Ketzer, W. Muller, W. Schmid, D. Horvath and J. Eades, Phys. Rev. Lett. **72**, 1180 (1994)
5. H. A. Torii, R. S. Hayano, M. Hori, T. Ishikawa, N. Morita, M. Kumakura, I. Sugai, T. Yamazaki, B. Ketzer, F. J. Hartmann, T. Von Egidy, R. Pohl, C. Maierl, D. Horvath, J. Eades and E. Widmann, Phys. Rev. A **59**, 223 (1999).
6. M. Hori, J. Eades, R. S. Hayano, T. Ishikawa, J. Sakaguchi, E. Widmann, H. Yamaguchi, H. A. Torii, B. Juhasz, D. Horvath and T. Yamazaki, Phys. Rev. Lett. **87** 093401 (2001).
7. M. Hori, J. Eades, R. S. Hayano, T. Ishikawa, W. Pirkl, E. Widmann, H. Yamaguchi, H. A. Torii, B. Juhasz, D. Horvath and T. Yamazaki, Phys. Rev. Lett. **91** 123401 (2003).

8. G. Gabrielse, A. Khabbaz, D. S. Hall, C. Heimann, H. Kalinowsky and W. Jhe , Phys. Rev. Lett. **82**, 3198 (1999).
9. S. Eidelman *et al.*,. Phys. Lett. B **592**, 1 (2004).
10. V.I. Korobov, Phys. Rev. A **54**, R1749 (1996).
11. V.I. Korobov and D.D. Bakalov, Phys. Rev. Lett. **79**, 3379 (1997).
12. V.I. Korobov, Phys. Rev. A **67**, 026501 (2003).
13. Y. Kino, M. Kamimura and H. Kudo, Nucl. Instr. Methods B **214**, 84 (2004).
14. A. M. Lombardi, W. Pirkl and Y. Bylinsky, in Proceedings of the 2001 Particle Accelerator Conference, Chicago, 2001 (IEEE, Piscataway, NJ, 2001), pp. 585–587.
15. M. Hori, J. Eades, R. S. Hayano, W. Pirkl, E. Widmann, H. Yamaguchi, H. A. Torii, B. Juhasz, D. Horvath, K. Suzuki, and T. Yamazaki, Phys. Rev. Lett. **94**, 063401 (2005).
16. J. Reichert, R. Holzwarth, Th. Udem and T. W. H?nsch, Opt. Commun. **172**, 59 (1999).
17. P. J. Mohr and B. N. Taylor, to appear in Review of Modern Physics **76** (2004); also available on web at http://physics.nist.gov/constants.
18. J. R. Sapirstein and D. R. Yennie, in Quantum Electrodynamics, ed. T. Kinoshita, (World Scientific, Singapore), pp 560–672 (1990).
19. D. Colladay and V. A. Kostelecky, Phys. Rev. **D55**, 6760 (1997).
20. R. Bluhm, V. A. Kostelecky and N. Russell, Phys. Rev. Lett. **79**, 1432 (1997).
21. R. Bluhm, V. A. Kostelecky and N. Russell, Phys. Rev. **D57**, 3932 (1998).
22. V. A. Kostelecky, Phys. Rev. Lett. **80**, 1818 (1998).
23. R. Bluhm, V. A. Kostelecky and Neil Russell, Phys. Rev. Lett., **82**, 2252 (1999).
24. M. Niering, R. Holzwarth, J. Reichert, P. Pokasov, Th. Udem, M. Weitz, T. W. H?nsch, P. Lemonde, G. Santarelli, M. Abgrall, P. Laurent, C. Salomon, and A. Clairon., Phys. Rev. Lett. 84 (2000) 5496.
25. S. G. Karshenboim, Simple atoms, quantum electrodynamics, and fundamental constants, in *Precision Physics of Simple Atomic Systems*, pages 142?162, Springer, Berlin, Heidelberg, 2003, hep-ph/0305205.
26. ASACUSA collaboration, CERN-SPSC 2005-02, available from CERN Document Server.
27. A. Mohri and Y. Yamazaki, Europhys. Lett. **63**, 207 (2003).

# Non-Neutral Plasma Confinement In A Cusp-Trap And Possible Application To Anti-Hydrogen Beam Generation

Akihiro Mohri*, Yasuyuki Kanai*, Yoichi Nakai*,
and Yasunori Yamazaki* [#]

* Atomic Physics laboratory, RIKEN, Wako 351-0198, Japan
# Institute of Physics, Gradate School of Arts and sciences,
University of Tokyo, Tokyo 153-8902, japan

**Abstract.** A new scheme for synthesizing antihydrogen by trapping positrons and antiprotons in a field consisting of a magnetic quadrupole and an electric octupole (cusp –trap) is now under investigation. The total electric field of the octupole with the space charge of a nonneutral plasma composed of particles of the same sign of charge, i.e., positrons or mixture of electrons and antiprotons, is expected to form a potential well for particles of the opposite sign of charge. Particles trapped in the well are mixed with the present dense particles, where positrons and antiprotons will combine to produce antihydrogen atoms. A considerable fraction of antihydrogen atoms in low-field seeking states will be transported outside as a beam.

Experiments on electron confinement in the cusp-trap were carried out in a strong magnetic quadrupole (3.8T at the maximum on the axis). The confinement time reached 400s for the trapped electron number $N_0= 3.6 \times 10^7$. The time decreased with $N_0$ but it was still about 100s for $N_0= 1.6 \times 10^8$.

An electron plasma initially formed around the zero-field point rapidly expanded and settled down onto a quasi-stable state. Cross-sectional density profiles had shapes like a high volcano with a big crater. Analysis of the density profile shows that a potential well for oppositely charged particles (positive ions in this case) is probably formed inside the trapped electrons.

**Keywords:** nonneutral plasma, cusp, octupole, confinement, antiproton, antihydrogen, positron
**PACS:** 36.10.-k, 29.27.Hj, 52.27.Jt

## INTRODUCTION

Antihydrogen ($\overline{H}$) has been synthesized by two working groups at CERN [1,2]. Research related to antihydrogen is now entering a new advanced stage. Many potential candidates for stringent CPT tests will be examined in their practical application, which are high-precision laser spectroscopy, high-precision measurement of hyperfine splitting with rf-cavity and so on. Such a high-precision measurement will be pursued either for $\overline{H}$ atoms trapped in a closed space or for $\overline{H}$ atoms constituting a beam. Antihydrogen atoms in low-field seeking states can be confined in a minimum-B magnetic field. Generation of $\overline{H}$ beams needs a proper spatial distribution of magnetic field gradient for guiding the atoms. In the both cases,

CP793 Physics with Ultra Slow Antiproton Beams
edited by Yasunori Yamazaki and Michiharu Wada
© 2005 American Institute of Physics 0-7354-0282-5/05/$22.50

confinement or guide of $\overline{\text{H}}$ atoms should be compatible with synthesis of them in the system.

A project named "ALPHA " [3] has recently started aiming at confinement of $\overline{\text{H}}$ atoms in a minimum-B field, which is produced by a magnetic multi-pole and a mirror field. $\overline{\text{H}}$ atoms are synthesized in a nested trap locating near the axis where the magnetic field is nearly uniform. This plan is one of solutions to make the synthesis and the confinement compatible together.

Several plans to produce beams of $\overline{\text{H}}$ atoms of low energies have been considered. A simple and certainly realizable plan [4] is to use $\overline{\text{H}}$ atoms leaking from slits of a nested trap in which $\overline{\text{H}}$ atoms are produced in the similar way taken by ATRAP and ATHENA groups. Charged particles can be confined in an effective potential well formed by ponderomotive forces of radio-frequency fields. A new concept to synthesize $\overline{\text{H}}$ atoms in a trap made by rf-fields is now under study at CERN [5]. In this case, $\overline{\text{H}}$ atoms are also taken out of the rf-trap through opening gaps of installed electrodes. Emerging $\overline{\text{H}}$ atom beams in above two cases are so divergent that collection of them may need large open areas on instruments.

The other scheme to produce $\overline{\text{H}}$ beams incorporates a magnetic quadrupole (cusp) and an electric octupole [6] in a trap. This arrangement, referred to as cusp-trap or MCEO, possibly allows extraction of almost fully polarized $\overline{\text{H}}$ beams with a considerable flux density. This intense $\overline{\text{H}}$ beam will make a high-precision determination of the antiproton magnetic moment feasible. Positrons and antiprotons recombine in the central region of the cusp-trap and synthesized $\overline{\text{H}}$ atoms in low-field seeking states are guided by the surrounding magnetic field with strong gradient. Realization of this scheme strongly depends on the dynamics of nonneutral plasma confined in the cusp-trap.

This report mainly describes recent experiments on confinement of nonneutral electron plasma in the cusp-trap.

## CUSP-TRAP

Combination of a magnetic quadrupole (cusp) and an electric octupole forms a trap of non-neural plasma. An axisymmetric magnetic quadrupole is simply expressed in terms of the vector potential in cylindrical coordinates $(r,\theta,z)$ as

$$A_\theta(r,z) = \frac{B_0}{L} rz, \tag{1}$$

and the magnetic field components are

$$\boldsymbol{B} = \left( -\frac{B_0}{2L} r, \ 0, \ \frac{B_0}{L} z \right) \tag{2}$$

Also, the electric potential of an axisymmetric octupole is given by

$$\phi(r,z) = \phi_0\left(\left(\frac{r}{L}\right)^2 + \left(\frac{z}{L}\right)^2\right)^2 P_4\left[\frac{\left(\frac{z}{L}\right)}{\sqrt{\left(\frac{r}{L}\right)^2 + \left(\frac{z}{L}\right)^2}}\right]. \tag{3}$$

Here, $L$ is the scale length, $B_0$ and $\phi_0$ are the magnetic field strength and the potential at $z=L$ and $r=0$, respectively, and $P_4$ is the Legendre function of the first kind and the fourth order. Figure 1 shows magnetic field lines of a magnetic cusp and equipotential surfaces of an electric octupole.

The magnetic cusp is representative of the minimum-B configuration in which a confined neutral plasma is magnetohydrodynamically stable. However, particles of confined neutral plasma are rapidly lost, since their orbital magnetic moments cannot be invariant around the null-field, i.e., the null-field acts as a scattering center for particles.

On the other hand, the cusp-trap, which incorpotates both a magnetic quadrupole and an electric octupole, perfectly confines a single charged particle [7]. That is, the Störmer region that constrains a charged particle is closed. Similarly, nonneutral plasmas are expected to be confined in this field configuration

Equilibrium state of nonneutral plasma in the cusp-trap has theoretically been found at the Brillouin limit for cold plasma [8]. Also, preliminary experiments using electrons in low magnetic fields have proved that cusp-traps are capable of nonneutral confinement [8,9]. However, plasma diffusion processes as well as formation of a well for particles of the opposite sign of charge have not been clear yet and remain to be studied.

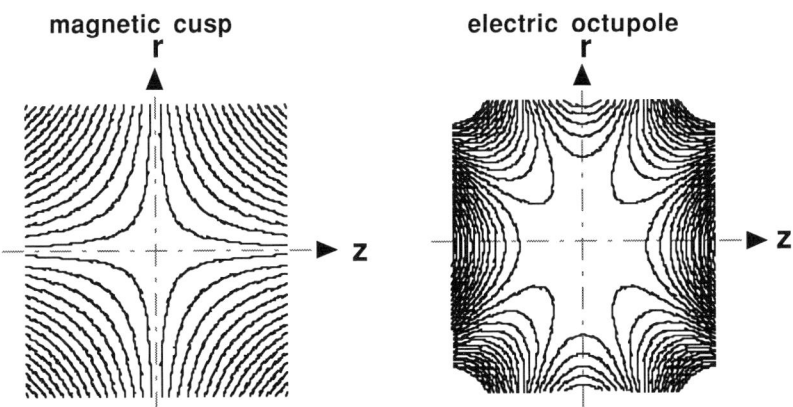

**FIGURE 1.** Field lines of a magnetic quadrupole( cusp) and equi-potential surfaces of an electric octupole.

# Application Possibility Of Cusp-Trap To Generation Of Spin-Polarized Antihydrogen Beam

We shall consider a magnetic field depicted in Fig.2. The field has a pure cusp in the central region and extends outwards having an appropriate field distribution. When a mixture of cold positrons and antiprotons condenses in the central region, recombination of antiprotons with positrons is expected to take place. Cold $\overline{H}$ atoms in the ground state enter the nonzero magnetic field region and their quantum states are split to four states as $(F, M_F) = (1,1), (1,0), (1,-1), (0,0)$ where $F$ and $M_F$ are the total spin and its magnetic quantum number, respectively. The magnetic moment in the state $(1,1)$ and $(1,0)$ is antiparallel to the magnetic field, so that $\overline{H}$ atoms in these two states prefer weaker fields (low-field seeker). In other words, low-field seekers feel backward force in magnetic field gradient $\nabla |B|$. When the magnetic field is designed to have the spatial distribution of $\nabla |B|$ as given in Fig.3, a considerable part of produced low-field seekers are transported outside as a spin-polarized beam. Trajectories of $\overline{H}$ atoms in $(1,1)$ state with the initial energy of 0.233meV are shown in Fig.2.

The magnetic fields shown in Fig.2 and Fig.3 are both the same. This field is produced by super-conducting coils and used for experiments described in this report. This magnet will be used for $\overline{H}$ beam generation when the concept is considered to be feasible.

To realize the concept, it becomes an essential theme how to make cold positrons and antiprotons coexist in the central region.

In a magnetic mirror, of which the minimum field is $B_L$ and the maximum one is $B_H$, the electric potential difference between these two extreme field points appears when a nonneutral plasma is confined therein [10]. This difference $\Delta\phi$ roughly amounts to

$$\Delta\phi \sim \left(\frac{kT}{e}\right) \cdot \beta \quad ; \quad \beta = \frac{B_H}{B_L} - 1, \tag{4}$$

where k is the Boltzmann constant and e is the unit of charge. Real nonneutral plasma is not perfectly cold in any case. Also, a magnetic cusp can be regarded as a bundle of magnetic flux tubes with high mirror ratios and the field strength increases in all directions. It is expected that, when a nonneutral plasma is confined in the cusp-trap, a potential well for particles of the opposite sign of charge is formed in the central region. That is, confined positrons (or mixture of electrons and antiprotons) digs a well for antiprotons ( or positrons). Therefore, the coexistence of positrons and antiprotons may become possible.

Formation of the well and its stable sustainment have not yet been studied. Those problems are key subjects of this experiment.

150

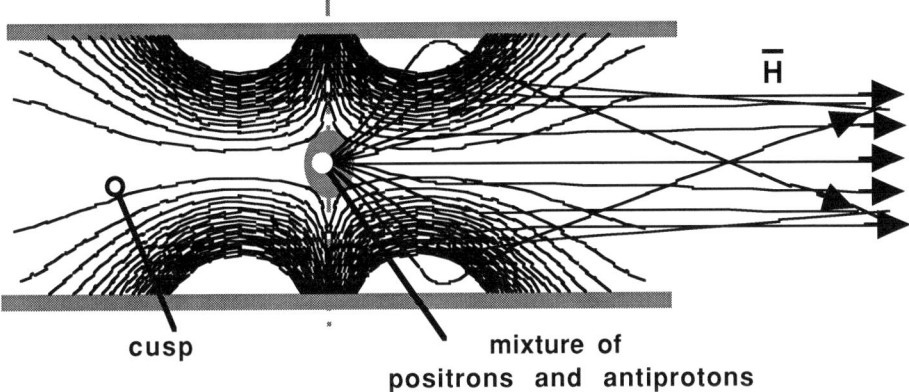

**cusp**              **mixture of
positrons and antiprotons**

Figure 2. Spin-polarized antihydrogen atoms $\overline{\mathrm{H}}$ flow out of the recombination region located around the center.

# EXPERIMENT

## Experimental Setup

The experiment is performed by using a trap consisting of a super-conducting quadrupole magnet and a vacuum tube which houses a set of electrodes for generating an octupole, an electron gun and a Faraday cup.

### *Super-Conducting Quadrupole Magnet*

Two pairs of super-conducting coils generate a magnetic cusp with a proper spatial distribution of field gradient necessary for intensity enhancement of outgoing spin-polarized antihydrogen atoms when they are synthesized.    These coils are cooled

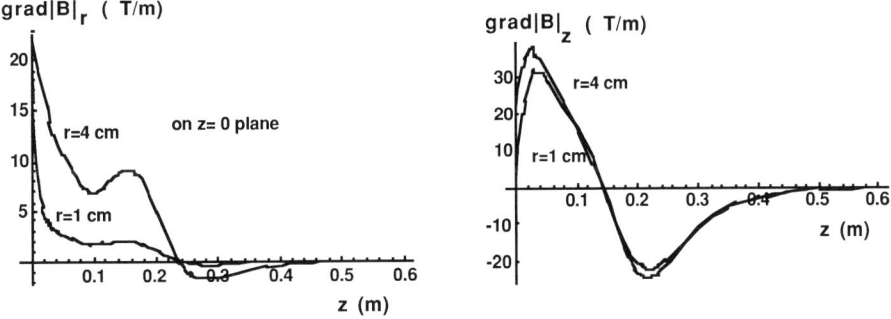

**FIGURE 3.** Radial and axial components of the magnetic field gradient in the quadrupole magnet.

down to 5K with a cryogenic refrigerator without the use of liquid helium. The magnetic field lines are shown in Fig.2 and the radial and axial components of field gradient at the radius of 1 and 4 cm are shown in Fig.3. The maximum magnetic field strength on the axis is 3.5T at 15cm from the plane of symmetry. Inside the warm bore of 16cm inner diameter is installed a vacuum tube of aluminium alloy.

## Generation Of Octupole And Equipments Inside Vacuum Tube

Figure 4 depicts the arrangement of electrodes producing an octupole, an electron gun and a segmented Faraday cup. The space enclosed with the electrodes is in the diameter of 9.2cm and in the axial width of 9.2cm. Shapes of the electrodes and the voltage allocation applied on them are optimized so as to make a wide region of octupole inside the trap region. Two electrodes on the both sides are copper-meshes with 80% transparency in order to enable the passage of electrons at the injection and also at the dump. The system is evacuated down to the vacuum pressure of $1.5 \times 10^{-7}$Pa.

A tiny electron gun with a cathode of Ba-sintered porous tungsten is set on the axis outside the trap as is shown in Fig.4. A burst of pulsed electron beams, each of which has the pulse width of 40μs typically, are injected into the trap.

The Faraday cup in the diameter of 6cm is segmented to eight parts in the radial direction with the pitch of 0.4mm. Line-density, that is integrated amount of density along magnetic field lines, can be inferred from charges collected on the segments. Also, the total number of the confined electrons, $N$, is determined by summing up all signals of the segment.

## Injection, Confinement And Dump Of Electrons

Synchronously to every pulsed electron beam, the potential on the electron injection side is made shallower to introduce the beam into the trap and, at the pulse end, the potential is again returned back to the former one to keep the injected electrons. Figure 5(a) and (b) show the potential distributions at the injection and the

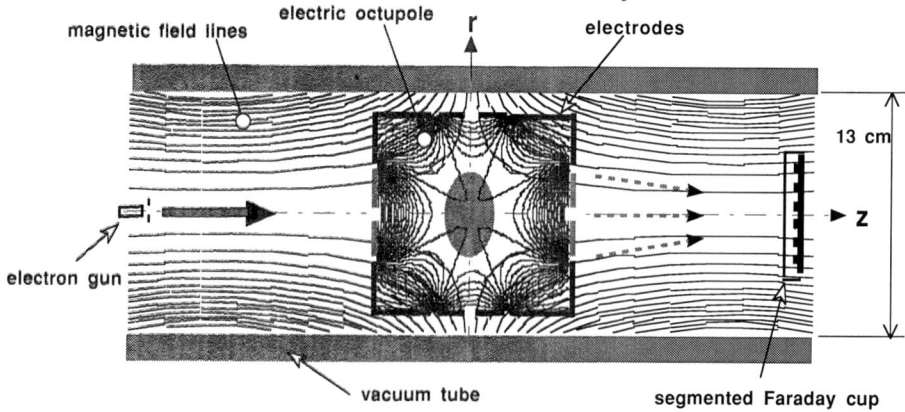

**FIGURE 4.** Experimental components inside the vacuum tube: the electrodes, electron gun and the segmented Faraday cup.

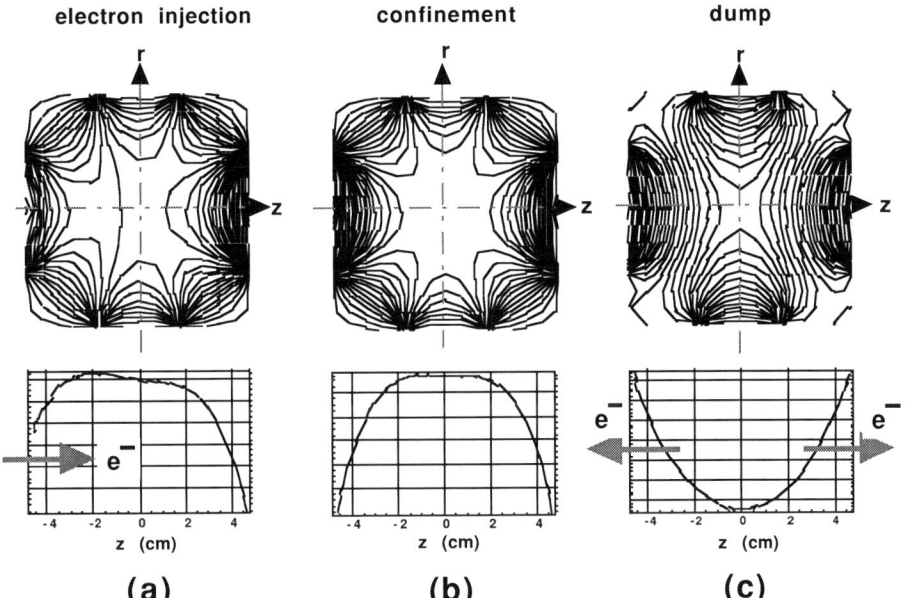

**FIGURE 5.** Equi-potential surfaces of the trap( upper figures) and axial change in the potential (lower figures) for the case (a) At electron beam injection. The potential barrier on the injection side ( left side) is lowered to allow the electrons come in, (b) At electron confinement. The octupole field is held during the confinement. (c) At dump of electrons. Electrons are forced to get out of the trap.

confinement, respectively. By repeating this cycle, electrons are stacked and confined. Stacked electron number can be adjusted by choosing the repetition number.

At an elapsed time after the end of electron stacking, the confined electrons are dumped out of the trap by changing the potential distribution as shown in Fig.5(c). In this moment, the electric field pushes and dumps the electrons outside the trap along magnetic field lines. A half of the dumped electrons are collected with the Faraday cup.

## Time Evolution Of Confined Electron Number

Trapped electrons are localized around the center of the trap in a quite earlier phase after the electron stacking. The field strength near the center is not so strong to withstand the expansion of the electrons due to their space charge, so that the electrons change their spatial distribution until they settle in a state where the space charge field is balanced with forces exerted by externally applied electric and magnetic fields. However, the plasma gradually expands across the magnetic field, being caused either by collisions with the residual molecules or by fluctuations in the plasma. Finally, the outward plasma edge touches the electrodes and the total number of electrons $N$ begins to decrease by the loss of electrons. Time dependence of $N$ strongly reflects

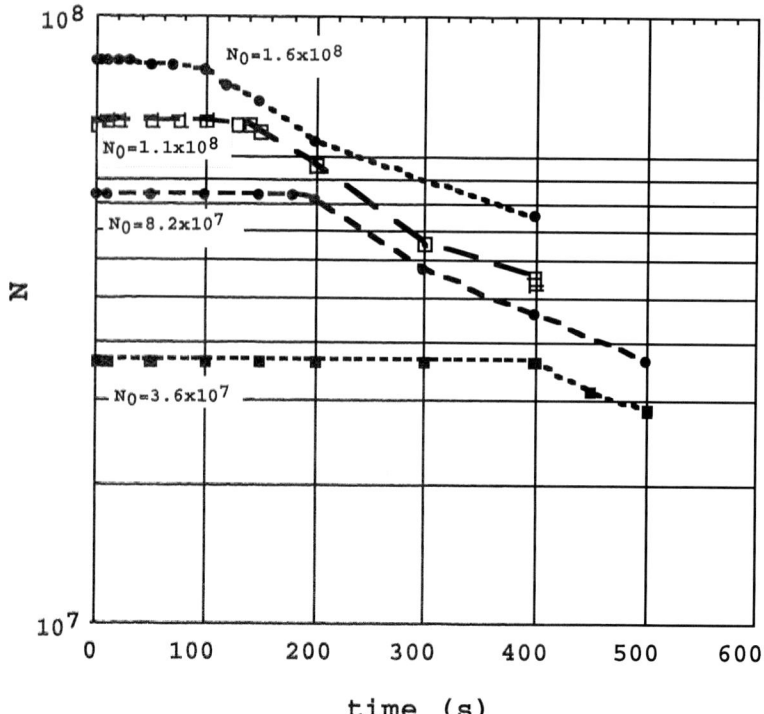

FIGURE 6. Time variation in the electron number $N$ for different initially stacked number $N_0$.

such a macroscopic behaviour of the plasma. Observed spatial change in the density distribution will be discussed later.

Figure 6 shows the change in the total trapped electron number $N$ with time for different initially stacked number $N_0$. Here, the magnetic field at the radial inner surface of the electrode, i.e., at $r$=4.6cm and $z$=0, is 0.9T and the well depth of the octupole is 33.7V. In the case of $N_0$ =3.6x10$^7$, $N$ is nearly constant until the time $t_c$~400s and then decreases faster. The turning time $t_c$ becomes shorter as $N_0$ increases, e.g., $t_c$~110s for $N_0$ =1.1x10$^8$. This trend suggests that the plasma periphery moves faster toward the electrode wall for larger $N_0$.

## Density Profile And Electric Potential Of Plasma

In order to determine the density distribution of the plasma $n(r,z)$ from experimentally obtained line-densities, it is necessary to assume a shape of constant density surface. In the equilibrium state of cold plasma at the Brillouin limit in this field configuration [8], the density is constant at a surface $\rho$=const as

$$n = n(\rho), \quad ; \quad \rho^2 = r^2 + 2z^2. \qquad (5)$$

154

Whence, each equi-density surface is the surface of an oblate spheroid with the aspect ratio of $\sqrt{2}$. This relation is used to determine density profiles although electron densities in this experiment are much less than that in Brillouin limit. Line densities estimated from a temporary function $n(\rho)$ are compared with experimentally obtained ones on the same field lines, and thereby a proper $n(\rho)$ that corresponds to the observation can be found.

## Density Profile At Early Phase After Electron Stacking

Electrons are injected into the trap and, at a time $t_D$ after the injection, trapped electrons are dumped. Time variation in the density distribution is inferred from the obtained line-densities, using the model eq.(5). Electrons during the injection and immediately after its end would be not so cold to allow the use of eq.(5) given for cold plasma and the practical profile itself would differ from it. However, by using the modeled profile, we may get a rough picture of cross-field motions of the trapped electrons.

Figure 7 shows examples of obtained density profiles. Here, the magnetic field at $r$=4.6cm and $z$=0 was 0.7T and the well depth of the octupole was 33.7V. The duration of the electron injection was 800µs. The total number of stacked electrons was $N_0$=6x10$^7$ at the end of the pulse. As noticed in Fig.7(a) ( $t_D$=0), expansion of electrons across the magnetic field had already occurred during the electron injection and the profile became like a crater where the density near the center was much less. This expansion rapidly proceeded thereafter as seen in Fig.7(b) ( $t_D$ =1ms). Though the expansion was fast at the early phase, it gradually slowed down as the plasma approached an equilibrium state.

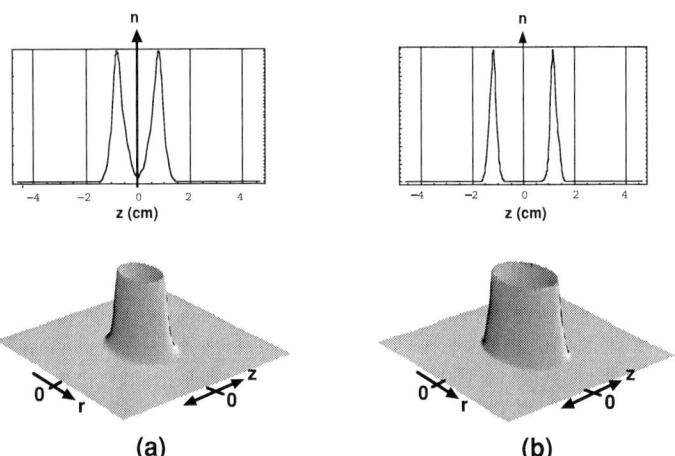

**(a)** **(b)**

**FIGURE 7.** Density profiles inferred from the relation eq.(5) at (a) the end of the electron injection : $t_D$ =0 and (b) $t_D$ =1ms. Upper figures are axial changes in the density and lower ones are density profiles on a $r$-$z$ plane. Shown densities are in an arbitrary scale.

# Calculation of Potential Distribution

When the plasma density distribution is expressed by eq.(5) , we can regard the plasma as a superposition of many spheroidal plasmas with the same aspect ratio, each of which has its own uniform density and size. To make such a spheroidal plasma, we divide the plasma density profile on a r-z plane into thin slices parallel to the plane. Each slice, which has a density and a size in the r- and z-directions, corresponds with a spheroidal plasma with a uniform density. Potentials produced by individual spheroidal plasmas $\phi_j^s(r,z)$ can be estimated analytically. If a slice is across the concave part of the profile as shown in Fig.7, the slice possesses an elliptic hole. This hole is considered as an additive slice having the same density but with the opposite charge. The potential for the hole is noted as $\phi_i^{sn}(r,z)$. Then, the electric potential of the plasma $\phi^s(r,z)$ is found by summing up all potentials of the sliced plasma components as

$$\phi^s(r,z) = \sum_j \phi_j^s(r,z) - \sum_i \phi_i^{sn}(r,z). \qquad (6)$$

The externally applied electric potential $\phi^{ex}(r,z)$ is numerically found from the voltages applied to the electrodes. Also, the potential caused by image charges on the electrode surfaces $\phi^i(r,z)$ is numerically obtainable using $\phi^s(r,z)$. Summing these potentials, we have the potential distribution inside the trap region $\phi^t(r,z)$ as

$$\phi^t(r,z) = \phi^s(r,z) + \phi^i(r,z) + \phi^{ex}(r,z). \qquad (7)$$

# Density Profile And Internal Potential Distribution Of Confined Plasma

The number of trapped electrons $N$ was nearly constant until a time $t_c$ as shown in Fig.6, while the plasma changed its shape in this period. Within a few seconds after the completion of electron stacking, the plasma rapidly expanded outwards. Exampled density profiles during this expansion have been shown in Fig.7. After then, the change in the density distribution became slow and the associated potential structure also gradually varied with time.

Figure 8 shows radial profiles of the density and the potentials in a typical case of a large $N_0$. The experimental conditions were as follows; the magnetic field at $r$=4.6cm and $z$=0 was 0.9T, $N_0$ =2.8x10$^8$, the potential well depth of the octupole was 33.7V. The data was taken at 10s after the electron stacking. Here, the referring potential is on the inner wall of electrode on the midplane, i.e., at $r$=4.6cm and $z$=0. It should be noted that all profiles shown in Fig.8 were obtained by using eq.(5). The plasma density $n$ radially increased till $r$=2.4cm and steeply fell. At this time the plasma periphery did not contact with the electrode surface. The plasma potential $\phi^s$ decreased from the electrode surface and also the potential due to the image charges $\phi^i$ did similarly. These two potentials $\phi^s$ and $\phi^i$ were negative to the external octupole $\phi^{ex}$, so that the resultant net potential $\phi^t$ became lower than $\phi^{ex}$. The figure on the right and the lower side in Fig.8 shows an enlarged one of $\phi^t$, where the

156

potential difference from the center $\Delta\phi^t$ is shown. There was a potential well, noted by $W^s$, for particles of a positive charge. Its well depth was about 40mV. This well formation is what we have expected. In this experimental condition the well continued for about 20s but the well depth became shallower with time.

Of course, the well: $W^s$ might be easily formed and its well depth would become deeper if $N_0$ is increased to make $\phi^s$ large enough. However, at larger $N_0$, the plasma diffuses faster across the magnetic field and the well: $W^s$ itself disappears in a shorter time. In the best case so far, the well: $W^s$ is maintained for 40s for $N_0 = 1.5\times10^8$.

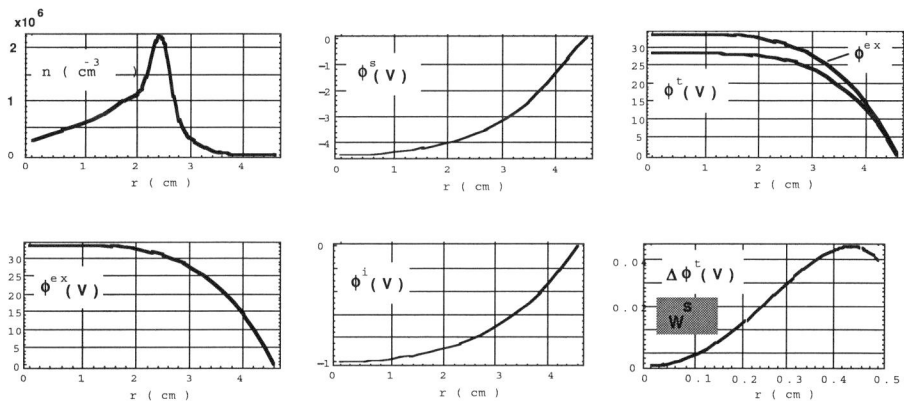

**FIGURE 8.** Radial density distribution and variation of electric potentials on the z=0 plane at 10s after the end of electron stacking, where $n$ is the density, $\phi^{ex}$ the external octupole, $\phi^s$ the self-field of the plasma, $\phi^i$ the potential caused by the image charge, $\phi^t$ the potential inside the trap region, and $\Delta\phi$ is the potential difference measured from the potential at the center. The experimental conditions are noted in the text.

# SUMMARY

Experiments on electron confinement in the cusp-trap with super-conducting coils were performed in order to see the possibility to generate a spin-polarized antihydrogen atomic beam.

Confinement of electron plasma continued for a long time, keeping the plasma isolated from the surrounding wall. There was dependence of the confinement time $\tau$ on the trapped electron number $N_0$. The time $\tau$ became shorter for larger $N_0$, e.g., $\tau\sim400$s for $N_0= 3.6\times10^7$ and $\tau\sim100$s for $N_0= 1.6\times10^8$. This dependence shows there are some mechanisms to enhance cross-field diffusion of plasma at larger $N_0$. Explication of the dependence remains for further study.

Formation of the well for positively charged particles in trapped electrons has been one of the key issues of antihydrogen synthesis in the cusp-trap. It is said from the data analysis that this potential well is probably formed inside electron plasma, although the applied model on the density distribution is an assumption. The well lasted for 40s in a proper condition. It is necessary to find out the condition for elongating the lasting time.

There are many subjects to be investigated further more. Electron plasmas in this experiment were confined in a weak magnetic field region. To realize efficient synchrotron cooling, the plasma should be extended to a higher magnetic field. This extension is accompanied with the deformation of plasma shape. Application of additive electric fields with quadrupole component may elongate the plasma shape in the axial direction. As pointed out by ATHENA group [11], sufficiently cold antiprotons should combine with positrons to make antihydrogen atoms cold enough to ease manipulation of them by realizable magnetic field gradient. Then, cooling of mixture of electrons and antiprotons may become necessary. How to load such different kinds of particles into the cusp-trap is also an important subject to be studied.

## ACKNOWLEDGMENTS

The authors are much indebted to Dr. T. Ikeda for his work to make a reliable data acquisition program used to collect multi-channel signals from the Faraday cup. This work is supported by Special Research Projects for Basic Science of RIKEN.

## REFERENCES

1. Amoretti M. et.al., Nature **419**, 456-459 (2002).
2. Gablrelse G. et.al., *Phys. Rev. Letters* **89**, 213401 (2002).
3. Hangst J., "Project ALPHA and Future of Antihydrogen Physics" in this AIP proceeding
4. Widmann E. Eades J., Hayano R.S. *et.al.,*, *"The hydrogen Atom: Precision Physics of Simple Atomic Systems"* , edited by Kahrsheinvoim S. G.. et al., Berlin-Heiderberg: Springer-Verlag, 2001, pp. 528-542..
5. Hori M. et.al., "Radio frequency Paul trap for antihydrogen Production" ( unpublished).
6. Mohri A, and Yamazaki Y., Euro. Phys. Letters **63**, 207-213 (2003).
7. Mohri A. et.al., Jpn. J. Appl. Phys. 37, L1553-L1555 (1998).
8. Tiouririne T. N., Turner L., and Lau A. W. C., *Phys. Rev. Letters* **72**, 1204-1207 (1994).
9. Mohri A. et. al., "Experiment on Non-Neutral electron Plasma Confinement in a Field Composed of a Magnetic Quadrupole and an Electric Octupole", in $27^{th}$ *EPS Conference on controlled and Plasma Physics*, edited by Szeo K. et.al., ECA **24B** pp.149-152 (2000).
10. Fajan J., *Phys.Plasma* **10**, 1209-1214 (2003).
11. Madsen N. et.al., *Phys. Rev. Letters* **94**, 0334031 (2005).

# Muonic Anti-Hydrogen
# -Formation and Test of CPT Theorem-

K. Nagamine[1]

*Muon Science Laboratory, Institute of Materials Structure Science,*
*High Energy Accelerator Research Organization (KEK), Oho, Tsukuba, Ibaraki 305-0801, Japan*
*Physics Department, University of California, Riverside,*
*Riverside, California 92521, U.S.A.*

**Abstract.** With a completion of high intensity hadron accelerator in the near future, there will be a realization of high intensity as well as high quality beam of secondary particles like positive muons, anti-protons, etc. Here, it is suggested that a formation of Muonic Anti-Hydrogen is not unrealistic at all. Once it is formed, a new type of testing of CPT theorem will be realized by comparing with Muonic Hydrogen; much powerful testing for a short-range CPT violating interaction in contrast to the case of usual Hydrogen versus Anti-Hydrogen comparison.

**Keywords:** Anti-Hydrogen, CPT, Muonic Atom
**PACS: 36.10.Dr, 36.90.+f**

## INTRODUCTION

Including anti-proton $\bar{p}$, as seen in Fig. 1, there are four types of hydrogen atoms with $e^+$, $e^-$, $\mu^+$ and $\mu^-$; these are the conventional H atom ($e^-p$), the corresponding antiparticle $e^+\bar{p}$ (known as anti-Hydrogen, $\bar{H}$), and the two muonic counterparts $\mu^-p$ and $\mu^+\bar{p}$. As it is well recognized, when a method of generating anti-hydrogen $\bar{H}$ ($e^+\bar{p}$) is established, a high precision spectroscopic measurement on $\bar{H}$, in comparison with the corresponding results for H, may contribute to the verification or falsification of the CPT conservation law (where CPT refers to the product of the three symmetry operations of charge transformation, parity-inversion, and time-reversal). In order to do the test of the CPT theorem, as it is already indicated in the earlier publication ( p.185 of [1] ), there is an obvious advantage for the use of the $\mu^-p$ and $\mu^+\bar{p}$ pair, in comparison with the use of usual H and $\bar{H}$ pair. Although a correct theory does not exist to the knowledge of the present author, when the CPT-violating interaction has a short-range nature with an extremely massive exchange boson, such an effect can more easily be seen in the ($\mu^-p$, $\mu^+\bar{p}$) comparison than in the ($e^-p$, $e^+\bar{p}$) comparison, since the atomic size becomes smaller by a factor of 1/207.

---

[1] Present Address in Japan
Atomic Physics Laboratory, RIKEN, Hirosawa, Wako, Saitama 351-0198, Japan
and J-PARC Project Office, KEK, Oho, Tsukuba, Ibaraki 305-0801, Japan

CP793 *Physics with Ultra Slow Antiproton Beams*
edited by Yasunori Yamazaki and Michiharu Wada
© 2005 American Institute of Physics 0-7354-0282-5/05/$22.50

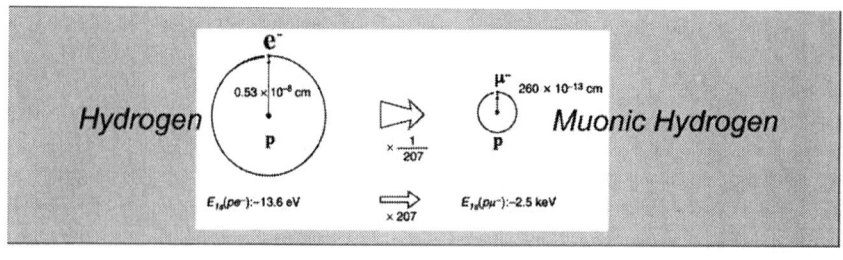

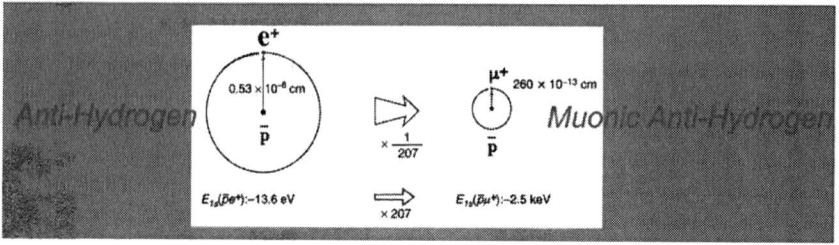

FIGURE 1. The possible four hydrogen atoms which are formed by proton, anti-proton, with electron and muon.

## POSSIBLE FORMATION

Since intense slow $\mu^+$/ Mu ( Muonium, hydrogen-like atomic state of $\mu^+$ and $e^-$ ) production as well as intense slow anti-proton beam will soon become available at the future high intensity hadron accelerator, it will be possible to produce $\mu^+\bar{p}$ through e.g. the following transfer reaction:

Mu $+\bar{p} \to \mu^+\bar{p} +e^-$, through a collision reaction by using a cloud target of thermal Mu and slow $\bar{p}$ projectile with the energy of $\bar{p}$ optimized for a capturing process to the new atomic state $\mu^+\bar{p}$ whose ground state has the binding energy of the $E_{g.s.}(\mu^+\bar{p}) \sim$ 2.528 keV. The exact transfer-reaction cross section has not been calculated yet. Some indication can be obtained by the similar reactions such as H $+\bar{p} \to p\bar{p} +e^-$ [2, 3], where the value of more than $10^{-16}$ cm$^2$ was indicated for the capturing stage at the c.m. energy of close to 1 eV. Correct theoretical studies for the transfer reaction including a population distribution of the initial states as well as the succeeding cascade transitions are called for.

Recent progress in the beam technology for the production of the ultra-slow $\bar{p}$ in the energy range of eV to 100 keV has been reported in the present workshop. In fact, a production of the intense slow $\bar{p}$ beam in the level of $10^8$/s is under planning at the new accelerator complex at GSI, where beam power of the primary proton is at the level of kW. Thus, if we install the same beam producing device at more intense hadron accelerator such as J-PARC (50 GeV x 15 μA), it is quite natural that one can expect the slow p̄ beam at the level of $10^{10}$/s.

At the same time, the production method of intense thermal Mu has been taking a significant progress. At KEK-MSL, the project of ultra-slow muon has been initiated in 1985, on the occasion of the discovery of the thermal Mu production from hot tungsten [4]. The work has been extended, via an initiation of the laser resonant spectroscopy of thermal Mu [5], to the generation of ultra-slow positive muon by laser resonant ionization of the Mu produced from tungsten placed at the position of primary proton beam line [6] and later placed at the secondary beam of intense 4 MeV surface muon at RIKEN-RAL [7]. At the same time, progress has been taken for the method of thermal Mu production; a series of experiments on $SiO_2$ powder [8] and systematic studies on noble metals [9] and recently on a porous tungsten [10]. As a summary of all of these historical developments, one can say that we have a method to convert a single MeV positive muon to thermal energy Mu in vacuum with a conversion efficiency of close to or more than 10 %.

In addition, in the fields of both the low-energy muon science and the particle physics of muon colliders and neutrino factories, the idea of a large acceptance pion and muon collection has recently become popular. One distinguished example is the Dai-Omega project at KEK-MSL, which is a large-acceptance axial-focussing superconducting surface muon channel and successfully collected 4 MeV surface muon with a solid angle of 1 Str [11]. Similar concept has been introduced at PSI by using a normal-conducting coil at the front-end [12]. More drastic improvement, close to a full-capture of all the produced muons, is employed for the forthcoming rare-decay measurement at BNL [13], based upon the design works on e.g. neutrino factories.

Combining all of these factors, as shown in Table 1, with a realization of 1 MW proton driver, one can expect at the level of more than $10^{10}$/s thermal Mu production in the vacuum space near e.g. hot porous tungusten. In the earliest stage of the formation experiment, a full stopping of the MeV positive muons in a multi-layer of the tungsten foil with optimized spacing will be applicable.

Thus, by injecting slow p̄ beam to the cloud of thermal Mu which are produced e.g. from the surfaces of a stack of hot porous tungsten foils, the formation of $10^4$/s muonic anti-hydrogen can be expected. The formation can easily be monitored by detecting the characteristic X-ray of 2 keV, a 2p-1s transition. At the discovery stage, in order to eliminate background due to annihilation gamma-rays, an employment of pulsed time structure for both Mu and anti-proton beam might be helpful.

Since the formation experiment is essentially a crossed beam experiment, depending upon the method of the ultra-slow anti-proton beam production, a space compression like a micro-beam in addition to a time-compression like a pulsed beam for the positive muons will contribute to the increase of the formation rate of the muonic anti-hydrogen.

**Table 1.** Intensity estimation for thermal muonium.

| | Expected Numbers ($s^{-1}$) | Conditions & Remarks |
|---|---|---|
| $N_p$ | $7.6 \times 10^{15}$ | $1 \text{ GeV} \times 1\text{mA}\,(1 \text{ Mw})$ |
| $N_{\pi^-}^{tot}$ | $4.8 \times 10^{13}$ | $\sigma_{\pi^-}^{tot} : 28\text{mb}$ |
| $N_{\mu+}$ | $4.8 \times 10^{12}$ | Pion Capture 2.9 T 20 cm Bore $\times$ 1.5m |
| | | Pion Decay 3.0 T, 25 cm Bore $\times$ 10 m |
| | | Full Reflection |
| $N_{\text{Th Mu}} \simeq 10^{12}$ | | Full Stopping($100 \times 10\,\mu$m Poruns W) |
| | | 20 % Conversion |

# HIGH PRECISION SPECTROSCOPY FOR CPT TEST

With keeping the aim of the test of CPT theorem in mind, it is somewhat a shame of the present muon physicist that there has been no successful high precision measurement like a laser resonance experiment on muonic hydrogen ($\mu^-$p). At the same time, other than the comparison to ($\mu^+$p) mentioned here, a high precition spectroscopy of the system ($\mu^-$p) presents many other fundamental physics interests such as the questions of QED testing of the vacuum polarization, proton polarizability, etc.

As for the energy levels of ($\mu^-$p) shown in Fig. 3, there have been several proposals for laser resonance spectroscopy.

One distinguished example is the measurement of the 2S Lamb shift ($2^5P_{3/2} - 2^3S_{1/2}$) of muonic hydrogen proposed at PSI [14]. There, laser resonance spectroscopy will be planned to a precision of 30 ppm by employing $\lambda \sim 6\ \mu$m laser produced by multiple Raman process of 708 nm laser. The detection of the resonance signal will be a change of the succeeding cascade X-ray; a missing $K_\alpha$ X-ray. Another example of the similar type of experiment is ($\mu^- p$) (3d-3p) with $\Delta E$ : 0.006 eV and $\lambda$ : 188 μm [15]. For this purpose, a variable-frequency free electron laser is required, together with intense $\mu^-$ stopping in low pressure (below 20 mbar) hydrogen gas. Since 88 % of the splitting is due to the vacuum polarization, one can determine vacuum polarization with an accuracy of 100 ppm which should be compared to the presently available value with an accuracy of 0.1 % from either $(g - 2)_\mu$ or $(g - 2)_e$.

The other example is the ($\mu^- p$) (n = 1, hfs) with $\Delta E$ : 0.183 eV and $\lambda$: 6.8 μm. As described by Kato [16], an appropriate laser source with frequency variability might be available. The major difficulty in this measurement is how to detect the resonance signal. A polarized ($\mu^- p$) (n = 1) state can in principle be produced by spin-exchange collision between a spin-polarized Kr atom and unpolarized ($\mu^- p$) as commonly used to obtain polarized radioactive nuclei, muonium, etc. The conditions, i.e. the Kr concentration in $H_2$ gas and the total pressure of the gas mixture, must be optimized so that $\mu^-$ transfer from p to Kr is minimized and spin polarization transfer maximized. Another possibility might be a brute-force low-temperature nuclear polarization of ortho $H_2$ (80% polarization under 15 T at 0.015 K). After generating the spin repolarized ($\mu^- p$) in its F = 1 state, the hfs resonance can easily be detected by observing the destruction of the decay electron asymmetry with reference to the spin polarization axis. Also, it is suggested that one can use a difference in the range as well as he interaction time between two hyperfine states in hydrogen [17].

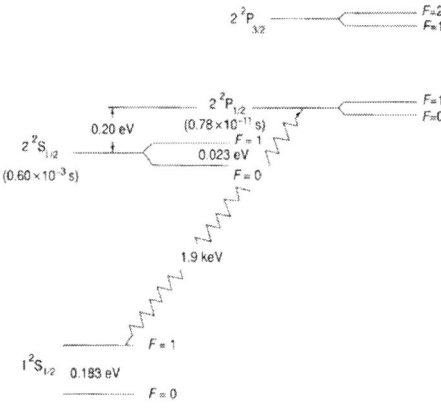

**FIGURE 2.** Energy level of Muonic Hydrogen and that of Muonic Anti-Hydrogen under CPT conservation.

# CONCLUSION

At the time of the beginning of the 21$^{st}$ century, the following concluding statements can be drawn.

1) Muonic Anti-Hydrogen is not a dream but a scientific subject to be realized within the 21st century. The technology of high-intensity and high-quality ultra-slow positive muon and $\bar{p}$ will take a rapid development *independently*, because of the needs of each community: surface-intersurface studies, nano-sciences, life-sciences, further higher energy projects like neutrino factories and muon colliders, etc. fot the positive mouns; anti-hydrogen spectroscopy and anti-particle gravity etc. for $\bar{p}$. Thus, good communications between low-energy muon-science community and anti-proton physics community should only be encouraged at the relevant timing of the future technical development of each community.

2) At the same time, theoretical works on CPT violation effect in Muonic Anti-Hydrogen taking into account the range difference effect in comparison with electronic system is called for.

# ACKNOWLEDGEMENTS

The author acknowledges helpful discussions with Professors Y. Yamazaki and J.S. Cohen. The work is partly supported by the JSPS Core-to-Core Program.

# REFERENCES

1. K. Nagamine *"Introductory Muon Science"*, Cambridge University Press, Cambridge (2003) 1-208.
2. J. S. Cohen, Rep. Prog. Phys. 67(2004) 1769 and private communication (2005)
3. K. Sakimoto, Phy. Rev. 65A (2001) 012706
4. A.P. Mills, Jr., J. Imazato, S. Saitoh, A. Uedono, Y. Kawashima and K. Nagamine, Phys. Rev. Lett. 56 (1986) 1463-1466.
5. S. Chu, A.P. Mills, Jr., A. Yodoh, K. Nagamine, Y. Miyake and T. Kuga, Phys Rev. Lett. 60 (1988) 101-104
6. K. Nagamine, Y. Miyake, K. Shimomura, P. Birrer, J.P. Marangos, M. Iwasaki, P. Strasser and T. Kuga, Phys. Rev. Lett. 74 (1995) 4811-4814.
7. Y. Matsuda, P. Bakule et al., J. Phys.: Nucl. Phys. 29(2003)2039.
8. G. Beer et al., Phys. Rev. Lett. 57(1995)671.
9. A. Matsushita and K. Nagamine, Phys. Lett. A 244 (1998) 174-178.
10. H. Miyadera, K. Nagamine, K. Shimomura, Y. Ikedo and K. Nishiyama, Preprint (2005)
11. H. Miyadera, K. Nagamine, K. Shimomura, K. Nishiyama, H. Tanaka, K. Fukuchi, S. Makimura and K. Ishida, Physica B 326 (2003) 265-269.
12. T. Proksha and E. Morenzoni, private communication (2005).
13. W. Molzen et al., UC-Irvine Phys. Tech. Report 96-30 (1996).
14. F. Kotttman et al., Hyperfine Interactions 138(2001)55.
15. P. Hauser, Hyperfine Interactions 103(1996)175.
16. K. Kato, IEEE J. Quantum Electron. QE-20(1984)698.
17. D.J. Abbott, V. Markusin and R.T. Siegel, Phys. Lett. A178(1993)398.

# A new path toward gravity experiments with antihydrogen

P. Perez and A. Rosowsky*

*DSM/Dapnia/SPP, CEA/Saclay, F-91191 GIF-SUR-YVETTE

**Abstract.** We propose to use a 13 KeV antiproton beam passing through a dense cloud of positronium (Ps) atoms to produce an $\overline{H}^+$ "beam". These ions can be slowed down and captured by a trap. The process involves two reactions with large cross sections under the same experimental conditions: the interaction of $\overline{p}$ with $P_S$ to produce $\overline{H}$ and the $e^+$ capture by $\overline{H}$ reacting on $P_S$ to produce $\overline{H}^+$. The decelerated $\overline{H}^+$ ions are captured and cooled in a trap. The extra $e^+$ is removed with a laser to measure the gravitational acceleration of neutral antimatter in the gravity field of the Earth.

**Keywords:** positron positronium matter antimatter symmetry antigravity
**PACS:** 41.75.Fr; 04.80Cc

Recently it has been proposed [1] to measure the gravity acceleration of antimatter using the $\overline{H}^+$ ion. This ion has the advantage that it can be cooled down to a few $\mu K$, a temperature suitable for a gravity experiment.

In order to create the $\overline{H}^+$ ion, we propose to launch 13 KeV antiprotons onto a dense cloud of positronium atoms. This Ps target is prepared with positronium atoms emitted from a metallic surface which is bombarded by a flux of positrons. The atoms are emitted from the surface with an energy $\leq 2$ eV and a speed $\leq 1$ mm/ns. The density of the Ps target is proportional to the positron flux. An amount of order $10^{11}$ positrons is needed in order to get the required density for the Ps target. However, the short Ps lifetime requires this amount to be delivered in a few nanoseconds while the highest foreseen rates of slow positrons are at best $10^{11}$ s$^{-1}$. The positrons have then to be accumulated in the metal vicinity and accelerated toward it. To counteract the effects of space charge, the positrons are accumulated into a small neutral $e^-e^+$ "plasma" for a short time ($t < 1$ s). The plasma size is $\sim 3$mm$^2 \times 1$cm. Then an electrostatic field accelerates the positrons toward an Aluminium crystal where they are converted into Positronium atoms.

This scheme is described in detail in [2].

## ACKNOWLEDGMENTS

We wish to thank Y. Yamazaki and A. Mohri.

## REFERENCES

1. J. Walz and T. Hänsch, *General Relativity and Gravitation* **36** 561 (2004); J. Waltz, H. Pittner, M. Hermann, P. Fendel, B. Henrich, T. W. Hänsch, *Appl. Phys.* **B77**, 713–717 (2003).
2. P. Perez and A. Rosowsky, *Nucl. Intr. Meth.*, **A** (2005) in press.

CP793 *Physics with Ultra Slow Antiproton Beams*
edited by Yasunori Yamazaki and Michiharu Wada
© 2005 American Institute of Physics 0-7354-0282-5/05/$22.50

# ANTIPROTON-NUCLEUS INTERACTION

# Light antiprotonic atoms

## Detlev Gotta

*Institut für Kernphysik, Forschungszentrum Jülich, D-52425 Jülich, Germany*

**Abstract.** The present knowledge on strong–interaction effects in light antiprotonic atoms is reviewed. Data were obtained during the LEAR era, where the high flux made possible the use of high–resolution devices like semiconductor detectors and a crystal spectrometer. Open questions and possible extensions at the future antiproton facilities are discussed.

**Keywords:** Exotic atoms, antiprotonic atoms, exotic–atom cascade, X–ray spectroscopy, crystal spectrometer, bound–state QED
**PACS:** 36.10.-k; 25.43.+t; 29.30.Kv; 07.85.Jy

## INTRODUCTION

The measurement of the characteristic X–radiation emitted from antiprotonic atoms constitutes an antinucleon–nucleus scattering experiment at relative energy zero. The strong interaction becomes manifest in an energy shift and broadening of the low-lying atomic states. Shift and broadening are directly related to the complex antiproton-nucleus scattering length and volume and are sensitive to the medium– and long range part of the antinucleon-nucleus potential. The hydrogen isotopes allow access to the elementary systems antiproton–proton and –neutron and the light nuclei serve as a testing ground to build up a consistent picture of the antinucleon–nucleus interaction. Furthermore, the study of the atomic cascade and its pressure dependence sheds light on the processes governing the de-excitation of the antiprotonic atoms [1].

For precision studies of the strong–interaction effects statistics and good energy resolution is essential. In order to achieve sufficiently high X–ray yields, antiprotonic hydrogen and helium must be formed in dilute gases to reduce the influence of non–radiative de–excitation processes occuring during collisions with other target particles. Therefore, low–pressure gas targets and thin windows must be used. At LEAR, the use of the cyclotron trap allowed to stop the antiproton beam quantitatively at such low densities [2].

Energies of the low–lying X–ray transitions in hydrogen and helium isotopes are in the range 2–15 keV. For hydrogen, the hadronic effects are of the order of 1 keV and 10–500 meV for the s–wave and p–wave interaction, respectively. Consequently, two different approaches are required: a direct measurement with semiconductor detectors, e. g., Si(Li)s or charge-coupled devices (CCDs) and to achieve ultimate resolution by using a Bragg crystal spectrometer. Whereas CCDs allow an efficient reduction of the annihilation induced background by the analysis of the hit pattern, a Bragg spectrometer is self collimating due to its narrow angular acceptance.

In this article, the present knowledge of light antiprotonic atoms is shortly reviewed. Facing the proposed low–energy antiproton facility FLAIR at GSI [3] an outlook to

CP793 *Physics with Ultra Slow Antiproton Beams*
edited by Yasunori Yamazaki and Michiharu Wada
© 2005 American Institute of Physics 0-7354-0282-5/05/$22.50

forthcoming developments for X–ray detection and the possible gain in quality and precision of data is discussed.

## ATOMIC CASCADE

In exotic atoms, the quantum cascade of a captured particle starts – as a rule of thumb – at a main principle quantum number $n_{\bar{p}}$ according to the outmost electron shell ($n_e$), i. e, at about $n_{\bar{p}} \approx n_e \sqrt{m_{\bar{p}}/m_e}$ (Fig. 1). Electrons are quickly removed by Auger emission and especially in the case of the electrically neutral hydrogen de–excitation is governed by collisional processes. When crossing other target molecules, fast transitions between angular momentum states of the same principle quantum number occur down to $n_{\bar{p}} \cong 9$ because of the Coulomb field (Stark–mixing). The induced s and p waves quickly lead to annihilation and the antiproton cannot reach the X–ray dominated part of the cascade. Stark–mixing essentially determines the density dependence of the line yields in exotic hydrogen (Fig. 2).

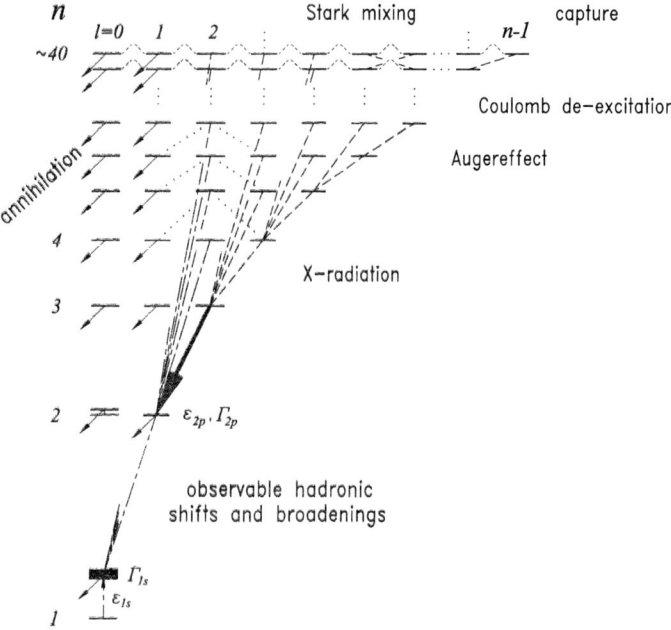

**FIGURE 1.** Atomic cascade in antiprotonic hydrogen. The sign of the level shift is defined by $\varepsilon \equiv (E_{experiment} - E_{QED})$, i. e., $\varepsilon$ is negative (positive) for a less (stronger) bound level.

For the most deeply bound atomic states radiative de–excitation dominates. As the size of the atomic orbit is inversely proportional to the reduced mass of the captured particle and the nucleus, the innermost levels are already affected by strong interaction. In the case of antiprotonic hydrogen, the Bohr radius is 58 fm only. Hence, X–ray transitions to the low–lying states are shifted and broadenend because of annihilation.

170

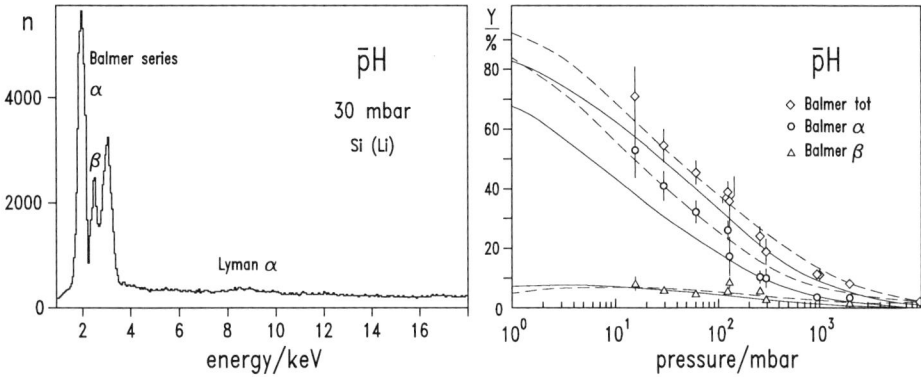

**FIGURE 2.** X–ray spectrum of antiprotonic hydrogen measured with a Si(Li) detector placed in the bore hole of the cyclotron trap (LEAR experiment PS175). The pressure dependence of L lines makes evident the strong influence of Stark–mixing (from [4]).

# EXPERIMENTAL APPROACHES

High–precision measurements of exotic hydrogen X-rays combine on principle conflicting items. On the one hand, the precise determination of energies and line shapes requires high statistics: this usually demands for an apparatus providing antiproton stops in high–density targets, because the exotic atoms must be formed at high rate in a small volume in order to achieve a bright X–ray source for devices of low efficiency like (rather small) solid–state detectors or high–resolution crystal spectrometers. On the other hand, a dense target leads to a drastic reduction of the X–ray intensities due to Stark–mixing (Fig. 2) which suggests to perform the measurements at the lowest possible density.

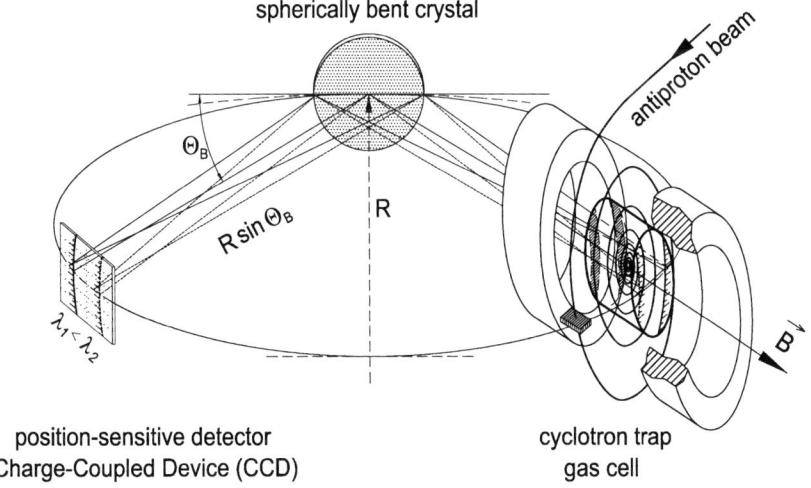

**FIGURE 3.** Principle setup of the cyclotron trap combined with a Bragg spectrometer (from [1]).

At LEAR finally up to $10^6$ antiprotons per second were available at a dedicated low–energy beam line of 105 MeV/c momentum. About 90% of the incoming beam could be stopped in hydrogen gas of 30 mbar pressure at room temperature in a volume of 2 cm in diameter by using the cyclotron trap [2], which allowed the use of a crystal spectrometer. Line yields close to saturation have been achieved by this method [4] (Fig. 2). The efficiency of the solid state detectors were about $10^{-4}$ and a few $10^{-6}$ only for the Bragg spectrometer. The crystal spectrometer was set up in Johann geometry equipped with spherically bent silicon and quartz crystals (Fig. 3). The Johann geometry allows the simultaneous measurement of an energy interval corresponding to the width of the X–source, which is an essential feature because of the low counting rates. Energy resolutions around 300 meV were achieved for the $L\alpha$ line of $\bar{p}H$ and $\bar{p}D$ at an energy of 1.7 and 2.3 keV, respectively [5].

## STRONG INTERACTION

At present microscopic calculations at the quark level do not yield a satisfactory description of the experimental results. Therefore, the low–energy nucleon–antinucleon interaction still has to be described by an optical potential ansatz. The real part is obtained by G–parity transformation from the nucleon–nucleon potential [6]. Annihilation is taken into account by an imginary or complex potential with a radial shape representing approximately the matter distribution of the nucleon or nucleus (Fig. 4). Usually no spin or isospin dependence is considered for the annihilation potential because of, amongst other things, the lack of high–precision data. A recent review of the low–energy $NN$ interaction may be found in [7].

An important feature of the G–parity transformation is the creation of a deep attractive potential, mainly by the sign change of the $\omega$ exchange contribution. It gave rise to a discussion of narrow nuclear bound states close below threshold (baryonium [8]). Evidence for such states is assumed to be identified, e. g., by an anomalous behaviour of the $\rho$ parameter. The $\rho$ parameter is given by the ratio of the real–to–imaginary part of the scattering amplitude and is determined at threshold by $2\varepsilon/\Gamma$ from shift and width as measured in antiprotonic hydrogen.

Furthermore, the coherences change drastically as compared to the $NN$ case. Roughly speaken, when changing from $NN$ to $\bar{N}N$ a strong spin–orbit and a weak spin–spin force turns into a rather weak spin–orbit and a strong spin–spin interaction (Fig. 5). A detailed comparison of the $NN$ and $\bar{N}N$ potentials is given by [6].

Antiprotonic atoms test the medium– and long–range part of the interaction, where pion and rho exchange dominate. Due to the strength of the annihilation the antiproton is not able to penetrate more than about 1.2 fm the nucleus' surface. A decisive question in testing the meson–exchange picture, which successfully describes the $NN$ interaction, is whether the predicted spin–spin and spin–orbit forces show up in the low–energy $\bar{N}N$ experiments. It is related to the search for baryonium, because a particular atomic level, i. e., of specific quantum numbers only, may have an induced width from such deeply bound states. The large spin–averaged broadening found for the 2p state (p–wave enhancement) is discussed to be an indication for baryonium states [9].

In light antiprotonic atoms, spin–spin and spin–orbit interaction are accessible by

172

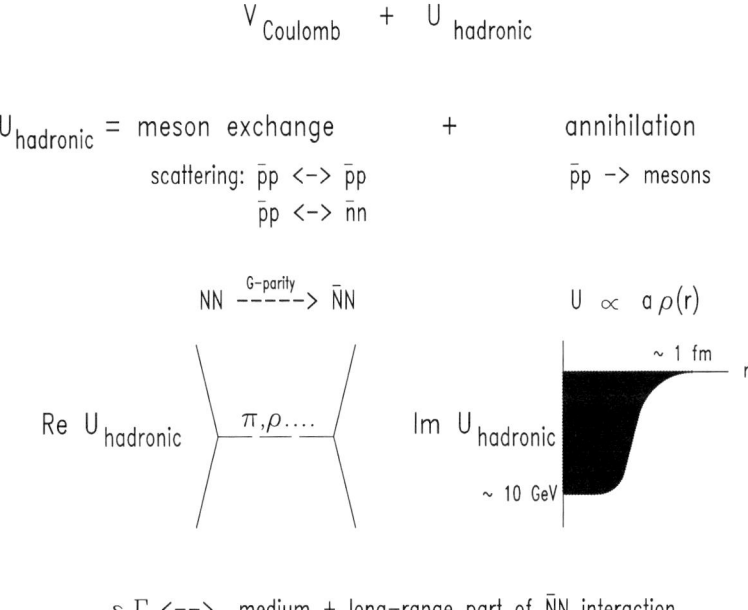

$$V_{Coulomb} \quad + \quad U_{hadronic}$$

$$U_{hadronic} = \text{meson exchange} \quad + \quad \text{annihilation}$$

scattering: $\bar{p}p \longleftrightarrow \bar{p}p$

$\bar{p}p \longleftrightarrow \bar{n}n$

$$NN \xrightarrow{\text{G-parity}} \bar{N}N \qquad\qquad U \propto a\,\rho(r)$$

Re $U_{hadronic}$ $\quad \pi,\rho\dots \quad$ $\sim 1$ fm

$r$

Im $U_{hadronic}$

$\sim 10$ GeV

$\varepsilon, \Gamma \longleftrightarrow$ medium + long–range part of $\bar{N}N$ interaction

**FIGURE 4.** Generation of the $\bar{N}N$ interaction from the $NN$ meson–exchange potential and an "ad hoc" short–range annihilation part.

measuring the hyperfine splitting of the 1s and the 2p state (Fig. 5). Hence, the precision spectroscopy of antiprotonic hydrogen is equivalent to a $\bar{p}p$ double polarisation experiment at relative energy zero. It is expected that the effects are largest for hydrogen, because here the level splitting is largest and annihilation is less dominant due to the comparatively small overlap of the antiproton and nucleus wave functions.

## STRONG–INTERACTION RESULTS

The setup of LEAR experiment PS207 allowed for a simultaneous measurement with CCDs and a (twin) crystal spectrometer (Fig. 6). The relative efficiencies, line yields and background requirements led to comparable time to measure the K (CCD) and L X–rays (crystal spectrometer). In a two week measurement, about 2000 $\bar{p}H$ L$\alpha$ events were accumulated in each branch of the crystal spectrometer. The statistics collected for the broad $\bar{p}H$ K lines is one order of magnitude larger [5, 10, 11].

A reasonable degree of suppression of the annihilation induced background is already obtained by guard–ring Si(Li) detectors [4]. It could be improved by the use of CCDs which allows an analysis of the hit pattern due to their granularity [10]. Few keV X–rays deposit the converted charge in one or two pixels, whereas background produces larger structures.

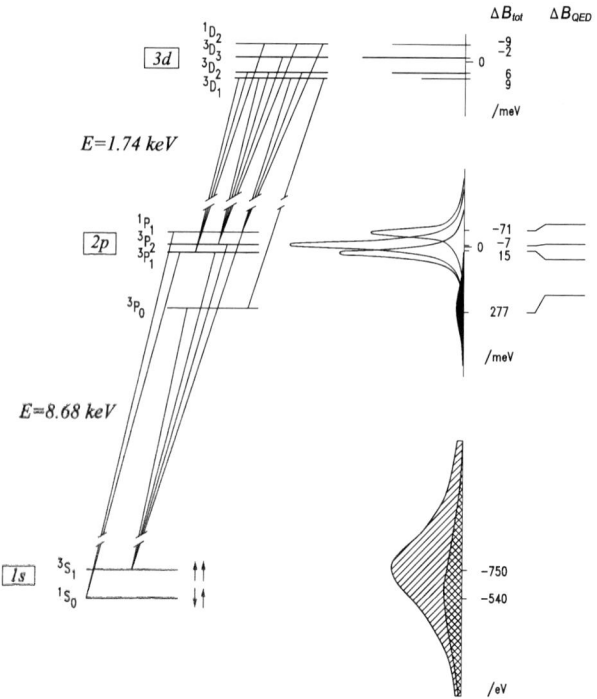

**FIGURE 5.** Hyperfine transitions between the low–lying states in antiprotonic hydrogen (from [5]).

**TABLE 1.** Spin–averaged hadronic shifts and broadenings in light antiprotonic atoms.

|  | $\varepsilon_{1s}$ / eV | $\Gamma_{1s}$ / eV | | $\varepsilon_{2p}$ / meV | $\Gamma_{2p}$ / meV | | $\Gamma_{3d}$ / meV | |
|---|---|---|---|---|---|---|---|---|
| $\bar{p}H$ | $-714 \pm 14$ | $1097 \pm 42$ | [10] | $+15 \pm 20$ | $38.0 \pm 2.8$ | [5] | – | |
| $\bar{p}D$ | $-1050 \pm 250$ | $1100 \pm 750$ | [11] | $-243 \pm 26$ | $489 \pm 30$ | [5] | – | |
| $\bar{p}^3He$ | – | – | | $-17 \pm 4$ | $25 \pm 9$ | [12] | $2.14 \pm 0.18$ | [12] |
| $\bar{p}^4He$ | – | – | | $-18 \pm 2$ | $45 \pm 5$ | [12] | $2.36 \pm 0.10$ | [12] |

# Hydrogen and Deuterium

In antiprotonic hydrogen and deuterium, only s and p states are affected by the strong interaction. Already the first measurements at LEAR proved that p–wave annihilation in $\bar{p}H$ is extra ordinarily strong. The broadening was determined from the intensity balance of the L series and the K$\alpha$ transition [4, 13]. The spin–averaged 2p–level broadening of 38 meV is almost a factor of 10 larger than expected from simple geometrical scaling. About 99% of the antiprotons reaching the 2p state annihilate, i. e., the line yield of the 1 keV broad K$\alpha$ is of the order of 1% only, which complicates a precise determination of the strong–interaction effects.

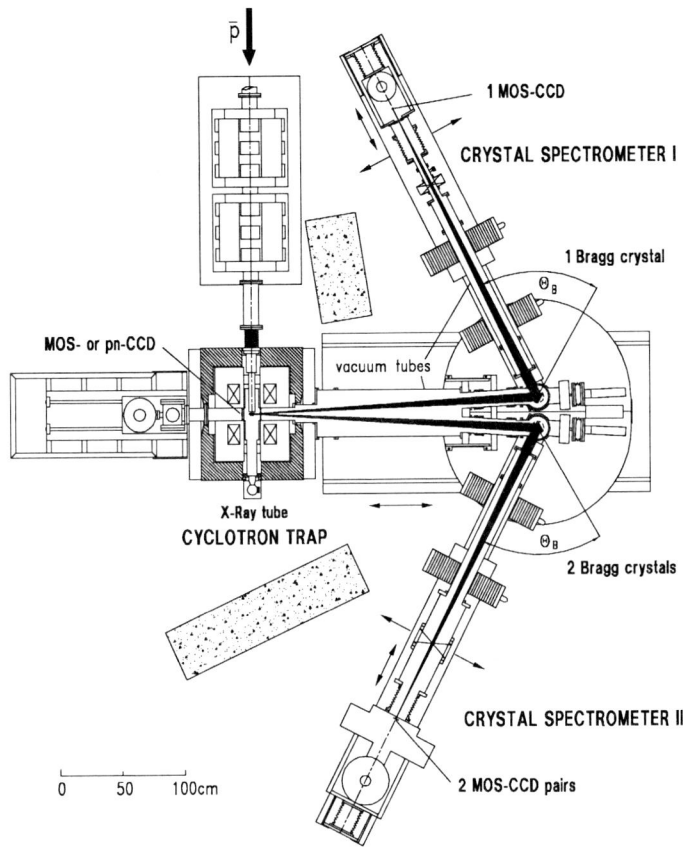

**FIGURE 6.** Experiment PS207 at LEAR using the cyclotron trap, CCD detectors placed in one of its bore holes for direct measurements and a twin crystal spectrometer equipped with spherically bent quartz and silicon Bragg crystals (from [5]).

The spin–averaged 1s–level shift and broadening could be determined at the few per cent level [10]. In addition, evidence was found for the ground–state transition in $\bar{p}D$ [11] (Fig. 7 and Table 1). Strong–interaction effects of the individual 1s hyperfine components in $\bar{p}H$ could not be obtained without additional information, because no clear separation of the ground–state doublet is achievable with the statistics accumulated at LEAR. The relative population of the singlet and triplet states has to be assumed as given by relative hadronic width of the statistically populated 2p sublevels to be about 2. Furthermore, the background beneath the $\bar{p}H$ $K\alpha$ had to be adapted from a $\bar{p}D$ measurement. Nonetheless, the order of magnitude could be fixed (Table 2). Simulations show, that an increase of statistics by a factor of 20 together with an improved background reduction results in an accuracy for the individual hyperfine states comparable to the accuracy for the spin–averaged quantities achieved at LEAR. Much progress in the quality of data is expected from the use of fast CCDs, where up to 1000 frames per second can

175

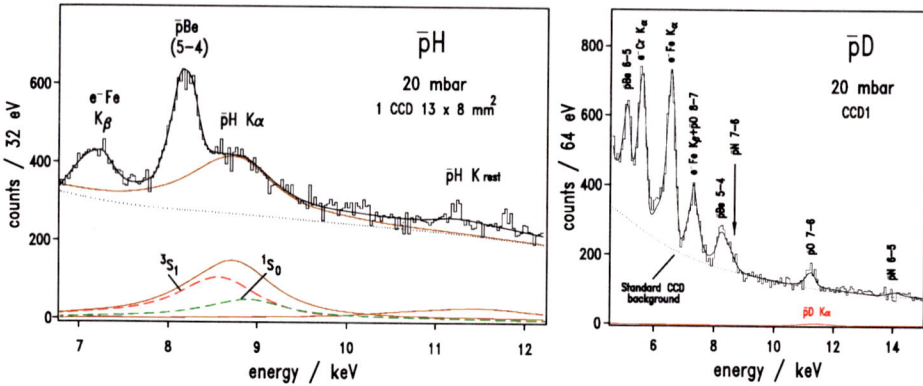

**FIGURE 7.** Ground–state transitions in antiprotonic hydrogen and deuterium measured with a CCD detector placed in the bore hole of the cyclotron trap (from [10, 11]).

be processed [14, 15, 16]. For MOS CCDs, used up to now, the read–out time is 10–20 s.

In the case of hydrogen, the 2p–level hyperfine pattern was partly resolved (Table 2). No structure was found in deuterium. However, the limited statistics did not allow to identify possible structures of the line shape. The spin-averaged broadening is found from the fit of a single Lorentzian to the $(3d - 2p)$ line shape. For a more detailed discussion of the $\bar{p}D(3d - 2p)$ line shape see below (bound–state QED). The large hadronic width in $\bar{p}D$ is consistent with geometrical scaling when compared to $\bar{p}H$ [5].

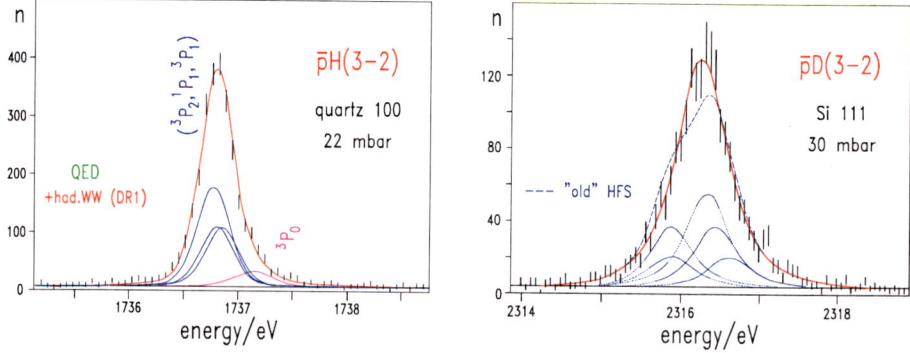

**FIGURE 8.** $\bar{p}H(3d - 2p)$ and $\bar{p}D(3d - 2p)$ transitions measured with a crystal spectrometer (from [5]).

**TABLE 2.** Strong–interaction effects in hyperfine states of $\bar{p}H$ as measured by LEAR experiment PS207.

| $\varepsilon\,(1^1S_0)$ / eV | $\Gamma\,(1^1S_0)$ / eV | $\varepsilon\,(1^3S_1)$ / eV | $\Gamma\,(1^3S_1)$ / eV | $\varepsilon\,(2^3P_0)$ / meV | $\Gamma\,(2^3P_0)$ / meV |
|---|---|---|---|---|---|
| $-440 \pm 75$ [10] | $1200 \pm 250$ [10] | $-785 \pm 35$ [10] | $940 \pm 80$ [10] | $+139 \pm 38$ [5] | $120 \pm 25$ [5] |

176

# Light nuclei

The isotope effect in antiprotonic helium was determined at LEAR using Si(Li) detectors [12]. The accuracy of the 2p–level widths, however, was limited by the resolution of the detector (Fig. 9 – left). For such experiments crystal spectrometers are better suited, but could not be used during the life time of LEAR. Such a measurement of the $(3d-2p)$ line shape can improve an order of magnitude on accuracy for the broadening.

An effective range analysis of low–energy annihilation cross sections and exotic atom data yields a consistent picture for the imaginary parts of the $\bar{p}A$ scattering lengths and volumes for hydrogen, deuterium and $^4He$ [17]. Especially, s– and p–wave parameters for the $\bar{p}p$ system are in striking agreement and a normal threshold behaviour is observed. However, exotic states for particular quantum numbers are not principally excluded because only spin–averaged quantities are obtained by such an analysis. Therefore, the measurement both of unpolarised and spin–dependent quantities in scattering experiments [18] as well as the resolution of hyperfine states in atoms is highly desirable.

An advantage of atom data is that they provide not only the absorptive part of the complex optical potential parameters because shift and width correspond approximately to real and imaginary part [19]. Furthermore, the different atomic levels yield an estimate for the strength of the different partial waves.

From the $A$ dependence a decrease of the s–wave annihilation strength is observed [17]. This behaviour is qualitatively understood from the increasing suppression of the wave function at the origin due to the strong annihilation [20, 21]. Similarly, evidence for a saturation of the p–wave annihilation strength is found (Fig. 9 – right).

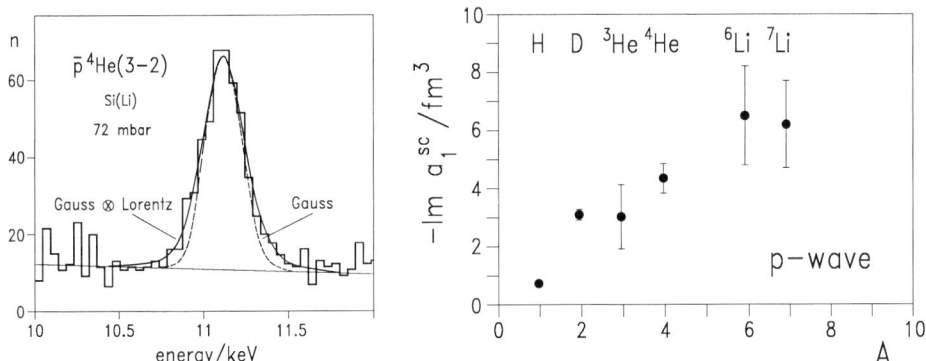

**FIGURE 9.** Left – Line shape of antiprotonic helium measured with a Si(Li) detector (from [12]). Right – Annihilation strength of the p wave exrtracted from light antiprotonic atoms (from [17].

# Annihilation on protons and neutrons

The relative annihilation strength of antiprotons on protons and neutrons gives access to the isospin dependence and is an important input for nuclear structure analyses like the determination of the neutron halo from antiprotonic atoms formed with heavy

**TABLE 3.** Relative annihilation strength on protons and neutrons from atom data and charge analysis of final states. For a comparison of the isotopes the measured broadening ($\Gamma$) has to corrected for the different overlap of the antiproton's wave function with the nucleus ($\tilde{\Gamma}$) [12].

| | $\bar{p}p$ and $\sigma_{\bar{n}p}$ | $\bar{p}d$ | $\bar{p}^3He$ | $\bar{p}^4He$ | $\bar{p}^3He + \bar{p}^4He$ average 2p and 3d width |
|---|---|---|---|---|---|
| $R^{bound}_{\bar{p}n/\bar{p}p}$ | $1.14 \pm 0.13$ | $0.78 \pm 0.03$ | $0.47 \pm 0.04$ | $0.42 \pm 0.05$ | $0.52^{+0.86}_{-0.41}$ |
| | [1] | [22, 23] | [24] | [25] | |

nuclei [21]. From $\bar{p}H$ atom data a mixture of isospin one and zero is obtained according to $a_{\bar{p}p} = (a_{I=0} + a_{I=1})/2$. After correcting for the strong–Coulomb interference $\Im a_{s,\bar{p}p} = -0.73 \pm 0.03$ fm is obtained [1]. Data for the isospin 1 channel are available from a measurement of the antineutron annihilation cross section close to threshold yielding $\Im a_{s,I=1} = -0.83 \pm 0.07$ fm [26]. The relative annihilation strength for free nucleons is then given by $R^{free}_{s-wave} = \Im a_{\bar{p}n}/\Im a_{\bar{p}p} = 2\Im a_{I=1}/(\Im a_{I=0} + \Im a_{I=1})$.

It can be assumed that annihilation occurs almost completely by a two–body process. Therefore, for bound nucleons the equivalent quantity $R^{bound}$ can be obtained by analysing the charge multiplicities after annihilation at rest (see Table 3). A second possibility is to compare the level broadening in isotopes of light nuclei. Annihilation on single nucleons yield

$$\frac{\tilde{\Gamma}_{3He}}{\tilde{\Gamma}_{4He}} = \frac{2 + R^{bound}}{2 + 2R^{bound}}. \tag{1}$$

Though qualitatively understood by the different partial wave composition, the quantitativ description of $R^{bound}$ with $A$ is not available yet (Tables 1 and 3). Obviously, precision atom data for the helium isotopes are essential.

## BOUND–STATE QED

The understanding of the electromagnetic level splitting forms the basis for the extraction of the strong–interaction effects. Though bound–state QED contains no unknown contributions at that level, a precise calculation involving the recoil and anomalous magnetic moment corrections requires the most recent calculation techniques. Because of the almost equal masses of "nucleus" and orbiting particle for the lightest antiprotonic atoms the hyperfine splitting has the same order of magnitude than the fine structure.

The measurement of the line shape of the $\bar{p}D(3d - 2p)$ transition led to an obvious contradiction, because the predicted level splitting should result in a doublet–like structure [1, 5] (Fig. 8). A critical review lead to a completely different result for the splittings [27]. A striking difference of the old and more recent calculations is obtained (Fig. 10). A further test of such calculations may come from the level splitting in higher–lying levels, where strong interaction is negligible.

At present, the $(5g - 4f)/(5f - 4d)$ splitting in $\bar{p}^3He$ of $\Delta E = 907 \pm 12$ meV [1] to be compared with the theoretical result of $\Delta E = 913$ meV [27] is the most precise experimental value obtained for antiprotonic atoms. A high–flux facility like FLAIR would allow to increase accuracy by one order of magnitude.

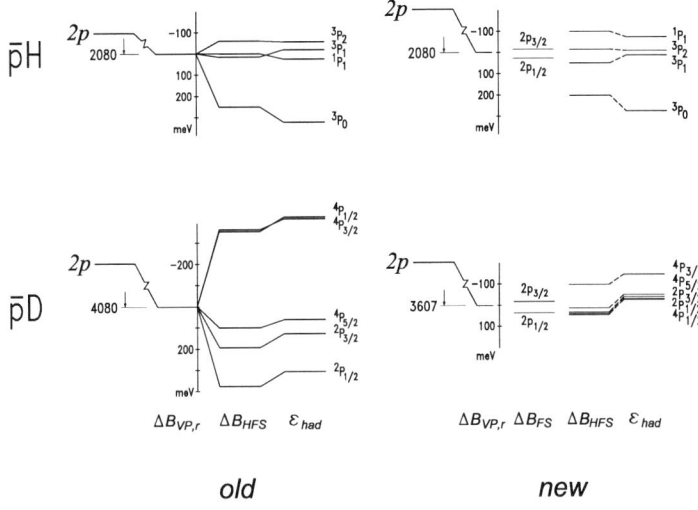

old                           new

**FIGURE 10.** Comparison of (former and recent) calculations for the electromagnetic $2p$ hyperfine splittings in antiprotonic hydrogen and deuterium. The calculations include vacuum polarisation and recoil $\Delta B_{VP,r}$, electromagnetic fine $\Delta B_{FS}$ and hyperfine structure $\Delta B_{HFS}$. To illustrate the magnitude of the strong interaction, the predicted hadronic shifts $\varepsilon_{had}$ are added in the rightmost columns. For a more detailed discussion see [1, 5].

# CAPTURE AND CASCADE

## Coulomb explosion

When captured in molecules the binding electrons are quickly emitted by fast Auger emission leaving back ions still at about a distance close to the original binding length. Coulomb repulsion leads to an acceleration (Coulomb explosion) of the ions which causes a Doppler broadening of the subsequently emitted X-rays. The captured particle itself will stick at one of the ions, there continuing its de–excitation cascade. First evidence for Coulomb explosion in exotic molecules was found from the pressure dependence of the K line yields in muonic atoms when comparing noble gases like neon and argon with binary molecules like $N_2$ or $O_2$ [28].

A direct observation of the Doppler broadening of X–rays was first achieved for pionic and muonic atoms in the (5–4) transitions of nitrogen and oxygen [29]. The Doppler width yields the velocity of the ion at the time of X–ray emission, which allows an estimate for the charge state at the time of separation. It was found that the charge state corresponds to about the number of electrons forming the covalent binding in these molecules. Similar processes should occur with antiprotonic atoms. Due to the larger mass of the antiproton, Auger emission out of molecular states may be more efficient than for pions and muons leading to higher charge states. This could be observed from the X–ray line width measured with a high–resolution crystal spectrometer.

179

# Capture

A significant fraction of the total K yield is accumulated in the not resolved transitions from high–lying $np$–states (series limit) when using solid state X–ray detectors like Si(Li)s or CCDs (see Fig. 2). The maximum principle quantum $n_{max}$ of the antiproton's orbit, for which such a transition is resolved, is given by

$$n_{max} = \sqrt[3]{\frac{2n_f^2}{\Delta E/E_{\infty-n_f}}}. \tag{2}$$

Assuming an energy resolution of $\Delta E = 300$ meV, the relative intensity of the L series transitions in $\bar{p}H$ and $\bar{p}D$ can be studied up to $n_{\bar{p}} \approx 40$. It provides the population of the high–lying $p$–states, which complements the information obtainable from laser spectroscopy for different $np$–states and constant $n_{\bar{p}}$.

# Electronic X–rays

It is known from the study of antiprotonic X-rays in dilute gases, that very early during de–excitation an almost saturation of the line yields is achieved. This is interpreted as a result of very efficient depletion of the electron shells by Auger emission leading quickly to an almost circular cascade. The non-occurance of strong electronic fluorescence lines immediately proves that K–Auger emission cannot contribute substantially to capture, but emission of the outer electron shells happens first. The antiprotonic argon spectrum shows the characteristic holes at the $\bar{p}Ar(17-16)$ and $\bar{p}Ar(16-15)$ transitions. Here the energy gain from $\Delta n = 1$ transitions is sufficient for Auger emission of the K electrons. The non zero, but small, intensity of the $\bar{p}Ar(17-16)$ and $\bar{p}Ar(16-15)$ yields the fraction of completely and all but one electron ionised argon [30] (Fig. 11–left).

Due to the possiblity to be captured at large $n_{\bar{p}}$ also in lower angular momentum states, non–radiative $\Delta n \gg 1$ transitions have enough energy to produce K or L holes. Such holes are quickly filled up by outer electrons. Here, amongst other processes X–ray emission occurs. A closer look to spectra formed with noble gases exhibits a rich structure attributed to electronic X–rays. The relative intensities are at about the 10% level of the saturated $\bar{p}A$ transitions in argon and krypton. In the case of xenon several L– and K–shell X–rays are emitted per formed antiprotonic atom. For xenon a complete ionisation is not longer possible as seen from the not saturated line yields. Two high intense and broad structures due to refilling of L and K holes from a large variety of configurations of the remaining electron shell and the antiproton's level are observed (Fig. 11–right).

A couple of contributions of the electronic X–ray complex could be tentatively identified as transitions occuring at a certain stage of the antiproton's cascade [31]. Not all contributions could be explained up to now. Resuming such studies at a high–flux antiproton facility will allow measurements using a crystal spectrometer. The much better energy resolution is ideally suited to resolve more details of the interplay of the antiproton with the remaining electrons during its descent.

180

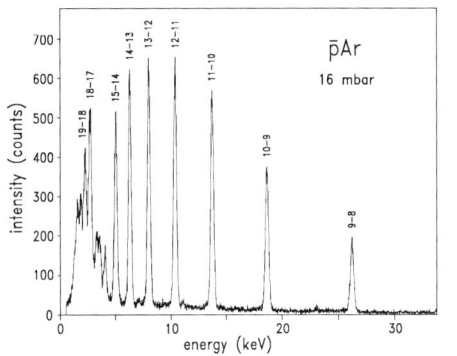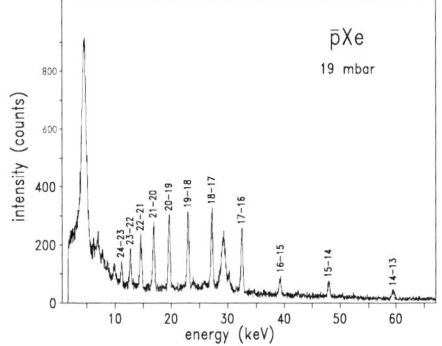

**FIGURE 11.** X–ray transitions in antiprotonic argon and xenon measured with a Si(Li)(PS175).

# OUTLOOK

Experiments performed at the high–flux antiproton beams provided at LEAR until 1996 constitute a major step forward in the understanding of the low–energy antinucleon–nucleon interaction. Continuation of such studies at a high flux low–energy antiproton facility will allow the determination of strong–interaction effects at a level of precision which became achievable for pionic atoms at the meson factories. There hadronic quantities are observed at the per cent level or better [32]. As an intermediate step, measurements at the AD facility at CERN may be considered [33]. The average intensity is sufficient for direct measurements using for example fast CCDs. Such measurements, however, require a continuous beam, which may be obtained by extraction from a trap at a rate of about $10^4$ to $10^5$ antiprotons per second [34].

The proposed FLAIR facility at GSI aims at low–energy antiproton beams exceeding the beam intensity achieved at LEAR by up to two orders of magnitude. Furthermore, ultra–slow antiproton beams of 100–300 keV are planned and ideally suited for resuming the measurements with stopped antiprotons. The high flux makes available outstanding possibilities for ultimate resolution devices like crystal spectrometers. The use of asymmetrically cut crystals can improve resolution by a factor up to two. Alternatively, the progress in trapping and storing antiprotons in electromagnetic devices opens new possibilities. A reverse arrangement of the usual setup, i. e., stopping an antiproton beam in a target may be considered. By using an atomic beam, crossing the antiproton plasma inside such a trap, the antiprotons are consumed by forming exotic atoms practically at zero density.

Antiprotonic atom data, which were not or inadequately provided by the LEAR experiments, and which should be remeasured to allow for decisive tests of the theoretical approaches are:

- determination of strong–interaction effects for the individual hyperfine states of the $\bar{p}H$ ground state at the per cent level,
- confirmation or disprove of the result for the $\bar{p}D$ K$\alpha$ line,

- study or search of the spin–orbit effects in $\bar{p}H$ and $\bar{p}D$ in high–resolution and high–statistics measurements,
- improving on the accuracy of the hadronic effects especially in the helium isotopes by using also in this case a crystal spectrometer, and
- exploiting the enhanced experimental techniques for a thorough study of the atomic cascade.

## ACKNOWLEDGMENTS

Support and assistance of the organizers allowing for the attendance of the workshop is warmly acknowledged.

## REFERENCES

1. D. Gotta, *Prog. Part. Nucl. Phys.* **52** (2004) 133.
2. L. M. Simons, *Physica Scripta* **T22** (1988) 90; *Hyperfine Interactions* **81** (1993) 253.
3. http://www.oeaw.ac.at/smi/flair/index.htm
4. K. Heitlinger et al., *Z. Phys.* **A 342** (1992) 359.
5. D. Gotta et al., *Nucl. Phys.* **A 660** (1999) 283.
6. W. W. Buck, C. B. Dover, and J. M. Richard, *Ann. of Phys. (NY)* **121** (1979) 47.
7. E. Klempt, F. Bradamante, A. Martin, and J. M. Richard, *Phys. Rep.*, 368 (2002) 119.
8. I. S. Shapiro, *Phys. Rep.* 35 (1978) 129, and references therein.
9. J. Carbonell, *Nucl. Phys.* **A 692** (2001) 11.
10. M. Augsburger et al., *Nucl. Phys.* **A 658** (1999) 149.
11. M. Augsburger et al., *Phys. Lett.* **B 461** (1999) 417.
12. M. Schneider et al., *Z. Phys.* **A 338** (1991) 217.
13. R. Bacher et al., *Z. Phys.* **A 334** (1989) 93.
14. A. Ackens et al., *IEEE transactions* vol. **46** (1999) 1995.
15. N. Meidinger et al., *Nucl. Instr. Meth.* **A 512** (2003) 341.
16. H. Gorke, this workshop.
17. K. V. Protasov, G. Bonomi, E. Lodi Rizzini, and A. Zenoni, *Eur. Phys. J.* **A 7** (2000) 429.
18. E. Lodi–Rizzini, this workshop.
19. T. L. Trueman, *Nucl. Phys.* 26 (1961) 57.
20. A. Gal, E. Friedman, and C. Batty, *Phys. Lett.* **B 491** (2000) 219.
21. S. Wycech, this workshop.
22. R. Bizarri et al., *Nuovo Cim.* **A 22** (1974) 225.
23. T. E. Kalogeropoulos et al., *Phys. Rev.* **D 22** (1980) 2585.
24. F. Balestra et al., *Nucl. Phys.* **A 491** (1989) 541.
25. F. Balestra et al., *Nucl. Phys.* **A 465** (1987) 714.
26. G. S. Mutchler et al., *Phys. Rev.* **D 38** (1988) 742.
27. S. Boucard and P. Indelicato, *priv. comm.*
28. P. Ehrhart et al., *Z. Phys.* **A 311** (1983) 259.
29. T. Siems et al.,*Phys. Rev. Lett.* **84** (2000) 4573.
30. R. Bacher et al., *Phys. Rev.* **A 38** (1988) 4395.
31. K. Rashid, D. Gotta, B. Fricke, P. Indelicato and L. M. Simons, *in preparation.*
32. http://pihydrogen.web.psi.ch.
33. The Antiproton Decelerator (AD) at CERN, http://psdoc.web.cern.ch/PSdoc/acc/ad.
34. *Atomic Spectroscopy and Collisions Using Slow Antiprotons* (LEAR/AD experiment PS205 – phase 3 experiments), http://asacusa.web.cern.ch/ASACUSA.

# Antiproton-nucleus annihilation at very low energies down to capture

E. Lodi Rizzini[*,†], A. Bianconi[*,†], M. Corradini[*,†], M. Leali[*,†],
L. Venturelli[*,†] and N. Zurlo[*,†]

*Department of Chemistry and Physics, University of Brescia, Via Valotti 9, 25133 Brescia Italy
†INFN Brescia group

**Abstract.** Results and opportunities at the low energies antiproton facilities at CERN (AD) are presented.

**Keywords:** Low energies antiproton interactions
**PACS:** 25.43.+t, 36.10-k,34.50.Bw

## ANTIPROTON-NUCLEUS ANNIHILATION

Recent measurements[1, 2] of $\bar{p}$ annihilation on proton, deuteron and $^4$He targets show an interesting "inversion" pattern at projectile momenta below 60 MeV/c: total annihilation cross sections on Deuteron and $^4$He are smaller than the same cross sections on free proton targets, see Fig.1. Such a behavior is not seen at larger momenta, where total annihilation cross sections increase with the target mass number $A$, accordingly to intuition. There are no low energy data on larger nuclei, to clarify whether the decrease of cross sections with $A$ is a systematic phenomenon or specifical of a few light nuclei. In the case of $^4$He target the phenomenon is particularly impressing if one takes into account that the Coulomb attraction introduces a further factor $\sim Z$ in any model for $\bar{p}$−nucleus annihilation, so a naive extimation of the ratio between low-energy $\bar{p}p$ and $\bar{p}^4$He annihilation cross section could be as small as $1/AZ = 1/8$. On the contrary, below 60 MeV/c the $\bar{p}$−D and $\bar{p}$−$^4$He annihilation cross sections are clearly smaller than $\bar{p} - p$ corresponding cross section. So, with $^4$He targets the most naive expectation largely overestimates the detected annihilation rate. A confirmation of these measurements comes from the recent data[3] on the widths and shifts of the electromagnetic levels of the antiprotonic deuterium. As demonstrated long ago[4] there is a relation between the width of a level in an antiprotonic atom and the $\bar{p}$ low energy annihilation rate on the corresponding nucleus. Although affected by large error bars, the comparison of the data on antiprotonic Hydrogen and Deuterium confirms the "inversion" behavior. For these last measurements a successful prediction existed[5]. In view of the possibility of a systematical analysis of $\bar{p}$ annihilation on complex nuclei at very low energies at the $\bar{p}$ accelerator AD at CERN[6] using the MUSASHI facility of the ASACUSA Collaboration [this Conference] it is desirable to look for an understanding of what could be the properties of the interactions between $\bar{p}$ and nuclear matter, and to try to organize some model or framework aimed at this analysis. A specific model for nuclear interactions, in this peculiar low energy domain, is still missing. From a fundamental point of view,

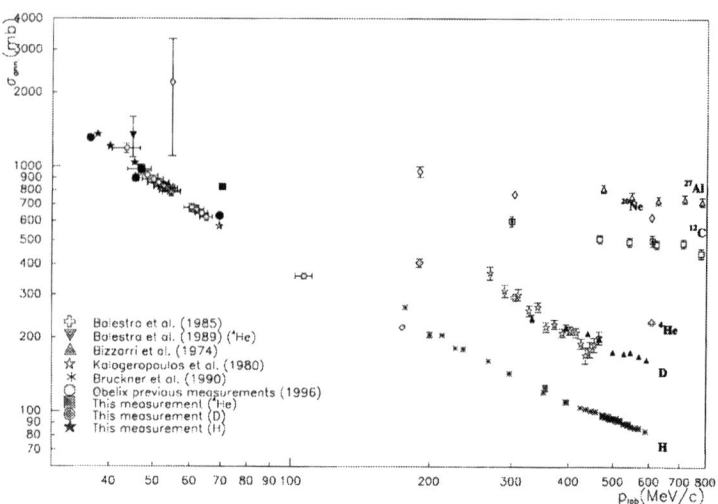

**FIGURE 1.**

it is not clear yet how much one can trust the concept of optical potential for describing hadronic annihilations. But even if one decides to trust it, it is desirable to have some physical connections between the nuclear potential and the underlying $\bar{p}$−nucleon interactions, not to reduce the physics of the problem just to a fit of radius, strength and a few more parameters. Moreover the traditional models used in nuclear physics for giving a meaning to the optical potential[7] can hardly be applied here. At large energies, short wavelengths allow for exploitation of the Glauber formalism[8]. The lowest $\bar{p}$ incident momentum at which data of $\bar{p}$−nucleus elastic and reaction cross sections were interpreted in a satisfactory way with a Glauber-style model was 600 MeV/c[9]. At lower momenta (300 and 200 MeV/c) such a model is only able to reproduce gross features of the differential cross sections, although it still revealed effective in reproducing integrated cross sections. We are now interested in a kinematical region where the Glauber model is surely not reliable anymore, i.e. with projectile momentum starting from around 100 MeV/c and going down to 1 MeV/c. At the lower end of this range, nonresonant annihilation processes are S-wave dominated with any nuclear target. At momenta around 200 MeV/c the S-wave contribution is not the most relevant one (with nuclear targets). The "inversion" pattern, which appears below 60 MeV/c, seems to be associated with the transition to the S-wave dominance region, where surely the Glauber model can't be applied. A simple black sphere potential model[10] shows that the general properties of low energy scattering on strongly absorbing optical potentials don't

184

contraddict the idea of "inversion": the S-wave contribution to the reaction cross section on a spherical imaginary potential ($V(r) = iW$ for $r < R$, and zero outside) is an increasing function of the strength parameter $W$ only at small $W$. When $W \to \infty$ the S-wave reaction cross section tends to zero. So the optical potential model does predict nuclear cross sections which are much smaller that naive geometrical expectations, with inversion or saturation behaviors. Actually, although first intuition suggests absorption cross sections $\sim R^3$ or $R^2$ for any strong absorber of radius $R$, the general properties of low energy scattering suggest an $R^2$ law for the elastic total cross sections only, while esoenergetic reactions should roughly follow a behavior $\sigma \sim R/k$[11]. The factor $1/k$ is a well known Bethe's prediction (which can be motivated by the idea that the probability for a spontaneous reaction is proportional to the time that any particle spends inside the interaction region), and $R$ can be put there to give the right dimension to a cross section. The $R/k$ law roughly works for any "moderately reacting sphere". The reaction properties of such a target can be expressed by an optical potential of the form $V = U - iW$. The imaginary part produces absorption of projectile flux according to the time dependence of the wavefunction $exp(-iEt)$ (since inside the potential region $E = T + V$, where $T$ is the initial kinetic energy of the projectile). The surprise is that with large values of $W$ (uncommon in traditional nuclear physics, but perhaps suitable to describe the violent annihilation phenomenon) low energy reaction cross sections predicted by the optical model become much smaller than $R/k$, and show no $R$ dependence. We analyze in detail the behavior of the wavefunction in the "strong optical model", to understand how much of its predictions is related to the use of an optical potential and how much is general. The general conclusion is that any model able to produce a "vacuum region" in the projectile flux will lead to a small reaction cross section, at the condition that the vacuum region surface diffuseness is small with respect to the vacuum region radius. The conclusion is more general than this model would allow for: it appears that whenever in a spherical region of space the projectile wavefunction is heavily suppressed almost all of the incoming flux must be elastically reflected. This is an obvious result when the suppression mechanism is elastic repulsion, but we find that the same must happen even when the mechanism which creates a "vacuum hole" in the projectile flux is inelastic. This phenomenon looks more suitable for application to the case of heavy nuclear targets (where the general nuclear matter properties are more relevant than the details of the structure) than for the case of the light nuclei used in the performed experiment[1]. In addition it would anyway suggest a value of the $\bar{p}$–$^4$He annihilation rate larger than the $\bar{p} - p$ one (because of the stronger Coulomb attraction) instead of the observed inversion between the two. So, although some explanations are available for the detected inversion mechanism, the problem is still open and need measurements.

## $\bar{P}$ STOPPING POWER IN HE

The energy release by a particle crossing matter is one of the most studied subjects of physics, starting with the seminal paper by Bohr in 1913 [12] and Bethe's theory [13] with its several extensions [14–22]. In Bethe's theory, the energy loss per unit path length (stopping power) is attributed to long range interactions with the target electrons leading to excitation and/or ionization of the target atoms or molecules. This mechanism will be named Electronic Stopping Power (ESP) in the following. This mechanism is supposed to present an adiabatic lower cutoff, which suppresses its effectiveness at decreasing projectile energies below a few keV. What exactly happens at such energies is not completely known. On the other hand, at decreasing energy the effect of collisions with the nuclei gives an increasing contribution to the energy loss of a heavy projectile. This mechanism has been called Nuclear Stopping Power (NSP in the following). Several models [23–27] and a few experimental measurements [28–30] exist for $\bar{p}$ NSP. In the OBELIX experiment [28] the antiproton beam produced at the low energy antiproton ring LEAR (CERN) was degraded by a suitable thickness of mylar, and entered a target with a continuous energy spectrum from a maximum to zero. Among the other ones, measures were taken with the cylindrical target (useful length = 75 cm, diameter = 22-30 cm) filled with hydrogen and deuterium at 0.2 mb and helium at 1 and 0.2 mb. As explained in refs. [28] and [29], one of the main peculiarities of our apparatus was the possibility of measuring for each event both the annihilation time (related to the incoming $\bar{p}$ signal) and the vertex coordinates, in particular the depth in the target $z$. In the following we name Scatter Plot the set of $(z,t)$ coordinates for annihilation events. This allows for a very detailed analysis of the slowing down and capture features of an antiproton beam in an extended target filled by low pressure gas, with events distributed in a 75 cm long region. In [29] ESP for $\bar{p}$ in gaseous hydrogen and helium was reported, for energies as small as 500 eV. Here ESP can be well fitted by the relation ESP $= \alpha E^{\beta}$, where E is the kinetic energy of the $\bar{p}$. An extrapolation of this relation at lower energies was not sufficient to explain the annihilation Scatter Plot. In [30] the Scatter Plot of $\bar{p}$ at rest annihilations in molecular hydrogen and deuterium gases at 0.2 mb pressure was presented, together with Monte Carlo reproductions of these data. The simulation included ESP as extrapolated from [29], and screened Rutherford $\bar{p}$−nucleus collisions that caused trajectory deflections and beam energy loss in the laboratory frame. To reproduce the Scatter Plot, a random decay process with life-times of $2.1\mu s$ for $H_2$ and $2.8\mu s$ for $D_2$ was added after the slowing-down and capture processes. A good reproduction of the data in the whole Scatter Plot confirmed that below 500 eV the stopping power is dominated by NSP, that this NSP can be reproduced by the simplest available model, and that a large part of the formed exotic atoms are characterized by decay processes with the above lifetimes. A Monte Carlo simulation based on the same technique is, in the He case, less effective. In fact, despite some features of the He data are reasonably similar to the $H_2$ and $D_2$ case, there are some peculiarities that cannot be interpreted as easily as there. The experimental Scatter Plot for the collection of $(z,t)$ annihilation points is presented in Figs.2a and Figs.2b, for helium at room temperature and at pressure 1 mb and 0.2 mb respectively. The $z \approx 0$ region is characterized by entrance wall annihilations. The data were collected using a temporal gate (above 300 ns and 450 ns respectively at 1 mb and 0.2 mb). For 1 mb data, the vertical belt at large $z$ in the target represents end wall annihilations. In the 0.2 mb case it is not present because in this case the target wessel was longer and the end wall was out of the region covered by the detectors. Apart from wall effects, most in-gas annihilations are concentrated in a reasonably defined belt which crosses the Scatter Plot from the origin to the right side. We name it MB, the "Main Belt". It includes those antiprotons that have been faster than the other ones in reaching a given $z_c$ capture point, and whose cascade process

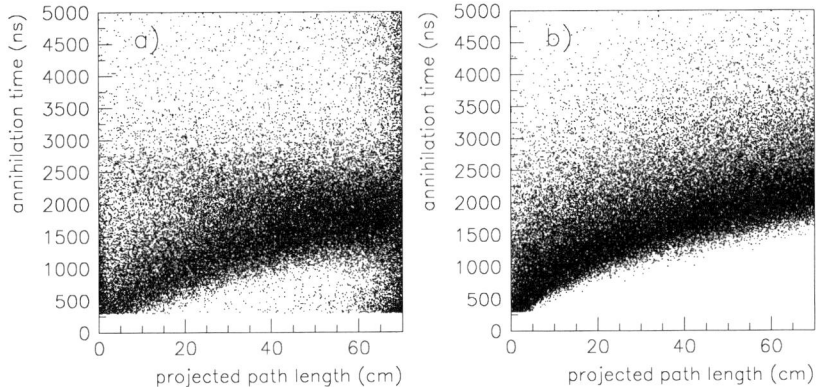

**FIGURE 2.**

has not been too slow. The MB, especially on the prompt rise, is qualitatively similar to the $H_2$ or and $D_2$ cases presented in [30]. Both in the 1 mb and 0.2 mb helium data, the most striking differences with respect to $H_2$ and $D_2$ data concern the region of annihilation times $t(z) > t_{peak}(z)$. The fraction of points that are present in this region is much larger in He than in $H_2$ and $D_2$. This is especially visible at intermediate times (i.e. not corresponding to long-time cascade tails). In the 1 mb case the intermediate time enhancement is particularly evident, since one can approximately identify a "Backward Belt" (BB). It amounts about $10 \div 20\%$ in respect to MB. The 0.2 mb Scatter Plot represents a sort of 5-times zooming a small region near the origin of the 1 mb Scatter Plot. Indeed, both the length of the pre-capture path, and the time needed for post-capture cascade are, roughly speaking, proportional to the helium density. For this reason the BB cannot be fully visible in the 0.2 mb Scatter Plot (the data taking was limited to $10\mu$s). Despite this, also in this case a large fraction of annihilations at intermediate and long times is present. Before going further, we want to remark that expressions like "long times" don't refer to metastable states. The fact that the $t-$distributions change at changing $z$ (apart for the time shift of the prompt rise due to obvious kinematical reasons) can perhaps be related with dominance of different cascade modes (i.e. different capture processes) at different $z$, or with multiple scattering. In both cases, however, this leads a $z-$dependence of the shape of the $t-$distributions only in presence of a different energy spectrum in the $\bar{p}$ beam at different $z-$ values. The actual entrance energy distribution is the one determined by degrading materials upstream the target. It has been seen to be regular, with a slow increase, on a keV energy scale, however the distribution in the $E < 100$ eV region is not known with high precision. This energy range is critical because large angle scattering is most effective here, and the cascade time is supposed to depend on the exact energy at which capture takes place. The clear $z-$dependence of the experimental Scatter Plots makes the analysis of the data complicate, but at the

187

same time it can potentially represent an open window on the energy dependence of $\bar{p}$−helium interactions near the capture region. A generalized adiabatic-ionization (AI) model [20] goes beyond the adiabatic model of Fermi-Teller by introducing collision broadening into the electronic potential curves of the "quasidipole" that is transiently formed in low-velocity collision of negative particles with atoms. The previsions of such a model are in impressive agreement with the recent data on single ionization in He [this Conference]. Moreover, the AI model is seen to provide a quite good description of the He data in the whole velocity range below the stopping power maximum. When the nuclear stopping power is added to the AI result for electronic stopping, the total stopping power exhibits a minimum at that position where indications for a minimum are found in our data. The deviation of the eveluted stopping power from the data is to be attributed to an underestimation of the calculated excitation cross section (not available). Clearly a better and more complete treatment of the two-electrons effect has to be evaluated. Moreover common sense suggests that a clear-cut distinction between electron and nuclear stopping power is not justified at low energies. Electron ionization in long-range interaction is adiabatically suppressed at $E \sim 100$ eV. Electron ionization is possible in collisions where the $\bar{p}$ penetrates deeply the atomic electron cloud. These ones are also those processes where $\bar{p}$−nucleus interactions are stronger. As an example of explanation of the Backward Belt (BB), let us assume the capture energy as distributed on a wide energy range, so that at $E < 40$ eV all $\bar{p}$ are captured, but capture is possible also at much larger energies, say at $E \approx 70$ eV. This may occur if double ionization of the target atom takes place in a collision [24]. In [24] a Monte Carlo calculations of antiproton collision with He were performed. The projectile $\bar{p}$ was launched at the target He atom with an initial energy of about 80 eV. About 23% of collision sequences ended with capture via single-electron ionization into the neutral $\alpha$ e $\bar{p}$ exotic, while the remain 77% resulted in capture via double ionization into a positively charged $\alpha$ $\bar{p}$ system. This capture energy spread has two consequences: (i) a wide (N,L)−distribution of the initial states of the exotic atom/ion $A$; (ii) a wide distribution of the recoil velocities $v_r$ of $A$. Concerning point (ii) we know that the average cascade time of a non metastable $A$ is approximately proportional to the number of $A$−atom collisions. So we have roughly $\tau_c \propto 1/v_r \propto 1/\sqrt{E_c}$. This phenomenon is enhanced if the $\bar{p}$ energy distribution at the entrance is populated by antiprotons with $E << 40$ eV. A quantum mechanical calculation of partial cross sections for $E < 30$ eV leading to different final states of antiprotonic helium atom was performed by J. Revai and N.V. Shevchenko [32], for the first time The results show, expecially in the low-energy region, the presence of strong resonance effects in connection with barrier penetration. The energy dependence of the calculated cross sections show that the different final states (J,v) are excited with a large probability in a relatively narrow window of the incident antiproton energy. We expect this to happen in the first part of the target. In the second part of the target the high helium capture cross section for $E < 40$ eV will not allow many such antiprotons to survive. The consequence would be a particularly wide distribution of cascade times as far as all $\bar{p}$ with $E < 30$ eV have not disappeared. If, on the contrary, the explanation of the BB had to be searched in the high energy capture events, a detail that should be taken into account is the entrance energy of the $\bar{p}$ involved in the BB. This energy ranges from zero to some keV. Rare fluctuations apart, larger entrance energy lead to $\bar{p}$ capture at $z$ values where the BB has been reabsorbed by the Main Belt. More difficult is to explain the BB in terms of trajectory spread, because one would expect this phenomenon to be more effective at increasing $z$, oppositely to what is seen. In a not too evident form, this effect can be seen in the higher pressure helium data, and in hydrogen data. However, a Backward Belt associated to relevant straggling phenomena at small $z$ would be possible, for example, if rare capture, no ESP and only very small impact parameter collisions were present at $E < 40$ eV (see e.g. [20]). This would imply large deflections accompanied by a long time delay before capture. To summarize, we have presented the $z, t$ Scatter Plots of $\bar{p}$–helium low-energy annihilations for helium pressure at 0.2 and 1 mb. Both the Scatter Plots and the time distributions show relevant qualitative differences with respect to the corresponding data in molecular hydrogen and deuterium. Annihilation points are more abundant than expected at intermediate times, where it is impossible for us to distinguish between post-capture cascade effects and pre-capture multiscattering effects. A previous Monte Carlo technique, successful in reproducing $H_2$, $D_2$ data, was employed here, with much smaller effectiveness.

# REFERENCES

1. A. Zenoni *et al*, Phys. Lett. **B 461** (1999) 405.
2. A. Zenoni *et al*, Phys. Lett. **B 461** (1999) 413.
3. M.Augsburger *et al*, Phys. Lett. **B 461** (1999) 417.
4. S.Deser *et al*, Phys. Rev. **96** (1954) 774; T.L.Trueman,
5. S.Wycech, A.M.Green and J.A.Niskanen, Phys.Lett. **B 152** (1985), 308.
6. S.Maury Design Report, CERN/PS 96-43 (AR).
7. H.Feshbach, C.E.Porter and V.F.Weisskopf, Phys. Rev. **96** (1954) 448.
8. V.Franco and R.J. Glauber, Phis.Rev. **142** (1966) 1195.
9. O.D.Dalkarov and V.A.Karmanov, Sov. Journ. Nucl. Phys. **45** (1987) 430; Nucl. Phys. **A 478** (1988) 635c;
10. K.V.Protasov, "Workshop on hadron spectroscopy 99", March 8-12, 1999, LNF, Frascati (Italy), 463.
11. E.g.: L.D.Landau and E.M.Lifshits: A course in theoretical physics, vol.III "quantum mechanics".
12. N. Bohr, Philos. Mag. 25, 10 (1913).
13. H. A. Bethe, Ann. Phys. (Leipzig) 5, 325 (1930).
14. E. Fermi and E. Teller, Phys. Rev. 72, 399 (1947).
15. F. Bloch, Ann. Phys. (Leipzig) 16, 287 (1933).
16. W. H. Barkas, J. N. Dyer and H. H. Heckman, Phys. Rev. Lett. 11, 26 (1963).
17. M. Inokuti, Rev. Mod. Phys. 43, 297 (1971).
18. J. F. Ziegler, J. P. Biersack and U. Littmark, *The Stopping and Range of Ions in Solids* (Pergamon, Oxford, 1985).
19. H. Knudsen and J. F. Reading, Phys. Rep. 212, 107 (1992).
20. G. Schiwietz, U. Wille, R. Diez Muino, P. D. Fainstein and P. L. Grande, J. Phys. B29, 307 (1996).
21. T. B. Quinteros and J. F. Reading, Nucl. Instrum. Methods B53, 363 (1991).
22. P. Sigmund and A. Schinner, Eur. Phys. J. D15, 165 (2001).
23. S. A. Wightman, Phys. Rev. 77, 521 (1950).
24. W. A. Beck, L. Wilets and M. A. Alberg, Phys. Rev. A48, 2779 (1993).
25. J. S. Cohen, Phys. Rev. A62, 022512 (2000).
26. G. Ya. Korenman, Nucl. Phys. A692, 145c (2001).
27. J.S. Briggs, P. T. Greenland and E. A. Solov'ev, Hyp. Int. 119, 235 (1999).
28. A. Adamo *et al.*, Phys. Rev. A47, 4517 (1993).
29. M. Agnello *et al.*, Phys. Rev. Lett. 74, 731 (1995).
30. A. Bertin *et al.*, Phys. Rev. A54, 5441 (1996).
31. A. Y. Voronin and J. Carbonell, Hyp. Int. 115, 143 (1998).
32. J. Revai, N.V. Shevchenko, private communication.

# Study of $S = -2$ baryonic states at FLAIR

D. Grzonka*, B. Bassalleck†, A. Gillitzer*, K. Kilian*, P. Kingsberry**,
W. Oelert*, J. Ritman*, T. Sefzick* and P. Winter*

*Institut für Kernphysik, Forschungszentrum Jülich GmbH, 52425 Jülich, Germany
†University of New Mexico, Department of Physics and Astronomy, Albuquerque, NM 87131, USA
**T.U.N.L., Duke University, Durham, NC 27708-0308, USA

**Abstract.** At the future FAIR project in Darmstadt/Germany [1] low energy antiprotons will be available at FLAIR, the Facility for Low energy Antiproton and Ion Research. Within the FLAIR LoI [2] it is proposed to study the production of strangeness $S = -2$ baryonic states based on ideas proposed for LEAR [3].

With stopped antiprotons a very efficient reaction chain for the production of slow $\Xi$ hyperons can be initiated. The special feature of this reaction channel is the low momentum of the produced $\Xi$ hyperon down to zero $MeV/c$ recoil momentum. These slow $\Xi$ particles can go into interacting $\Xi N$ systems which can couple to $YY$ or might also directly connect to the $H$ particle, if it exists.

From the experimental point of view the delayed decays of the strange exit particles allows a highly selective trigger on these reaction channels and the event reconstruction is relatively simple. A non magnetic detection system with track reconstruction ability is sufficient for the complete kinematical reconstruction.

**Keywords:** Xi production, hyperon nucleon interaction, recoil-less kinematics
**PACS:** 25.43.+t

## INTRODUCTION

Studies of baryon-baryon systems are a basic tool for investigations of the strong interaction. Limited to the light quarks (u,d,s) the baryons are grouped within $SU(3)$ multipletts where the lowest mass states build the octet of $1/2^+$ baryons shown in fig. 1. The coupling strengths between the different baryons are related by the corresponding Clebsch-Gordon coefficients if $SU(3)$ symmetry holds which is generally applied in hyperon-nucleon potential models. The experimental determination of the baryon-baryon interaction within the baryon octett will allow to develop improved hyperon-nucleon potentials and to proof the degree of $SU(3)$ symmetry which is of general importance for questions concerning the mechanism of symmetry violation.

For the NN-system an extended data base exists due to the stability of protons and neutrons which allows to prepare proton and neutron (by use of deuterons) beams and targets for NN scattering experiments. Hyperons are unstable with decay lengths $c\tau$ of up to a few $cm$ which complicates drastically conventional scattering experiments. A hyperon target is impossible and a hyperon beam is only available as secondary beam resulting from production reactions which is connected to much smaller beam intensities compared to NN scattering experiments. Nevertheless there are some elastic scattering cross section data available in the $\Lambda N$ [4], [5],[6],[7],[8],[9] and the $\Sigma N$ channel [9], see fig. 2. The majority of the data were taken with $\Lambda$ hyperons. To extract the scattering length which is related to the $YN$ potential the data have to be extrapolated to zero energy

CP793 Physics with Ultra Slow Antiproton Beams
edited by Yasunori Yamazaki and Michiharu Wada
© 2005 American Institute of Physics 0-7354-0282-5/05/$22.50

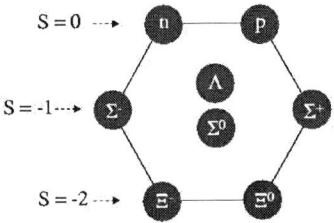

**FIGURE 1.** The $SU(3)$ $1/2^+$ baryon octett with proton, neutron, and the $\Lambda$, $\Sigma$, $\Xi$ hyperons.

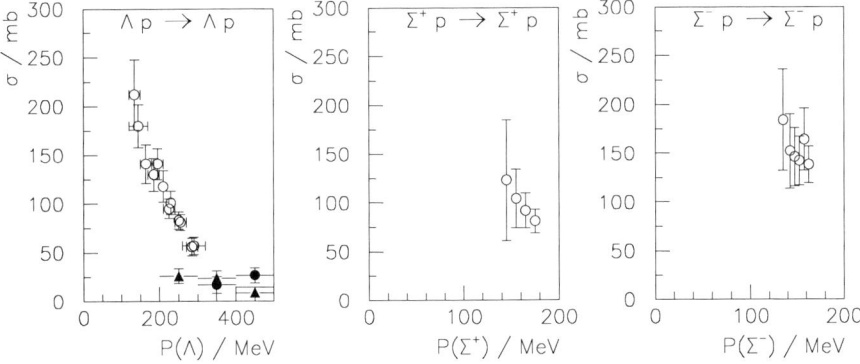

**FIGURE 2.** Total cross section data of hyperon-nucleon elastic scattering for $\Lambda$, $\Sigma^+$, and $\Sigma^-$.

where data are missing. By reducing the primary $\Lambda$ energy (velocity) the decay length goes down which makes secondary scattering experiments more difficult.

Another approach to get low energy scattering data are hyperon productions in multi particle exit channels like $pp \rightarrow pK^+\Lambda$. With a kinematical complete event reconstruction event samples with low $\Lambda N$ relative momenta can be selected from which the scattering length can be extracted. Such type of studies have been successfully performed for the $p\Lambda$ system in $K^-d \rightarrow \pi^- p\Lambda$ [10] and $pp \rightarrow pK^+\Lambda$ [11].

In the $S = -2$ sector tha data situation is even worse. Practically no data are existing concerning the $\Xi N$ scattering. The available data are limited to the search for the $H$ dibaryon, a (uuddss) quark configuration which was first predicted by Jaffee [12] to result in a bound state below the $\Lambda\Lambda$ threshold. The search for the $H$ particle was conducted in various reaction channels without any confirmation of such a state. Overviews of this field can be found e.g. in ref. [13], [14]. The cascade hyperons $\Xi$ are normally used as entrance into $B = 2$ systems with $S = -2$. Slow $\Xi$ particles can go into interacting $\Xi N$ systems which can couple to $YY$ systems. These $\Xi N$ might also connect directly to the $H$ particle or might be relevant in double $\Lambda$ hyper-nuclei, if produced in a larger nucleus. The most common way of $\Xi$ production is a double strangeness exchange reaction $(s, \bar{s})$ where a $K^-$ secondary beam induces a reaction like $K^- p \to \Xi^- K^+$ With a 2 $GeV/c$ kaon beam momentum $\Xi$ hyperons with momenta above 500 $MeV/c$ result which are used as secondary 'beam' to induce $\Xi N$ reactions.

The facility for low energy antiproton and ion research (FLAIR) at the new FAIR project in Darmstadt will deliver low energy phase space cooled antiproton DC beams which will allow $\Xi$ production in recoil-free kinematics. This production scheme via stopped antiproton annihilation was already proposed to be studied at LEAR [3] but could not be realized before its closing.

## Production of $S = -2$ baryonic states via antiproton annihilation

It is proposed to produce $S = -2$ systems via the $\Delta S = 2$ exchange reaction $(\bar{K}^*, K)$ [3]. The $\bar{K}^*$ is produced in the $\bar{p}$ annihilation reactions $\bar{p}N \to \bar{K}^* K^*$ and $\bar{p}N \to \bar{K}^* K$ with branching ratios BR [15],[16], $\bar{K}^*$ momenta $P_{\bar{K}^*}$ (for the nominal $K^*$ mass of 892 $MeV$ [17]) and decay lengths $\lambda$ of:

$\bar{K}^* K^*$ :  BR $= 5 \cdot 10^{-3}$  $P_{K^*} = 285 MeV/c$  $\lambda = 1.2 fm$
$\bar{K}^* K$ :  BR $= 1 \cdot 10^{-3}$  $P_{K^*} = 616 MeV/c$  $\lambda = 2.6 fm$.

The decay lengths $\lambda$ correspond to the momentum from $\bar{p}p$ annihilation and the decay width of $\sim 50$ $MeV$. Due to the small decay length both, $K^*$ production and the double strangeness exchange reaction have to take place in the same nucleus. The exchange reactions can proceed via: $\bar{K}^* N \to K\Xi$, e.g. $K^{*-} p \to K^+ \Xi^-$. That reaction is expected to be a very efficient source of $\Xi$ production. The first step in these studies would be the measurement of the branching ratio for the $\Xi$ production per stopped $\bar{p}$ annihilation in a deuteron target leading to the following two step processes:

1. step: $\bar{p}N \to \bar{K}^* K^{(*)}$ 2. step: $\bar{K}^* N \to \Xi K$ giving the overall reaction : $\bar{p}d \to \Xi K K^{(*)}$, with the possible exit channels: $\Xi^0 K^0 K^0$, $\Xi^0 K^0 K^0 \pi^0$, $\Xi^0 K^0 K^+ \pi^-$, $\Xi^- K^0 K^+$, $\Xi^- K^0 K^0 \pi^+$, $\Xi^- K^0 K^+ \pi^0$, and $\Xi^- K^+ K^+ \pi^-$.

Concerning the kinematics of the second step, there is a 'magic' $\bar{K}^*$ momentum at which the $\Xi$ can be produced at rest in the laboratory system. This magic momentum only depends on the masses of the reaction particles and is given by:

$P_{magic}^2 = (((m_1^2 + m_3^2) - (m_2 - m_4)^2)^2 - 4 m_1^2 m_3^2)/(4(m_2 - m_4)^2)$,
for the reaction: $m_1 + m_2 \to m_3 + m_4$.

Concerning e.g. the reaction $K^{*-} p \to K^+ \Xi^-$ a magic momentum is only possible if the $K^*$ mass is above 880 $MeV/c^2$. In fig. 3 in the left part the magic $K^*$ momentum is plotted as a function of the $K^*$ mass. The dashed line indicates the $K^*$ mass with a width of 50 $MeV$ which covers the region of the magic momentum.

The magic momenta match well with the momentum of the $\bar{K}^*$ listed above and shown as a function of the $K^*$ mass in fig. 3 in the right part for the $\bar{p}p \to \bar{K}^* K^*$ reaction. The cross section of the $(\bar{K}^*, K)$ reaction in the second step should be rather high since it is strongly exothermic and proceeds at low energies. In fact, it should be substantially higher than that of the endothermic reaction $K^- p \to K^+ \Xi^-$, which reaches a maximum

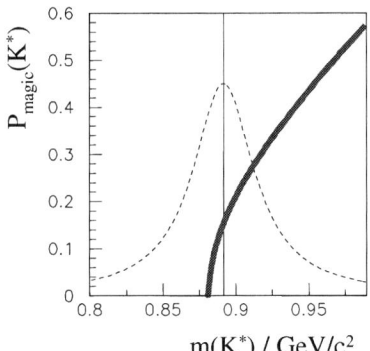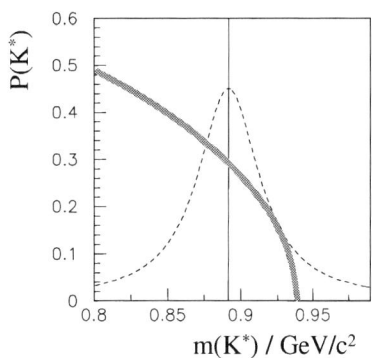

**FIGURE 3.** Magic $K^*$ momentum needed for a recoil-free $\Xi$ production (left) and $K^*$ momentum resulting from the $\bar{p}p \to \bar{K}^*K^*$ annihilation channel (right) as a function of the $K^*$ mass.

cross section of about 180 $\mu$b at 1.6 $GeV/c$ [18] corresponding to a branching ratio of $6 \cdot 10^{-3}$.

For the study of $\Xi N$ interaction, $\Lambda\Lambda$ (or $YY$) production and possible $H$ particle production the antiproton annihilation has to take place in a $^3$He target. The above mentioned $\Xi$ production reaction is now the entrance and we consider with an additional third reaction step: 1. step: $\bar{p}N \to \bar{K}^*K^{(*)}$, 2. step: $\bar{K}^*N \to \Xi K$, 3. step: $\Xi N \to \Xi N$, $\Xi N \to YY$ or $\Xi N \to H$. The first two steps are the same as for the deuterium target. The second and third step might proceed already directly together as an off shell reaction. In fig. 4 the reaction chain is shown schematically with the three steps.

The probability $P_H$ of producing a $(B = 2, S = 2)$ system in a $\Delta S = 2$ $(\bar{K}^*, K)$ re-action on two nucleons depends exponentially on the ratio of the recoil energy on the $\Xi$, $E_R = q_\Xi^2/2m_\Xi$, and the binding energy of the $(B = 2, S = 2)$ system, $E_B$ : $P_H \sim exp{-E_R/E_B}$. This probability is almost 100% if the recoil energy is sufficiently small. Therefore small momentum transfers $q_\Xi$ are important and should enhance the chance for the creation of a $(B = 2, S = 2)$ system. Compared to the $K^-$ and proton induced reactions [19],[20] a much smaller momentum transfer is possible for this sug-gested $\bar{p}$ induced process.

## EXPERIMENTAL REALIZATION

In order to discuss the experimental setup a specific reaction channel is considered now. Well suited is the channel $\bar{p}d \to \Xi^- K_s^0 K^{*+} \to (p\pi^-)_\Lambda \pi^- (\pi^+\pi^-)_{K_s} (\pi^+\pi^-)_{K_s} \pi^+$ with its three delayed decays. Both the trigger and the event reconstruction are particularly simple when using an apparatus which triggers on delayed decays and records charged

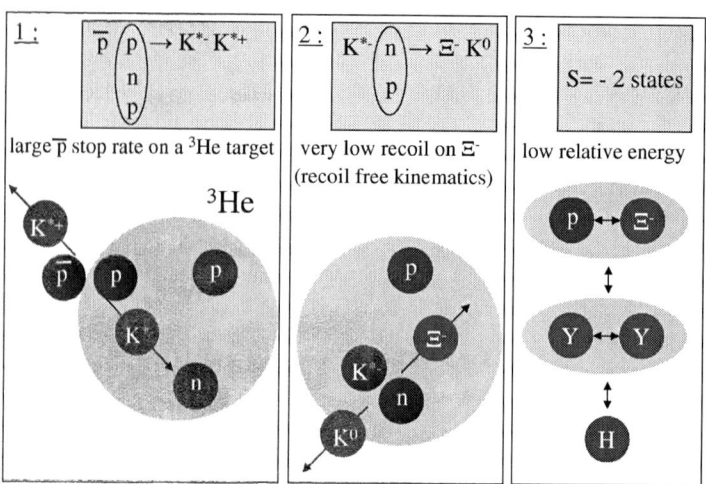

**FIGURE 4.** Sketch of the reaction steps from antiproton anihilation in the first step with a $K^*$ generation to the recoil-free $\Xi$ production in the second step and the interacting $\Xi - N$ system in the third step.

particle tracks. The same is valid for the $^3$He target resulting in $(B = 2, S = 2)$ systems: $\bar{p}\,^3\text{He} \to p\Xi^- K_s^0 K^{*+}$ or $\Lambda\Lambda K_s^0 K^{*+}$ or $HK_s^0 K^{*+}$ where the $H$ decays e.g. into $\Sigma^- p$.

For the experimental apparatus to be considered it is important that normal associated strangeness production in $\bar{K}K$ or $KY$ can lead to only two delayed decays. Those are frequent and have to be distinguished from the cases of $\Xi KK$ or $YYKK$ or $HKK$ production. A perfect signature for all $S = -2$ particle productions is the observation of at least three delayed decays in an event. An efficient and simple experimental and trigger setup can be envisaged using the multiplicity increase which occurs after the delayed decay of strange neutral particles sketched in fig. 5. On the right side in fig. 5 an event is sketched to illustrate the requirements on the detection system. In the final state of the reaction chain 8 charged particles result.

The event identification is possible via the geometry of the particle tracks. From the charged pion tracks the decay vertices of the $K_S^0$ can be reconstructed and with the known target vertex their tracks are determined. The $\Lambda$ track is given from the vertices reconstructed from the $(p, \pi^-)$ and $(\Xi^-, \pi^-)$ tracks. With this information all momentum directions of all particles are known and applying a mass hypothesis for the detected particles the event can be completely reconstructed if a high resolution tracking system and a well defined target vertex is used.

As target, a very small high pressure deuteron or $^3$He gas cell will be used where the full stop distribution is in the gas. This requires a low emittance antiproton beam with an energy around 1 $MeV$ which will be delivered by the FLAIR facility.

A clean trigger for this type of reaction channels is the high multiplicity of charged particles in the final state and the multiplicity increase. By measuring the number of

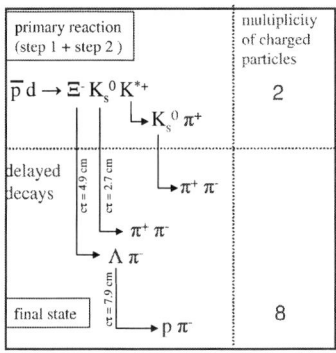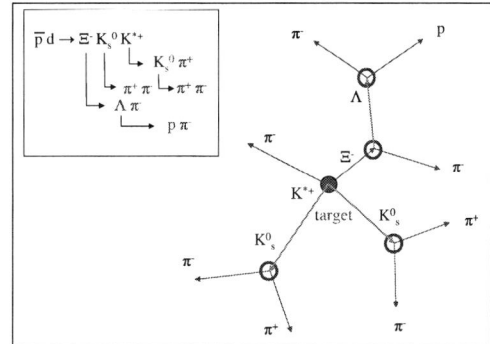

**FIGURE 5.** Sketch of the $\bar{p}d \rightarrow \Xi^- K_s^0 K^{*+} \rightarrow (p\pi^-)_\Lambda \pi^- (\pi^+\pi^-)_{K_s} (\pi^+\pi^-)_{K_s} \pi^+$ reaction indicating the reaction chain with the three delayed decays and the resulting particle tracks to be detected.

charged particles close to the target which is only 2 in comparison to the multiplicity at a reasonable distance a very efficient trigger signal can be generated due to the delayed decays in this multi-strangeness production. The next step in the experimental programme will be the study of the $\Xi - N$, $Y - Y$ or $H$-particle system. Here $^3$He has to be used as target gas. The particle configuration is similarly, only an additional proton appears in the exit channel. Therefore the arguments given above for the pure $\Xi$ production are the same. A further extension of the programme could be the production of double hypernuclei. With the technique of recoil-free kinematics the $\Xi$ can also be produced and deposited in more extended nuclei. A high efficient production of double hypernuclei is expected with this method.

## Detection system

The detector must deliver multiplicity signals and tracking information of the charged particles. In fig. 6 a sketch of the proposed setup is shown. The target is surrounded by a segmented scintillation detector. Here plastic scintillators readout by mirror foils or a cylinder out of scintillating fibres could be used. For charged tracks with a short decay length like the $\Xi^-$ in the reaction channel sketched in fig.6 , a vertex detecor very close to the target is necessary. Two layers of Si-microstrip detectors are forseen. The charged particle tracking will be done by straw tubes. Several layers of straw tubes in different orientations will allow a 3 dimensional track reconstruction. The detailed configuration has to be determined from Monte Carlo studies. The closure of the detection volume will be done by plastic scintillator hodoscopes, a barrel type combined with two endcaps. First simulation studies to check the performance of the proposed detector system have been started using GEANT 3 where the reaction: $\bar{p}d \rightarrow \Xi^- K_s^0 K^{*+}$ was generated. With a detector configuration sketched in fig. 6 about 90 % of the events produce a multiplicity increase of two or more which result in a sufficient background reduction at the main

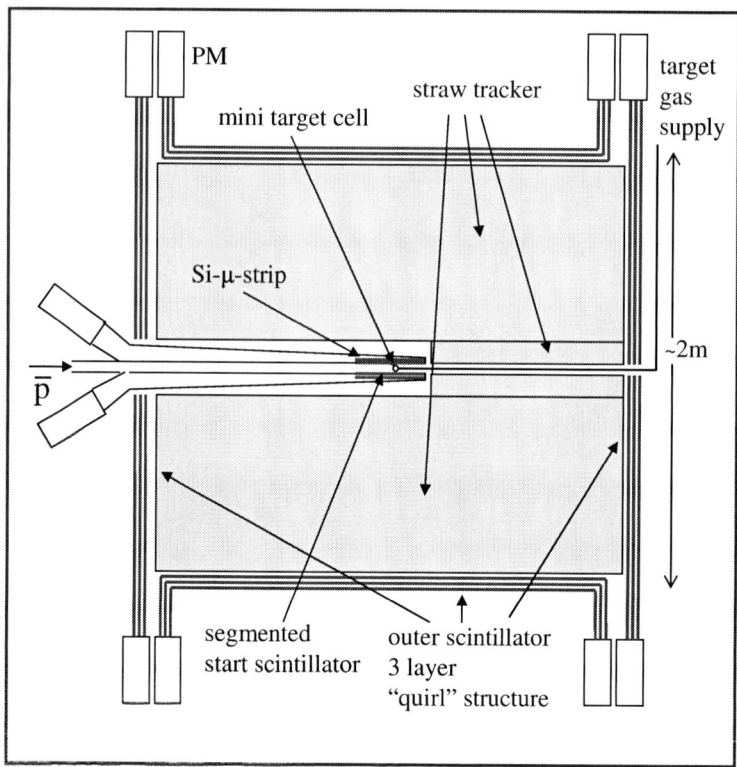

**FIGURE 6.** Sketch of the detection system with two scintillator layers to trigger on the multiplicity increase and a tracking system for the charged particles.

trigger level. The efficiency for the $\Xi^-$ detection in a $Si\text{-}\mu$-strip detector would reach about 35 % for a quite conservative assumed distance of $1\,cm$ from the target which guarantees the possibility of a sufficiently precise $\Xi^-$ track determination. The reaction particles with mean momenta around 200 - 300 $MeV/c$ traverse the straw detector and the multiple scattering within the expected detector material is low enough with a mean around 5 - 10 $mr$ that a reasonable track reconstruction is expected. More detailed MC studies with the final detector configuration will be performed to determine the event reconstruction efficiency.

## SUMMARY

The low energy antiproton facility FLAIR will deliver phase space cooled antiproton beams with low emittance which can be stopped within a small target volume.

The annihilation of these stopped antiprotons will be a very efficient method to produce $S = -2$ systems via the double strangeness exchange reaction $(\bar{K}^*, K)$. With a sizeable branching ratio the annihilation of antiprotons results in the production of a $\bar{K}^*$ "beam" which interacts with another nucleon via $\bar{K}^* N \rightarrow K\Xi$. The momenta of the $\bar{K}^*$ are well matched for the production of slow $\Xi$ particles which undergo efficient $\Xi N$ interactions. The three delayed decays in this reaction chain allows a very effective trigger and selection of the $S = -2$ production events and the event reconstruction via geometry of the particle tracks requires a relatively simple detection system.

The proposed studies will result in detailed information of $S = -2$ baryonic and possible dibaryonic states.

# REFERENCES

1. FAIR, Facility for Antiproton and Ion Research, *http://www.gsi.de/fair*.
2. FLAIR, Facility for Low-Energy Antiproton and Ion Research, *http://www.oeaw.ac.at/smi/flair*.
3. K. Kilian, *Proc. 4th LEAR Workshop*, Villars-sur-Ollon 1987, C. Amsler et al., ed., Harwood Academic Publishers, Chur, 529 (1988), K. Kilian et al., *Memorandum PSCC CERN* 1990.
4. G.Alexander,U.Karshon,A.Shapira,G.Yekutieli, *Phys.Rev.* **173**,1452 (1968).
5. J.A.Kadyk,G.Alexander,J.H.Chan,P.Gaposchkin,G.H.Trilling, *Nucl. Phys.* **B27**, 13 (1971).
6. R.P.Mount,R.E.Ansorge,J.R.Carter,W.W.Neale,J.G.Rushbrooke,D.R.Ward, *Phys. Lett.* **58B**, 228 (1975).
7. K.J.Anderson,N.M.Gelfand,J.Keren,T.H.Tan, *Phys.Rev.* **D11**, 473 (1975).
8. J.M.Hauptmann,J.A.Kadyk,G.H.Trilling, *Nucl. Phys.* **B125**, 29 (1977).
9. Eisele *Phys. Lett.* **37B**, 205 (1971).
10. H.H. Tan, *Phys. Rev. Lett.* **23**, 395 (1969).
11. J.T. Balewski, A. Budzanowski, C. Goodman, D. Grzonka, M. Hofmann, L. Jarczyk, A. Khoukaz, K. Kilian, T. Lister, P. Moskal, W. Oelert, I.A. Pellmann, C. Quentmeier, R. Santo, G. Schepers, T. Sefzick, S. Sewerin, J. Smyrski, A. Strzalkowski, C. Thomas, C. Wilkin, M. Wolke, P. Wüstner, D. Wyrwa *Eur. Phys. J.*, **A2**, 99 (1998).
12. R.L. Jaffe, *Phys. Rev. Lett.*, **38**, 195 (1977).
13. T. Sakai, K. Shimizu, K. Yazaki, *Prog.Theor.Phys.Suppl.*, **137**, 121 (2000).
14. B. Bassalleck, *Nucl. Phys.*, **A639**, 401c-406c (1998).
15. N. Barash, L. Kirsch, D. Miller, T. H. Tan, *Phys. Rev.*, **145**, 1095-1102 (1966).
16. L. Mandrup, A.S. Jensen, A. Miranda, G.C. Oades, J. Carbonell, *Nucl.Phys.*, **A512**, 591-625 (1990).
17. Review of particle physics, *Phys. Lett.*, **B592** (2004).
18. A. de Bellefon et al., *Nuovo Cimento*, **7A**, 567 (1972).
19. R.W. Stotzer, T. Burger, P.D. Barnes, B. Bassalleck, A.R. Berdoz, M. Burger, R.E. Chrien, C.A. Davis, G.E. Diebold, H. En'yo, H. Fischer, G.B. Franklin, J. Franz, L. Gan, D.R. Gill, T. Ijima, K. Imai, P. Koran, M. Landry, L. Lee, J. Lowe, R. Magahiz, A. Masaike, R. McCrady, F. Merrill, C.A. Meyer, John M. Nelson, K. Okada, S.A. Page, P.H. Pile, B.P. Quinn, W.D. Ramsay, E. Rossle, A. Rusek, R. Sawafta, H. Schmitt, R.A. Schumacher, R.L. Stearns, I.R. Sukaton, R. Sutter, J.J. Szymanski, F. Takeutchi, W.T.H. van Oers, D.M. Wolfe, K. Yamamoto, M. Yosoi, V.J. Zeps, R. Zyber, *Phys. Rev. Lett.*, **78**, 3646-3649 (1997).
20. K. Nakazawa, *Nucl. Phys.*, **A639**, 345c-353c (1998).

# NUCLEAR STRUCTURE

# Nuclear structure studies with low energy $\bar{p}$

Sławomir Wycech

*Sołtan Institute for Nuclear Studies, Warsaw, Poland*

**Abstract.** The nuclear $\bar{p}$ capture from atomic states is briefly reviewed and several capture modes are compared. All these modes may test neutron density distributions in different regions of nuclei and yield complementary information on the *Rms* and higher moments of the neutron density profiles. Some advantages and difficulties of experimental methods are indicated and a special attention is paid to the $\pi^{\pm}$ emission following $\bar{p}$ annihilation. It is shown that this useful method may become very powerful if it determines more than the separate $\pi^+$ and $\pi^-$ multiplicities. Two specific questions are analysed: the ratio of $\bar{p}n$ and $\bar{p}p$ annihilation rates and the reabsorption of $\pi$ mesons in the residual nuclei.

**Keywords:** Antiprotonic atoms, nuclear radii
**PACS:** 13.75.-n;21.10Gv;36.10.-k

## INTRODUCTION

About half century ago it was realized that kaonic and antiprotonic atoms provide a way to study the tail of the nuclear density distribution, its isospin structure and nuclear correlations there [1]. Three methods have been used so far, which give information in the regions roughly 1 fm to 3 fm beyond the half density radius.

- The observation of X-ray cascades in hadronic atoms and extraction of the atomic level shifts and widths [2, 3, 4, 5, 6, 7]. These level widths, related to the nuclear absorption of hadrons may determine high moments of the nuclear density distributions. Information obtained in this way is limited to one, in some special cases to two level widths and one level shift. The latter is in general difficult to interpret and provides a check on interaction models.

- Studies of the nuclear absorption products related to $\bar{p}$ annihilation

$$\bar{p} + (N, Z) \longrightarrow (N, Z - 1) + mesons \qquad (1)$$

$$\bar{p} + (N, Z) \longrightarrow (N - 1, Z) + mesons \qquad (2)$$

where a nucleus of $N$ neutrons and $Z$ protons is denoted by $(N, Z)$. Thus (1) involves predominantly the $\bar{p}p$ and (2) the $\bar{p}n$ annihilation. In this way one can discriminate captures on protons from captures on neutrons and obtain more nuclear information. However, these results are also more difficult to describe since the initial states of capture are not known directly and the final mesons may exchange charge or become absorbed. Some additional information is required.

  - One way to study such reactions is the detection of mesons. This research was initiated in the nuclear emulsion with $K^-$ atoms [8, 9] and in bubble chambers with the $\bar{p}$ atoms [10, 11]. Scintillator detectors and fixed targets have later

CP793 *Physics with Ultra Slow Antiproton Beams*
edited by Yasunori Yamazaki and Michiharu Wada
© 2005 American Institute of Physics 0-7354-0282-5/05/$22.50

been used [12]. Recently, a novel method to study the pion emission from trapped $\bar{p}$ atoms of unstable nuclei has been proposed [13].

– Complementary measurements detect residual nuclei instead of the mesons. The radiochemical method was proven to be successful [14],[15]. In particular the final nuclei which may be detected in this way are "cold" Z-1, N-1 nuclei of excitation energies less than the neutron emission thresholds.

The basic $\bar{p}N$ interactions required for this research are limited to several phenomenological parameters: range of the $N\bar{N}$ annihilation, absorptive parts of the scattering amplitudes, pion production multiplicities, pion momentum distributions. These may be taken from other experiments while effects of uncertainties must be quantified.

## ATOMIC ANTIPROTONS

Antiprotons bound into atomic orbits cascade down to annihilate within the nucleus. The annihilation happens at the extreme nuclear surface due to the two effects: the atomic cascade populates states of high centrifugal barrier and the free path of antiprotons in nuclear matter is less than 1 fm. The peripherality of capture allows for low density approximations: quasi-free scattering and single-particle picture of the nucleus. It simplifies the description of final mesons, an important point for understanding the absorption experiments. On the other hand the difficulty inherent in the surface studies is related to its sensitivity to $N\bar{N}$ force range.

The tool to describe the antiprotonic atomic level shifts and widths is an optical potential $V^{opt}$. The simplest one is assumed usually [4, 5, 6, 7, 16] in the form

$$V^{opt}(\mathbf{R}) = \frac{2\pi}{\mu_{N\bar{N}}} \Sigma_s t^s_{N\bar{N}} \rho^s(\mathbf{R}) \tag{3}$$

where $\mu_{N\bar{N}}$ is the $N\bar{N}$ reduced mass, $\rho^s(\mathbf{R})$ are nuclear densities at a radius $\mathbf{R}$, index $s$ denotes protons or neutrons. The $t^s_{N\bar{N}}$ are complex "effective" scattering lengths. The finite range in the $N\bar{N}$ interaction requires folded densities

$$\rho^s(R) = \int d\mathbf{u} \rho^s_0(\mathbf{R} - \mathbf{u}) \upsilon(\mathbf{u}) \tag{4}$$

instead of "bare" nucleon densities given in terms of single nucleon wave functions $\varphi^\alpha_N$, as

$$\rho^s_0(R) = \Sigma_\alpha [\varphi^\alpha_N(R)]^2 \tag{5}$$

The form-factor $\upsilon(u)$ represents $N\bar{N}$ force range $r_o$. For the absorptive part of $V^{opt}$, the annihilation range of 1 fm is expected from models of $N\bar{N}$ annihilation [24], and nuclear $\bar{p}$ scattering experiments [18]. The potential itself is highly absorptive Im $t^s_{N\bar{N}} \approx 2$ fm and at the nuclear centre Im$V^{opt}$ would be 200 MeV strong. The corresponding free path length is well below 1 fm. However, it should be remembered that the form and the strength of $V^{opt}$ is tested only in the surface region. In particular, Im $V^{opt}$ is determined

202

by the atomic level widths via

$$\frac{\Gamma}{2} = \frac{2\pi}{\mu_{N\bar{N}}} \Sigma_s \, \mathrm{Im} \, t_{N\bar{N}}^s \int d\mathbf{R} \rho^s(\mathbf{R}) \, | \, \Psi_{\bar{N}}(\mathbf{R}) \, |^2 \tag{6}$$

where $\Psi_{\bar{N}}(\mathbf{R})$ is the atomic wave function. Since $\Psi_{\bar{N}} \approx R^l$ and only high angular momenta $l$ are available, the absorption strength is peaked at the surface. In a crude approximation, the atomic widths are determined by $< R^{2l} >$-th moments of the nuclear density. Precise atomic wave functions, corrected for nuclear interactions, indicate the dominant moment to be $< R^{2l-2} >$.

The form of optical potential is not fully settled, in particular the relation of the effective amplitude $t_{N\bar{N}}^s$ to the low energy expansion parameters

$$t_{\bar{p}N} = a_N + 3b_N \mathbf{p}\mathbf{p}' \tag{7}$$

where $a$ are scattering lengths, $b$ are scattering volumes and $\mathbf{p}$ are momenta. Such an amplitude generates the optical potential composed of central and gradient terms

$$V^{opt} = \Sigma_s \, V_s = \frac{2\pi}{\mu_{N\bar{N}}} \Sigma_s \, [a_s \rho^s + 3b_s \nabla \rho^s \nabla] \tag{8}$$

Overall, the X ray data indicates $t_{N\bar{N}}^{protons} \approx t_{N\bar{N}}^{neutrons}$, [20], but other experiments show more complicated structure. These questions are discussed later in the text.

## STUDIES OF ANTIPROTON ANNIHILATION PRODUCTS

In this section we discuss nuclear reactions induced by the annihilation. The point of special interest is the effect of the final state mesons upon the formation of final states detected in specific experimental conditions.

The partial widths $\Gamma_s$ for reactions (1,2) leading to a final state $s$ may be described by a formula analogous to eq. (6)

$$\frac{\Gamma_s}{2} = \frac{2\pi}{\mu_{N\bar{N}}} \mathrm{Im} t_{N\bar{N}}^s \int d\mathbf{R} \, | \, \Psi_{\bar{N}}(R) \, |^2 \, \rho_s(R) P^s(R) \tag{9}$$

where some functions $P^s$ are introduced in order to describe the formation of required final states. For an isolated $N\bar{N}$ system or in case of full atomic level width eq.(9), summed over all states $s$, reduces to the unitarity condition for the absorptive part of the elastic scattering amplitude $\mathrm{Im} t_{N\bar{N}}^s$. Inside a nucleus this condition is modified by the meson interactions and the selection of final states. A reliable description of $\Gamma_s$ is facilitated by large multiplicity of mesons ($\approx 5$) and large ($\approx 2$ GeV) deposit of the annihilation energy. On the nuclear scale, and on the scale of each individual meson, the closure approximation is justified, and it leads to the simple formula (9). The calculation of $P^s$ is now briefly outlined. Finer details: finite range of interactions, meson multiplicities and correlations, unitarity, applicability of the closure approximation and other corrections are described at some length in ref. [22].

Let the antiproton in a state denoted by $n$ and described by a wave function $\Psi_{\bar{N}}^n(\mathbf{R})$ annihilate on a nucleon in a single particle state $\alpha$ of wave function $\varphi_N(\mathbf{R})^\alpha$. The outcome is $k$ pions with momenta $\mathbf{p_i}$ and other quantum numbers denoted jointly by $\xi_i$, $i = 1...k$. The transition amplitude for this process, is

$$f_{n,\alpha} = \int \Psi_{\bar{N}}^n(\mathbf{R}) \varphi_N^\alpha(\mathbf{R}) t_{N\bar{N}\to M}(\xi) \prod_i \bar{\varphi}_M(\mathbf{R}, \xi_i). \tag{10}$$

where $t_{N\bar{N}\to M}$ is a transition matrix that describes the annihilation. All wave functions in eq. (10) involve nuclear interactions. In the bulk of the phase space, pions are fast enough to allow an eikonal description. Following this, the wave function $\varphi_M$ for each pion is taken in the form of ingoing waves

$$\bar{\varphi}_M(\mathbf{R}, \xi_i) = \exp(i p \mathbf{R} - i S_i(\mathbf{p}, \mathbf{R})) \tag{11}$$

with $S$ calculated in terms of a pion-nucleus potential $U_i$

$$S_i(\mathbf{p}, \mathbf{R}) = \int_0^\infty dt [\sqrt{(p^2 - U_i(\mathbf{R} + \hat{\mathbf{p}}t))} - p] \tag{12}$$

by integrating the local momentum over a linear meson trajectory. Due to nuclear excitations and pion absorptions this wave is damped with a rate described by Im$S$.

To calculate the absorption probability, amplitudes (10) are squared, summed over the final pionic channels and integrated over the phase space. The latter includes also the total momentum of the pions $\mathbf{P} = \Sigma_i \mathbf{p_i}$ equal to the recoil momentum of the residual nucleus. The nuclear recoil and excitation energy is negligible with respect to the total energy release and the closure approximation is applicable. That simplifies the mesonic part of the bilinear expression $f \bar{f}$ summed over pionic states

$$\Sigma_k \int d\mathbf{P} \int d\xi t_{N\bar{N}\to M}^s(\xi) \bar{t}_{N\bar{N}\to M}^s(\xi) \varphi_M(\mathbf{R}, \xi) \bar{\varphi}_M(\mathbf{R}', \xi) \approx$$
$$\text{Im} t_{N\bar{N}}^s \ \delta(\mathbf{R} - \mathbf{R}') \ < \prod_i |\exp(-S(\mathbf{p_i}, \mathbf{R}))|^2 > \tag{13}$$

Here, $d\xi$ denotes integrations over the relative pion momenta restricted by the energy conservation. The integration over momentum $\mathbf{P}$ generates (approximately) $\delta(\mathbf{R} - \mathbf{R}')$ and the sum covers pion multiplicities. For an isolated $N\bar{N}$ system eq. (13) reduces to the unitarity condition for the absorptive part of the elastic scattering amplitude Im$t_{N\bar{N}}^s$ Inside nuclei this condition is modified by a product of functions

$$P_{miss}(\mathbf{R}) = < \prod_i |\exp(-S(\mathbf{p_i}, \mathbf{R}))|^2 > \tag{14}$$

which describe final state interactions of each meson, averaged over the multiplicities and phase space.

In the case of radiochemical measurements the summation and averaging is done over all final mesons. In addition there are some limitations on the final nuclear states detected in the experiment. That leads to two final state functions describing different physical effects, $P^s = P_{miss}(\mathbf{R})P^s_{dh}(\mathbf{R})$. The chance that *all* mesons miss the residual nucleus is given by the $P_{miss}$ of eq. (14) which describes final state interactions of mesons. The closure approximation materialized in eq. (13) allows the transition from eq. (6) to eq. (9) in the text, as the squares of $\varphi^\alpha_N$ reduce to the nuclear density. Most of the final meson interactions remove the residual nucleus from the "cold" $A-1$ state. Thus Im $S$ is generated by Im $U^{opt}$ which includes all final meson interactions: the true meson absorption and the scattering which is predominantly an inelastic scattering [30]. Under these conditions the optical potential Im $U^{opt}$ is determined mainly by the absorptive $\pi - N$ scattering amplitude in the $\Delta$ and higher resonance region. The other factor $P_{dh}$ represents a correction for the $\bar{p}$ annihilation on "deeply bound" nucleons. A fast annihilation of a nucleon leaves the nucleus with a hole in one of nuclear shells. Later, the nucleus rearranges into an excited system. If the related excitation energy exceeds $\approx 8$ MeV a neutron is rapidly emitted. Such annihilations lead to "hot" $A-1$ nuclei, and the true residuals become $A-2$ nuclei. One should remove those "deep hole" contributions, and a good estimate for $P_{dh}$ is

$$P_{dh} = \Sigma_{\alpha,upper}[\varphi^\alpha_N(R)]^2/\Sigma_{\beta,all}[\varphi^\beta_N(R)]^2 \qquad (15)$$

where the first sum is limited, by the value of the neutron emission threshold, to the upper nucleon orbitals. At the surface this correction is small, but it becomes relevant for more central absorptions.

In experiments that detect a final meson of charge $s$ the states of residual nuclei are irrelevant. Hence, $P_{dh} = 1$ and

$$P^s(\mathbf{R}) = <| \exp(-S_s(\mathbf{p_i}, \mathbf{R})) |^2> \equiv P_\pi \qquad (16)$$

describes the chance of survival for each charged meson. The relevant calculation should involve only true $\pi^+, \pi^-$ absorptions as well as $\pi^+ \to \pi^0$ , $\pi^- \to \pi^0$ charge exchange reactions. From the scattering cross sections, the absorptive and charge exchange one [30], the first process is expected to dominate as 4:1. The averaging in eq. (16) involves meson multiplicities and the meson phase space. Some results are discussed in the last section.

The localisation of antiproton capture events is given by a product of the densities and the final state factors $| \Psi_{\bar{N}}(R) |^2 \rho_s(R)P^s(R)R^2$. The relevant regions are indicated in fig.1. The X-ray measurements offer an advantage of well defined initial and final states, but cannot distinguish the proton contribution from the neutron one. The other methods can do that, but the knowledge of initial $\bar{p}$ states is limited. In principle the composition of capture states may be obtained in cascade calculations, but additional data are required to check these. In the cold captures the necessary constraints are given by the ratios $\Gamma(A-1)/\Gamma(total)$ which amount to $\approx 10\%$. In the meson emission studies, equivalent constraints are given by the meson absorption rate $P_\pi$. Both these quantities are obtained in the relevant experiments. Figure 1 shows that the three discussed methods offer complementary information on different regions of the nuclear densities. The X-ray data involve densities of $\approx 5\%$ of the central one around $c + 1$ fm, where $c$ is the half density radius. The mesonic measurements are slightly more peripheral while the radiochemical method tests the most extreme region. These differences reflect the

**TABLE 1.** Brief comparison of the three methods using antiprotons to study the neutron skin.

| Method | measured quantity | advantages | difficulties |
|---|---|---|---|
| X-rays | level widths $\Gamma$<br>level shifts $\rightarrow$ | $\bar{p}$ state known<br>check on $V_{\bar{p}}^{optical}$ | no p,n separation |
| Cold capture | $\Gamma(\bar{p}n)/\Gamma(\bar{p}p)$<br>$\Gamma(A-1)/\Gamma(total) \rightarrow$ | n,p separation<br>check on $\bar{p}$ states | $\bar{p}$ state unknown |
| Meson emission | $N(\pi^-) - N(\pi^+)$<br>$N(\pi^-) + N(\pi^+) \rightarrow$ | n,p separation<br>check on $\pi$ absorption | $\bar{p}$ state unknown |

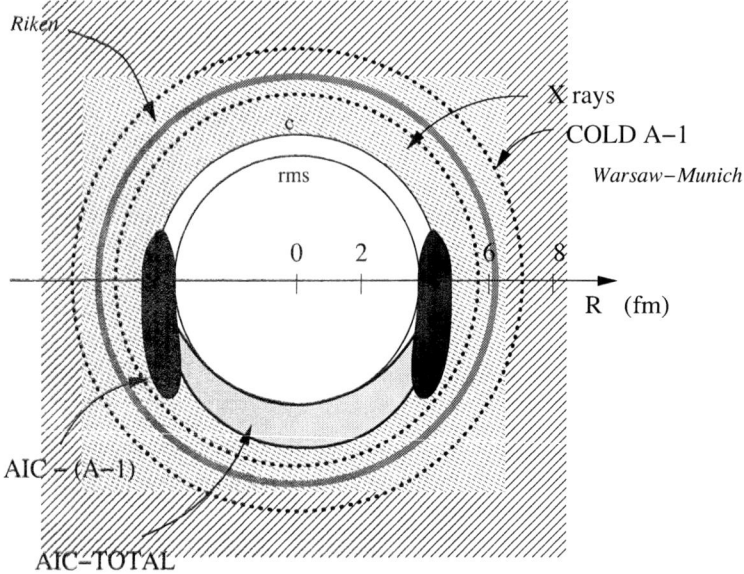

**FIGURE 1.** A schematic representation of the antiproton capture regions in $^{112}$Sn that could be tested in three atomic experiments: X-rays, cold A-1 capture *Warsaw − Munich* and charge meson detection, *Riken*. AIC - denotes the in flight capture proposal [17].

chances for mesons to leave the nucleus. The $P^s(R)$ are very small at central densities and rise only in the nuclear surface region. All characteristic features of the three methods are briefly summarised in table 1.

All methods require additional data: the strengths of the capture rates $\sigma(\bar{p}n)$ and $\sigma(\bar{p}p)$ or their ratio $R_{n/p} = \sigma(\bar{p}n)/\sigma(\bar{p}n)$. The experiments which measure the capture rates on neutrons $\Gamma(\bar{p}n)$ and protons $\Gamma(\bar{p}p)$ determine a "halo factor" [10] defined as

$$\frac{\Gamma(\bar{p}n)}{\Gamma(\bar{p}p)} = \frac{Z}{N} R_{n/p} f^{halo}. \tag{17}$$

With the $R_{n/p}$ known, $f^{halo}$ represents the neutron excess in the capture region. It is left to the nuclear theory to disclose what region of the nuclear surface is involved and what is the physical origin of the halo. In the next sections the $R_{n/p}$ is extracted from several experiments and the capture site in the mesonic measurements is discussed.

## The $\sigma(\bar{p}n)/\sigma(\bar{p}p)$ ratio

The $R_{n/p}$ may be extracted from the low energy in-flight captures, $\bar{N} - N$ interaction models and the antiproton optical potential fitted to the X-ray data.

The atomic widths for partial absorptions $\Gamma(\bar{p}n), \Gamma(\bar{p}p)$ are related to the absorptive amplitudes $\mathrm{Im}\, t_{N\bar{N}}$ for the $\bar{p}n$ and $\bar{p}p$ pairs. In the $\bar{N}N$ scattering states these amplitudes are given by the total cross sections $\sigma(\bar{p}n), \sigma(\bar{p}p)$. For bound nucleons such relations are not possible since there are no cross sections. One has to extrapolate the amplitudes to negative energies and check the results against other available experiments. The amplitudes determined in this way allow to calculate the antiproton optical potentials. Next, these potentials could be used in the analysis of the nuclear capture data.

The number $R_{n/p}$ required in the neutron halo studies has been obtained in several experiments listed in table 2. The numbers seem random but, as shown below, there is some simplicity behind these results. Let us notice that the $\bar{p}N$ scattering amplitudes $t$ describing the scattering on a bound nucleon enter the formalism as $t(E_{\bar{p}} - E_B - E_{recoil})$ where $E_B$ is the nucleon binding and $E_{recoil}$ is the recoil energy of the $\bar{p}N$ pair with respect to the residual system. In atomic states the energies $E_{\bar{p}}$ are slightly negative while in the "stopped" antiproton states $E_{\bar{p}} \approx 1$ MeV, [29]. The tested quantity is the scattering amplitude

$$\bar{t} = \int t(E_{\bar{p}} - E_B - \frac{p^2}{2m_R}) \mid \tilde{\phi}(p) \mid^2 d\vec{p} \qquad (18)$$

weighted by the recoil momentum distribution. The latter is determined by $\phi(p)$, the Fourier transforms of $\Psi^n_{\bar{N}}(\mathbf{R})\varphi^\alpha_N(\mathbf{R})$. Typical energies involved in eq.(18) are (-2,-8) MeV in D, (-8,-40) MeV in heavy nuclei, and (-20,-40) MeV in $^4$He. These energies cover the unphysical subthreshold region and a model is required for such an extrapolation. Here, the amplitudes averaged over spin states are calculated in terms of Paris potential [24]. This potential has been fitted to the 3400 $\bar{p}p$ and $\bar{n}p$ scattering data and the recent $\bar{n}p$ scattering measurements ref. [25] are essential for our purpose.

There are four basic antiproton amplitudes of interest in eq. (7). Figure 2 gives $R_{n/p} = \mathrm{Im}\, a_n/\mathrm{Im}\, a_p$ calculated for the S wave and compared to the data. Similar calculations for P waves can be compared only to the atomic data as shown in fig. 3. The following procedure was adopted to obtain these rsults.

(1) The stopped $\bar{p}$ experiments involve essentially the S wave capture. For He targets the initial experimental momenta of 10 MeV/c up to 50 MeV/c [29] indicate pure S wave capture. On the other hand, the deuteron data [28] may contain some higher momenta and a P wave component. In the figure, the experimental ratio is compared to the ratio calculated at the threshold. The calculated $R_{n/p}$ compare well with the chamber data indicated in the figure. For P waves the calculated $R_{n/p} \approx 1$.

**TABLE 2.** The experimental antiproton capture ratios $R_{n/p} = \sigma(\bar{p}n)/\sigma(\bar{p}p)$ extracted in atomic states and in states of flight. In C and Ni nuclei, the relation $\rho_p/Z = \rho_n/N$ was assumed.

| Element | $R_{np}$ | method | state | reference | remarks |
|---------|----------|--------|-------|-----------|---------|
| D | 0.81(3) | chamber | stopped | [28] | |
| $^3$He | 0.47(4) | chamber | stopped | [29] | |
| $^4$He | 0.48(3) | chamber | stopped | [29] | |
| $^{12}$C | 0.63 | $\pi^+, \pi^-$ | atoms | [10] | $\rho_p = \rho_n$ |
| $^{58}$Ni | 0.8 | cold A-1 | atoms | [22] | $\rho_p = \rho_n$ |
| Z=8-90 | $\approx 1$ | X-rays | atoms | [20] | |
| Z=8-90 | $\leq 1$ | X-rays | atoms | [21] | |
| Z=50-90 | $\approx 1$ | X ,cold A-1 | atoms | [19] | |

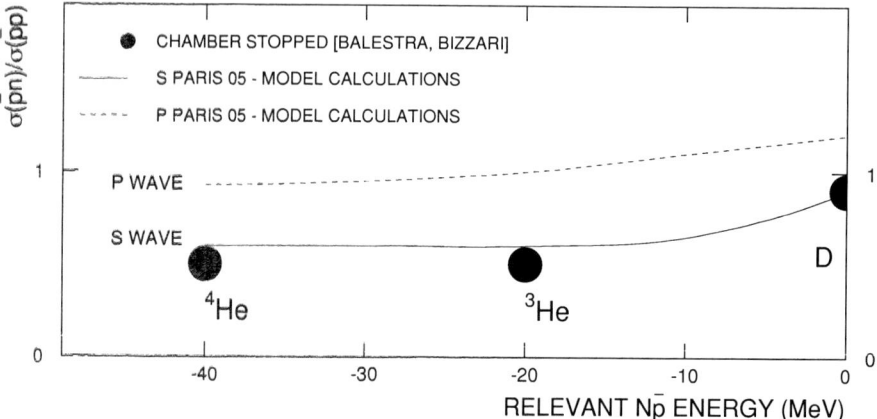

**FIGURE 2.** The $R_{n/p}$ ratio obtained in low energy in-flight capture experiments, [29, 28]. The horizontal scale indicates the energies involved in the $\bar{p}N$ CM systems, relevant to these experiments. The two curves correspond to the Paris potential calculation of the ratio in S and P wave $\bar{p}N$ interactions.

(2) In the case of atoms the $R_{n/p}$ involves mixtures of the S and P vave interactions. Thus, it depends on the state and cannot be presented as a single curve. A good quidance is given in terms of the optical potential for antiprotons $V^{opt}$ of eq. (8). The strengths of $\bar{p}n$ and $\bar{p}p$ capture rates are determined by average values of $V^{opt}$ in atomic states characterized by the principal quantum number $n$ and angular momentum $l$. The ratio becomes

$$R_{n/p} = < n,l \mid ImV_n \mid n,l > / < n,l \mid ImV_p \mid n,l > . \qquad (19)$$

Roughly one has

$$< n,l \mid V_s \mid n,l > \sim < \rho^s(R)[a_s + 3b_s \frac{l(2l+1)}{R^2}] >_{n,l} \qquad (20)$$

where $R$ is the radial coordinate. The contribution from the P wave term $b$ dominates and its role increases with the increasing atomic number. That happens for the orbitals tested

208

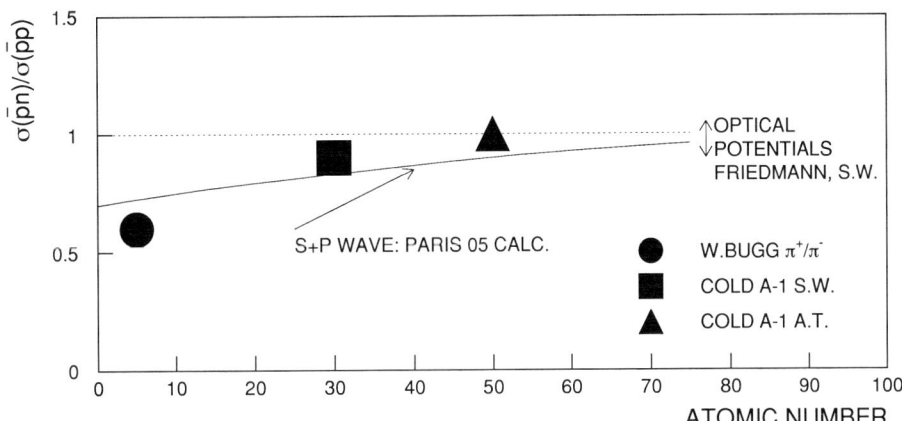

**FIGURE 3.** The $R_{n/p}$ ratio obtained in atomic states. The continous line gives the expectation ( crude with respect to the capture orbitals ) based on the Paris potential calculations including the S and P wave $\bar{p}N$ interactions. The dotted line indicates typical optical potential results obtained with the assumption of constant effective parameters.

in the X-ray cascade since the average ratio of the antiproton angular momentum to the annihilation distance $< l^2/R^2 >$ increases slowly with the nuclear size. With the values of $a, b$ obtained in the Paris model one obtains $R_{n/p} \approx 2/3$ in Carbon, and this number increases slowly to $R_{n/p} \approx 1$ in heavy nuclei. The trend is indicated in fig. 3 but the real numbers must be calculated specifically for each state of interest. The trend compares well with the results given in table 1.

(3) An optical potential [21] indicates a fair consistency with the $\bar{N}N$ model calculations. In particular the best fit parameters extracted from the X-ray data Im $a = -0.85$ fm, Im $b = -0.80$ fm$^3$ are fairly close to the Paris model averages calculated at a characteristic subthreshold energy of - 20 MeV. These are not final results, both the optical potentials and the $\bar{N}N$ potential models are being improved. In particular one needs to understand the important deuteron and helium atoms as well as the real parts of $a$ and $b$.

One has reached an almost quantitative agreement of the elementary $\bar{N}N$ model based upon the $\bar{N}N$ data and the nuclear $\bar{p}$ capture experiments. It gives a good prospect for a reliable calculations of $R_{n/p}$.

# THE ANALYSIS OF $\pi^+$ AND $\pi^-$ MULTIPLICITIES

This section is devoted to the analysis of the RIKEN proposal [13] and its potential advantages. The analyses of the mesonic measurements, aimed to extract the halo factor, consist of two steps [10, 11]. First, one finds the pion emission probability $P_\pi$ given by the total number of the emitted pions $N(\pi^-) + N(\pi^+)$ and the total capture events number $N_{events}$. These define the charged pion multiplicity $M(\pi^{\pm})$

$$M(\pi^{\pm}) = \frac{N(\pi^-) + N(\pi^+)}{N_{events}} = M^{prim}(\pi^{\pm})P_{\pi} \tag{21}$$

in terms of the pion emission probability $P_{\pi}$ and a primary multiplicity $M^{prim}(\pi^{\pm})$. The latter is taken from the deuteron. The second step is to determine $N(\bar{p}n)$, the number of events due to $\bar{p}n$ captures

$$N(\bar{p}n) = \frac{N(\pi^-) - N(\pi^+)}{P_{\pi}}. \tag{22}$$

Next, one obtains $N(\bar{p}p) = N_{events} - N(\bar{p}n)$ and the halo factor follows from eq. (17) $[N \sim \Gamma]$. In those analyses it has been assumed that $N(\pi^-) \pm N(\pi^+)$ are not much affected by the absorption of several mesons and by the charge exchange $\pi^0 \to \pi^+$, $\pi^0 \to \pi^-$ in the final states. These corrections will be studied later, let us first look into the consequences of Eqs.(21,22).

The early experiments yield $M(\pi^{\pm})$ in 5 elements C, N, Ti, Ta, Pb [10, 11]. Later, more extensive measurements found $M(\pi)$ for 15 elements, albeit with no $\pi^+$ and $\pi^-$ distinction [12]. These measurements and the related emission probabilities allow to locate the $\bar{p}$ annihilation site. Calculations of ref. [23] by the way of eq. (16) prove a simple, but quite precise, geometric interpretation of $P_{\pi}$ in terms of a solid angle $\Omega$ under which the nucleus is seen from the annihilation point. A simple geometry yields

$$P_{\pi} = 1 - \Omega/4\pi, \quad \frac{\Omega}{2\pi} = 1 - \sqrt{1 - (c/(c+\delta))^2} \tag{23}$$

where $\delta$ is the radial distance from the nuclear half density radius $c$ to the average capture radius $c + \delta$. Two factors generate such a simple picture. First the $\pi$ absorption is very strong and second it involves two (or more nucleons). The related $\rho^2$ absorption profile narrows the surface region and the pions see the nucleus as a black sphere. The best fit to the available $P_{\pi}$ and the nuclear pion absorption cross sections [30] produces $\delta = 1.35(0.1)$ fm [23]. The real absorption region has some 2 fm radial depth centered at $c + \delta$ as given by eq. (9) and indicated in figure 1.

However, such analyses have not used the full power of the experiment. In particular the bubble chamber measurements produce 7-8 multiplicities for the total mesonic charges. Two examples are given in table 3. To understand these charge multiplicities a simple phenomenological calculation is now performed. Let the primary meson multiplicities be $W[n,k]$ where $n$ is the total number of mesons and $k$ is the number of charged $\pi^+, \pi^-$ pairs. The multiplicities $W$ are well known from $\bar{p}$ stopped in hydrogen and deuterium followed by the charged pion emissions [26]. The inclusion of $\pi^0$ may be done fairly precisely with the inclusion of single $\pi^0$ production data, total meson production data and the statistical model [27].

Consider the process initiated by a primary $\bar{p}p$ annihilation in a nucleus. The observed final multiplicities differ from the primary multiplicities because of two effects: meson absorptions and meson charge exchanges. Let us discuss the absorption first. and denote

210

**TABLE 3.** The experimental and calculated in a phenomenological way pion charge multiplicities following the $\bar{p}$ absorption in C and Pb nuclei. The unit is %. The C nucleus: the data of ref. [11] are used with the errors calculated in ref. [26]. The other data of Bugg [10] were corrected for the hydrogen contamination by a simple subtraction of the hydrogen events in the charge 0 sector. No errors are attributed to these results. The parameters used in this calculation $R_{n/p} = 0.66, P_\pi = 0.13$ $\omega = 0.2, \lambda = 0.1, f^{halo} = 1$ should be compared to $R_{n/p} = 0.63(6)$, $P_\pi = 0.13(1), f^{halo} = 1$ from ref. [11]. The Pb nucleus: data from ref. [10], with no available experimental errors. The parameters used in this calculation $R_{n/p} = 0.96$, $P_\pi = 0.235$, $f^{halo} = 1.8$. The charge symmetric version c.s with $\omega = 0.29, \lambda = 0.12$ misses the charge $(+1)$ and $(-2)$ sectors. The charge non symmetric results n.c.s. with $\omega_+ = 0.26, \omega_- = 0.32, \lambda = 0.12$ are given in the last column. These numbers should be compared to $R_{n/p} = 0.63(6)$, $P_\pi = 0.221(0.014), f^{halo} = 2.34(50)$ from ref. [10].

| charge | exp. | calc | exp | exp | calc c.s. | calc n.c.s. |
|---|---|---|---|---|---|---|
| - | C [11] | C | C [10] | Pb [10] | Pb | Pb |
| + 2 | 1.8(2) | 1.3 | 2.1 | 2.6() | 1.3 | 1.7 |
| + 1 | 12.5(4) | 13.0 | 17.5 | 15.1() | 11.6 | 14.3 |
| + 0 | 43.0(8) | 43.1 | 38.3 | 30.6() | 30.5 | 32.3 |
| - 1 | 34.5(7) | 34.1 | 33.7 | 37.7() | 39.2 | 37.0 |
| - 2 | 6.5(3) | 8.0 | 7.8 | 12.3() | 16.2 | 13.8 |
| - 3 | 1.0(1) | 0.4 | 0.6 | 1.7() | 1.0 | 0.8 |

the probabilities for a charged meson absorption by $\omega_+$ and $\omega_-$. The probability that all charged mesons are emitted and the total charge 0 is observed equals to

$$P[0] = \Sigma_{n,k} W[n,k](1 - \omega_-)^k (1 - \omega_+)^k, \tag{24}$$

the probability that one $\pi^-$ is absorbed giving total meson charge 1

$$P[1] = \Sigma_{n,k} W[n,k](1 - \omega_-)^{k-1} \omega_- (1 - \omega_+)^k, \tag{25}$$

in a similar way a double $\pi^-$ absorption gives

$$P[2] = \Sigma_{n,k} W[n,k](1 - \omega_-)^{k-2} \omega_-^2 (1 - \omega_+)^k \tag{26}$$

and so on. Similar analysis may be done for the primary $\bar{p}n$ annhilations but it turns out that the final multiplicities are not well reproduced. The charge exchange processes $\pi^0 \to \pi^+$ and $\pi^0 \to \pi^-$ must be taken into account. Let the probability for $\pi^0 \to \pi^\pm$ be $\lambda_\pm$. With these processes the primary $\bar{p}p$ annihilation contributes to the final charge 1 state

$$\Delta P[1] = \Sigma_{n,k} W[n,k] P[n,k](1 - \omega_-)^k \lambda_+ (1 - \omega_+)^k \tag{27}$$

where $P[n,k]$ is the fraction of $\pi^0$ in the $n,k$ states [27]. The expression given above assumes also that all the charged pions are identified. With our phenomenological approach the inverse processes $\pi^+ \to \pi^0$ constitutes some fraction of the $\pi^+$ absorption

211

probability already included in the parameter $\omega_+$. This procedure is continued and parameter $\lambda$ is extracted from the data. The results are shown in table 3. For C nucleus a charge symmetric solution with two parameters gives good fit to the data of ref. [11]. In this fit the overall absorption parameter $P_\pi$ was fixed to the experimental result given by eq.(21). The charge exchange parameter $\lambda$ turns out sizable. It reflects the effect of $\Delta(1233)$ resonance which produces charge exchange cross section of $30mb$ strength. The outcome in terms of $R_{n/p}$ is essentially the same as in the standard approach. This happens since the two uncertain processes $\pi^\pm \to \pi^0$ and $\pi^0 \to \pi^\pm$ balance each other. The essential fact is that the measurements of the pion multiplicities allow a direct experimental control over both the absorption and the charge exchange processes. Let us also comment that the results from hydrogen chamber [10]which require uncertain background subtraction, offer much worse comparison with this phenomenological calculation. This underlines the need to remove the $\bar{p}p$ background.

Similar analysis made for Pb nucleus indicates that in the neutron excess situation the absorption of final $\pi^-$ is larger than that of $\pi^+$. As seen from table 3 the effect is small but noticeable. It is, yet again, related to the neutron halo but it is not a trivial one. The explanation may be given in terms of the $\pi NN \to \Delta N \to NN$ absorption model. The dominant mode involves $np$ pair with the deuteron quantum numbers. Thus the pion absorptive cross section is not much affected by the neutron excess [30]. However, the geometry of the surface capture of antiprotons makes the difference. Thus, the $\pi^-$ meson born at distant surface is much more likely to collide with a neutron than with the proton. With the $\pi^- n$ collision the $\Delta$ resonance is more likely to be formed. Next, the $\Delta$ propagates by about 1 fm and interacts with a more central proton by a long range pion exchange force. This sizable range in the absorption mechanism, indicated in ref. [31], generates the asymmetry in the $\pi^-$ and $\pi^+$ behavior. An estimate of the propagation ranges and the geometry involved makes the ratio $\omega_-/\omega_+$ to rise up to about 4/3. Similar observations are also made in other (Ti,Ta) targets studied in the W.Bugg's experiment.

Finally let us notice that such an extended analysis does not change the $f^{halo}$ results given by Bugg, within the error limit given by the standard method of eqs. (21) and (22). Some reduction obtained in the Pb case is due to the updated value of the $R_{n/p}$.

# CONCLUSIONS

There are several methods which use antiprotons to study the neutron distribution in nuclei. These methods are complementary as they provide information on different nuclear regions. The pion emission experiments, which are of our special interest, may be a very useful tool to test the densities at nuclear radii about 1 fm beyond the nuclear half density radius. In particular:

• The pion emission measurements may provide not only the "halo factors" but also probabilities of the final pion absorption and charge exchange.

• The parallel theoretical calculations for the pion absorption and charge exchange probabilities will pinpoint the properties the initial $\bar{p}$ capture state.

• The essential primary rates of captures on neutrons and protons and their ratio $R_{n/p}$ are now fairly well understood as an interplay of the S and P wave absorptions. The S wave is tested in the low energy absorption in flight, the interplay of S and P is tested in the antiprotonic X-rays.

## ACKNOWLEDGMENTS

This work has been initiated by Yasunori Yamazaki and Michiharu Wada whose kind help and collaboration is acknowledged.

## REFERENCES

1. D.H. Wilkinson, Phil. Mag. **4**, 215 (1959). P.B. Jones, Phil. Mag. **3**, 33 (1958).
2. C.E. Wiegand and G.L.Godfrey, Phys. Rev. **A9**, 2284 (1974). R. Seki and C.E. Wiegand, Ann. Rev. Nucl. Sc. **25**, 241 (1975).
3. P. Roberson et. al, Phys. Rev. **C5**, 1954 (1977).
4. Th. Köhler et. al, Phys. Lett. **B176**, 327 (1986).
5. D. Rohmann et. al, Z. Phys. **A325**, 261 (1986).
6. A Trzcinska Nucl. Phys. **A692**, 176c(2001).
7. C.Batty, E.Friedman and A. Gal, Phys.Rep. 287(1997)385.
8. D.H. Davis et.al, Nucl. Phys. **B1**, 434 (1967).
9. E.H.S. Burhop, Nucl. Phys. **B1**, 438 (1967); Nucl. Phys. **B44**, 445 (1972).
10. W.M. Bugg et.al, Phys. Rev. Lett. **31**, 475 (1973).
11. M. Wade and V.G.Lind, Phys. Rev. **D14**, 1182 (1976).
12. D.Polster et.al, Phys. Rev. **C51**, 1167 (1995).
13. M. Wada and Y. Yamazaki, Nucl. Instr. Meth. **B**214(2004)196.
14. J. Jastrzębski Nucl. Phys. **A558**, 405c (1993).
15. P. Lubiński et. al, Phys. Rev. Lett. **73**, 3199 (1994).
16. C.J. Batty, Phys. Lett. **B189**, 393 (1987).
17. P. Kienle, Nucl. Instr. Meth. **B**214(2004)191.
18. E. Friedman and J. Lichtenstadt, Nucl. Phys. **A455**, 573 (1986).
19. A. Trzcinska et. al, Nucl. Inst. Meth.**B**214(2004)157.
20. E.Friedman and A.Gal, Nucl. Inst. Meth. **B**214(2004)160.
21. S.Wycech, Nucl. Inst. Meth. B Phys. **B**214(2004)164.
22. S.Wycech et.al, Phys. Rev. **C54**,1832(1996).
23. J.Cugnon et al, Phys.Rev. **C**63(2001) 027301
24. M.Lacombe et. al, to be published, B.El-Bennich et. al, Phys.Rev.**C**59(1999)2319.
25. F.Iazzi et. al, Phys. Lett. **B**475(2000) 378 .
26. J.Riedlberger et. al, Phys. ReV. **C**40(1989)2717.
27. P.D.Zemany, J.M.Muntz and Z.Ming Ma Phys. ReV. **D**23(1981)1473.
28. R. Bizzari et. al, Nuovo Cim. **22A**, 225 (1974); T.E. Kalogeropoulos, Phys. Rev. **D22**, 2585 (1980).
29. F.Balestra et.al, Nucl. Phys. **A491**, 541 (1989).
30. D. Ashery, Nucl. Phys. **A355**, 385 (1980); Phys. Rev. **C23**, 2173 (1981).
31. H.Kim and H.Toki, Progr.Th. Phys. (Lett) 72,1050(1984).

# Antiprotonic atoms - a tool for the investigation of the nuclear periphery

A. Trzcińska*, J. Jastrzębski*, P. Lubiński*, B. Kłos†, F.J. Hartmann**, T. von Egidy** and S. Wycech‡

*Heavy Ion Laboratory, Warsaw University, PL-02-093, Warsaw, Poland
†Physics Department, Silesian University, PL-40-007, Katowice, Poland
**Physics Department, Technische Universität München, D-85748, Germany
‡Sołtan Institute for Nuclear Studies,
PL-00-681 Warsaw, Poland

**Abstract.** Antiprotonic X rays were used to investigate the nuclear matter densities. Neutron densities in 26 isotopes were determined using this method. The information on the nuclear matter density at relatively large radii was then converted to rms radii by the use of a two-parameter Fermi-function profile. The obtained systematics of differences of the neutron and proton rms radii is in a fair agreement with theoretical calculations and results of other experimental methods.

**PACS:** 21.10.Gv, 13.75.Cs, 27.60.+j, 36.10.–k

## INTRODUCTION

The antiprotonic atoms are the good probe to study the nuclear periphery. Antiprotons captured into atomic orbits cascade down emitting X rays. In the vicinity of a nucleus the strong interaction reveals its presence: antiprotonic levels are shifted if compared to the pure electromagnetic energy and become broadened. The width and shift of the levels give an information on the nuclear matter density at the nuclear periphery.

The antiprotonic cascade ends with the annihilation on a peripheral nucleon: a neutron or a proton. If all produced pions miss the target nucleus two kinds of residues can be created with one neutron or one proton removed from the target nucleus. The yields of these products are proportional to the neutron and the proton densities at the nuclear surface.

## THE EXPERIMENT

The experiment was performed at CERN (LEAR facility) within the PS209 collaboration in two runs: 4 weeks in 1995 and 3 weeks in 1996.

Two experimental methods based on the phenomena described above were used to study distribution of nuclear matter at relatively large radii in nuclei.

The first method is based on the measurements of antiprotonic X-rays. The line shapes, intensities and energies allow to determine the level widths and shifts induced by the strong interaction. At present, 44 level shifts and 62 level widths for 34 isotopes (see Fig. 1) have been evaluated [1, 2, 3].

CP793 *Physics with Ultra Slow Antiproton Beams*
edited by Yasunori Yamazaki and Michiharu Wada
© 2005 American Institute of Physics 0-7354-0282-5/05/$22.50

In the second method the yield of the radioactive antiproton annihilation residues one mass unit lighter that the target was measured. The yield of products with one neutron less than the target neutron number $N_t$ (annihilation on a neutron) and those with one proton less than the target proton number $Z_t$ (annihilation on a proton) were determined [5, 6, 7]. The halo factor was calculated using these yields [5]. This factor gives the normalized neutron to proton density ratio at a distance about 2.5 fm larger than half density charge radius [8].

## RESULTS AND DATA INTERPRETATION

The obtained in this experiment halo factors were compared with the results of other experiments examining a difference in neutron and proton density distributions, i.e. the difference of the neutron and proton rms radii $\Delta r_{np}$. This comparison shows that $\Delta r_{np}$ result mainly from differences of the surface thickness in neutron and proton densities and not from differences of the half density radii [3]. In the data analysis density distributions were assumed in the form of two-parameter-Fermi distribution for both, the protons and neutrons. The proton densities were adapted from the compilations [9, 10].

The simple optical potential of the antiproton-nucleus interaction of the form: $V_{opt} \sim \bar{a} \cdot (\rho_p + \rho_n)$ ($\bar{a}$ – antiproton effective scattering length, $\rho_p$ and $\rho_n$ proton and neutron densities) was used in the X-ray data analysis. The potential was taken from [11]: $\bar{a} = 2.5 + i \cdot 3.4$. The details of this analysis and the way of neutron density extraction from the level widths and shifts was described in several articles [3, 12, 13, 14].

Figure 2 shows 3 examples of the neutron to proton density ratios. Neutron densities were deduced from the X-ray data analysis. In the same figure the theoretical calculations with the Hartree-Fock-Bogoliubov (HFB) method using SkP Skyrme force are presented jointly with the halo factors. The good agreement of both experimental results and theoretical predictions can be observed. This is the case in the most of examined isotopes except of [106]Cd, [112]Sn and [144]Sm [13]: the isotopes with weakly bound proton. The effect of quasi-bound $\bar{p}p$ states was proposed [16] as an explanation of this discrepancy.

Using neutron densities deduced from the X-ray data and proton densities from the compilation [9, 10], the difference of neutron and proton root mean square radii $\Delta r_{np}$ was deduced for the studied isotopes under the assumption of 2pF proton and neutron density distributions. The calculated $\Delta r_{np}$ values are presented in the Table 1 and in Fig. 3 (as a function of the asymmetry parameter $\delta = (N - Z)/A$). The linear relationship $\Delta r_{np}$ as a function of $\delta$ was fitted. The function $\Delta r_{np}(\delta)$ was finally parametrized using following coefficients: $\Delta r_{np} = (-0.03 \pm 0.02) + (0.90 \pm 0.15) \cdot \delta$ fm.

**TABLE 1.** Neutron and proton root mean square radii differences $\Delta r_{np}$, obtained from antiprotonic X-ray data.

| Isotope | $\Delta r_{np}$ [fm] | Isotope | $\Delta r_{np}$ [fm] | Isotope | $\Delta r_{np}$ [fm] |
|---------|------------------|---------|------------------|---------|------------------|
| $^{40}$Ca | $-0.08^{+0.05}_{-1.0}$ | $^{59}$Co | $0.00^{+0.08}_{-0.13}$ | $^{122}$Te | $0.08^{+0.04}_{-0.05}$ |
| $^{48}$Ca | $0.09^{+0.05}_{-0.05}$ | $^{90}$Zr | $0.09^{+0.02}_{-0.02}$ | $^{124}$Te | $0.06^{+0.04}_{-0.04}$ |
| $^{54}$Fe | $0.04^{+0.06}_{-0.08}$ | $^{96}$Zr | $0.12^{+0.03}_{-0.03}$ | $^{126}$Te | $0.11^{+0.03}_{-0.05}$ |
| $^{56}$Fe | $0.03^{+0.08}_{-0.11}$ | $^{106}$Cd | $0.10^{+0.10}_{-0.14}$ | $^{128}$Te | $0.11^{+0.04}_{-0.05}$ |
| $^{57}$Fe | $0.07^{+0.05}_{-0.05}$ | $^{116}$Cd | $0.15^{+0.04}_{-0.04}$ | $^{130}$Te | $0.15^{+0.06}_{-0.08}$ |
| $^{58}$Ni | $-0.09^{+0.09}_{-0.16}$ | $^{112}$Sn | $0.07^{+0.02}_{-0.02}$ | $^{208}$Pb | $0.15^{+0.02}_{-0.02}$ |
| $^{60}$Ni | $-0.01^{+0.08}_{-0.15}$ | $^{116}$Sn | $0.10^{+0.03}_{-0.03}$ | $^{209}$Bi | $0.18^{+0.04}_{-0.06}$ |
| $^{64}$Ni | $0.04^{+0.07}_{-0.08}$ | $^{120}$Sn | $0.08^{+0.03}_{-0.04}$ | $^{232}$Th | $0.21^{+0.07}_{-0.07}$ |
| | | $^{124}$Sn | $0.14^{+0.03}_{-0.03}$ | $^{238}$U | $0.21^{+0.07}_{-0.07}$ |

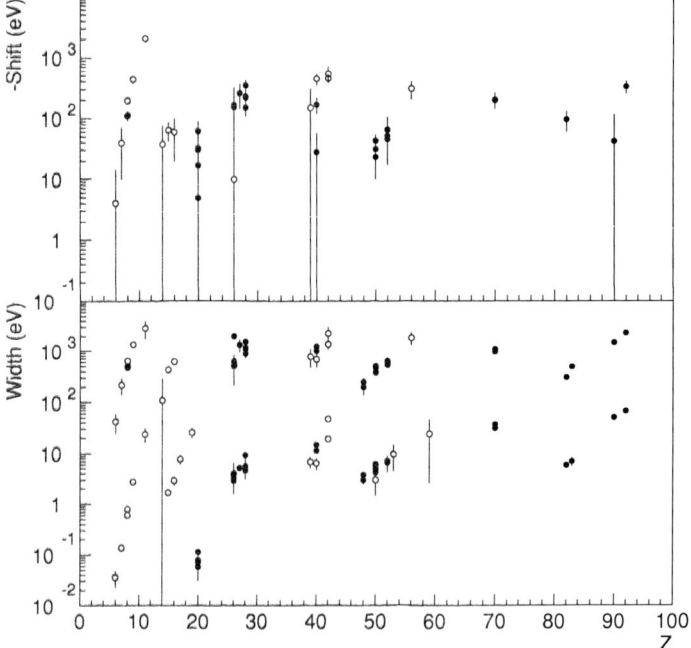

**FIGURE 1.** The antiprotonic strong interaction level widths and shifts plotted as a function of the atomic number Z. Full points – the values determined in the PS209 experiment; open points – previous data [4].

216

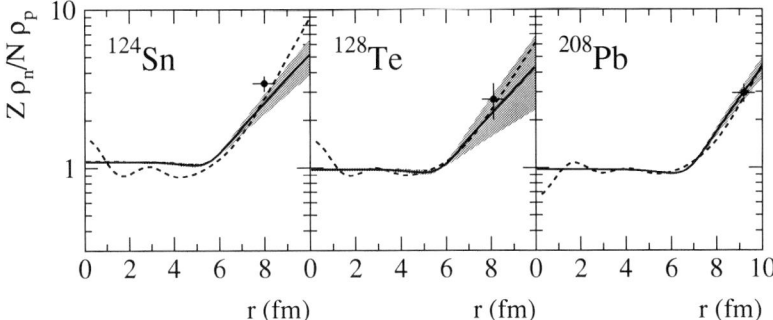

**FIGURE 2.** The normalized neutron to proton density ratio $(Z \cdot \rho_n/N \cdot \rho_p)$ deduced from the level widths and shifts (solid lines with errors indicated by grey bands) and the charge distributions given in Ref. [10] (Sn) and Ref. [9] (Te and Pb). They are compared to halo factors measured in the radiochemical experiments (full points) and with the HFB model calculations (with SkP Skyrme force) [15] (dashed lines). The halo factors are marked at radial distances corresponding to the most probable annihilation sites.

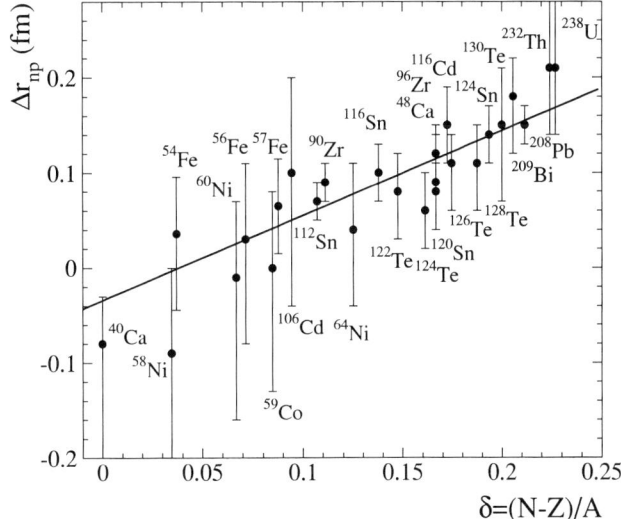

**FIGURE 3.** Difference $\Delta r_{np}$ between the rms radii of the neutron and proton distributions deduced from the antiprotonic atom X-ray data as a function of $\delta = (N-Z)/A$. The proton distributions were obtained from electron scattering data [10] (Sn nuclei) or from muonic atom data [9, 17, 18] (other nuclei). The full line represents the linear relationship between $\delta$ and $\Delta r_{np}$ as obtained from a fit to the experimental data.

# COMPARISON WITH OTHER EXPERIMENTS AND THEORETICAL CALCULATIONS

The $^{208}$Pb nucleus is one of the most intensely studied isotopes by many experiments as well as by a theory and because of this is a good case for comparisons. Figure 4 shows compilation of $\Delta r_{np}$ in $^{208}$Pb determined in hadron scattering experiments and the average value is $\Delta r_{np} = 0.17 \pm 0.02$ fm. The results are compared with the value $\Delta r_{np} = 0.15 \pm 0.02$ fm obtained from antiprotonic X-rays. The theoretical predictions for $\Delta r_{np}$ are also shown. All presented results agree very well.

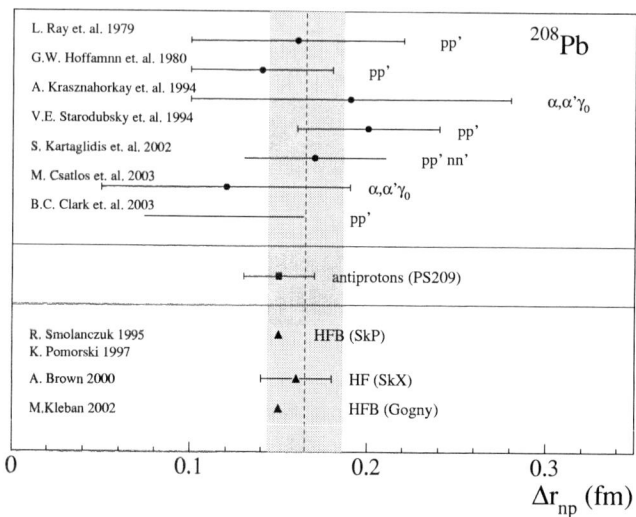

**FIGURE 4.** The difference $\Delta r_{np}$ between the rms radii of the neutron and proton distributions in $^{208}$Pb determined in hadron scattering experiments [19, 20, 21, 22, 23] compared with the antiprotonic atoms result and several mean field calculations [15, 24, 25].

The results of these comparisons can be used as an additional confirmation of the assumption presented in the previous section that the neutron and proton distributions in a given isotope differ mainly by the surface thickness and not by the half density radius. Figure 5 shows the results of $\Delta r_{np}$ obtained from the analysis of X-ray data in $^{208}$Pb with the assumption that $c_n \neq 0$. One can see that the average $\Delta r_{np}$ from various experiments allows $c_n \leq 0.1$ fm only.

In Fig. 4 a good agreement of $\Delta r_{np}$ for $^{208}$Pb obtained in the hadron scattering and in the $\bar{p}$ experiments is evident. However, the situation is not so good if one compares $\Delta r_{np}$ obtained for other isotopes [27] – the old data from different experiments are much more scattered. The scatter of the theoretical values is significantly smaller. Examples of HF and HFB calculations are shown in Fig. 6, while the predictions of a Droplet Model were presented in Ref. [28].

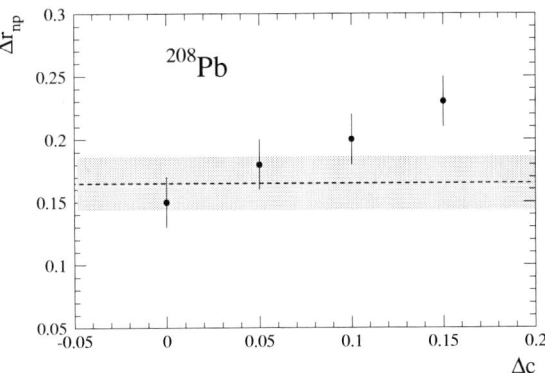

**FIGURE 5.** The difference of neutron and proton rms radii in $^{208}$Pb obtained in the X-ray data analysis as a function of $\Delta c_{np}$. The dashed line with a grey band describes an average $\Delta r_{np}$ value for $^{208}$Pb obtained from hadron scattering experiments – see Fig. 4.

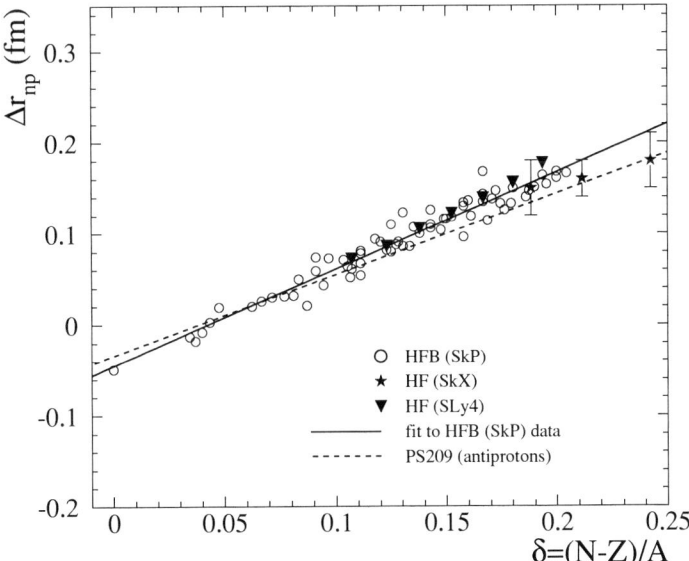

**FIGURE 6.** The difference $\Delta r_{np}$ between the rms radii of the neutron and proton distributions as a function of $\delta = (N-Z)/A$ calculated by mean field models. Open points Ref. [15] (HFB model with SkP force); full triangles - Ref. [29] (HF model, SLy4 force); asterisks - Ref. [24] (HF model, SkX force). The full line is the linear relationship between $\delta$ and $\Delta r_{np}$ as obtained from the fit to the HFB model calculations (open points). The dashed line is the fit to the antiprotonic atom data as presented in Fig. 3.

# SUMMARY

A rich set of data concerning antiprotonic atoms was gathered in the PS209 experiment. The data were analyzed using a simple optical potential and density distributions in a form of 2pF function. The neutron density distributions and neutron-proton rms radii differences were determined. Despite the model dependence of data treatment a good agreement with various theories (HF and HFB calculations, Droplet Model predictions) was obtained for density distributions and rms radii differences. Also a fair agreement with other experimental results of $\Delta r_{np}$ was obtained.

The presented experimental data make an interesting material for theory –the search of a new, more sophisticated optical potential [30, 31, 32].

The PS209 experiment of course didn't exhaust all the interesting topics in the medium and heavy antiprotonic atoms. For example: measurements of X rays for deformed even-A nuclei and for odd-A nuclei or the investigation of the properties of deeply bound $\bar{p}$ states via E2 resonance would be of great interest. The FLAIR (Facility for Low-Energy Antiproton and Ion Research) planned at Darmstadt will allow a continuation of the presented studies.

# REFERENCES

1.  A. Trzcińska, J. Jastrzębski, T. Czosnyka, T. von Egidy, K. Gulda, F.J. Hartmann, J. Iwanicki, B. Ketzer, M. Kisieliński, B. Kłos, W. Kurcewicz, P. Lubiński, P.J. Napiorkowski, L. Pieńkowski, R. Schmidt, E. Widmann, *Nucl. Phys.* **A692** (2001) 176c.
2.  A. Trzcińska, J. Jastrzębski, P. Lubiński, F.J. Hartmann, R. Schmidt, T. von Egidy, B. Kłos, *Acta Phys. Pol.* **32** (2001) 917.
3.  A. Trzcińska, J. Jastrzębski, P. Lubiński, F.J. Hartmann, R. Schmidt, T. von Egidy, B. Kłos, *Phys. Rev. Lett.* **87** (2001) 082501.
4.  Batty, C. J., Friedman, E., Gils, H. J., and Rebel, H., *Advances in Nuclear Physics*, J. W. Negele and E. Vogt, Plenum Press, New York, 1989.
5.  P. Lubiński, Jastrzębski, A. Grochulska, A. Stolarz, A. Trzcińska, W. Kurcewicz, F.J. Hartmann, W. Schmid, T. von Egidy, J. Skalski, R. Smolańczuk, S. Wycech, D. Hilscher, D. Polster, I. Rossner, *Phys. Rev. Lett.* **73** (1994) 3199.
6.  P. Lubiński, J. Jastrzębski, A. Trzcińska, W. Kurcewicz, F.J. Hartmann, W. Schmid, T. von Egidy, R. Smolańczuk, S. Wycech, *Phys. Rev.* **C57** (1998) 2962.
7.  R. Schmidt, F.J. Hartmann, B. Ketzer, T. von Egidy, T. Czosnyka, J. Jastrzębski, M. Kisieliński, P. Lubiński, P. Napiorkowski, L. Pieńkowski, A. Trzcińska, B. Kłos, R. Smolańczuk, S. Wycech, W. Poschl, K. Gulda, W. Kurcewicz, E. Widmann, *Phys. Rev.* **C60** (1999) 054309.
8.  S. Wycech, J. Skalski, R. Smolańczuk, J. Dobaczewski, J.R. Rook, *Phys. Rev.* **C54** (1996) 1832.
9.  G. Fricke, C. Bernhardt, K. Heilig, L.A. Schaller, L. Schellenberg, E.B. Shera, C.W. De Jager, *At. Data Nucl. Data Tables* **60** (1995) 177.
10. H. de Vries, C.W. De Jager, C. de Vries, *At. Data Nucl. Data Tables,* **36** (1987) 495.
11. C. J. Batty, E. Friedman, and A. Gal, *Nucl. Phys.* **A592** (1995) 487.
12. F.J. Hartmann, R. Schmidt, B. Ketzer, T. von Egidy, S. Wycech, R. Smolańczuk, T. Czosnyka, J. Jastrzębski, M. Kisieliński, P. Lubiński, P. Napiorkowski, L. Pieńkowski, A. Trzcińska, B. Kłos, K. Gulda, W. Kurcewicz, E. Widmann, *Phys. Rev.* **C65** (2001) 014306.
13. R. Schmidt, A. Trzcińska, T. Czosnyka, T. von Egidy, K. Gulda, F. J. Hartmann, J. Jastrzębski, B. Ketzer, M. Kisieliński, B. Kłos, W. Kurcewicz, P. Lubiński, P. Napiorkowski, L. Pieńkowski, R. Smolańczuk, E. Widmann, S. Wycech, *Phys. Rev.* **C67** (2003) 044308.
14. B. Kłos, S. Wycech, A. Trzcinska, J. Jastrzebski, T. Czosnyka, M. Kisielinski, P. Lubinski, P. Napiorkowski, L. Pienkowski, F.J. Hartmann, B. Ketzer, R. Schmidt, T. von Egidy, J. Cugnon, K. Gulda, W. Kurcewicz, E. Widmann, *Phys. Rev.* **C69** (2004) 044311.

15. R. Smolańczuk, private communication, 1995.
16. S. Wycech, *Nucl. Phys.* **A692** (2001) 29c.
17. J. D. Zumbro, E. B. Shera, Y. Tanaka, C. E. Bemis, Jr., R. A. Naumann, M. V. Hoehn, W. Reuter, R. M. Steffen, *Phys. Rev. Lett.* **53** (1984) 1888.
18. J. D. Zumbro, R.A. Naumann, M.V. Hoehn, W. Reuter, E.B. Shera, C.E. Bemis Jr., Y. Tanaka, *Phys. Lett.* **B167** (1986) 383.
19. L. Ray, *Phys. Rev.* **C19** (1979) 1855.
20. G.W. Hoffmann, L. Ray, M. Barlett, J. McGill, G. S. Adams, G. J. Igo, F. Irom, A. T. M. Wang, C. A. Whitten, Jr., R. L. Boudrie, J. F. Amann, C. Glashausser, N. M. Hintz, G. S. Kyle, G. S. Blanpied, *Phys. Rev.* C21 (19880) 1488.
21. A. Krasznahorkay, A. Balanda, J.A. Bordewijk, S. Brandenburg, M.N. Harakeh, N. K alantar-Nayestanaki, B.M. Nyako, J. Timar, A. Van der Woude, *N*ucl. Phys. **A567, 521** (1994).
22. V. E. Starodubsky, N.M. Hintz, *Phys. Rev.* **C49** (1994) 2118.
23. S. Karataglidis. K. Amos, B.A. Brown, P.K. Deb, *Phys. Rev.* C65 (2002) 044306.
24. A. Brown, *Phys. Rev. Lett.* 85 (2000) 5296.
25. M. Kleban, B. Nerlo-Pomorska, J.F. Berger, J. Decharge, M. Girod, S. Hilaire, *Phys. Rev.* **C65** (2002) 024309.
26. M. Csatlos, Z. Gacsi, J. Gulyas, A. Krasznahorkay, A. Krasznahorkay Jr., Z. Mate, J. Timar, M. Heil, F. Kaeppeler, R. Reifarth, M. Fayez-Hassan. *Nucl. Phys.* **A719** (2003) 304c.
27. J. Jastrzębski, A. Trzcińska, P. Lubiński, b. Kłos, F.J. Hartmann, T. von Egidy, S. Wycech, *Int. J. of Mod. Phys.* E 13 (2004) 343.
28. W. Swiątecki, A. Trzcińska, J. Jastrzębski, *Phys. Rev.* C 71 (2005) 047301.
29. F. Hofmann and H. Lenske, *Phys. Rev.* **C57** (1998) 2281.
30. S. Wycech, *Nucl. Instrum. & Methods*, **214** (2004) 164.
31. E. Friedman, A. Gal, *Nucl. Instrum. & Methods*, **214** (2004) 160.
32. E. Friedman, A. Gal, J. Mareš, nucl-th/0504030

# An Antiproton Ion Collider (AIC) for Measuring Neutron and Proton Distributions in Stable and Radioactive Nuclei

Paul Kienle*

*Technische Universität München (Germany) and Stefan Meyer Institut, Wien (Austria)*

**Abstract.** An antiproton-ion collider is proposed to independently determine mean square radii for protons and neutrons in stable and short lived nuclei by means of antiproton absorption at medium energies. The experiment makes use of the electron ion collider complex (ELISE) of the GSI FAIR project with appropriate modifications of the electron ring to store, cool and collide antiprotons of 30 MeV energy with 740A MeV energy ions

The total absorption cross-section of antiprotons by the stored ions will be measured by detecting their loss by means of the Schottky noise spectroscopy method. Cross sections for the absorption on protons and neutrons, respectively, will be studied by detection of residual nuclei with A-1 either by the Schottky method or by analysing them in recoil detectors after the first dipole stage of the NESR following the interaction zone. With a measurement of the A-1 fragment momentum distribution, one can test the momentum wave functions of the annihilated neutron and proton, respectively. Furthermore by changing the incident ion energy the tails of neutron and proton distribution can be measured.

The absorption cross section is at asymptotic energies in leading order proportional to the mean square radius of the nucleus. Predicted cross sections and luminosities show that the method is applicable to nuclei with production rates of about $10^5 s^{-1}$ or lower, depending on the lifetime of the ions in the NESR, and for half-lives down to 1 second.

**Keywords:** Antiproton Absorption, Antiproton Ion Collider, RMS neutron and proton radii of stable and radioactive nuclei.
**PACS:** 21.10Fr, 21.10Gv. 25.43.+t 25.60.-t 29.20Dh

* For the AIC Collaboration: :P. Beller, F. Bosch, M. Cargnenelli, L. Fabbietti, Th. Faestermann, B. Franzke, H.Fuhrmann, R.S. Hayano,, J.Homolka, Ch. Kozhuharov, R. Krücken, H. Lenske, Y. Litvinov, J. Marton, F.Nolden, P.Ring, Y.Shatunov, A.N. Skrinsky, K. Suzuki, V.A. Vostrikov, T. Yamaguchi, E. Widmann, S. Wycech, J. Zmeskal.

CP793 *Physics with Ultra Slow Antiproton Beams*
edited by Yasunori Yamazaki and Michiharu Wada
© 2005 American Institute of Physics 0-7354-0282-5/05/$22.50

# MOTIVATION

The very limited knowledge of the nuclear neutron compared with the proton distribution is an old unsolved problem of nuclear structure physics. Whereas the proton distribution is well known using electromagnetic probes [1], there is until now no equivalent probe for the neutrons. The total nucleon distribution has been studied with hadronic probes, but with rather limited accuracy because the interaction T-matrix is poorly known Only at this workshop, Terashima reported 295 MeV proton scattering data on Sn and Ca nuclei, analyzed with in medium modified coupling constants and masses of the exchanged mesons, calibrated with $^{58}$Ni, and using relativistic impulse approximation to deduce neutron density distributions [2] Root mean square radii differences for neutrons and proton indicate small errors, and an unexplained small value for $^{120}$Sn. Trzcinska presented at this workshop antiproton atom data, which are aimed at investigating the nuclear periphery which is dominated by the neutron distribution[3]. Finally Wada proposed a new method to extend the study of the neutron to proton density ratio in the nuclear periphery to short lived nuclei by measuring the charged pion multiplicity of radioactive antiproton atoms in a trap [4].

At LEAP02, we proposed to measure directly the difference of the mean square neutron and proton radii ($<r^2_n>-<r^2_p>$) in stable as well radioactive nuclei by observing medium energy antiproton absorption on neutrons and protons simultaneously in an antiproton-ion collider (AIC) [5]. This new method allows also measuring the momentum distribution of the neutron and proton before annihilation, thus giving direct evidence of their spatial distributions. In paragraph 2 we sketch the main features of medium energy antiproton absorption as worked out recently [6]. In paragraph 3 we present the lay out of the AIC as given in a technical proposal [7] to the FAIR project of GSI Darmstadt [8]. In paragraph 4 we describe the methods to measure total and partial antiproton absorption cross sections using Schottky noise spectroscopy and magnetic spectrometry which allow also observing the momentum distribution of the recoil nuclei. In the final paragraph the analysis of the data is given including main physics goals.

# Medium Energy Antiproton Absorption

An optical potential for the antiproton-nucleus interaction is derived using the *impulse* $t\rho$-folding approximation with the $t$-matrix determined from the known energy dependent antiproton-N cross section and the density $\rho_{n/p}$ for neutrons ($n$) and protons ($p$) derived from HFB calculations [6]. A comparison of early antiproton-deuteron and -proton cross section measurements showed that the antiproton-neutron interaction is within 10% equal to the antiproton-proton interaction in the medium energy range [9]

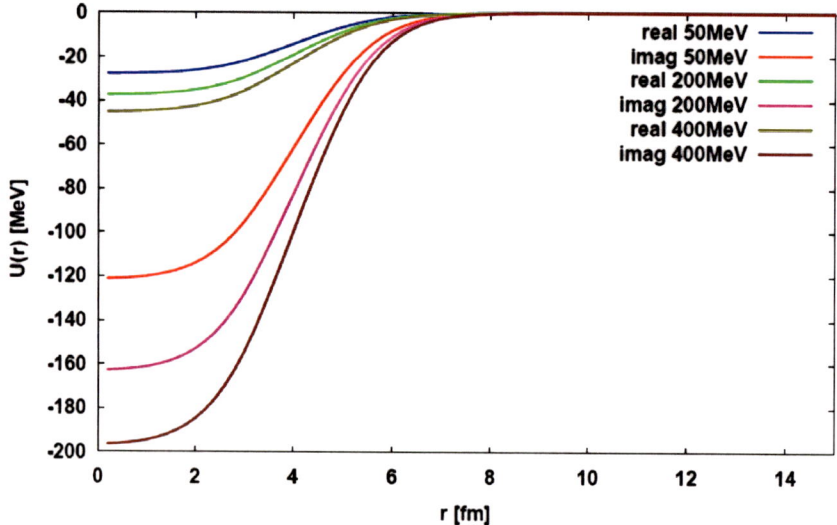

**FIGURE 1** Energy dependent optical potential U(r) of antiprotons in $^{58}$Ni at 50, 200, 400 MeV energy as indicated.

Figure 1 show optical potentials for $^{58}$Ni and antiprotons of 50, 200, and 400 MeV separated in real and imaginary parts. Notice that the imaginary part is about 4 times deeper than the real part, thus absorption of the antiprotons is dominating the reaction. Using these energy dependent optical potentials, which also reflect the nucleon distribution, the absorption cross sections can be calculated for various energies and nuclei. In the following the main features of antiproton absorption is illustrated with the Ni isotopic chain. Figure 2 shows the impact parameter dependence of the antiproton absorption of $^{78}$Ni at 50, 100, 200, 300 and 400 MeV. The cross section rises linearly with the impact parameter and drops reaching the edge of the nuclei. With increasing bombarding energy it reaches an asymptotic impact parameter dependence which reflects the radial extension of the absorbing nucleus, with an effective area corresponding to its mean square radius ($<r^2>$). At lower bombarding energies the outer parts of the diffuse nuclear surface contributes more and more. This suggests that a measurement of the energy dependence of the absorption is suited to determine both the radius parameter c and the diffuseness a of an assumed Fermi distribution. Figure 4 shows for energy of 400 MeV the calculated total antiproton absorption cross sections for Ni isotopes. It is shown that the total absorption cross section, $\sigma_{abs} = \alpha <r^2>$, with $\alpha$ denoting an energy dependent constant, normalized with the mean square radius $<r^2>$ of $^{58}$Ni in our presentation. Note that the proportionality of the total absorption cross section to $<r^2>$ holds extremely well along the whole Ni isotopic chain. Figure 4 also shows the calculated partial neutron and proton absorption cross sections which scale with $\alpha(N/A) <r^2_n>$ and $\alpha(Z/A) <r^2_p>$ respectively (not explicitly shown). So in the asymptotic limit, which is reached at 400 MeV bombarding energy, measurements of the total and partial antiproton absorption cross section open a direct way to determine the respective <r²> values. Details concerning calibration of α and determination of loss factors due to pion absorption in the residual nuclei will be discussed in paragraph 5.

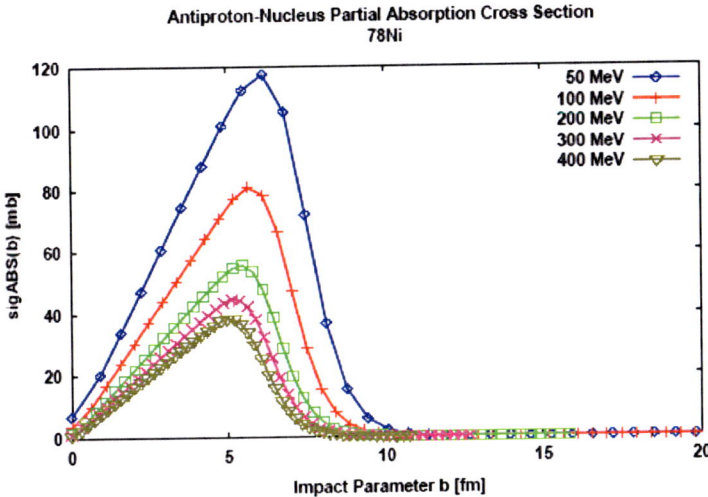

**FIGURE 2** Antiproton absorption cross sections as function of the impact parameter b for $^{78}$Ni at 50, 100, 200, 300, and 400 MeV energy. Note the asymptotic property at higher energies and the contribution from the tail of the nucleus at lower energies.

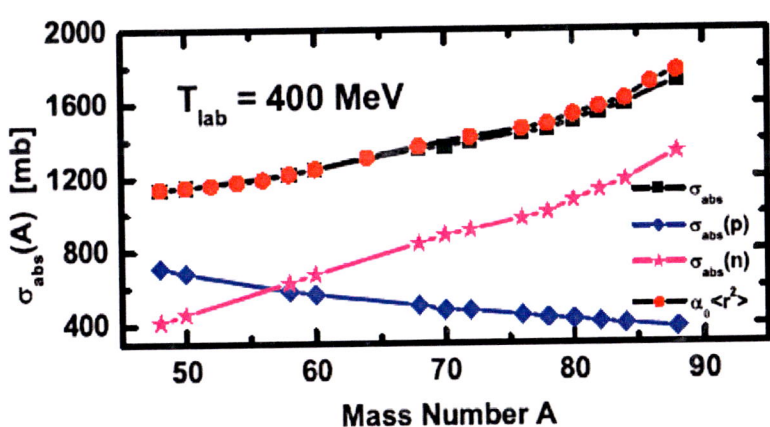

**FIGURE 3** Antiproton absorption cross sections σ, σ(n), σ(p) as function of A for Ni isotopes with mass number A at $T_{lab}$ = 400 MeV energy. The scaling of the total absorption cross section σ with α<r²>, normalized at $^{58}$Ni, is also shown.

# Antiproton Ion Collider

The antiproton ion collider proposed [5,7] makes use of the electron-ion collider complex ELISE [10] with appropriate modifications of the electron ring to store, cool and collide antiprotons with energies up to 100 MeV with up to 740A MeV ions coasting in the NESR. The collider is aimed at studying the fragment production in medium energy antiproton ion collisions under favorable kinematical conditions. The fragments preferentially emitted within the magnetic rigidity and angular acceptances of the NESR are analyzed using their Schottky noise frequency signal. Alternatively they can be detected and analyzed in a even wider magnetic rigidity, (p/m), range with the magnetic system following the collision zone using conventional charged particle spectroscopy and detection techniques (Figure 4). Both antiproton and ion beam are operated in the coasting mode, so no RF for bunching is needed so far.

Radioactive beams, produced by projectile fragmentation, are separated in the fragment separator Super-FRS, accumulated in a collector ring CR, precooled by fast stochastic cooling (<1s) at a fixed energy of 740A MeV, and fed via the RESR into the NESR. For short lived nuclei continuous accumulation using longitudinal phase space stacking is foreseen to increase the luminosity. The ions can also be decelerated in the RESR by ramping the magnetic field and the RF in an inverse synchrotron mode with a ramp rate of 1 T/s. This feature allows studying an excitation function for the antiproton absorption. In the NESR the ions are further cooled by electron cooling with cooling times of less than 0.3s. Stable nuclear beams can be stored in a similar way in the NESR.

Antiprotons are produced by 30 GeV protons from the SIS100 synchrotron in a special target arrangement, collected and cooled in the collector ring CR at 3GeV energy, .and transferred into the RESR storage ring. The RESR is then used for deceleration of the antiprotons and cooling with an electron cooler with up to 70 keV electrons with a cooling time of about 12s, before transfer into the modified electron ring.. With a similar electron cooler added to the electron ring, $10^9$ antiprotons, accumulated in about a minute, are cooled to their space charge limit and collided in a collision zone, as in the electron ion collider, with the ions coasting in the NESR. With $10^6$ radioactive ions accumulated in the NESR and $10^9$ antiprotons at 30 MeV in the electron ring a luminosity of $10^{23}$ cm$^{-2}$s$^{-1}$ is reached. By increasing the antiproton energy to 100 MeV and thus decreasing the space charge limitation, the luminosity rises to $5.10^{23}$.cm$^{-2}$s$^{-1}$.

As the antiproton absorption cross section is in the order of one barn, and the detection efficiency for the reaction is very high, the reaction rate will allow to measure accurate absorption cross sections, enabling e.g. systematic studies of changes of $<r^2_n>$ and $<r^2_p>$ along isotopic and isotonic chains in an acceptable length of beam time. Detailed rate estimates, including the transport efficiency from the production target to the NESR and the beam life times in the storage ring are given for key nuclei in the technical proposal of the AIC [7]. The luminosity can be raised by continuous stacking the ions in the longitudinal phase space as indicated before. There is however for long lived nuclei the space charge limit and for short lived ones the overall cooling time limit, which is about one second. For nuclei of 1s half-life the production limit is $\sim 10^5$ s$^{-1}$ and for nuclei with longer half lives correspondingly lower.

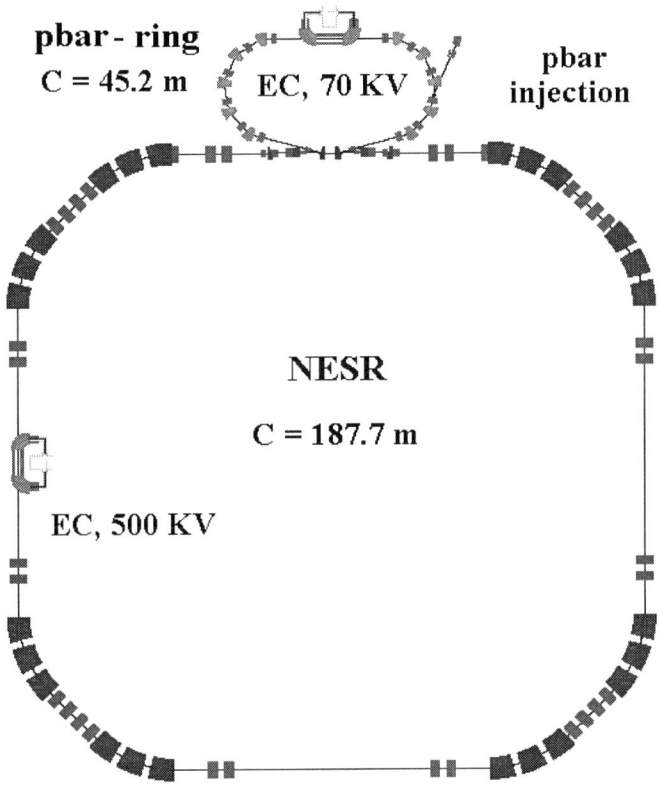

**FIGURE 4** Magnetic lattice of the Antiproton Ion Collider (AIC).. The NESR stores and cools the coasting ions which collide head on with cooled antiprotons coasting in the pbar ring.. The reaction fragments are cooled by a 500 kV electron cooler (EC) in the NESR and are detected using Schottky noise frequency spectroscopy. Those with larger magnetic rigidity deviations are analyzed by conventional charged particle spectroscopy after the first dipole magnet section following the collision zone.

# Antiproton Absorption Measurement

Schottky noise frequency analysis has been originally developed for beam diagnostics in storage rings. At the ESR storage ring of GSI, Darmstadt, the reaction product analysis was developed using Schottky noise Fast Fourier Transform (FFT) frequency analysis as a nondestructive tool for mass and lifetime measurements of reaction products. Figure 5 shows a schematic set up with the Schottky noise pick up plates traversed by the circulating ion beam and fragments. The signals of both plates are summed, amplified and frequency analyzed in a FFT. Its frequency spectrum shows the fundamental as well as the harmonics of the revolution frequencies of the fragments coasting in the ring.

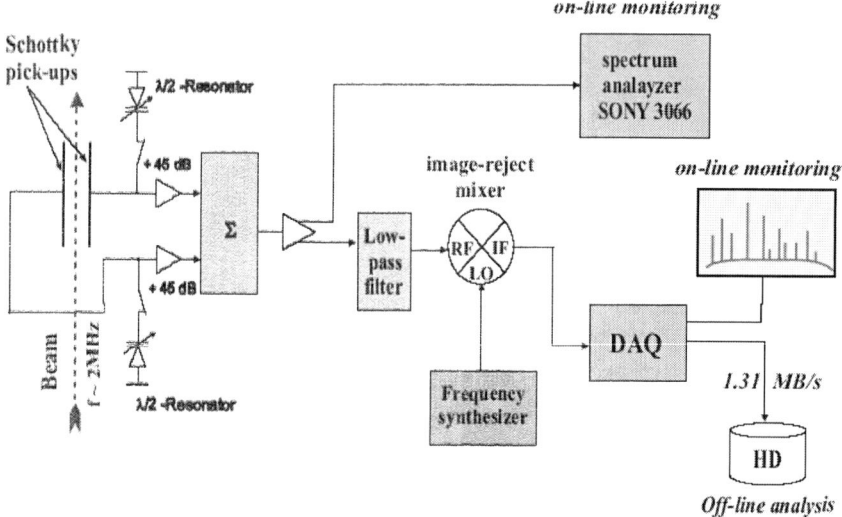

**FIGURE 5** Schematic diagram of the Schottky noise frequency analysis system, as used at the ESR storage ring @ GSI.

As an example, Figure 6 shows a frequency spectrum of a coasting $^{187}Re^{75+}$ beam with $2.10^8$ ions stored for 5.7 h after passing for 100s through a $10^{12}$ atoms/cm² thick argon jet target to strip from the β-bound state decay product $^{187}Os^{75+}$ the K-electron for the detection of this decay process of $^{187}Re^{75+}$. The frequency spectrum shows in addition weak lines assigned to Re and W isotopes, reaction products produced by knock out reactions of neutrons, and one proton plus neutrons, respectively, produced in collisions while the argon target was on for the short time.

It is proposed to use this nondestructive method to detect the antiproton annihilation products circulating in the NESR after the collision The revolution frequency of the fragments is ~p/q with p denoting the momentum of the fragment and q its charge. After cooling, which equalizes the fragment velocities, the frequency is ~m/q, with m denoting its mass. The high p/q-resolution power of $2.10^6$ allows detecting the spread of the magnetic rigidities and thus the momentum p of the

228

fragments directly after production and the observation of the cooling process to fixed p. The peak area is ~$q^2n$, with n being the number of circulating ions. The high sensitivity for highly charged ions allows to detect few ions and even single ones with q>40 in addition to a large dynamic range as demonstrated in Figure 6.

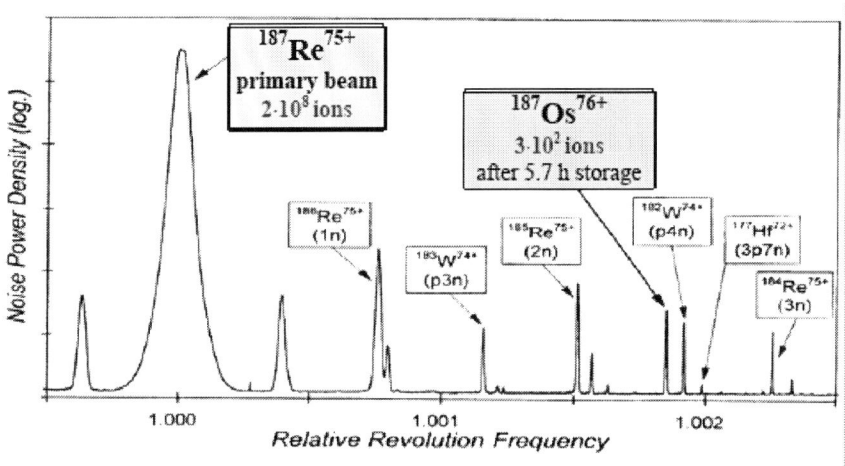

**FIGURE 6** Schottky noise frequency spectrum of a coasting $^{187}Re^{75+}$ beam and its bound β-decay daughter $^{187}Os^{75+}$, after storage of 5.7 h and stripping in an argon jet target to $^{187}Os^{76+}$..In addition one sees Re and W fragments from knock out reactions while the target was on for 100s.

The p/q distribution of the fragments with atomic numbers (A-1) after antiproton annihilation in a nucleus with atomic number A and magnetic rigidity ($p_0$/q) was simulated, assuming quasi free absorption of the antiproton by a bound neutron and proton, respectively, with momentum given by a Fermi distribution. Figure 7 shows the relative deviation of the magnetic rigidities from that of a $^{132}Sn$ beam for its antiproton annihilation fragments $^{131}Sn$ and $^{131}In$ produced in head on collisions of 740A MeV $^{132}Sn$ ions with 30 MeV antiprotons, $\varepsilon_z$= ($p_z$/q-$p_0$/q)/($p_0$/q). The broadening of the fragment lines is a result of the Fermi momentum distribution of the annihilated neutron ($^{131}Sn$) and proton ($^{131}In$), respectively. It can be measured by Schottky spectroscopy, as well as conventional magnetic rigidity analysis using the magnetic system behind the collision zone (see Figure 4). There is an acceptance limitation for the (A-1) fragments because the NESR magnetic rigidity acceptance is limited to ± 2.1% at most, which means that only A>60 fragments are accepted with one magnetic field setting of the NESR. For masses A lower than ~60 the (A-1) fragments from antiproton annihilation have to be detected by recoil detectors after the third magnetic dipole of the NESR behind the interaction zone (see Figure 4). Suitable detectors have been developed and successfully used in the ESR. This technique can also be used to measure the recoil momentum of the fragments precisely and deduce the momentum distribution of the annihilated neutron and proton, respectively.

229

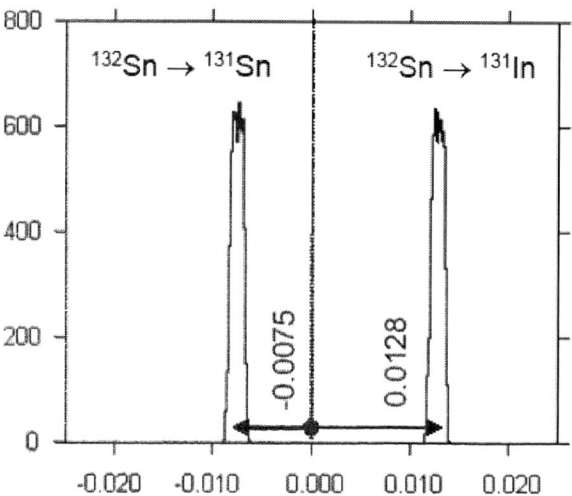

**FIGURE 7** Magnetic rigidity spectra of $^{131}$Sn and $^{131}$In as function of $\varepsilon_z$, the relative change of magnetic rigidity of the fragments with respect to the $^{132}$Sn beam, following antiproton annihilation of a neutron and a proton, respectively.

## Cross Section and $<r^2>$ Determination

The total absorption cross section, $\sigma$, can be determined from the decrease of the number of circulating ions, measured by the decrease of their Schottky line, normalized to the time-integrated luminosity, determined by detecting the Rutherford scattering of antiprotons into small forward angles relative to the direction of the antiproton beam. The (A-1) fragment production cross section, following antiproton annihilation of a neutron (n) or a proton in a nucleus, $\sigma_{n,p}$, is determined from the number of observed fragments with (N-1) and (Z-1) respectively, and normalization with the measured time integrated luminosity. A small correction for acceptance efficiency $\varepsilon$(A-1) of the NESR may have to be applied.

In paragraph 2 we have shown that at asymptotic high energies (T>400A MeV) the absorption cross section is in good approximation given by $\sigma = \alpha<r^2>$ with $<r^2>$ denoting the mean square radius of the nucleon distribution, and $\alpha$ being a proportionality constant, which has to be calibrated or determined by theory. Further more $<r^2>= (N/A)<r^2_n> +(Z/A)<r^2_p>$ is the sum of the mean square radii of the neutron, $<r^2_n>$, and the proton, $<r^2_p>$, distribution including the normalization factors (N/A) and (Z/A), respectively. The measured neutron and proton annihilation cross section $\sigma_{n,p}$ have to be corrected with a loss factor x, which takes into account the destruction of the (A-1) fragments by absorbed pions. It can be determined using the relation $x = [(N/A)\sigma_n + (Z/A)\sigma_p]/\sigma$, experimentally. Using the relation $x\sigma=[(N/A)\sigma_n + (Z/A)\sigma_p] =x\alpha<r^2> = x\alpha[(N/A)<r^2_n> +(Z/A)<r^2_p>]$, we finally get $<r^2_n> = \sigma_n/(x\alpha)$ and

$<r^2_p> = \sigma_p/(x\alpha)$. The normalization constant $\alpha$ can be calibrated with a stable nucleus, the $<r^2_p>$ of which is known from electron scattering, while the loss factor x, which is estimated to be ~0.3 for Ni and 0.6 for Sn nuclei is a measurable quantity, as indicated.

We aim at an accuracy of a few percent for the measurement of the mean square radii of the nucleon, neutron, and proton distribution of long chains of nuclei, such as from neutron poor $^{104}$Sn to neutron rich $^{134}$Sn to study the development of the mean square radii differences between neutrons and protons with increasing isospin. This allows a systematic study of changes of the nuclear structure caused by shell and collective effects and on the influence of the asymmetry energy. Whereas for heavy nuclei we focus on the built up of neutron skins with increasing proton-neutron number difference, for light nuclei the onset of neutron halos is in the center of our interest. A new class of measurements accessible with or technique is the study of odd-even effects of neutron mean square radii in analogy to the poorly understood odd-even isotope shifts.

A precise knowledge of the difference of neutron-proton density distribution is also needed to extract the iso-vector pion- nucleus scattering length from the binding energies and widths of deeply bound pionic s-states in heavy nuclei, such as Sn and Pb isotopes.. We introduced this new method recently to study the chiral dynamics of pions in a nuclear medium [10, 11]. A first indication of the partial restoration of spontaneously broken chiral symmetry has been seen. The systematic uncertainties of the restoration effect, depends critically on the accuracy of the knowledge about neutron- proton distribution differences. Odd-even effects may also play an important role.

Finally antiproton absorption measurements on light iso- scalar and neighboring nuclei may be favorably used to study t-matrix differences of antiproton-neutron and –proton interaction with high accuracy and systematically.

# ACKNOWLEDGMENTS

I'd like to acknowledge the contribution of the whole AIC collaboration to the development of this work, which led within less than half a year to a Technical Proposal of the AIC project. I appreciated the support of the Austrian Academy of Sciences during the whole development period of the project and thank all the members of the Stefan Meyer Institute for helping me to make my stay in Vienna in many respects .very enjoyable.

# REFERENCES

1.  Fricke, G., et al. At. Data Nuclear Data 60 (1995) 177.
2.  Terashima, S. this workshop
2.  Trczinska, A. this workshop. Trczinska, A. et al. Phys. Rev. Letters 87 (2001) 082501

4.  Wada, M. this workshop. Wada, M. and Yamazaki, Y. NIM B214 (2004) 196.
5.  Kienle, P. NIM B214 (2004) 191
6.  Lenske, H. and Kienle, P. Phys. Letters B (in process of publication)
7.  Technical Proposal of the AIC Project, GSI Darmstadt, Jan. 2005.
8.  An International Accelerator Facility for Beams of Ions and Antiprotons, Conceptual Design Report. GSI Darmstadt 2004.
9.  Elioff, T. et al. Phys. Rev. 128 (1962) 869.
10. Kienle, P. and Yamazaki, T. Progress Part. Nuclear Phys. 52 (2004) 85.
11. Suzuki, K. et al Phys. Rev. Letters 92 (2004) 082501.

# Antiprotonic Radioactive Atom for Nuclear Structure Studies

M. Wada* and Y. Yamazaki*,†

*Atomic Physics Laboratory, RIKEN 2-1 Hirosawa, Wako, Saitama 351-0198, Japan
† Institute of Physics, Graduate School of Arts and Sciences, University of Tokyo, 3-8-1, Komaba, Meguro, Tokyo 153-8902, Japan

**Abstract.** A future experiment to synthesize antiprotonic radioactive nuclear ions is proposed for nuclear structure studies. Antiprotonic radioactive nuclear atom can be synthesized in a nested Penning trap where a cloud of antiprotons is prestored and slow radioactive nuclear ions are bunch-injected into the trap. By observing of the ratio of $\pi^+$ and $\pi^-$ produced in the annihilation process, we can deduce the different abundance of protons and neutrons at the surface of the nuclei. The proposed method would provide a unique probe for investigating the nuclear structure of unstable nuclei.

**Keywords:** antiprotonic atom, nuclear structure, neutron halo, radioactive beam
**PACS:** 29.25.Rm 41.85.Ar 21.10.Gv 25.43.+t 36.10.-k

## INTRODUCTION

Exotic properties of nuclei, such as halo and skin, have been investigated in nuclei far from the stability line [1]. Various experimental methods have been applied or proposed to investigate the nuclear size and form of unstable nuclei as listed in Table 1. Precision optical spectroscopy have been used to determine the charge radii of many unstable nuclei. Bohr-Weisskopf effect [2], which can be derived by precision hyperfine structure spectroscopy, is sensitive to the distribution of magnetism in a nucleus (DNM). Muonic X-ray measurement is more sensitive to the nuclear volume effect in such atomic spectroscopy. Although the short lifetime of muons makes experiments difficult, a new experimental scheme for unstable nuclei was proposed and tested with some stable nuclei [3]. Electron elastic scattering is an ideal probe to determine nuclear form factors; however, the small cross section has limited its applications to unstable nuclei. Recently, a new scheme—SCRIT (self-confining radioactive isotope target)— where unstable nuclear ions are trapped in an electron beam was proposed for electron scattering experiments for unstable nuclei [4]. We expect to obtain accurate charge radii and charge form factor of some unstable nuclei using such new experimental methods based on the electromagnetic interaction at the new radioactive ion beam facilities.

The most successful experiment to investigate the halo structure of unstable nuclei is the interaction cross section measurement in the nuclear fragmentation reaction of intermediate-energy radioactive nuclear beams [5]. Although a nuclear model is required to determine the matter radii of nuclei, many exotic nuclei have been investigated by this highly sensitive experimental method. The distribution of nuclear matter can be measured by proton elastic scattering wherein solid hydrogen can be used as the target

CP793 *Physics with Ultra Slow Antiproton Beams*
edited by Yasunori Yamazaki and Michiharu Wada
© 2005 American Institute of Physics 0-7354-0282-5/05/$22.50

**TABLE 1.** Experimental methods for determining the nuclear size and form of unstable nuclei, based on the electro-magnetic (top) and the strong (bottom) interactions.

| Method | Quantity | For RIB | |
|---|---|---|---|
| Optical spectro-scopy (IS) | $\langle R^2_{\text{charge}} \rangle^{1/2}$ (Relative) | Possible ($> 10^2/\text{s}$) | Many unstable nuclides of good elements. Only sensitive to protons. |
| Microwave spectro-scopy (hfs) | $\langle R^2_{\text{mag}} \rangle^{1/2}$ (Relative) | Possible ($10^2$ in trap) | Bohr-Weisskopf effect (DNM), Only for a few elements such as alkaline-earth. |
| Muonic X-ray | $\langle R^2_{\text{charge}} \rangle^{1/2}$, $\langle R^2_{\text{mag}} \rangle^{1/2}$ (Absolute) | [3] ($10^9$ atoms) | Absolute charge radii, DNM, short $T_{1/2}$ of $\mu$ limits possible nuclides |
| Electron scattering | $\rho_{\text{c}}(r), (\rho_{\text{mag}}(r))$ | [8] ($> 10^6$ atoms) | Charge and magnetic form factors, universal electro-magnetic probe. Small cross section limits possible nuclides |
| Interaction cross section | $\langle R^2_{\text{matter}} \rangle^{1/2}$ | Possible ($0.01/\text{s}$) | Nuclear matter size. Model is required to determine nuclear rms radii. |
| Proton elastic scat-tering | $\rho_{\text{matter}}(r)$ | Possible ($10^3/\text{s}$) | Nuclear matter distribution. Model is required. |
| Antiproton scatter-ing | $\langle R^2_{\text{p}} \rangle^{1/2}, \langle R^2_{\text{n}} \rangle^{1/2}$ | Possible [7] | Independent determination of proton & neutron rms radii. |
| Antiprotonic atom | $\frac{\rho_{\text{p}}}{\rho_{\text{n}}}|_{\text{surface}}$ | Possible ($> 10^3/\text{s}$) | Different abundances of of protons and neutrons at nuclear surface. |

and an intermediate-energy radioactive ion beam hits the target. Using this scheme even very short-lived nuclei, i.e., $T_{1/2} < 1$ ms, can be investigated.

Recently, two new experiments for unstable nuclei using antiproton were proposed [6, 7]. One is the antiproton ion collision experiment in a storage ring; in this experiment, the absorption cross section of an antiproton by a nucleon of the unstable nucleus is measured at intermediate energy. Since particle identification of the residual nuclei allows us to distinguish proton absorption events and neutron absorption events, the root mean square radii of protons and neutrons in the nucleus can be determined independently. The other experiment is the antiprotonic radioactive atom experiment, which is discussed in this paper.

Different abundances of protons and neutrons at the surface of the nuclei are an important concern for nuclear structure studies. Antiprotonic atoms would be excellent probes for such different nucleon abundance at the surface; this is because annihilation of an antiproton dominantly occurs with a nucleon at the surface of a nucleus and the vanished nucleon can be identified by the total charge of the emitted pions or the residual nucleus.

Antiprotonic atoms have been studied exclusively for stable nuclei by using various experimental methods as shown in Table 2. Antiprotonic atoms were produced by irradiating an antiproton beam on a fixed target material. When an antiproton is captured in an electronic orbital of an atom, it decays to lower levels by radiating Auger electrons and X-rays. The lowest X-ray transition level indicates the matter radius of the nucleus [9]. At a certain level where a sizeable overlapping of the wavefunctions of the antiproton

**TABLE 2.** Experimental methods for antiprotonic atom and obtainable physical quantities

| Physical quantity | Observable | Method | For RIB | previous studies |
|---|---|---|---|---|
| Nuclear size | X-ray | | | Trzcinska et al [9] |
| p, n abundance at nuclear surface | Pion's net charge | Calorimetric | | Bugg et al [10] |
| | | Statistical | Possible | |
| | Cold residue | $\gamma$-ray | | Jastrzebski et al [11] |
| | | Recoil momentum | Possible | |
| Surface nucleon's momentum | Cold residue | Recoil momentum | Possible | |

and the nucleons exists, annihilation occurs between the antiproton and a nucleon of the nucleus. The implication of these studies is that the annihilation dominantly occurs with a nucleon at the surface of the nucleus and that it can be determined whether the vanished nucleon is a proton or a neutron based on the following phenomena. One is that p̄-n and p̄-p annihilations produce charged pions with a net charge of $-1$ and 0, respectively. Bugg et al. used a bubble chamber to detect charged pions and identified the annihilated nucleons [10]. The other is the fact that the "cold" residual nucleus $^A_N Z$ becomes $^{A-1}_{N-1}Z$ and $^{A-1}_N(Z-1)$, as consequences of p̄-n and p̄-p annihilations, respectively. The Warsaw group detected $\gamma$-rays to identify the cold residues [11].

An important advantage of the pion detection method is its universality. In particular, when nuclei close to the drip line are concerned, other methods are often inapplicable. One or both of residual nuclei, $^{A-1}_{N-1}Z$ and $^{A-1}_N(Z-1)$, often become particle-unbound nuclei. Pions can be observed even if the residual nucleus does not exist.

## FORMATION AND DETECTION OF ANTIPROTONIC RADIOACTIVE ATOMS

All the previous experiments used a fixed target and a fast antiproton beam. However, such methods are impossible to be applied to radioactive nuclei since the number of radioactive nuclei is always much smaller compared to that of stable nuclei and they decay within finite lifetimes. The sensitivity of the detection methods should also be high for such a limited number of radioactive species. In our proposal, therefore, we use a cloud of antiprotons trapped in a Penning trap as the target and singly ionized slow radioactive nuclear ions as projectiles [6]. Figure 1 shows a schematic potential diagram of the nested Penning trap. In this figure, negatively charged antiprotons are confined in a positive well at the center of a nested trap while positively charged slow ions are bunch-injected into the negative outer well of the trap.

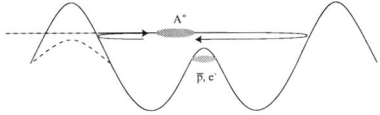

**FIGURE 1.** Potential diagram of the nested Penning trap for the p̄-RI atom.

Among several detection schemes, statistical analysis of the ratio of the detected $\pi^+$ to $\pi^-$ is the most feasible and universal method for antiprotonic radioactive atoms. The detection efficiency for $\gamma$-rays is much lower than that for charged pions and the radiochemical method is not effective for nuclei close to the drip line. The detection of X-rays enables us to investigate the matter radii of nuclei, however, the detection efficiency is also limited if all possible geometries are considered. Furthermore, the matter radii can be measured quite easily, for instance, by the measurements of the total interaction cross sections in intermediate energy reactions. The particle identification of recoiling residual nuclei is also a potential method for antiprotonic radioactive atoms. Details of the detection scheme will be discussed in a later section. Here, we focus on the detection of charged pions, so far.

## Production rate

The antiproton capture cross sections of slow singly charged ions were theoretically estimated by Cohen [12] and they are as large as those of neutral atoms (Fig. 2). A typical value is $4 \times 10^{-16}$ cm$^2$ for $^{11}$Li$^+$ ions when the relative energy is 0.1 atomic units which corresponds to a $^{11}$Li$^+$-beam energy of 33 eV in the antiproton rest frame. Assuming that the number of trapped antiprotons is $5 \times 10^6$ [13] and that they are confined within 1 mm$^2$, the target density is $N(\bar{\text{p}}) = 5 \times 10^8$ cm$^{-2}$. Slow RI ions are bunch-injected into a nested trap and they pass through the antiproton cloud for $5 \times 10^5$ s$^{-1}$ if the ions are 33-eV $^{11}$Li$^+$ and the trap length is 4 cm. Since we are mainly interested in very short-lived nuclei, we assume a short measurement cycle of 10 ms, in which only 10 RI ions are involved when the RI beam intensity is $10^3$ s$^{-1}$. Then, the production rate per cycle is $Y = 4 \times 10^{-16} \cdot 5 \times 10^8 \cdot 10 \cdot 5 \times 10^3 = 1 \times 10^{-2}$. Thus, in total one antiprotonic- $^{11}$Li$^{++}$ ion can be produced per second. This is quite a feasible number.

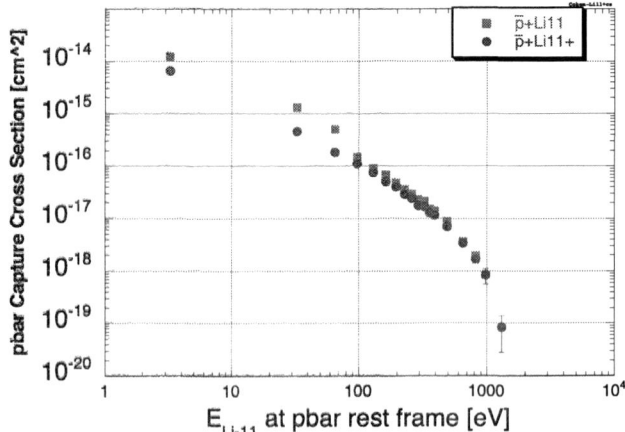

**FIGURE 2.** Theoretically evaluated capture cross section of antiproton in Li$^+$ ion and neutral atom [12].

# Detector setup

The purpose of the detectors is to identify the polarity of the charged pions from annihilation events. A stack of position sensitive detectors (PSDs) cylindrically covers the nested trap in order to track the path of the charged pions. The polarity can be deduced from the deflection direction in a high magnetic field ($\sim$5 T) and noise events from the annihilations that are not caused in the center of the trap can be discriminated. For the charged pions with an energy between 100 MeV to 400 MeV, the curvatures of the paths under 5 T are $\sim$100 to $\sim$200 mm. If three layers of detectors are located 10 mm apart from one another, the position at the central layer is shifted by more than 0.5 mm from the other two layers. A relatively simple PSD can be used for this purpose and several types of detectors are candidates—silicon PSD, multi-wire gas counter, fiber scintillators and so on.

Discrimination of the background events is an important concern for the detector setup. The major background events are: 1) energetic protons directly emitted from the annihilated nucleus, 2) $\gamma$-rays due to uncharged pions and X-rays from cascade, and 3) pions produced in the annihilation events with background gas. Protons and $\gamma$-rays can be discriminated by the response of the detectors and also by the curvature of the paths in the tracking detectors. Note that the multiplicity of direct proton event is as small as one [14]. Pions originating to the background gas would have the biggest effect, since $10^7$ antiprotons in a trap with a long storage lifetime of one day already annihilate at a rate of 100 Hz, which is much higher than the expected true event rate. However, the residual gas is almost purely $H_2$ at cryogenic temperature, i.e., such an annihilation does not produce any recoil nuclei. If we detect a recoil nucleus coincident with pions, we can clearly identify a true event. Figure 3 shows the detection efficiencies of recoil ions obtained by using a detector with a diameter of 40 mm located at a distance of 400 mm from the center of the trap under a magnetic field of 5 T. Even if the solid angle is as small as 0.06%, high efficiencies are expected—especially for heavier ions—due to the presence of the strong magnetic field.

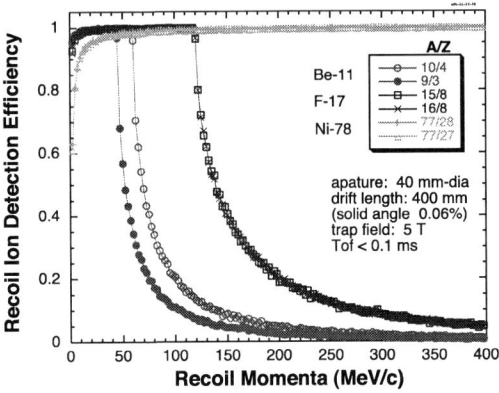

**FIGURE 3.** Detection efficiencies of recoil ions from annihilation events.

The detector of recoil ions can be used not only to eliminate background events but also to identify the recoil particles, if the position of the detected ions in the recoil ion detector and the position of the annihilation point are accurately measured. The relation of the longitudinal distance between the two positions, $L$, and the radial distance between the two positions, $d$, is expressed as

$$d = 2\frac{m}{qB}\sqrt{\frac{2E}{m}\left(\frac{L}{tof}\right)^2}\left|\sin\left(\frac{qB}{2m}tof\right)\right|$$ (1)

where $E$ is the kinetic energy of ions; $m$, mass; $q$, charge; $B$, magnetic field and $tof$, time-of-flight. Here we see that even the kinetic energy of recoil ion can be determined from the measured quantities. However, it should be noted that accurate measurement of the annihilation position to determine the kinetic energy is difficult by means of the present technique. Further development is required for this particular purpose.

## SIMULATION AND ANALYSIS

The simplest statistical analysis of the pion detection method is to compare the ratio of the total numbers of detected $\pi^-$ and $\pi^+$ throughout a measurement. This yields the nucleon density ratio $\rho(n)/\rho(p)$ at the surface of a nucleus, $R_{n/p}$, even if the detection efficiency of charged pions is low. Note that if all the charged pions can be detected, a single event can determine the net charge of an annihilation event; therefore $\bar{p}$-n or $\bar{p}$-p annihilation events can be distinguished each other. However, the detection efficiency cannot be unity in usual cases. A simulation (Fig. 4) shows that the nucleon density ratio at the surface $R_{n/p}$ can be obtained with a 5% accuracy from $5\times10^5$ antiprotonic atoms.

If the detection efficiency is reasonably high, the multiplicity of detected charged pi-ons ($M$) and their total charge per event ($\Sigma_c$) can be measured; this provides useful infor-

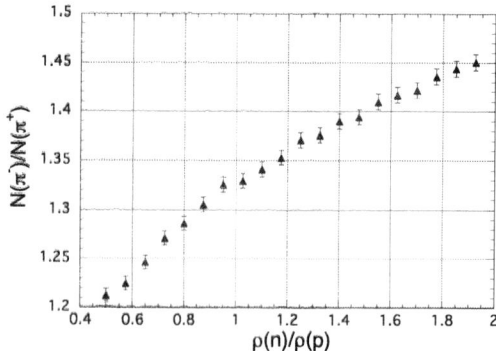

**FIGURE 4.** Simulated ratio of the total number of $\pi^-$ to that of $\pi^+$ plotted as function of $R_{n/p}$. The error bar shows the statistical error when $5\times10^5$ antiprotonic atoms are formed and charged pions produced by annihilation are detected with a 50% efficiency.

TABLE 3. Simulated histogram of the multiplicity of charged pions (row) and the total charge per event (column) under a typical experimental condition: $R_{n/p} = 1.0$, $\varepsilon_+ = \varepsilon_- = 0.6$, $\omega_+ = \omega_- = 0.1$, $\lambda_+ = \lambda_- = 0.1$. The number of total annihilation event is $10^5$.

| $M \backslash \Sigma_c$ | -5 | -4 | -3 | -2 | -1 | 0 | +1 | +2 | +3 | +4 | |
|---|---|---|---|---|---|---|---|---|---|---|---|
| 0 | 0 | 0 | 0 | 0 | 0 | 0 | 0 | 0 | 0 | 0 | (11386)* |
| 1 | 0 | 0 | 0 | 0 | 17223 | 0 | 11233 | 0 | 0 | 0 | 28456 |
| 2 | 0 | 0 | 0 | 7530 | 0 | 21437 | 0 | 2844 | 0 | 0 | 31811 |
| 3 | 0 | 0 | 1029 | 0 | 11901 | 0 | 6591 | 0 | 179 | 0 | 19700 |
| 4 | 0 | 44 | 0 | 1904 | 0 | 4394 | 0 | 519 | 0 | 5 | 6866 |
| 5 | 1 | 0 | 99 | 0 | 979 | 0 | 451 | 0 | 13 | 0 | 1543 |
| 6 | 0 | 2 | 0 | 75 | 0 | 133 | 0 | 14 | 0 | 0 | 224 |
| 7 | 0 | 0 | 1 | 0 | 7 | 0 | 3 | 0 | 0 | 0 | 11 |
| 8 | 0 | 0 | 0 | 1 | 0 | 1 | 0 | 0 | 0 | 0 | 2 |
| 9 | 0 | 0 | 0 | 0 | 1 | 0 | 0 | 0 | 0 | 0 | 1 |
| | 1 | 46 | 1129 | 9510 | 30111 | 25965 | 18278 | 3377 | 192 | 5 | 88612 |

* unobservable events

mation [15]. The observation probabilities of charged pions are functions of the charge of annihilated nucleons, the initial multiplicities of $\pi^{0,\pm}$, the absorption or charge exchange probabilities ($\omega_\pm$) of $\pi^\pm$ by the residual nucleus, the charge exchange probabilities ($\lambda_\pm$) of $\pi^0$, and the detection efficiencies ($\varepsilon_\pm$). The double charge exchange probabilities can be ignored. Assuming 60% detection efficiencies of charged pions and other relevant experimental parameters, observed events are simulated and histogrammed in terms of $M$ and $\Sigma_c$ (Table 3). The initial multiplicities of p̄-p and p̄-n annihilations are obtained from literature [16].

The count distribution of each cell depends not only on $R_{n/p}$ but also on $\varepsilon_\pm$, $\omega_\pm$, and $\lambda_\pm$. If these parameters are sufficiently orthogonal to each other, they can be obtained by fitting the measured values like Table 3. While $\varepsilon_\pm$ and $\omega_\pm$ are strongly correlated, $R_{n/p}$ and $\lambda_\pm$ cause reasonably independent effect to the values of the histogram. Figure 5 is an example simulation result of their dependences. Here, we evaluate only the projected histograms of $M$ and $\Sigma_c$, which correspond to the rightmost column and the bottom row of Table 3—not each cell of the table. Evidently, $M$ has small dependence on $R_{n/p}$ while $\Sigma_c$ is strongly dependent on $R_{n/p}$. Increases in $\lambda_-$ and $\omega_+$ or decreases in $\lambda_+$ and $\omega_-$ cause a similar global effect of lowering the average charge; however, the trend of each values shows a different dependence.

## EXPERIMENTAL LOCATIONS

The proposed experiment requires the simultaneous availability of slow radioactive nuclear beams and trapped antiprotons. To date, a large number of trapped antiprotons have only been available at the CERN antiproton decelerator facility (AD). If a beam transport line can be build from CERN ISOLDE to AD, it can be a possible location for the experiment proposed in this study. The other possibility at CERN is to bring trapped antiprotons from AD to ISOLDE by using a portable trap. In this scheme, a long beam transport line for slow RI-beams from ISOLDE is not needed. On the other hand, if such

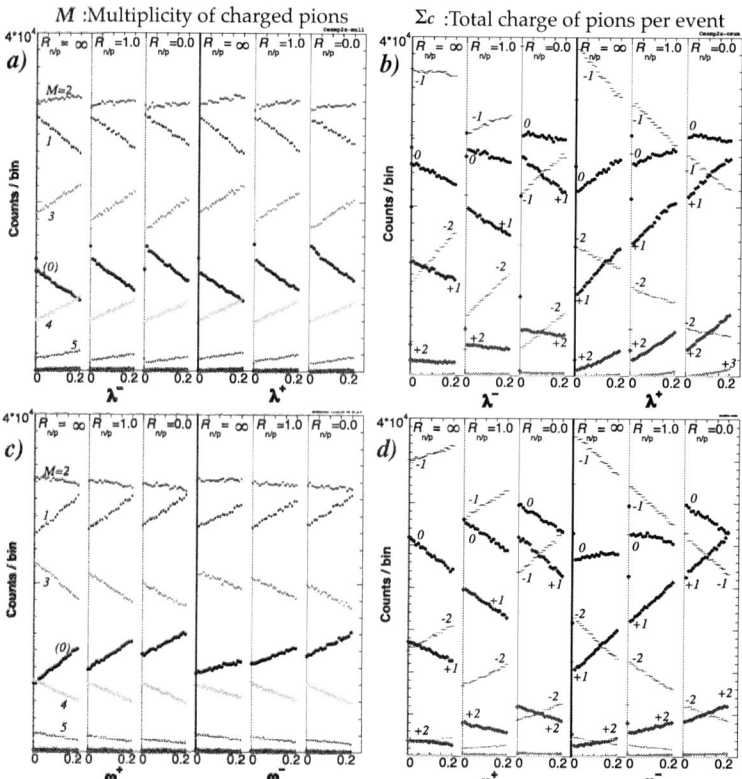

**FIGURE 5.** Simulation of the dependence of the detected charged pion's multiplicity (left (a, c)) and the total charge per event (right (b, d)) on the charge exchange probability $\lambda_{\pm}$ (top (a, b)) and on the absorption probability $\omega_{\pm}$ (bottom (c, d)). For a) and b), $\lambda_-$ and $\lambda_+$ are scanned from 0.0 to 0.2 while the other parameters are fixed: $\varepsilon_{\pm} = 0.6$, $\omega_{\pm} = 0.1$ and $\lambda_+ = 0.1$ for the three left scans ($\lambda_- = 0.1$ for the three right scans). For c) and d), $\omega_+$ (three left scans) and $\omega_-$ (three right scans) are scanned from 0.0 to 0.2 when $\lambda_{\pm} = 0.1$. $R_{n/p}$ is set for three extreme cases: $\infty$ (neutron only), 1.0 (equal distribution) and 0.0 (proton dominant).

a portable trap is realized for antiprotons and the trapping lifetime is sufficiently long, trapped antiprotons can be brought to RI beam facilities around the world. One candidate is SLOWRI at RIKEN RIBF [17]. This facility in Japan, which is under construction, aims to provide slow and trapped RI beams of thousands of unstable isotopes of all atomic elements using a superconducting heavy ion cyclotron, the projectile fragment separator BigRIPS, and an RF ion guide. One of the planned experiments at SLOWRI is p̄-RI. Since the trapping lifetime of the portable trap must be more than one week, technical evaluation of the vacuum technologies for such a trap is currently being done. As regards other future facilities, FAIR at GSI is an important candidate. This facility will provide us trapped antiprotons at FLAIR, and the low-energy branch of NUSTAR will provide us universal slow RI beams. Thus we propose antiprotonic radioactive atom experiments at the FAIR project by the Exo+pbar collaboration.

# CONCLUSION

Antiprotonic radioactive nuclear atoms would be new and unique probes for nuclear structure studies of unstable nuclei. Among several experimental schemes of antiprotonic atoms, the pion detection method is advantageous in terms of the universality and richness of the obtainable information. Charged pion emissions followed by annihilation are observable even if the residual nuclei are unbound or fragmented. The statistically acquired matrix of the charged pion multiplicity $M$ and the total charge per event $\Sigma_c$ allows us as to fit not only for the different distributions of protons and neutrons at the nuclear surface, $R_{n/p}$, but also the pion absorption probabilities $\omega_\pm$ and the charge exchange probabilities $\lambda_\pm$. The experimental technologies in the detectors and the trap should be improved further in order to realize the proposed experiments in the coming decade.

# ACKNOWLEDGMENTS

This report describes the proposed experiments by the SLOWRI collaboration at RIKEN RIBF and by the Exo+pbar collaboration for the GSI-FAIR project. The authors acknowledge the members of these two collaborations, especially, Prof. Wycech for helpful discussions.

# REFERENCES

1. I. Tanihata, et al., *Phys. Rev. Lett.* **55** (1985) 2676.
2. A. Bohr and V. F. Weisskopf, *Phys. Rev.* **77** (1950) 94.
3. P. Strasser, these proceedings.
4. W. Wakasugi, T. Suda and Y. Yano, *Nucl. Instr. Meth.* **A532** (2004) 216.
5. T. Ozawa, these proceedings.
6. M. Wada and Y. Yamazaki, *Nucl. Instr. Meth.* **B214** (2004) 196.
7. P. Kienle, these proceedings.
8. T. Suda, these proceedings.
9. Trzcinska et al., *Phys. Rev. Lett.* **87** (2001) 82501.
10. W. M. Bugg, G. T. Condo, E. L. Hart, H. O. Cohn and R. D. McCulloch, *Phys. Rev. Lett.* **31** (1973) 475.
11. Jastrzebski et al., *Nucl. Phys.* **A588** (1993) 405c.
12. J. S. Cohen, *Phys. Rev.* **A 69** (2004) 022501.
13. N. Kuroda, H. A. Torii, K. Yoshiki Franzen, Z. Wang, S. Yoneda, M. Inoue, M. Hori, B. Juhasz, D. Horvath, H. Higaki, A. Mohri, J. Eades, K. Komaki and Y. Yamazaki, *Phys. Rev. Lett.* **94** (2005) 023401.
14. D. Polster et.al, *Phys. Rev.* **C51** (1995) 1167.
15. S. Wycech, these proceedings.
16. J. Riedlberger et. al, *Phys. Rev.* **C40**(1989) 2717.
17. M. Wada, Y. Ishida, T. Nakamura, Y. Yamazaki, T. Kambara, H. Ohyama, Y. Kanai, T. M. Kojima, Y. Nakai, N. Ohshima, A. Yoshida, T. Kubo, Y. Matsuo, Y. Fukuyama, K. Okada, T. Sonoda, S. Ohtani, K. Noda, H. Kawakami and I. Katayama, *Nucl. Instr. Meth.* **B204** (2003) 570.

# Muonic Atoms of Unstable Nuclei

P. Strasser[*†], K. Nagamine[*!], T. Matsuzaki[†], K. Ishida[†], Y. Matsuda[†] and M. Iwasaki[†]

[*]Muon Science Laboratory, Institute of Materials Structure Science, High Energy Accelerator Research Organization (KEK), 1-1 Oho, Tsukuba, Ibaraki 305-0801, Japan
[†]Advance Meson Science Laboratory, RIKEN, 2-1 Hirosawa, Wako-shi, Saitama 351-0198, Japan

**Abstract.**
New intense muon beams with flux several orders of magnitude higher than at present muon facilities would allow many novel experimental studies that were until now statistically not feasible. The investigation of the nuclear properties of short-lived nuclei using muonic atom spectroscopy would become possible. A feasibility study at RIKEN-RAL muon facility using the cold hydrogen film method to produce radioactive muonic atoms is in progress. This method would allow studies of the nuclear properties and nuclear sizes of unstable nuclei by means of the muonic X-ray method at facilities where both intense negative muon and RI beams would be available. Encouraging experimental results were obtained with stable argon ions implanted in solid deuterium films. In this paper, we present a compilation of the recent progress of this project that were previously reported in several shorter publications, and discuss future perspectives at the new Muon Experimental Facility in the J-PARC project.

## INTRODUCTION

Muonic atom spectroscopy [1] has played an important role in establishing and refining nuclear structure models since many years, and has been successfully used to study stable nuclei. New intense muon beams, with fluxes several orders of magnitude higher than at present muon facilities, would allow many novel experimental studies that were statistically not feasible until now. The investigation of the nuclear properties of short-lived nuclei using muonic atom spectroscopy would become possible. This would be a unique tool to increase our knowledge of the nuclear structure far from stability where new effects may be expected, in particular, the nuclear charge distribution and the deformation properties of nuclei.

Muonic X-ray measurements can yield very precise and absolute values for the charge radii and other ground state properties. It would usefully complement the knowledge obtained from electron scattering and laser spectroscopy, since in the past calibration data were used from muonic atom measurements with stable nuclei. Also, measurements of the quadrupole hyperfine splittings of muonic X-rays could provide precise and reliable absolute values of nuclear quadrupole moments, which are sensitive to the deviation of the shape of a nuclear charge distribution from spherical symmetry.

---

[1] Present address: University of California, Riverside, California, USA and Atomic Physics Laboratory, RIKEN, 2-1 Hirosawa, Wako-shi, Saitama 351-0198, Japan.

CP793 *Physics with Ultra Slow Antiproton Beams*
edited by Yasunori Yamazaki and Michiharu Wada
© 2005 American Institute of Physics 0-7354-0282-5/05/$22.50

We proposed the cold hydrogen film method [2] to extend muonic atom spectroscopy to the use of nuclear beams including, in the future, radioactive isotope (RI) beams, and to produce radioactive muonic atoms. This method would allow studies of unstable nuclei by means of the muonic X-ray method at facilities where both intense negative muon and RI beams would be available.

Indeed, this project has been strongly motivated by the advances of the recent RI beam projects that are in operation, or being planned, at facilities where negative muon beams are also available, e.g., the ISAC project at TRIUMF [3], and the E-arena in the J-PARC project (in a later phase) [4]. The RI Beam Factory project at RIKEN [5] is also of interest, provided intense negative muon beams can be produced there.

Furthermore, the new neutrino factory concept to produce unprecedented high flux of muons could benefit many muon related research topics, including the proposed study. Actually, it is a very attractive concept because the same proton driver beam could also be used simultaneously for second generation RI beam facilities. This would be a unique opportunity to combine massive amounts of muons with very intense RI beams. Already in 2001, two workshops on radioactive antiprotonic and muonic atoms (RAMA) [6, 7] took place in connection with a possible neutrino factory at CERN [8]. Strong efforts were reported towards possible formation and investigation of radioactive muonic atoms and even possibly radioactive antiprotonic atoms, which could be produced similarly and promise measurements of the neutron distribution in nuclei. Several experimental approaches other than the cold hydrogen film method have been envisaged to combine muons and antiprotons with radioactive nuclei, including a beam merging scenario and a cyclotron trap [9].

Very recently, muon capture on short-lived neutron-rich nuclei has also been proposed as a tool to study highly excited states in the daughter nucleus, with the additional attractive feature that the resulting system is one step further away from the line of stability [10].

## FUTURE PERSPECTIVES AT J-PARC

The new Muon Experimental Facility that is now under construction at the Materials and Life Science Facility (MLF) in the J-PARC project [11] is also a very attractive alternative to realize the proposed study. Besides conventional surface muon and decay muon beams that will be produced using a 1-MW (3 GeV, 333 $\mu$A) proton beam on a 2-cm thick carbon target, there is also a proposal for the world highest intensity surface muon channel called "Super Omega" [12]. This advanced muon channel consists of a double normal conducting solenoid lens to accept muons in a solid angle of nearly 400 msr, a curved superconducting solenoid for transportation, and a set of superconducting coils to focus the beam at the exit of the solenoid. According to Monte Carlo simulations, a surface muon intensity of $2$–$5 \times 10^8$/sec is expected. Since this muon beam channel uses only solenoids and coils, cloud negative muons can also be extracted simultaneously. Furthermore, by placing a magnetic bend right at the end of this channel, both negative and positive muon beams with the same momentum could be produced in principle. The cloud negative muon intensity at 27 MeV/c is roughly 1–2% of the surface muon intensity, and a negative muon beam of $10^{6-7}$/sec could be obtained. This is more than 2–3 orders of magnitude higher than presently available at RIKEN-RAL Muon Facility, and therefore very attractive to perform the proposed study. However, at present, this advanced muon channel "Super Omega" has not been funded yet.

# MUONIC ATOM FORMATION WITH UNSTABLE NUCLEI

## The Cold Hydrogen Film Method

The basic concept of the cold (solid) hydrogen film method is to stop both a negative muon ($\mu^-$) beam and a radioactive ion ($Z^+$) beam simultaneously in a solid hydrogen ($H_2/D_2$) film, followed by the direct muon transfer reaction to higher Z nuclei to form $\mu Z$ radioactive muonic atoms [2, 13].

It is well known that when hydrogen ($H_2$) or deuterium ($D_2$) is enriched in elements with atomic number $Z > 1$, the transfer of a muon from hydrogen to nuclei of heavy elements will inexorably take place [14], i.e., $H\mu + Z \rightarrow \mu Z + H$, with $H \equiv$ p, d. This direct transfer reaction is irreversible, since the increase in binding energy greatly exceeds any attainable target temperature. It proceeds mainly to highly excited levels of the muonic atom $\mu Z$, being accompanied by emission of characteristic X-rays and Auger electrons in the deexcitation of the $\mu Z$ muonic atom. The transfer rate of this process in liquid-hydrogen is very fast (except for transitions to helium) and, as a rule, can be expressed as $\lambda_Z \approx Z\,C_Z \times 10^{10}$ s$^{-1}$ [14], where Z is the atomic number and $C_Z$ is the concentration of higher Z nuclei (i.e., $C_Z = N_Z/N_X$, with $X \equiv$ H, D).

Figure 1 shows the proposed scheme of muonic atom formation with implanted ions in a solid deuterium film. It should be noted that the idea which makes use of the transfer mechanism is not new. It was already used 30 years ago to measure transfer rates and rms radii of rare isotopes with high-pressure gas target (see, e.g., ref. [15]). The advantage of using a pure $D_2$ film for the formation of $\mu Z$ atoms, compared to a pure $H_2$ film, is that

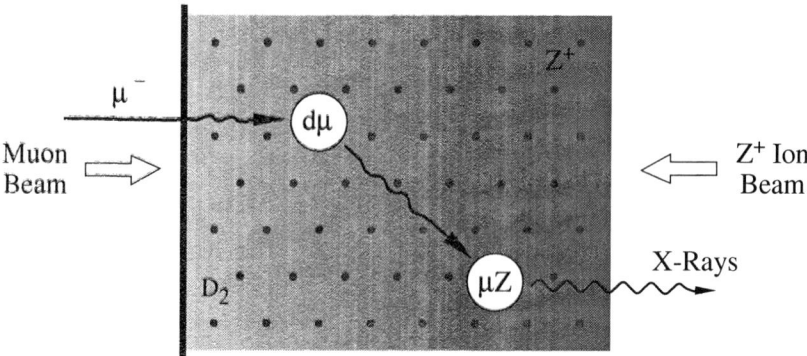

**FIGURE 1.** Simplified scheme of muonic atom formation from implanted ions in a solid deuterium film.

244

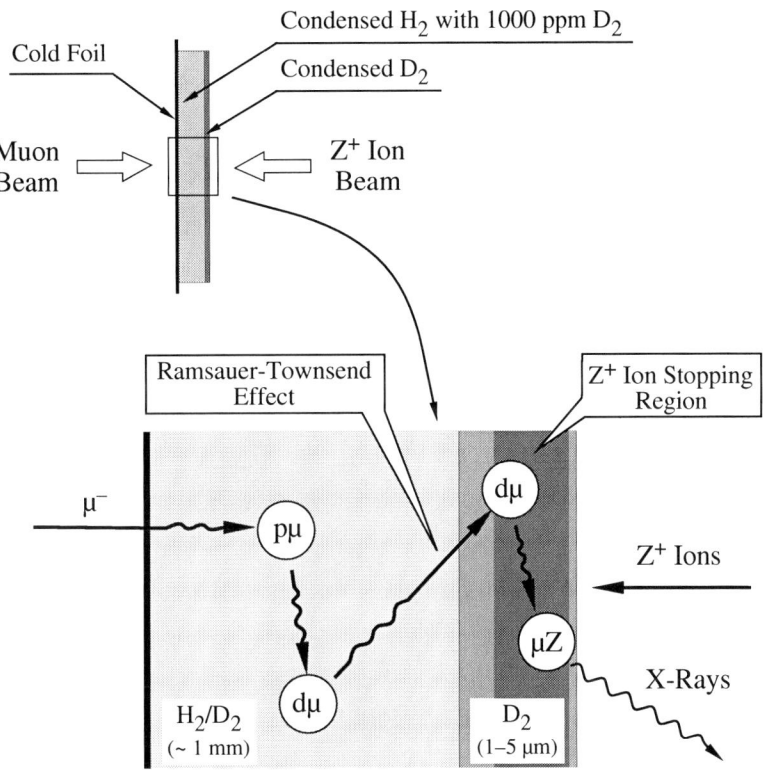

**FIGURE 2.** Simplified scheme of muonic atom formation from implanted ions in solid hydrogen films using the two-layer arrangement.

after dd$\mu$ muonic molecule formation the $\mu^-$ is mostly released, while it is lost after pp$\mu$ is formed.

The two-layer arrangement, which was first proposed in [2], seems well appropriate to current muon beams that have a relatively large momentum spread. It uses a separate layer to stop efficiently $\mu^-$, and, with the help of the Ramsauer-Townsend effect in the scattering of muonic deuterium atoms in $H_2$, confine the production region of $\mu Z$ atoms to an optimized film thickness. Figure 2 shows the proposed scheme of muonic atom formation with implanted ions in a solid hydrogen film using the two-layer arrangement. The first layer – consisting of a hydrogen-deuterium ($H_2/D_2$) mixture (1-mm thick with a $D_2$ concentration of 0.001) – is used for $\mu^-$ stopping followed by muonic protium (p$\mu$) atom formation, muon transfer to form muonic deuterium (d$\mu$) atoms, and d$\mu$ diffusion with the help of the Ramsauer-Townsend effect in the scattering of d$\mu$ atoms. The second layer – pure deuterium ($D_2$) with implanted $Z^+$ ions as impurities – is used for the formation of $\mu Z$ atoms. Experimental results showed that a sizable fraction of incident $\mu^-$ (as high as 0.01, depending on the incoming $\mu^-$ beam momentum spread) are released as $\sim$ eV d$\mu$ atoms from the first $H_2/D_2$ layer surface. Also, it was found that those energetic $d\mu$ atoms are stopped within a thickness of 5 $\mu$m of the added $D_2$ layer [16, 17]. Consequently, if impurity ions are implanted into this $D_2$ layer, muon transfer from $d\mu$ atoms to higher Z nuclei will occur. This scheme would require fewer implanted ions because it confines the production region of $\mu Z$ atoms to a much thinner $D_2$ film thickness.

Several other production schemes can be also considered. For instance, if very narrow momentum spread negative muon beams would become one day available, either by phase-space rotation or cooling, a single $D_2$ layer can be used to stop $\mu^-$ in a much thinner region and more efficiently. The ion beam could then be stopped precisely at the same position increasing the transfer yield. In this case, the minimum possible thickness will be mainly determined by the diffusion process of muonic deuterium ($d\mu$) atoms in solid. As shown in ref. [18], the calculated total cross-section for $d\mu$ scattering in solid deuterium has a low-energy minimum due to a strong reduction of the Bragg cross-section, which strongly increases the mean-free-path of $d\mu$ atoms.

It should be noted that the present method may be the only one capable to study very short-lived isotopes (below a few milliseconds). Other methods usually require RI beams to be cooled down. But with this method it is possible to use a thick $D_2$ layer to stop directly high energy RI beams. For very short lived nuclei, however, very intense muon beams will be required.

Detailed Monte Carlo simulations are still in progress to examine these different schemes and optimize the formation yield.

## Transfer Yield Estimation

The total transfer yield to implanted ions in a $D_2$ layer can be approximated by neglecting the muon sticking after $dd\mu$ fusion and other losses as follows:

$$Y_Z = \frac{\phi \lambda_Z}{\lambda_0 + \phi \lambda_Z} Y_\mu, \tag{1}$$

where $\lambda_0$ is the muon decay rate, $\phi$ is the number density of the $D_2$ layer (normalized to liquid-hydrogen density $N_0 = 4.25 \times 10^{22}$ cm$^3$), and $Y_\mu$ is the $\mu^-$ stopping yield in the target. Presently, at RIKEN-RAL Muon facility port 4, we can obtained a stopping yield of $\sim 0.6$ in a 1-mm $D_2$ layer with a 27 MeV/c backward decay negative muon beam. For the two-layer arrangement case, $Y_\mu$ needs to be multiplied by the fraction of stopped $\mu^-$ entering the $D_2$ layer as $d\mu$ atoms. In a previous experiment, we obtained a measured value of $\sim 0.04$ using a 1-mm $H_2/D_2$ emission layer [17]. If only a pure $H_2$ layer is used, the $pp\mu$ muonic molecule formation rate needs also to be taken into account.

Figure 3 shows the estimated transfer yield as a function of $Z \cdot N_Z$ (the atomic number Z multiplied by the implanted ion number $N_Z$) assuming an uniform stopping region of implanted ions in a 1-mm pure $H_2$ layer (dotted line), in a 1-mm pure $D_2$ layer (dashed line), and in a 5-$\mu$m $D_2$ layer for the two-layer arrangement case (solid line), respectively. The same $\mu^-$ stopping yield $Y_\mu$ was assumed in all three cases. We also supposed for the two-layer arrangement case that all emitted $d\mu$ atoms emitted from the first 1-mm $H_2/D_2$ layer are stopped in the 5-$\mu$m $D_2$ layer (no $d\mu$ loss). Almost one order

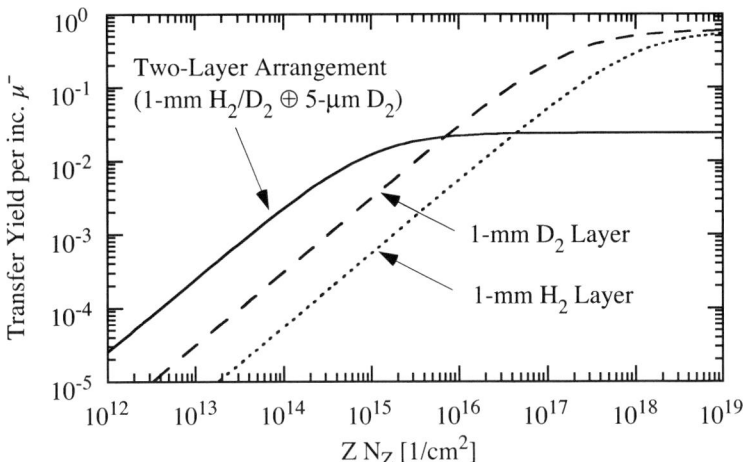

**FIGURE 3.** Comparison of estimated transfer yield to implanted ions uniformly stopped in a 1-mm pure $H_2$ layer (dotted line), in a 1-mm pure $D_2$ layer (dashed line), and in a 5-$\mu$m $D_2$ layer (two-layer arrangement case) (solid line), respectively.

of magnitude higher transfer yield could be achieved in principle by using the two-layer arrangement compared to a pure $D_2$ layer, if d$\mu$ loss from the second $D_2$ layer is found to be negligible. In the case of a pure $H_2$ layer, the yield becomes smaller because of the losses due to pp$\mu$ muonic molecule formation.

As an example, let us consider a target of 1 cm$^2$ with an incident muon beam intensity ($N_\mu$) at 30 MeV/c of $10^8$ $\mu^-$ per second, which could be achieved at a new generation muon beam facility and under a superconducting Helmholtz coil confinement field, and $N_Z = 10^{12}$ implanted Sn ions (Z=50). With a 1-mm pure $D_2$ layer, we could expect $\sim 1.5$ $10^4$ muonic tin ($\mu$Sn) atoms to be formed per second, and $\sim 200$ $\mu$Sn 2p$\to$1s transition photons (3.4 MeV) to be detected per hour, assuming an X-ray detection efficiency of 0.005, a detector solid angle of 0.001 and a branching ratio of 0.7. The same using the two-layer arrangement, we could get $\sim 1500$ events per hour.

The main uncertainty in this preliminary yield estimation is the d$\mu$ loss from the $D_2$ layer that was neglected. It is expected to reduce the transfer yield for low ion concentration $C_Z$. Furthermore, solid state effects that appear in the d$\mu$ thermalization in solid hydrogen [18] may also affect directly this loss. More detailed investigations are required to take into account all these effects, and optimize the total transfer yield.

## Experimental Considerations

There are several advantages in using the solid hydrogen ($H_2/D_2$) film method to generate and investigate radioactive muonic atoms. The system is a windowless target in vacuum, with a well-defined interaction region between $\mu^-$ and radioactive nuclei to set a compact detector system around the target. The $H_2/D_2$ film can easily be evaporated and replaced by a fresh one when needed. RI beam characteristics are not critical in terms of beam emittance and energy spread, since the beam will be stopped completely in the target.

Because conventional $\mu^-$ beams have large sizes ($\sim$ 4 cm in diameter), the use of a magnetic confinement field around the target would constrain incoming $\mu^-$ within a small target area ($\sim$ 1 cm$^2$), while reducing the background of e$^-$ associated with $\mu$-decay. It would also reduce the number of radioactive nuclei needed to be implanted in the solid hydrogen target.

The activity deposited by the RI beam in the solid hydrogen target will need to be carefully considered, because the beam will be stopped completely. Indeed, the proposed experiment is quite the opposite compared to nuclear astrophysics experiments that are planned with RI beams, where the beam only interacts with the target without actually stopping inside (only a small faction will be scattered and stopped inside the vacuum chamber vessel). It will reach a maximum value at the steady state equal to the beam intensity $\Phi$. This will result in a high radiation background making the measurement of muonic X-rays difficult. A pulsed $\mu^-$ beam will definitely help to reduce this constant background noise by several orders of magnitude. When using pulsed $\mu^-$ beams and measuring X-rays and $\gamma$-rays, it is important that the counting rate per detector be kept well below one event per pulse, otherwise the detector will be "blind" to delayed events. A large detector array with small individual solid angles will be required. In that regard, we would benefit enormously from a proton driver beam that would produce $\mu^-$ beams of the same intensities but with a much higher repetition rate such as 1–10 kHz. Each pulse would contain 2–3 orders of magnitude fewer muons, making measurements more accessible. Also, the detection system will have to be designed specifically depending on the decay mode of the unstable nuclei to be studied to further minimize this background (e.g., active background suppression, ...).

The decay product, or daughter nuclei, will act as impurities in the hydrogen target and affect the total transfer yield to implanted unstable nuclei. The transfer rates being quite similar, the ratio between both population will be the critical parameter. It should be noted that the limitation comes from the transfer yield that will be decreasing with time, and not from the X-ray measurement itself, since muonic transition energies of unstable nuclei and daughter nuclei are quite distinct (because of different atomic number) and can be easily distinguished using standard high resolution semiconductor detectors. As soon as the implantation of unstable nuclei starts with a beam intensity $\Phi$, the number of unstable nuclei ($N_A$) and daughter nuclei ($N_B$) in the hydrogen target will increase. Both populations will be the same at around 1.6 times the decay time $\tau$, and when the steady state is reached, $N_A$ will saturate at a value of $\Phi\tau$ ($= N_Z$), while $N_B$ will continue to increase with time, i.e., $\Phi t$. The ratio between unstable nuclei and daughter nuclei ($N_A/N_B$) will be decreasing as $\tau/(t - \tau)$. At five times the decay time, this ratio will be nearly 0.25. As an example, Fig. 4 shows the time dependence of each population, considering a decay rate $\tau$ of 1000 s ($T_{1/2} = 11$ min.) and a beam intensity $\Phi$ of $10^9$ particles per second. The effect of a beam impurity of 1 % is also shown. We can notice that even if a small amount of contaminants are contained in the primary beam (below a few percent), the effect of the accumulation of daughter nuclei in the target is much more significant. Therefore, high purity radioactive isotope beams are not essential to

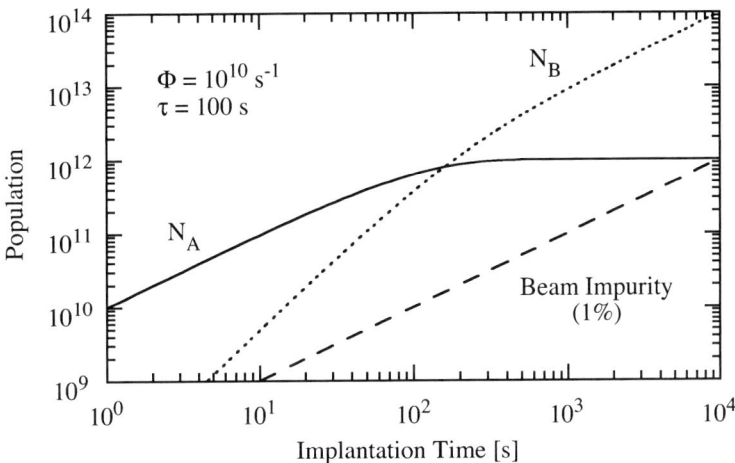

**FIGURE 4.** Time dependence of implanted unstable nuclei ($N_A$, solid line) and daughter nuclei ($N_B$, dotted line) in the solid hydrogen target, considering a decay rate $\tau$ of 1000 s and a beam intensity $\Phi$ of $10^9$ particles per second. The accumulation of a beam impurity of 1% is also shown (dashed line).

this experiment.

With a standard target system, after a certain period of time when the accumulation of daughter nuclei becomes significant, the hydrogen target can be easily evaporated and replaced by a new one. A few hours are needed at the moment to perform this operation. Therefore, we are practically limited with this method to study unstable nuclei with a half-life of a few tens of minutes, or longer. But we could in principle use a rotating target, where the implantation region is changing with time, so that the area contaminated with daughter nuclei is regularly moved away from the $\mu^-$ beam stopping region. This method could help gain a factor of roughly ten or more, and allow unstable nuclei with a half-life of a few minutes to be studied. An other idea would be to use the associated sputtering of the solid $D_2$ to change the ion stopping region with time, so that daughter nuclei are continuously left behind in the target.

## RECENT EXPERIMENTAL PROGRESS

### Test Experiment

An experimental setup ($\mu A^*$) has been constructed to implant stable ions in solid hydrogen films using a cryogenic target combined with an ion-implantation system. The objective of this project is to establish experimentally the feasibility of using the cold hydrogen film method, and determine the optimum conditions to perform muonic atom X-ray spectroscopy with implanted ions.

The layout of this experimental setup installed at RIKEN-RAL muon facility [19]

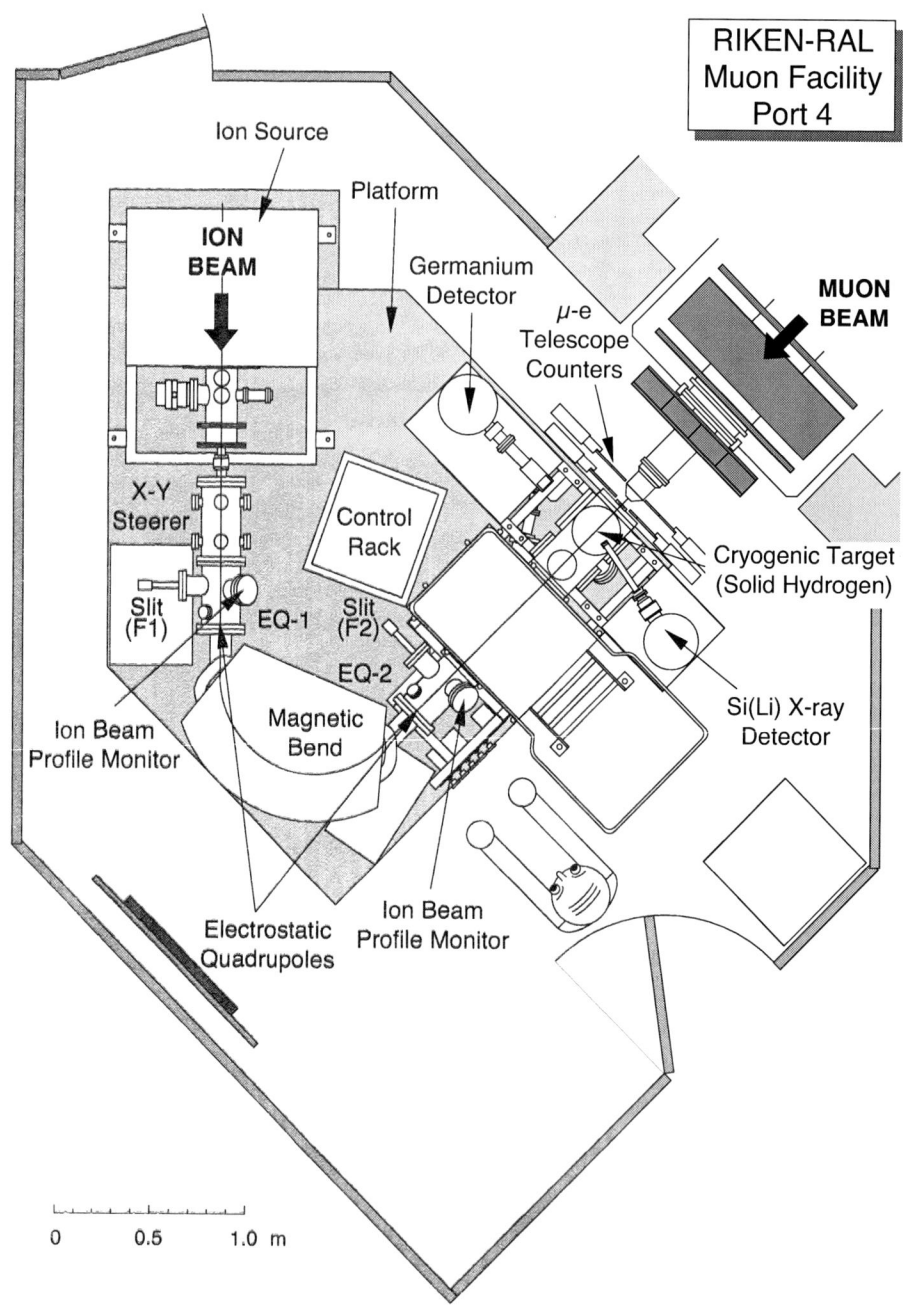

**FIGURE 5.** Layout of the experimental setup for muonic atom spectroscopy with implanted ions in solid hydrogen films installed at the new Port 4 of the RIKEN-RAL muon facility.

port 4 is shown schematically in Fig. 5. The design and construction was already reported in detail in ref. [20]. In brief, it comprises mainly two parts: (1) a cryogenic target system to cool a silver target foil (cold foil) down to 3K onto which a thin target layer of solid hydrogen is formed by freezing hydrogen gas using a diffuser, and (2) an ion source with an analyzing magnet to accelerate (up to 33 kV/q), select and implant stable ions into this solid hydrogen film. Three beam profile monitors are used to to determine the ion beam profile and current.

The key feature of the cryogenic target system is the liquid helium reservoir of the cryostat being at atmospheric pressure, so that it can be refilled without any disturbance to the good operation of the system. This allows a solid hydrogen target to be kept indefinitely as long as the LHe reservoir is refilled on a regular basis (typically once a day). It was also designed to sustain the additional power load deposited by the ion beam, ensuring a temperature at the center of the target below 4K (e.g., a 10-$\mu$A 50-keV ion beam would deposit 0.5 W on the target). The same technique as employed in previous experiments [16, 17] is used to deposit hydrogen films.

All the components of this setup were mounted on a platform 700 mm high for quick installation and removal from the experimental area, this one being used by many other users.

## Experimental Results and Discussion

As already reported in several short publications [21, 22, 23, 24, 25, 26], an experimental program has been initiated at the RIKEN-RAL muon facility to establish experimentally the feasibility of this method. We summarize here the present status and the progress achieved so far.

Two different target configurations were investigated. At first, the two-layer arrangement was used. The primary layer made of 0.5-mm $H_2/D_2$ is used to efficiently stop 27 MeV/c $\mu^-$ followed by p$\mu$ formation, p$\mu$ to d$\mu$ transfer and d$\mu$ emission with the help of the Ramsauer-Townsend effect in the diffusion of the d$\mu$ atoms, while the second layer made of pure $D_2$ (several $\mu$m thick) with implanted Ar ions is used to confine the production region of muonic argon ($\mu$Ar) atoms to an optimized film thickness. This scheme would require fewer implanted ions if d$\mu$ loss from the second layer is found to be negligible.

The very first targets with implanted Ar ions that were measured, only faint $\mu$Ar X-rays could be detected. It was then realized that the sputtering effect of solid $D_2$ was underestimated when making a very thin target with implanted ions. Indeed, too many Ar ions were actually being implanted at once, and due to the large sputtering yield of solid $D_2$ [27], the region with implanted Ar ions was being removed by the argon beam a while later. According to Monte Carlo simulations performed with SRIM [28], the range of 33-keV Ar ions in solid $D_2$ is only about 250 nm with a range straggling of roughly 100 nm (FWHM). This sputtering yield was measured with our system by depositing a thin $D_2$ film ($\sim$ 1 $\mu$m) directly on the cryostat silver foil and measuring the time needed by the argon beam to remove it. When the ions reach the silver foil, the interaction between the beam and the metallic substrate produces an enhanced sputtering and the pressure in the vacuum chamber increases momentarily. A preliminary sputtering yield of 300–400 $D_2$ molecules per incoming Ar ions was extracted. This value is consistent with those published in ref. [27] for hydrogen ions with comparable energies.

The subsequent target was carefully made by taking into account this sputtering effect, and alternate implantation and $D_2$ deposition were used to built a $D_2$ layer with a larger thickness.

At ISIS-RAL, the proton beam has a two-bunch pulse structure (70 ns wide, separated by 320 ns), and both of these beam bunches in a spill were used. Raw X-ray time spectra had therefore two prompt peaks, and, since the delayed time range very close to those two prompt peaks were examined, the delayed component (from +75 to +250 ns after the muon pulse) after the second prompt region was added to the one after the first prompt region. Those events are labeled as "short" delayed events. Data from +250 ns to 32 $\mu$s after the second muon pulse are labeled as "long" delayed events.

The germanium energy spectra shown in Fig. 6 indicates very clearly $\mu$Ar 2p→1s transition X-rays at 644 keV from the muon transfer transfer reaction. Around $10^{16}$ Ar ions per $cm^2$ were implanted in a deuterium thickness of 5-$\mu$m, corresponding to an average concentration of 500 ppm. At this concentration, all $d\mu$ atoms emitted from the primary 0.5-mm $H_2/D_2$ layer are expected to transfer very rapidly. The "short" delayed events in Fig. 6b are mainly due to the diffusion time of the $d\mu$ atoms before reaching the added $D_2$ layer. Almost no "long" delayed events can be seen in Fig. 6c, either because muon transfer already occurred or $d\mu$ atoms escaped into vacuum ($d\mu$ loss).

The triangular peaks observed at 596 keV and 691 keV, respectively, result from inelastic neutron excitation of the Ge isotope nuclei within the detector itself [29]. The neutrons are produced following $\mu^-$ capture on nuclei and consist mainly of evaporation neutrons from de-excitation of the nucleus formed after the capture.

Later, a different target configuration made of a pure $D_2$ layer was used. Since the range of 33-keV Ar ions in solid $D_2$ is only about 0.25 $\mu$m, with a range straggling of 0.1 $\mu$m, it is very difficult to perform an uniform implantation in a thick $D_2$ layer. Consequently, each Ar implantation was separated from the next by depositing about 20 $\mu$m of $D_2$ to make a total $D_2$ layer thickness of 0.5 mm. Every implantation region had a

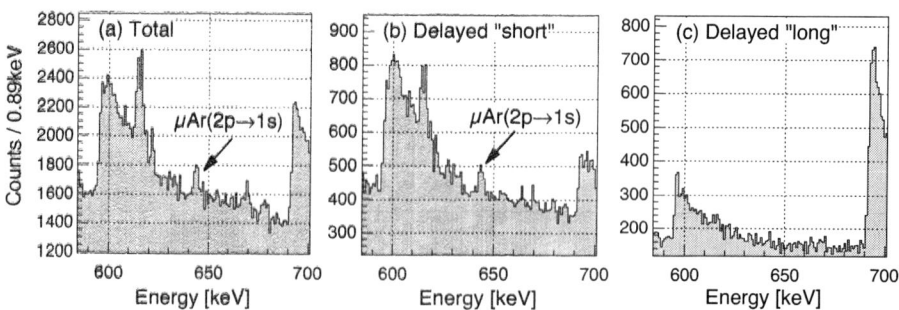

**FIGURE 6.** Germanium energy spectra measured with 0.5-mm $H_2/D_2 \oplus$ 7-$\mu$m $D_2$(Ar): (a) total, (b) delayed events from +75 to +250 ns after each muon pulse and (c) delayed events from +250 ns to 32 $\mu$s after the second muon pulse, respectively.

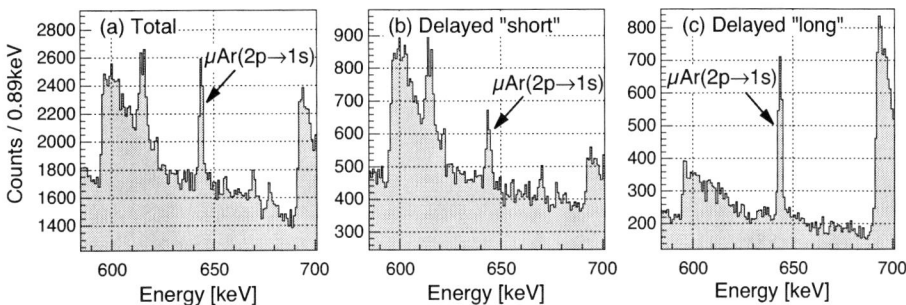

**FIGURE 7.** Germanium energy spectra measured with 0.5-mm $D_2$(Ar): (a) total, (b) delayed events from +75 to +250 ns after each muon pulse and (c) delayed events from +250 ns to 32 $\mu$s after the second muon pulse, respectively.

thickness of only 0.1 $\mu$m with a local concentration of 1000 ppm. The average Ar concentration throughout the $D_2$ layer was around 5 ppm, corresponding to approximately the same amount used in the previous target (i.e., $10^{16}$/cm$^2$). Figure 7 clearly shows that even with Ar ions implanted non-uniformly throughout the target, very strong muon transfer events can be detected. Due to the strong reduction of the Bragg cross-section at low $d\mu$ collision energy in solid $D_2$ [18], the $d\mu$ atom mean-free-path is strongly increased to nearly 10–20 $\mu$m below $\sim$ 1 meV, resulting in a high $d\mu$ atom mobility and a very long diffusion length. Preliminary analysis shows that the 644-keV X-ray disappearance rate is consistent with an average Ar concentration of 5 ppm. At this concentration, most of the muon transfer occurs delayed, and the S/N ratio can be greatly improved by selecting only delayed events as shown in Fig. 8, demonstrating the advantage of using a pulsed muon beam. Higher transition X-rays ($\mu$Ar 3p$\rightarrow$1s and $\mu$Ar 4p$\rightarrow$1s) from the muon transfer transfer reaction can be observed.

These first test experiments were performed with a rather large amount of implanted argon ions, because of the relatively low $\mu^-$ beam intensity on the target. At present, we are using a backward decay $\mu^-$ beam with a momentum of 27 MeV/c. The $\mu^-$ flux on the target through a $\phi$40-mm collimator is about 5000/sec with a repetition rate of 50 Hz. A stopping yield of nearly 60% in a 1-mm thick $D_2$ layer was achieved. With higher muon momentum, the beam intensity increases but as the energy spread of the beam. Therefore, the total number of muons stopping in the target does not improve much, while the overall S/N ratio is decreasing. Future intense muon beams with higher muon flux would require fewer implanted muons.

Further measurements were performed with reduced Ar concentrations to study the muon transfer reaction and also investigate the diffusion process of $d\mu$ atoms in solid $D_2$ films with Ar ions implanted non-uniformly. This effect will seriously limit the minimum layer thickness that can be used in the two-layer arrangement without being affected by the $d\mu$ loss.

A series of targets was measured with each Ar implantation separated from the next by depositing a fixed amount of $D_2$ to make a total $D_2$ layer thickness of 1 mm. As before, every implantation region was identical, with a local concentration of nearly

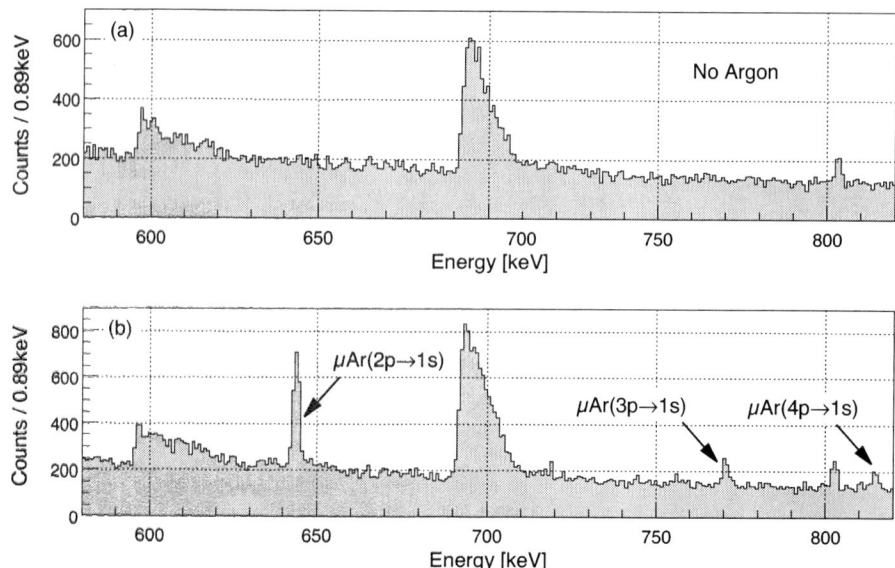

**FIGURE 8.** Delayed germanium energy spectra measured with (a) 1-mm $D_2$ only, and (b) 0.5-mm $D_2(Ar)$. Delayed events are from 250 ns to 32 $\mu$s after the second muon pulse.

1000 ppm, while the distance between each implantation was changed for each target. Figure 9 shows the total germanium energy spectra measured with 1-mm pure $D_2$ and (a) 20 implantations spaced 50 $\mu$m apart, (b) 10 implantations spaced 100 $\mu$m apart, and (c) 5 implantations spaced 200 $\mu$m apart, corresponding to an average Ar concentration of about 2, 1, and 0.5 ppm, respectively. Even with only 5 implantations, very strong muon transfer $\mu$Ar 2p→1s X-rays at 644 keV can still be detected. The amount of argon implanted in this target was $3 \times 10^{16}$ ions in $\sim$ 4-cm diameter and 1-mm thick $D_2$ film, or $\sim 3 \times 10^{15}/cm^2$. Unfortunately, due to a compressor trouble of the muon decay channel solenoid, only 12 hours of data could be taken.

The data analysis is now in progress and will be presented in a later publication. We can preliminarily conclude that an uniform implantation is not critical to observe muon transfer reaction in solid $D_2$. A d$\mu$ atom diffusion length of nearly 20 $\mu$m could be extracted, which is consistent with the predicted value based on calculated cross-sections. This effect limits seriously the minimum layer thickness that can be used in the two-layer arrangement, and makes the use of very thin layers not possible because of the large d$\mu$ loss.

## FUTURE PLANS

After the feasibility of this method has been clearly demonstrated, we plan to go one stage closer towards the generation and investigation of radioactive muonic atoms. As

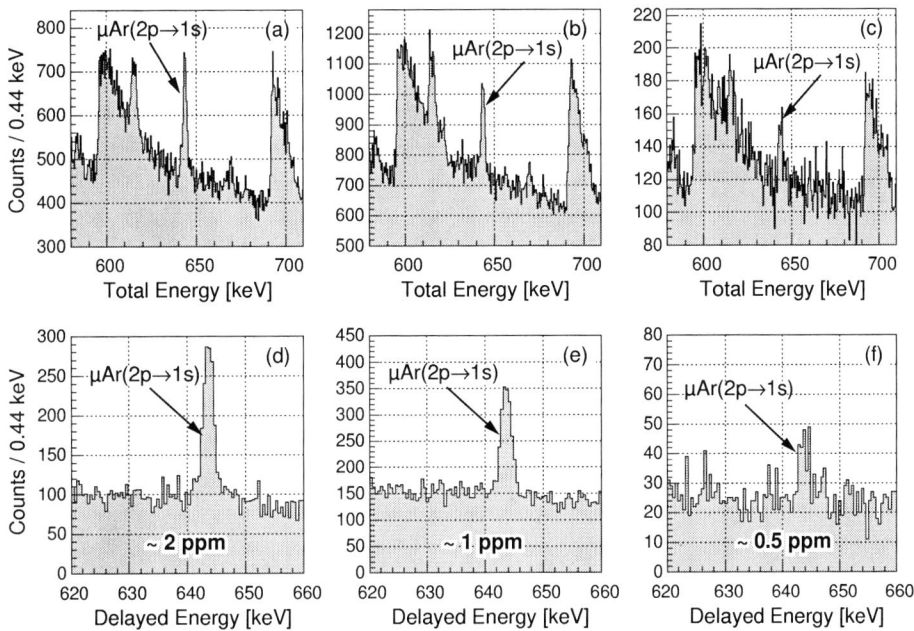

**FIGURE 9.** Germanium energy spectra measured with 1-mm pure $D_2$ and about 2, 1 and 0.5 ppm of argon ions implanted non-uniformly. The figures at the top (a, b and c) show the total energy spectra, while the ones at the bottom (d, e, and f) show delayed events only from 250 ns to 32 $\mu$s after the second muon pulse.

an intermediate step towards muonic spectroscopy with unstable nuclei, an experiment using long-lived isotopes is already under consideration.

An element selective ion source is now under development by Miyatake et al. [30]. One candidate element is $^{56}$Ni (neutron and proton magic, $T_{1/2} = 6.1$ d). Other interesting elements considered include $^{146}$Gd (neutron magic, $T_{1/2} = 48.3$ d), $^{148}$Gd (systematics, $T_{1/2} = 74.6$ y), and $^{126}$Sn (proton magic, $T_{1/2} = 10^5$ y).

Furthermore, radium isotopes are also of strong interest, since there are no stable isotopes for good measurements of nuclear parameters like the nuclear charge radius. These parameters would be urgently needed to exploit the full potential of the radium atom for atomic parity non-conservation studies [9].

A new surface ionization type ion source is been constructed with the aim of using in the future radioactive isotopes. This type of ion source is capable of producing ions from alkali and alkaline-earth metals with high efficiency. It is now being tested and will be installed on the existing experimental setup at port 4 at the end of this summer. At the moment, it will be used only to produce stable beams in order to optimize the new ion source and tune the beam transport optics. It will also be used to asses to the implantation system safety for future use of radioactive isotopes and determine the necessary modification that will be required. In the case of radium, since there are no stable isotopes, the best candidate is barium, which is in the same group and has a similar ionization energy. We propose to conduct muon transfer reaction studies with barium isotope ions implanted in solid deuterium films. Different targets with implanted barium isotopes will be measured in order to determine the muon transfer rate and the isotope shift. Transfer rate results will be compared to the data already obtained with argon ions. These measurements are crucial to this project in order to plan a further experiment using long-lived isotopes.

## ACKNOWLEDGMENTS

One of the authors (PS) acknowledge beneficial discussions with Prof. Y. Miyake and Dr. K. Shimomura concerning the "Super Omega" project. Gratitude is also extended to Profs. A. Taniguchi, S. Ichikawa, H. Miyatake and T. Shinozuka for their helpful collaboration in designing the new surface ionization ion source. This research was partially supported by the Japanese Ministry of Education, Science, Sports and Culture, Grant-in-Aid for Scientific Research (B), 2003, 15340091.

## REFERENCES

1.  L. Schaller, Z. Phys. C 56 (1992) S48.
2.  P. Strasser, T. Matsuzaki and K. Nagamine, Hyp. Int. 119 (1999) 317.
3.  P. Bricault, R. Baartman, M. Dombsky, C. Mark, L. Moritz, G. Stanford and P. Schmor, Nucl. Phys. A 701 (2002) 49c.
4.  H. Miyatake, S.C. Jeong, H. Ishiyama, Y. Ishida, H. Kawakami, N. Yoshikawa, I. Katayama, M. H. Tanaka, E. Tojyo, M. Oyaizu, S. Arai, S. Tomizawa, K. Niki, Y. Arakaki, M. Okada, Y. Takeda, M. Wada, P. Strasser, S. Kubono and T. Nomura, Nucl. Phys. A 701 (2002) 62c.
5.  I. Tanihata, Nucl. Phys. A 616 (1997) 56c.
6.  *Workshop on Radioactive Antiprotonic and Muonic Atoms (RAMA), February 23, 2001, CERN, Geneva, Switzerland.*
7.  *ETC\* Workshop on Radioactive Antiprotonic and Muonic Atoms (RAMA), May 22-26, 2001, European Centre for Theoretical Studies in Nuclear Physics and Related Areas (ETC\*), Trento, Italy.*
8.  R. Garoby, in *Proc. 18th Int. Conf. on High Energy Accelerators (HEACC 2001), March 26-30, 2001, Tsukuba, Japan*, CERN-NuFact-Note-74 (2001), e-proceedings available at http://conference.kek.jp/heacc2001/ProceedingsHP.html.
9.  K. P. Jungmann, Hyp. Int. 138 (2001) 463.
10. T. Nilsson, J. Äysto, K. Langanke, K. Riisager, G. Martinez-Pinedo and E. Kolbe, Nucl. Phys. A 746 (2004) 621c.
11. Y. Miyake et al., Y. Miyake, N. Kawamura, S. Makimura, P. Strasser, K. Shimomura, K. Nishiyama, J.L. Beveridge, R. Kadono, N. Sato, K. Fukuchi, K. Nagamine, W. Higemoto, K. Ishida, T. Matsuzaki, I. Watanabe, Y. Matsuda, M. Iwasaki and S.N. Nakamura Nucl. Phys. B (Proc. Suppl.) 149 (2005) 393.
12. K. Shimomura, K. Ishida, H. Miyadera and K. Nagamine, AIP Conf. Proc. 721 (2004) 346.
13. P. Strasser, T. Matsuzaki and K. Nagamine, Eur. Phys. J. A 13 (2002) 197.
14. S.S. Gershtein and L.I. Ponomarev, in *Muon Physics*, Vol. 3, edited by V.W. Hughes and C.S. Wu (Academic Press, New York, 1975), p. 151, 163-165, 190-194.
15. G. Backenstoss, H. Daniel, K. Jentzsch, H. Koch, H. P. Povel, F. Schmeissner, K. Springer and R. L. Stearns, Phys. Lett. B 36, 422 (1971).
16. P. Strasser, K. Ishida, S. Sakamoto, K. Shimomura, N. Kawamura, E. Torikai, M. Iwasaki and K. Nagamine, Phys. Lett. B 368, (1996) 32.

17. P. Strasser, K. Ishida, S. Sakamoto, K. Shimomura, N. Kawamura, E. Torikai, M. Iwasaki and K. Nagamine, Hyp. Int. 101/102 (1996) 539.
18. A. Adamczak, Hyp. Int. 119 (1999) 23.
19. T. Matsuzaki, K. Ishida, K. Nagamine, I. Watanabe, G. H. Eaton and W. G. Williams, Nucl. Instr. and Meth. A 465 (2001) 365.
20. P. Strasser, T. Matsuzaki and K. Nagamine, Nucl. Instr. and Meth. A 460 (2001) 451.
21. P. Strasser, T. Matsuzaki and K. Nagamine, Hyp. Int. 138 (2001) 497.
22. P. Strasser, K. Nagamine, T. Matsuzaki, K. Ishida, Y. Matsuda, K. Itahashi and M. Iwasaki, J. of Phys. G 29 (2003) 2047.
23. P. Strasser, K. Nagamine, T. Matsuzaki, K. Ishida, Y. Matsuda, K. Itahashi and M. Iwasaki, Nucl. Phys. A 722 (2003) 523c.
24. P. Strasser, K. Nagamine, T. Matsuzaki, K. Ishida, Y. Matsuda, K. Itahashi and M. Iwasaki, AIP Conf. Proc. 721 (2004) 309.
25. P. Strasser, K. Nagamine, T. Matsuzaki, K. Ishida, Y. Matsuda, K. Itahashi and M. Iwasaki, Nucl. Phys. A 746 (2004) 621c.
26. P. Strasser, K. Nagamine, T. Matsuzaki, K. Ishida, Y. Matsuda, and M. Iwasaki, Nucl. Phys. B (Proc. Suppl.) 149 (2005) 390.
27. B. Stenum, O. Ellegaard, J. Schou and H. Sørensen, Nucl. Instr. and Meth. B 48 (1990) 530.
28. J.F. Ziegler, J.P. Biersack, U. Littmark, in *The Stopping and Ranges of Ions in Matter*, Vol. 1, edited by J.F. Ziegler (Pergamon Press, New York, 1985); for further information and download see also http://www.srim.org/.
29. E. Gete, D. F. Measday, B. A. Moftah, M. A. Saliba and T. J. Stocki, Nucl. Instr. and Meth. A 388 (1997) 212.
30. H. Miyatake (private communication).

# Nuclear matter radii determined by interaction cross sections

A. Ozawa

*Institute of Physics, University of Tsukuba*
*Tsukuba, Ibaraki 305-8571, Japan*

**Abstract.** Experimental studies on nuclear matter radii determined by the interaction cross sections ($\sigma_I$) are reviewed. In particular, the procedure to determine the root-mean square matter radii from the measured $\sigma_I$ by Galuber model analysis is described. Future $\sigma_I$ measurements at the RI beam factory (RIBF) in RIKEN are introduced. As new calculations, the sensitivity of the skin is discussed in the case with a proton target based on Glauber-model calculations. In the energy region of RIBF, $\sigma_I$ is sensitive for the skin; however, measurements with high accuracies are needed.

**Keywords:** Measured interaction $\sigma$; Deduced r.m.s. matter radii; Glauber-model analysis
**PACS: 25.60.Dz**

## INTRODUCTION

The nuclear size and density distribution are important bulk properties of nuclei. To study the size and densities for unstable nuclei, measurements of the interaction cross sections ($\sigma_I$) and the reaction cross sections ($\sigma_R$) are indispensable [1]. Recently, $\sigma_I$ have been extensively measured at the FRS facility in GSI, where RI beams with relativistic energies (~1 $A$ GeV) are available. Using Glauber-model analysis, root-mean square (RMS) matter radii of unstable nuclei can be determined from the measured $\sigma_I$. We have determined RMS matter radii in the *p-sd* shell region [1] and proton-rich Cl and Ar isotopes [2]. In the near future, measurements of $\sigma_I$ will be performed at the RIKEN RI beam factory (RIBF), where the RI beam energies are around 300 $A$ MeV [3]. In RIBF, we will determine RMS matter radii for more neutron-rich nuclei and much heavier nuclei up to $Z$~40.

## GLAUBER MODEL CALCULATIONS

The most-used model for $\sigma_I$ is the Glauber model [1]. In the Glauber model, $\sigma_R$, which is given by $\sigma_I$ plus the inelastic cross sections, can be calculated by using a transmission function,

$$\sigma_R = 2\pi \int_0^\infty [1 - T(b)] b\, db \, , \tag{1}$$

*CP793 Physics with Ultra Slow Antiproton Beams*
edited by Yasunori Yamazaki and Michiharu Wada

where $T(b)$ is the transmission for the impact parameter, $b$. At relativistic energy, the inelastic cross sections are fairly small, and thus in measurements with relativistic energy we can ignore the inelastic cross sections and take into account $\sigma_I$ as $\sigma_R$ [2]. One of simplest approximations to calculate the transmission function is the optical limit. In this approximation, the profile function is replaced by the nucleon-nucleon cross sections ($\sigma_{NN}$) under the zero range limit. $T(b)$ can then be calculated from the nucleon density distribution and $\sigma_{NN}$ as

$$T(b) = \exp\left\{-\sum_{ij}\sigma_{ij}\int \rho_{Ti}^{z}(s)\rho_{Pj}^{z}\left(|\boldsymbol{b}-\boldsymbol{s}|\right)d\boldsymbol{s}\right\}, \qquad (2)$$

where $\rho_{ki}^{z}(s)$ is a $z$-direction integrated nucleon-density distribution,

$$\rho_{ki}^{z}(s) = \int_{-\infty}^{\infty}\rho_{ki}\left(\sqrt{s^2 + z^2}\right)dz . \qquad (3)$$

The index k=P (projectile) or T (target) and $\sigma_{ij}$ are $\sigma_{NN}$ in which indices $i, j$ are used to distinguish a proton and a neutron. The nucleon density distribution in the nucleus is written as $\rho_{ki}(r)$. $\sigma_{NN}$ has an energy dependence, as shown in Fig. 1. We took into account the energy dependence and difference of $\sigma_{pp}$ and $\sigma_{np}$ in our calculations.

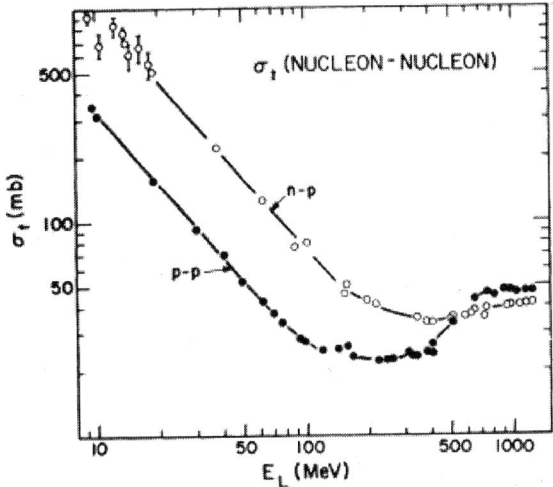

**FIGURE 1.** Energy dependence of the intrinsic nucleon-nucleon cross sections [4].

The choice of the shape of the density is an important issue. For light nuclei ($A<20$), a harmonic-oscillator (HO) type of density is sufficient to describe the experimental data. HO-type proton density-distributions, including contributions up to the $p$-shell, are given by the following equations:

$$\rho(r) = \frac{2}{\pi^{3/2}Z\lambda^3}\left\{1 + \frac{Z-2}{3}\left(\frac{r}{\lambda}\right)^2\right\}\exp\left(-\frac{r^2}{\lambda^2}\right). \qquad (4)$$

In this case, the parameter is only a size parameter, denoted by $\lambda$, if we assume the same parameter for proton and neutron distributions.

For heavier nuclei, good fits of the experimental data can be obtained if one assumes a distribution of the form

$$\rho(r) = \frac{\rho_0}{1 + \exp\left(\dfrac{r - r_0 A^{\frac{1}{3}}}{a}\right)} \quad . \tag{5}$$

In this case, there are two parameters to be determined: one is the radius parameter ($r_0$) and the other is the diffuseness ($a$). The RMS matter radii can be deduced by integrating the obtained density distributions. The presently deduced RMS matter radii are visualized in Fig. 2.

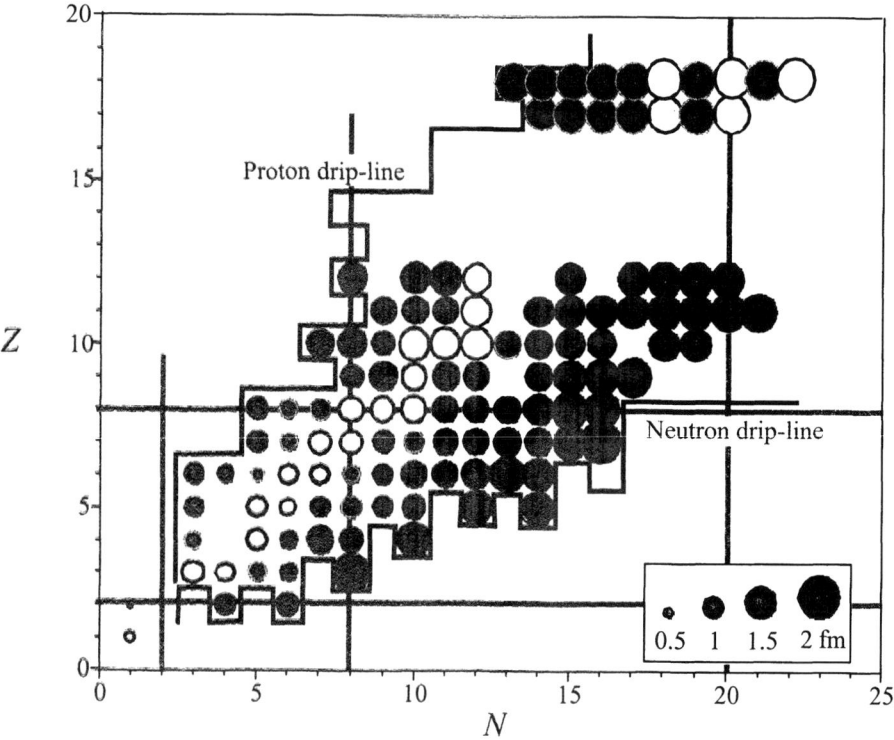

**FIGURE 2.** Nuclear matter radii determined from the interaction cross sections. The radius of $^4$He (1.47 fm) is subtracted. The grey (open) circles show unstable (stable) nuclei, respectively.

## FUTURE MEASUREMENTS OF $\sigma_I$ IN THE RI BEAM FACTORY AT RIKEN

The construction of RIBF is progressing at RIKEN. In 2006, construction of the superconducting ring cyclotron and the RI beam generator, BigRIPS [5], will be completed. Thus, in 2007, the first physics experiments can be performed in BigRIPS. We are planning to perform $\sigma_I$ measurements in BigRIPS. Since BigRIPS is a tandem fragment separator, as shown in Fig. 3, in the first stage up to F2, separation and purification of RI beams will be performed. Thus, we can use the second half of BigRIPS as a spectrometer to measure $\sigma_I$. In our setup, the reaction target will be located at the intermediate focusing point at F5 in BigRIPS, as shown in Fig. 3. The main difficulties in RIBF will be mixing different charge states in heavier nuclei, since the energy of the RI beam is around 300 $A$ MeV. Since, by preliminary calculations, at $Z\sim40$ with an energy of 300 $A$ MeV, the RI beam is not fully ionized and other 1 % charge states will be mixed, complete particle identification can not be performed. Thus, $\sigma_I$ will be easily measured in BigRIPS at $Z<40$. Measurements of the neutron-rich side will be limited by the RI beam intensity in BigRIPS. Our $\sigma_I$ measurements can be performed down to ~0.01 cps RI-beam intensity.

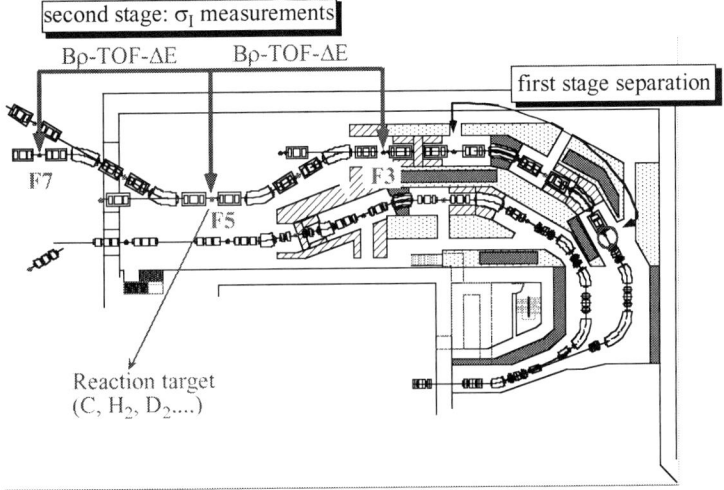

**FIGURE 3.** Experimental setup of BigRIPS [5] in RIBF.

# SENSITIVITY OF THE SKIN IN $\sigma_I$ WITH THE PROTON TARGET

According to Fig. 1, $\sigma_{pp}$ and $\sigma_{pn}$ are almost the same above 500 MeV. However, they are minimized at around 300 MeV, where $\sigma_{pp}$ and $\sigma_{pn}$ are different by a factor of 2. Thus, at this energy, $\sigma_I$ with the proton target is sensitive only to the neutron densities. On the other hand, $\sigma_I$ with a carbon target is not sensitive to neutron and proton densities, since the numbers of protons and neutrons are the same in the carbon target. Thus, we may deduce the skin thickness by measuring $\sigma_I$ with proton and carbon targets, respectively, at 300 $A$ MeV, which corresponds to the energy of RI beams produced at RIBF.

We checked the sensitivity of the skin in $\sigma_I$ with a proton target by using our Glauber model. Our calculation did not allow a proton target; thus instead of the proton target, $^2$He was the target in our calculations. We assumed $^{30}$Mg as a projectile with HO-type density. Thus, the parameters in our calculations were only the size parameters of the projectile density, $\lambda(p)$ and $\lambda(n)$, respectively. We assumed

$\lambda(p)=\lambda(n)=1.5$ fm, at first. In this case, there is no skin, and $\sigma_I$ with a carbon target at 300 $A$ MeV is 938 mb. To reproduce this number, we changed $\lambda(p)$ and $\lambda(n)$, respectively. Finally, we prepared three sets of $\lambda$'s. We calculated the RMS radii for protons and neutrons from the prepared sets, and calculated the skin, which corresponds to the difference of the RMS proton and neutron radii. Next, we changed the target to $^2$He and calculated $\sigma_I$ at 300 $A$ MeV using the prepared sets. The results of $\sigma_I$ with a $^2$He target are shown in Fig. 4. There is a sizeable slope in Fig. 4. Thus, it is shown that $\sigma_I$ with the proton target has some sensitivity to the skin. However, the sensitivity is smaller than the experimental uncertainty that we expected. In Fig. 4, 1 % experimental uncertainty is taken into account, which is a typical experimental error for $\sigma_I$ measurements [1]. If we measure $\sigma_I$ with the proton target precisely (<1 % accuracy) in RIBF, we can deduce the skin thickness by using $\sigma_I$ measurements alone. These measurements may be useful to study the skin thickness in unstable nuclei, since independent measurements for charge radii are not necessary.

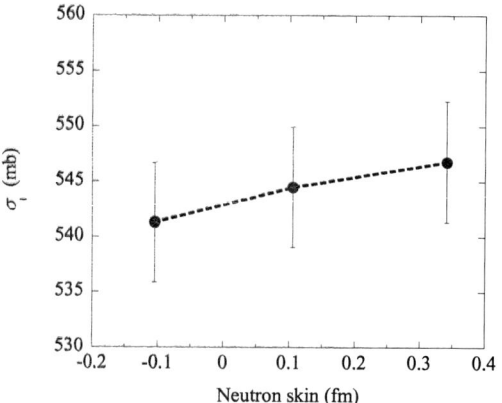

**FIGURE 4.** Calculated skin-dependence of $\sigma_I$ for $^{30}$Mg with a $^2$He target at an energy of 300 $A$ MeV. The expected uncertainty with 1% is shown by the error bars.

## SUMMARY

We introduced a Glauber model calculation with the optical limit approximation to determine matter radii from $\sigma_I$ measured at relativistic energies. Up to now, we have determined matter radii of the *p-sd* shell region and proton-rich Cl and Ar isotopes. Intrinsic $\sigma_{NN}$ have an energy dependence, and are different from $\sigma_{pp}$ and $\sigma_{pn}$ at ~300 MeV, which corresponds to the energy region of RI beams at RIBF. According to our Glauber model calculations, $\sigma_I$ with a proton target at ~300 $A$ MeV is sensitive to the thickness of the skin. However, measurements with high accuracies are necessary. Measurements of $\sigma_I$ will be performed at BigRIPS in RIBF. We expect that we can easily measure $\sigma_I$ up to $Z$~40.

262

# REFERENCES

1. A.Ozawa et al., *Nucl. Phys.* A 693 (2001) 32.
2. A.Ozawa et al., *Nucl. Phys.* A 709 (2002) 60.
3. Y.Yano et al., in: F.Marti (Ed.), Proceedings of XXth International Conference on Cyclotrons and their Applications, East Lansing, Michigan, American Institute of Physics, 2001, p. 161.
4. R.M.De Vries and J.C.Peng, *Phys. Rev.* C 22 (1980) 1055.
5. T.Kubo et al., *Nucl. Instr. Meth.* B 204 (2003) 97.

# Structure Studies of Unstable Nuclei by Electron Scattering

Toshimi Suda

*Heavy Ion Nuclear Physics Laboratory, RIKEN*
*Hirosawa, Wako, 351-0198, JAPAN*

**Abstract.** Electron scattering provides essential information on the internal structure of atomic nuclei. A new scheme, SCRIT(Self-Confining Radioactive Isotope Target), is proposed to study the internal structure of short-lived nuclei by electron scattering. Numerical simulations reveal that SCRIT provides high enough luminosity to determine the charge form factor by elastic scattering. A feasibility study of this novel scheme is now underway at an existing electron storage ring.

**Keywords:** electron scattering, SCRIT, Unstable nuclei, charge distribution
PACS: 21.10.Ft; 25.30.Bf; 29.25.-t

## INTRODUCTION

The size and shape are one of the most basic quantities of the atomic nucleus. The precise determination of the size and shape (i.e. nucleon density distributions) provides a critical test of our theoretical understanding of the nuclear structure. Indeed, the precise charge distributions determined by electron scattering for stable nuclei played an essential role to establish the nuclear structure models currently available.

The discovery of halo nuclei by pioneering work by I. Tanihata *et al.*, [1], however, raised a question about our understanding on the structure of the atomic nucleus. Halo nuclei, such as $^{11}Li$, have been discovered to have much larger spatial distributions than expected from the well-established $A^{1/3}$ rule of the nuclear radius for stable nuclei. This fact may indicate that our understanding of even the nuclear size is limited to stable nuclei, and the nuclear structure theory that we presently have is applicable only to a very limited number of nuclei near the stability line.

Today, a considerably broader range of $\beta$-unstable nuclei far from stability is accessible owing to successive technical improvements. It is now possible to probe the limits of nuclear existence, especially for light elements. Many new notable structures near the drip line have been discovered. They are, for example, the disappearance of the well-established magic numbers [2, 3], and the appearance of a new one close to the neutron drip line [4].

The root-mean-square (rms) nuclear-matter radii of many unstable nuclei have been deduced from interaction cross-section data measured using a high-energy radioactive isotope beam. Today, matter radii over a wide range of isotopes are available[5]. The rms charge radius can be accessed by isotope-shift measurements. By combining the matter and charge radius of an isotope, one can deduce the neutron radius. One example is Na, whose charge and neutron radii are both available over a wide range of neutron number. A growth of the neutron skin is observed when the Na isotope becomes more neutron rich [6].

The size and shape of a nucleus can not be, however, uniquely determined from the rms radius data only. The density distribution is necessary to be experimentally measured for the internal structure study. Little is still experimentally known about unstable nuclei.

It is of great importance to study how protons and neutrons distribute in an asymmetric system, such as proton-rich and neutron-rich nuclei, and how the distribution changes when the system becomes more asymmetric. In order to separately determine the proton and neutron distributions, one needs to determine the nuclear matter distribution and the charge distribution. Recently it has been demonstrated that the matter distribution can be rather accurately extracted from intermediate-energy proton elastic scattering data [7]. The elastic scattering cross section and the analyzing power for several (stable) isotopes of Sn have been measured using a polarized proton beam of 400 MeV, and their matter distributions have been extracted. Combining the matter distribution and the proton distribution from the electron-scattering data, the neutron distributions in Sn isotopes were extracted. A change in the neutron distribution has been as the number of neutron changes in Sn isotopes.

The charge distribution is best determined by electron scattering. Due to an obvious lack of targets, no charge distribution of short-lived unstable nuclei has ever been determined by electron scattering. In this report, a new scheme, which may make electron scattering from unstable nuclei possible, is discussed.

# ELECTRON SCATTERING

## Elastic Scattering and Charge Form Factor

Since the charge distribution of unstable nuclei is the main subject, we focus in this report only on *elastic* electron scattering. Because of a coherent contribution of all protons in a nucleus to elastic scattering, its cross section is the largest, at least up to a moderate momentum transfer, among those for the other scattering processes.

The differential cross section for elastic scattering from a spin-less nucleus under a Plane-Wave Impulse-Approximation (PWIA) is given as [8],

$$\frac{d\sigma}{d\Omega} = \frac{d\sigma_{Mott}}{d\Omega} |F_c(q)|^2, \qquad (1)$$

where $d\sigma_{Mott}/d\Omega$ is the Mott cross section, and $F_c(q)$ is the charge form factor. The Mott cross section is the elastic-scattering cross section from a point particle of charge $Z$,

$$\frac{d\sigma_{Mott}}{d\Omega} = \frac{(Z\alpha)^2 \cos^2 \frac{\theta}{2}}{4e^2 sin^2 \frac{\theta}{2}}, \tag{2}$$

where $e$ is electron energy, $\theta$ the scattering angle and $\alpha$ the fine-structure constant. The form factor is a Fourier component of the charge distribution, $\rho_c(r)$, at momentum transfer, $q$,

$$F_c(q) = \frac{1}{(2\pi)^{\frac{3}{2}}} \int \rho_c(r) e^{-i\vec{q}\cdot\vec{r}} d\vec{r}. \tag{3}$$

One determines $\rho_c(r)$ of a target nucleus by an inverse Fourier transformation of the experimentally determined charge form factor. There is a comprehensive data compilation of the charge form factor for stable nuclei obtained by electron scattering [9].

Figure 1 shows the results of a toy-model calculation for the charge form factor of Sn to demonstrate the sensitivity of the form factor to a change in the charge distribution. A Fermi-type distribution is assumed, and the parameters for the size and diffuseness are changed for ±5% and ±10%, respectively. The dip position and the height of the diffraction maxima are found to be separately sensitive to the change of the distribution parameters.

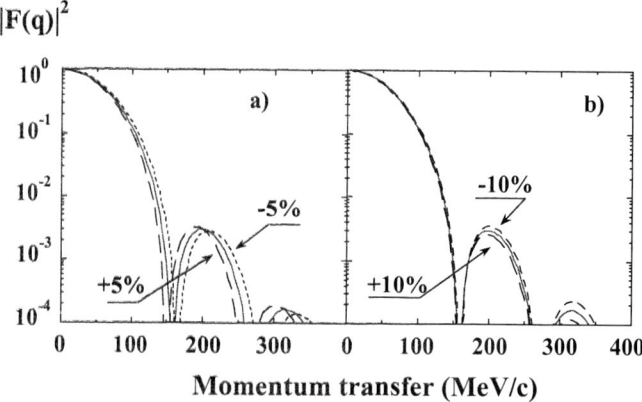

**FIGURE 1.** Result of a model calculation of the charge form factor for Sn. The parameters for the size and diffuseness in the Fermi distribution are changed for ±5% and ±10%, respectively.

For larger $Z$ nuclei, such as Sn isotopes, the PWIA framework may not be adequate due to serious distortion of the incoming and outgoing electron waves. A distorted-wave calculation was carried out with a calculation code, DREPHA [10], to check how the cross section behaves against changes of the size and diffuseness parameters.

Figure 2 shows the results for Sn. In the figure, the cross sections divided by the Mott cross section is plotted as a function of the momentum transfer. Although the diffraction dip is found to become less clear owing to the Coulomb distortion, the cross section is confirmed to still have the same sensitivity to changes in the size and diffuseness, as seen in the PWIA case.

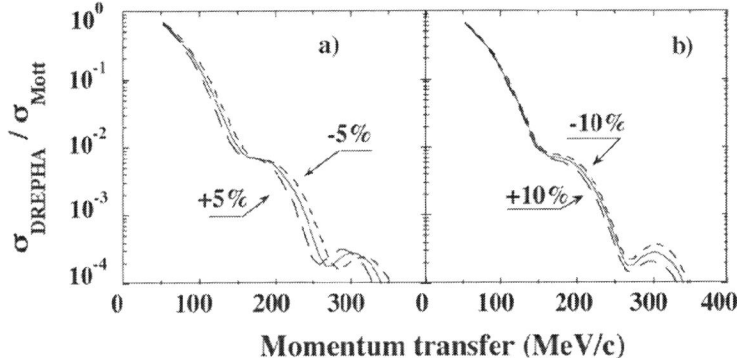

**FIGURE 2.** Result of a distorted wave calculation, DREPHA, of the elastic scattering cross section for Sn. The parameters for the size and diffuseness in the Fermi distribution are changed for ±5% and ±10%, respectively.

By definition, the charge distribution, $\rho_c(r)$, can be determined model-independently, if one measures the form factor up to an infinite $q$. Practically, however, the form factor can be measured only in a finite $q$ range. The low luminosities, unavoidable at electron-RI scattering facility, will seriously hinder the measurement up to a high-momentum transfer region due to a nearly logarithmic reduction of the cross section as a function of $q$. Since our goal is to determine at lease the size (radius) and shape (surface diffuseness) of the charge distribution, it is necessary to determine, at least, the dip position and the first diffraction maximum of the form factor.

## Required Luminosity

The required luminosity for an electron-RI scattering experiment is determined from the requirement that the charge form factor is necessary to be measured at least up to the first maximum. It is, therefore, strongly coupled to the solid angle of an electron spectrometer for electron-RI scattering experiments. Due to the facts that a quite low luminosity is expected and an identification of the elastic scattering is essential, both a large acceptance, such as a $2\pi$ azimuthal angular coverage, and a high resolution of $\Delta p/p \sim 10^{-4}$, are demanded for an electron spectrometer. Since such a spectrometer is not practical, a high-resolution spectrometer currently in operation is assumed in the following discussion. At present, this may be a sensible choice in our mission, since a high resolution instead of a large acceptance is essential for the charge form-factor measurement, where identification of the elastic scatting process is indispensable.

Figure 3 shows the expected yields of the elastic scattering events from $^{132}$Sn under the assumption of a luminosity of $10^{28}$/cm$^2$/s. The measurement is assumed to continue for a week. In the calculation, the electron energy is 300 MeV, and an electron spectrometer having an azimuthal acceptance of ± 70 mrad is assumed.

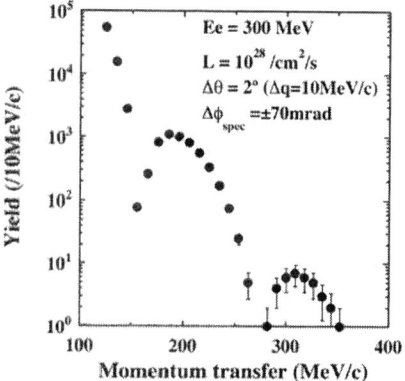

**FIGURE 3.** Expected yields for $^{132}$Sn(e,e') in one-week measurement. It is assumed that the luminosity is $10^{28}$ /cm$^2$/s, and an electron spectrometer covers an azimuthal angle of ± 70 mrad.

It is clear from the figure that the luminosity of an order of $10^{28}$ /cm$^2$/s is required for electron scattering from Sn isotope to cover the diffraction maximum in the charge form factor with a reasonable measuring time of one week, as long as a high-resolution spectrometer having smaller acceptance is employed.

## SCRIT: A NOVEL METHOD FOR E-RI SCATTERING EXPERIMENT

There may be several ways to realize not-yet-performed electron-scattering experiments off radioactive isotopes (RI). One way is to employ an electron-RI (eRI) collider, where two storage rings for electron and RI beam have a colliding section. Due to poor emittance of the secondary RI beam produced by fragmentation, a cooling process is necessary before sending the RI beam to an eRI collider in order to achieve a higher luminosity. An eRI collider, thus, is an accelerator complex of an RI generator, a cooler ring and a collider consisting of two storage rings for RI and electrons.

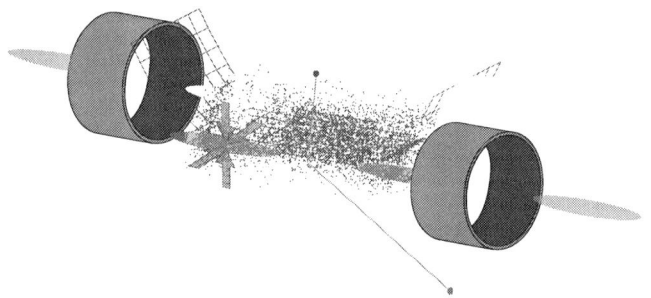

**FIGURE 4.** SCRIT concept. Electron bunches give focusing kicks to the ions, which results in a transverse confinement. By placing electrodes along a circulating electron beam for longitudinal confinement, one can form a localized trapped target.

We once carried out feasibility studies intensively on the collider scheme at RIBF [11]. Due to a duty factor mismatching between our RI-beam generator, cyclotrons, and storage rings in the collider, a sufficiently high luminosity was found to be unachievable. As an example, the expected luminosity for $^{132}$Sn was $\sim 10^{23}$/cm$^2$/s, which is far too low compared with the required luminosity of $\sim 10^{28}$/cm$^2$/s. A eRI collider is now planned in the GSI future project.

Instead of employing a costly collider configuration, we are proposing a completely new method to realize the eRI scattering experiment in much more cost effective way. This novel method, named SCRIT (Self-Confining RI Target), is based on the ion-trapping phenomena, known at electron storage rings, such as synchrotron-radiation facilities.

Ion trapping is a phenomenon observed at electron storage rings: circulating electron bunches ionize residual gases, and the bunches give a focusing kicks to the ionized gas ions to be trapped. Consequently the ionized residual gas distributes on the circulating electron beam. Since the ion trapping reduces the performances of electron-storage rings, such as shortening beam lifetime and introducing even beam instability, many efforts have been made so far to either reduce or remove the trapped ions from the circulating electron beam.

SCRIT uses this ion-trapping phenomenon to trap ions of interest transversely on a circulating electron beam. By placing electrodes in the longitudinal direction to form a mirror trapping potential, one may form a localized RI target on the electron beam, as schematically shown in Fig. 4. Since an eRI scattering facility based on this SCRIT concept requires only an electron storage ring and an ion source; this is very cost effective compared with an eRI collider where large accelerator complex is needed. Instead, very low-energy RI will be injected to SCRIT. A low-energy RI generator, such as ISOL or a gas-catcher facility which follows a fragment separator, is necessary.

Since the SCRIT concept is completely new, we started to perform a numerical simulation to study whether a high enough luminosity for eRI scattering can be achievable. The conclusion of the numerical simulations is that a luminosity of an order of $10^{27}$/cm$^2$/s can be expected when the number of injected ions is $10^7$/s. The electron beam current is assumed to be 500 mA. The details of the numerical simulations are described elsewhere [12].

# SCRIT Feasibility Study

Encouraged by the results of the numerical simulation, we constructed a SCRIT prototype in order to experimentally investigate its feasibility. The goals of this study are to confirm the following issues:

1. Is a trapped ion target really formed inside the circulating electron beam ?
2. Is a high luminosity really achievable, as expected by the numerical simulation?
3. Is it possible to detect elastically scattered electrons close to a circulating electron beam of a few 100 mA ?

Issue 1 is to confirm that ions of interest can be injected on an electron beam from an external ion source without giving any disturbance to the electron ring performance. To check issue 2, we measured the luminosity experimentally by observing bremsstrahlung which is a direct consequence of an overlap of ions with an electron beam. Issue 3 is to demonstrate the feasibility of the SCRIT system for an electron-scattering experiment. A SCRIT prototype, including a luminosity monitor and an electron detection system, was constructed, and then installed at an existing storage ring, KSR, of Kyoto University, Japan [13]. The ring has a 2-m straight section, which fits our requirement for the SCRIT installation, as shown in Fig. 5. An electron beam of 100 MeV with a maximum current of 100 mA is to be stored. Under this condition, a luminosity of an order of $10^{26}$/cm$^2$/s is expected with $10^7$/s ion injection.

## Electron Storage Ring, KSR

The KSR (Kyoto Storage Ring) is a race-track type electron-storage ring, whose circumference is 25.6 m. The maximum electron energy is 300 MeV. The electron beam of 100 MeV is injected to KSR from an electron linac. The RF frequency is 116.7 MHz. The beam size at the 2-m straight section is $\sigma_x = 0.8$ mm and $\sigma_y = 0.18$ mm, respectively. A typical beam life-time at 100 MeV is about 100 s for 100 mA, which is mainly determined by the vacuum of KSR, $10^{-8}$ Pa.

## The SCRIT Prototype

The constructed SCRIT system consists of an ion source, a transporting system, a deflector and a trapping electrode system, as shown in Fig. 5. Figure 6 shows photos of the SCRIT system, and close views of the trapping electrode system.

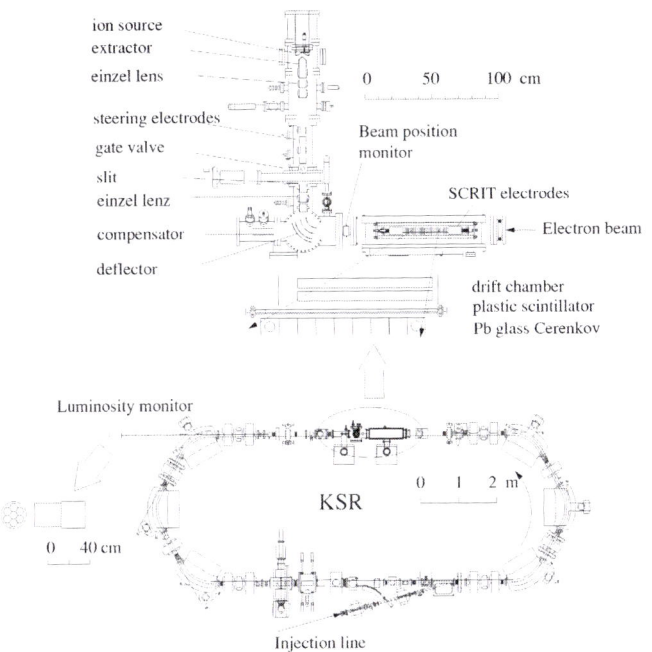

**FIGURE 5.** SCRIT prototype installed at KSR, Kyoto University.

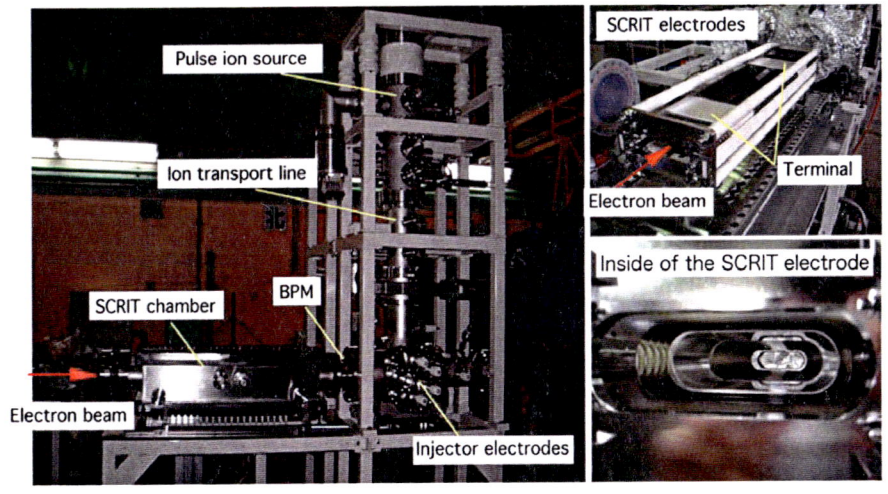

**FIGURE 6.** SCRIT prototype and SCRIT electrodes.

A surface ionizing ion source is employed for this SCRIT prototype, which is simple and provides a high quality ion beam to SCRIT. $Cs^{1+}$ ions pulsed by a grid were extracted from the ion source by an extractor, where 10 kV was applied. The extracted ions were transported by two Einzel lenses and four steering electrodes to the deflector electrode. At the deflector, the ions were merged with a circulating electron beam. A transverse kick, that the electron beam experienced at the deflector, was always compensated by a compensator electrode installed on the electron beam line. Once the ions are merged with the electron beam, they are automatically transported to the SCRIT electrodes with 10 keV kinetic energy. When they reach the trapping electrodes, they start to 'claim up' the longitudinal potential. The potential barrier height was adjusted so as to introduce 10 keV ions inside the trap. When the pulsed ions were completely inside, the potential barrier was switched on to longitudinally trap the ions inside. By adjusting the potential height inside the trapping barrier, one could freely control the kinetic energy of ions inside the trap.

The trapping electrode system consists of 42 isolated electrodes including two terminal electrodes. They are stacked every 11 mm. Each electrode has 100x35 mm$^2$ of aperture for the electron beam. The electrodes have a mesh structure for the path of scattered electrons to minimize multiple scattering. Since all of the electrodes are electrically independent, one can form any potential shape by applying a high voltage for each electrode from outside through a feedthrough.

## Luminosity Monitor

We attempt to determine the luminosity of trapped ions and electron beam by observing bremmstrahlung at downstream of the straight section of the KSR. The luminosity monitor consists of optically-isolated seven BaF$_2$ crystals, as shown in Fig. 5, with a thin plastic scintillator in front of the BaF$_2$ detector for charged-particle veto. Each crystal has a hexagonal cross section with a diameter of 7 cm, and a length of 20 cm. The detector is surrounded by Pb blocks. A pair of collimators with 10-mm diameter is placed in front of the BaF$_2$ detector to remove background, and to restrict the γ-ray irradiation only to the central crystal. The BaF$_2$ detector is energy-calibrated by using an extracted 60-MeV electron beam from the KSR, and has been confirmed to have an energy resolution of 7% (FWHM) for 100-MeV γ-rays. The expected counting rate for the luminosity of $10^{26}$/cm$^2$/s is 500 Hz for $E\gamma \geq 20$ MeV.

## Electron Detection System

An electron detector consists of a drift chamber, 12 plastic scintillators and 9 calorimeters, as shown in Fig. 5. Due to a geometrical limitation, the detection system has been installed below the SCRIT chamber. It is designed to cover scattering angles of 30-80°. With the beam energy of 100 MeV, the corresponding momentum transfer range is 50-130 MeV/c, as shown in Fig.7, where the elastic cross section dominates.

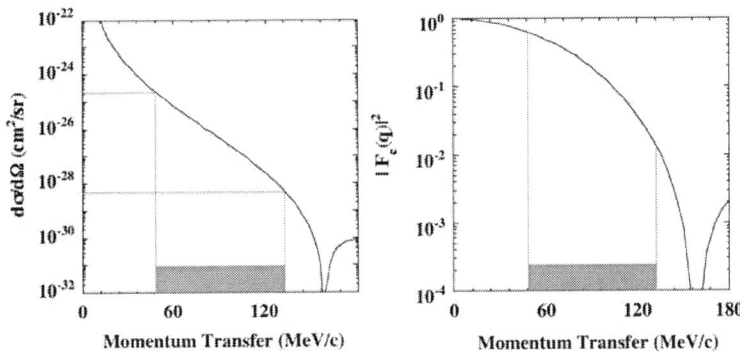

**FIGURE 7.** Momentum transfer region covered by the SCRIT prototype for $E_e$ = 100 MeV.

All electrons scattered at SCRIT are designed to pass through a 1-mm Be window of the SCRIT chamber to minimize multiple scattering. The drift chamber has 128 hexagonal-shape drift cells, and the drift length is 18 mm. These cells are arranged so that at least 4 cells are fired for an electron track to reconstruct its trajectory. A premixed He+$C_2H_6$ (50:50) gas is used to make the chamber be insensitive as possible to low-energy photons originated from electromagnetic showers. The performance of the drift chamber has been studied using cosmic rays, where a position resolution of 200 $\mu$m has been achieved. During test runs with a stored electron beam of 80 mA, the drift chamber was confirmed to function normally.

Two layers of six plastic scintillators are placed under the drift chamber as a trigger system. The size of each scintillator is 200 x 300 x 10 $mm^3$. A coincidence between two layers triggers data processing. The total energy of the scattered electron is to be measured by 9 calorimeters. A Pb-glass Cerenkov detector is employed as a calorimeter. Prior to installation, all detectors were energy-calibrated using a mono-energetic electron beam. Although the Pb-glass Cerenkov detector has a poor energy resolution, scattered electrons from the trapped ions under the present kinematics can be safely assumed only from elastic scattering. The expected counting rate integrated over an angular range of $\theta$=30-40°, where the elastic cross section is about 100 mb/sr, is about 10 Hz for the expected luminosity of $10^{26}$/$cm^2$/s.

## RESULTS AND DISCUSSIONS

## SCRIT at KSR

A series of test experiments were started just after installing the SCRIT prototype in KSR. Special attention has been paid to examine any negative influence on the KSR performances, such as the beam lifetime and the maximum storage beam current, as a result of the installation. This is because the following facts may shorten the beam lifetime or limit the maximum storage beam current. The SCRIT prototype has electrodes at very close to circulating electron beam. They are for producing a longitudinal trapping potential, and for deflecting the ion beam to merge with the electron beam. A high voltage on the order of 10 kV is applied for all electrodes. For the electrodes to form the trapping potential, a high voltage is quickly switched on/off to inject and trap the ions, and to release them from SCRIT. In addition, the deflector electrode always gives a transverse kick to the electron beam, although it is canceled, in principle, by a compensator.

The results of the measurements showed that the SCRIT operation does not make any serious influence on the KSR performance. Most of the R&D efforts made so far were to observe evidence of ion trapping in this novel SCRIT system. We employed two ways to examine the ion trapping:

1) extracting ions from SCRIT
2) measuring bremsstrahlung γ-rays with the luminosity monitor.

## Extracted Ions from SCRIT

The trapped ions were extracted from SCRIT by opening the confinement barrier potential. When the potential gate is opened, the trapped ions fall through the (longitudinal) trapping potential. They return back the injection line through the deflector with the 10-keV kinetic energy followed by detection at an electrode placed at the double collimator installed in the ion-transport line.

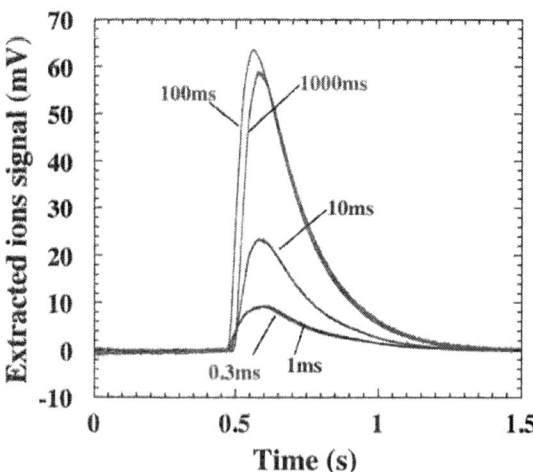

**FIGURE 8.** Signals due to the extracted ions from SCRIT. The trapping time is indicated.

Figure 8 shows signals of the electrodes for detecting the extracted ions, which should be proportional to the detected total charge. The trapping duration in SCRIT is indicated inside the figure. In these measurements, the $Cs^{1+}$ ions were injected as a pulsed beam of 100 μs. The electron beam was continuously injected to KSR in order to keep the averaged beam current to be 80 mA.

One notices that the detected total charge rapidly increased, after 10 ms, as a function of the trapping time. This may be due to continuous ionization of the trapped ions by the electron beam. Because of this ionization, the averaged charge state of the trapped ions is expected to increase as a function of the trapping duration.

Figure 9 is a plot of the detected total charges (solid circle) as a function of the trapping duration. In the figure, the result of a numerical calculation for estimating the total charge in SCRIT is also shown, which is normalized at the trapping time of 300 $\mu$s. The calculation is found to reproduce the time dependence of the detected total charges up to one second. Since the calculation assumed that all ions distribute inside the electron beam, this agreement indicates that the most of detected ions are really trapped inside the electron beam. Since most of the trapped ions at 300 $\mu$s after injection can be safely assumed to be still singly charged, we determined the number of detected ions to be 4 x $10^6$ at 300 $\mu$s. Correcting for the inefficiency of the released ion collection, more than $10^7$ Cs ions were finally concluded to be trapped in SCRIT. It is very close to our expectation from the numerical simulation [12]. The number of trapped ions in SCRIT was determined as a function of time using the calculated charge state of the trapped ions, as shown in Fig. 9. As a conclusion, an order of $10^7$ Cs ions are stably trapped in SCRIT for an order of a few 100 ms. The number of trapped ions, then, start to decrease after one second, where highly charged ions can not be trapped anymore due to the Coulomb repulsion force.

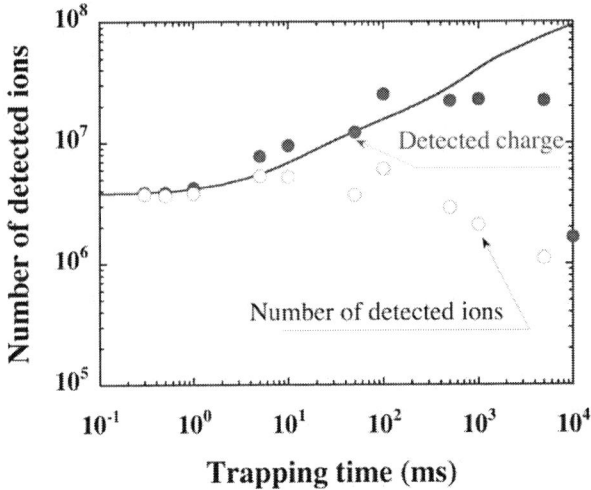

**FIGURE 9.** Extracted total charges and calculated number of the trapped ions as a function of the trapping duration. The solid and open circles show the detected total charges and the calculated number of ions, respectively. The solid line is the result of a calculation on the time dependence of the total charges of the trapped ions.

## Luminosity Monitoring

The growth of the detected total charge of the extracted ions from SCRIT as a function of time, as shown in Fig. 9, strongly supports that the trapped ions are well overlapped with circulating electrons in KSR. This is, however, still not a sufficient proof of the practicability of the SCRIT system for eRI scattering. It is essential to determine the luminosity, and to confirm that the expected luminosity by the simulation is really achieved.

In parallel to the measurements of ions extracted from SCRIT, we have tried to determine the luminosity by the $BaF_2$ detector to measure bremmstrahlung. Since a luminosity of an order of $10^{26}/cm^2/s$ was expected, an enhancement of the $\gamma$-ray flux of $\sim 500$ Hz above $E\gamma \geq 20MeV$ was foreseen. In addition to bremsstrahlung from the trapped CS ions, bremsstrahlung from residual gas ions trapped by electron beam contributes. Since the $BaF_2$ luminosity monitor was placed downstream of the straight section of KSR, the monitor also counts bremsstrahlung from residual gas in the vacuum pipe of the straight section. The integrated luminosity over a straight section of 8.5 m under the average vacuum pressure of $10^{-8}$ Pa is an order of $10^{27}/cm^2/s$, which results in a $\gamma$-ray flux of $\sim 1.5$ KHz.

It has been turned out, however, that the counting rate of the luminosity monitor at a storaged beam current of 80 mA was very high, $\sim 10$ KHz. No enhancement of $\gamma$-flux due to Cs ion injection has been observed so far.

It is still not yet clear where the origin of the huge number of $\gamma$ comes from. One of the reasons for the high $\gamma$-flux may be attributed to beam loss at the straight section, which initiates an electromagnetic shower. Due to relatively a short beamlife time of 100 s, the beam loss rate of KSR is rather high, approximately $10^7/s$ in total. In order to reduce the beam loss at the straight section, the beam pipe, where $\beta$-function becomes maximum, is to be modified.

As an additional monitor, the detection of characteristic $X$-ray (30 keV for Cs ion) is planned. The detection of the characteristic $X$-rays is also direct evidence that the Cs ions are overlapped with the electron beam. The characteristic $X$-ray is emitted when an innermost electron of Cs ion is knocked out by the electron beam. Since the Moeller cross section for this process is on the order of 100 b, one can expect $\sim 10^3$ characteristic $X$-rays. A Ge detector will view the target region through a Be window.

## CONCLUSION

We have been proposing a novel scheme to realize electron-scattering experiments for internal-structure studies of short-lived radioactive isotopes (RI). This scheme, SCRIT, is based on well-known ion-trapping phenomena at electron-storage rings. SCRIT forms a localized RI target on a circulating electron beam in a storage ring, and is found by a numerical simulation to provide a luminosity of more than $10^{27}/cm^2/s$. This luminosity is enough to measure the elastic scattering cross section to determine the charge distribution.

In order to demonstrate its feasibility, a SCRIT prototype has been constructed, and installed at an existing electron storage ring, KSR, Kyoto University. A series of test measurements is underway. The test experiments conducted so far have already revealed that the SCRIT prototype does not have any negative influence on the storage-ring performances, such as the lifetime and maximum storage beam current.

By means of extracting ions from SCRIT, we have clearly confirmed that externally injected $Cs^{1+}$ ions were stably trapped in SCRIT about one second, as expected in a numerical simulation. The SCRIT system has been demonstrated to work as a new type of ion trap. The number of trapped ions was estimated to be an order of $10^7$, which is also in good agreement with the expectation. The growth of the charge states of detected ions strongly indicates that all trapped ions distribute inside the electron beam, since the ions inside the electron beam are continuously further ionized by the electron beam.

In this R&D experiment, the luminosity is expected to be on the order of $10^{25}$/cm$^2$/s, whereas no direct evidence of this luminosity has yet been observed in the bremmstrahlung measurements. A modification of the beam transport is to be made in order to reduce the huge background for the bremsstrahlung detector. In addition, another detection system is to be installed to obtain direct evidence for the overlapping of trapped ions with the electron beam.

## ACKNOWLEDGMENTS

This work is financially supported by Grants-in-Aid for Scientific Research from Japan Society for the Promotion of Science.

## REFERENCES

1  Tanihata, I. *et al. Phys. Rev. Lett.* 55 (1985) 2676
2  Motobayashi, T. *et al.*, *Phys. Lett.* B346 (1995) 9
3  Navin, A. *et al.*, *Phys. Rev. Lett.* 85 (2000) 266
4  Ozawa, A. *et al..*, *Phys. Rev. Lett.* 84 (2000) 5493
5  Ozawa, A., Suzuki, T. and Tanihata, I., *Nucl. Phys.* A 693 (2001) 32
6  Suzuki, T. *et al.*, *Phys. Rev. Lett.* 75 (1995) 3241
7  Takeda, H. *et al.*, AIP Conference Proceedings Vol.675, 720 (2003)
8  deForest, T. Jr. and Walecka, J. D., *Adv. in Phys.* 5 (1966) 1
9  de Vriese, H., de Jager, C.W. and de Vries, C., *Atomic Data and Nuclear Data Tables* 36 (1987) 495
10  DREPHA, Phase shift code for the calculation of elastic electron scattering,
     B. Dreher and J. Friedrich, private communication.
11  Katayama, T., Suda, T. and Tanihata, I. , *Physica Scripta* T104 (2003) 129
12  Wakasugi, M., Suda, T. and Yano, Y., *Nucl. Instrum. and Meth.* A532 (2004) 216
13  Shirai, T. *et al.*, Proc. of the 7th European Acc. Conf. (2000) 442.

# Highly charged ions at rest: The HITRAP project at GSI

F. Herfurth, T. Beier, L. Dahl, S. Eliseev, S. Heinz, O. Kester, H.-J. Kluge, C. Kozhuharov, G. Maero, W. Quint and the HITRAP collaboration

*GSI, Planckstr. 1, 64291 Darmstadt, Germany*

**Abstract.** A decelerator will be installed at GSI in order to provide and study bare heavy nuclei or heavy nuclei with only few electrons at very low energies or even at rest. Highly-charged ions will be produced by stripping at relativistic energies. After electron cooling and deceleration in the Experimental Storage Ring the ions are ejected out of the storage ring at 4 MeV/u and further decelerated in a combination of an IH and RFQ structure. Finally, they are injected into a Penning trap where the ions are cooled to 4 K. From here, the ions can be transfered in a quasi dc or in a pulsed mode to different experimental setups. This article describes the technical concepts of this project as well as planned key experiments.

**Keywords:** highly charged ions, QED test
**PACS:** 07.90.+c,12.20.Fv

## 1. INTRODUCTION

The theory that is normally regarded as best understood is quantum electrodynamics (QED). This theory is, for instance, able to predict with remarkable accuracy atomic level energies in the hydrogen atom. However, in the very high electromagnetic fields of highly-charged heavy ions the usual perturbative approach is no longer applicable. Experimentally, such high fields become accessible by stripping heavy atoms off all or most of their electrons. There are two approaches: stripping of the projectile at high energy or collisions with high-energy electrons in an ion source. If moderately charged ions are accelerated and sent through a target they will lose all or most electrons in collisions with the target atoms if the ion energy is high enough. Similarly, atoms placed in an electron beam will be stripped sequentially off their electrons, if the energy of the electrons in the beam is high enough. It has been demonstrated that both approaches can produce heavy highly-charged ions (HCI) [1, 2]. The most intense source of bare uranium ions is at GSI where the high-energy stripping approach is used.

For precision experiments the HCI must be prepared at low energy, i.e. at nearly rest in space or with only a few keV total energy with a very small energy spread. For this the produced HCI will be injected into the Experimental Storage Ring (ESR) [3], electron cooled and decelerated to an energy of 4 MeV/u. After ejection out of the ring the ions are further decelerated by a combination of IH and RFQ structures and injected into a Penning trap. Here the ions are accumulated and cooled to 4 K. For this purpose the decelerator-cooler facility, called HITRAP, has been designed and is now under construction at GSI.

In addition to HCI of stable nuclei, those of radioactive ones will become available.

Radionuclides are produced and separated by use of the fragment separator (FRS) [4] and then injected into the ESR. In the more distant future, HITRAP will be a part of the Facility for Low Energy Antiproton and Ion research (FLAIR) at the planned international accelerator Facility for Antiproton and Ion Research (FAIR). There, HITRAP will not only provide low-energy highly-charged stable and radioactive ions but also low energy antiprotons.

## 2. EXPERIMENTS WITH HIGHLY-CHARGED, HEAVY IONS

### 2.1. Precision experiments with single ions

Precision measurements of basic quantities have been conducted successfully using HCI. Examples are the electronic g-factor, the binding energy of the electrons, and the atomic mass. The electronic g-factor of hydrogen-like ions can be used to test the theory of quantum electrodynamics in strong fields. In turn, fundamental constants like the electron mass or the fine-structure constant $\alpha$ can be determined by these measurements if the results of reliable quantum electrodynamics calculations are used as input [5]. The mass of HCI is another important quantity. A determination of the mass of the ion in different charge states gives a new access to the binding energy of the inner electrons. Measured for different isotopes, nuclear binding effects can be determined with high precision.

The g-factor measurement is based on the continuous Stern-Gerlach effect [6] where the g-factor of the bound electron

$$g = 2 \frac{\omega_L}{\omega_C} \frac{mq}{Me} \tag{1}$$

is determined by the ratio of the Lamor frequency $\omega_L$ of the electron to the cyclotron frequency $\omega_C$ of the ion [7, 8]. Additional quantities needed are the ion mass $M$, its charge state $q$ and the mass $m$ and charge $e$ of the electron. Remarkable accuracy can be reached since the frequency of the oscillations of an ion confined in a Penning trap can be determined with extremely high accuracy.

If the electronic g-factor of the bound electron can be calculated reliably to an accuracy that matches that of the experiment, then the electron mass can be extracted from equation (1). Using the g-factor measurements of $O^{7+}$ and $C^{5+}$ ions [9, 10] the electron mass has been obtained with a four-fold improved accuracy [11]. When these measurements can be extended to heavier ions and the QED calculations can be further improved, then the fine structure constant $\alpha$ can be determined in a way completely independent from present procedures [5].

Some of the most precise mass measurements have been performed using HCI [12]. For this the ion is stored in a Penning trap and its cyclotron frequency $\nu_C$ is measured. Then the mass $M$ can be determined using the relation

$$\nu_C = \frac{1}{2\pi} \frac{q}{M} B. \tag{2}$$

The magnetic field strength $B$ is determined via an ion with a well known mass or even using $^{12}C$ which is the microscopic mass standard. Since the cyclotron frequency depends linearly on the charge state $q$ of the ion, a higher resolving power and hence accuracy can be reached [13]. The pioneering experiment for mass measurements on HCI [12] reaches a relative mass uncertainty of about $10^{-10}$ by use of a trap at room temperature and a destructive detection technique. Mass measurements with a precision of $2 \cdot 10^{-11}$ have been performed by the use of singly-charged ions in a cryogenic Penning trap [14]. Here, the motion of only a single ion stored in the trap has been detected directly and non-destructive. Combining both methods, as possible at HITRAP, will push the relative mass uncertainty even below $1 \cdot 10^{-11}$.

By this, it will be possible to measure the mass differences between bare, H-like, He-like and so on uranium ions to 2 eV or better. Such measurements would provide the most accurate measurement of the 1s Lamb shift in hydrogen-like heavy ions and provide a stringent test of atomic theory including electron-electron correlations in strong fields. Furthermore, the possibility to produce radioactive nuclei using the FRS will enable for the first time mass measurements with such high precision on radioactive nuclei. Since these measurements are limited by the half-life of the nucleus of interest and the cooling and preparation will take about 10 seconds, the half-life of the isotope under investigation should be longer than 10 seconds.

## 2.2. Experiments with beams of slow heavy highly-charged ions

Atomic collisions at low energy provide valuable and indispensable data for many areas of physics, e.g. plasma physics, accelerator physics, and astrophysics. Atomic structure and collision dynamics can be studied by kinematically complete collision experiments using the COLTRIMS technique [15, 16]. For HCI charge exchange is the dominating process at energies of less than a few keV/u. Single or multi-electron capture occurs into high-lying states of the highly charged ion which forms strongly inverted systems, so-called hollow atoms. The initial capture process as well as the deexcitation via x-rays or Auger electrons can be monitored in detail by detecting in coincidence the recoil ion, the emitted electrons and photons (in the x-ray as well as in the optical spectral region) and by determining their energies.

Electron capture processes will be studied with the available beam intensity of about $10^4$ HCI per second. For this the beam of HCI is directed onto a target of, for instance, helium atoms in a supersonic jet. The determination of the recoil ion momentum along with the projectile properties gives informations about the energy difference between initial and final state, i.e. the Q-value. If performed for different states this enables high-resolution spectroscopy of the HCI energy levels. The presently achieved resolution of 0.7 eV and accuracy between 3 and 300 meV [17] is already competitive with the methods of conventional spectroscopy. In the future, an improved momentum resolution can be envisaged and hence a reduction of the uncertainty towards the sub-meV level.

Highly charged ions can be used to deposit large amounts of potential energy on surfaces. For instance, the potential energy of several hundred keV of a single bare uranium ion is deposited in a very small area of just a few square nanometers. This high energy density leads in general to a highly non-linear response of matter. For experiments it is, however, important to distinguish between processes induced by the potential energy of the ion due to the high charge state and those induced by kinetic energy. This requires well defined beams of very low kinetic energy and very small energy spread.

The measured quantity is the number of electrons emitted from the surface during the collision. If this is investigated for different charge states information can be gained on the non-linear response of semi-conductors and insulators to strong Coulomb fields and on defect-induced mobility in the limit of high defect density. The main parameter is the speed at which the insulating material can replenish electrons in the nanometer-sized impact region. A wide range of semiconducting and insulating surfaces produced by thin-film coating shall be studied using this method. In first experiments ions in low charge states were impinging on thin films of LiF [18] and $C_{60}$ [19] evaporated on Au. These experiments indicate that electron spectra depend strongly on the properties of the thin film.

For a certain range of kinetic energy it might happen that the impinging ion is repelled from the surface. This so-called trampoline effect occurs because the positive charges, created on the surface by the first electrons transfered to the incoming ion, are not compensated fast enough. So far experimental evidence for this effect could not be found [20]. Up to now the experimental search was limited by the low charge states available and the large energy spread for beams of only a few eV. Both limitations will be eliminated by use of the low-energy beam available at HITRAP.

## 2.3. Photon spectroscopy of highly charged ions at rest

Generally electronic transitions in HCI are in the far ultraviolet or X-ray spectral range, beyond the access by laser light in the visible spectral region. However, for very highly charged ions, for instance hydrogen-like $^{207}$Pb, the ground-state hyperfine transition (21 cm microwave radiation for atomic hydrogen) is moved into the spectral region accessible by conventional lasers. Since also the lifetime of these states (proportional to $Z^{-9}$) lies in the microsecond or millisecond rangelaser spectroscopy can be performed with high precision. In this way, the magnetic sector of high-field QED can be tested and, if successfully done, the distribution of the nuclear magnetization can be extracted.

At the HITRAP facility laser spectroscopy will be performed in a cryogenic Penning trap kept at 4 K. This offers several advantages: The ions are stored as a dense and well localized cloud (up to 100 000 ions stored within a few cubic millimeters) enhancing this way the overlap with the laser beam. Due to the low temperature the resonance is barely Doppler broadened. In combination with an easily achievable high collection efficiency for the fluorescence photons these properties will allow for an excellent signal-to-noise ratio. Furthermore, the long storage time enables one to optically pump the ions even using the rather weak M1 transition and to reach a high degree of electronic and nuclear polarization.

281

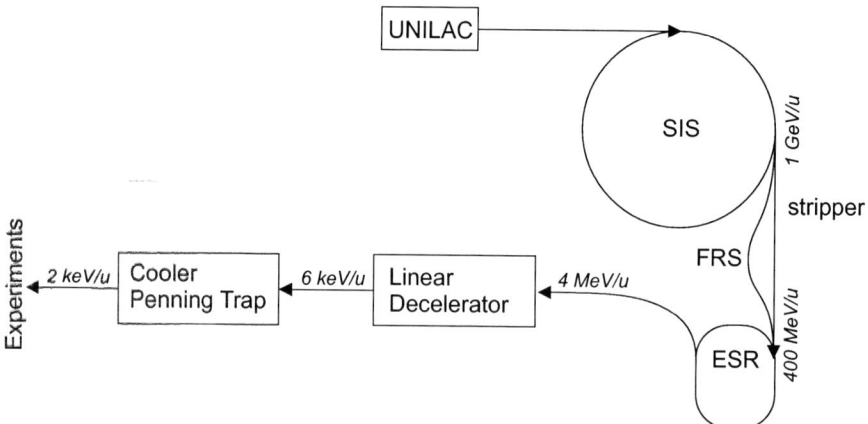

**FIGURE 1.** Overview of the planned HITRAP facility.

# 3. THE PLANNED HITRAP FACILITY

An overview of the planned facility is presented in Fig. 1. After production the ions will be decelerated down to 4 MeV/u using the Experimental Storage Ring (ESR) at the present GSI facility. Similarly, at the future FAIR facility the New Experimental Storage Ring (NESR) followed by the Low Energy Storage Ring (LSR) will be used for HCI and antiprotons deceleration. This will be accompanied by electron cooling in the storage ring, such that the emittance of the beam does not grow. Finally, the ions will be collected into a single bunch, extracted from the ESR and injected into a linear decelerator. After re-bunching with a double drift buncher (DDB) the beam will enter an interdigital H-type (IH) structure and will be decelerated to 0.5 MeV/u. Then it will be matched in longitudinal and transverse direction to a four-rod Radio Frequency Quadrupole (RFQ) decelerator. The final energy after the RFQ will be as low as 6 keV/u.

The beam properties of the different structures are listed in Tab. 1. Since there is no cooling applied in the linear decelerator section the absolute emittance increases considerably with decreasing energy. Further emittance growth occurs due to non-

**TABLE 1.** Beam properties along the decelerator

| Part of the decelerator | Energy keV/u | Emittance* $\pi$ mm mrad | Energy spread % | Remaining intensity[†] % |
|---|---|---|---|---|
| ESR - IH | 4 000 | 1 | $1 \cdot 10^{-1}$ | 100 |
| IH - RFQ | 500 | 13 | 4 | 67 |
| RFQ - cooler trap | 6 | 100 | 8 | 50 |
| cooler trap - experiment | $\approx 2$ | 1 | $1 \cdot 10^{-1}$ | 45 |

* not normalized
[†] simulated

282

linearities in the RF-gap fields and due to ion optical aberrations. The beam after the RFQ will have about $100\,\pi$ mm mrad transversal emittance and an energy spread calculated to be in the order of $\pm 4\%$. In order to further slow down the antiprotons or highly-charged ions, the beam is captured in a Penning trap. There, electron cooling and, subsequently, resistive cooling will be applied to cool up to $10^5$ HCI.

A cylindrical Penning trap in a magnetic field of 6 T will be kept at the temperature of liquid helium, 4 K. This will ensure the best possible vacuum that is needed to store antiprotons or HCI for an extended period. When injected into the strong magnetic field the ions or antiprotons will be decelerated further to energies below 2 keV/u. The strong magnetic field prohibits the transversal blow up of the beam in this phase. After in-flight capture by closing the trap after the ion bunch entered, the ions or antiprotons will be first cooled by interaction with a dense electron plasma. After about 1 s the ions or antiprotons will be separated from the electrons and stored in a harmonic electric field region. Here, it is planned to apply resistive cooling in order to cool the particles to final temperatures close to 4 K, equivalent to energies below 1 meV. Then the ions or antiprotons will be ejected and sent to experiments in either of two ways, as a quasi dc beam of $10^5$ particles distributed over up to 10 s or as a pulse with a bunch length of a few microseconds.

## 3.1. The IH structure and the double-drift buncher

H-type structures are the most efficient ion accelerator structures in the low-$\beta$ range due to their high shunt impedance. For heavy ions in the low energy range ($< 10$ MeV/u) the IH-structure is the best choice within drift tube structures since effective acceleration voltages up to 6 MV/m can be established. The HITRAP IH-structure operates at 108.408 MHz and 0.2% duty cycle. It has 25 gaps, one triplett inner-tank quadrupol lens and is 2.7 m long. An overall effective acceleration voltage of 10.5 MV is required to decelerate ions with a mass-to-charge ratio of 3 to 500 keV/u, the energy required for the injection into the RFQ [21]. The cavity design is optimized in order to reach an effective shunt impedance of 255 MΩ/m. In that case 165 kW power would be sufficient to reach the required tank voltage and a 180 kW power amplifier can be used which is available at GSI.

The ions extracted from the ESR are continuously distributed over a macro bunch with a length of 1.2 $\mu$s [22]. Thus the ions spread out over 360° RF-phase. The typical phase acceptance of the IH-decelerator is about 10°-15°. Therefore a buncher is required, which matches a larger fraction of the particles into the IH phase acceptance. A single harmonic buncher has a maximum phase acceptance of 130°, which corresponds to 37% of the particles bunched into the 10° phase acceptance of the HITRAP-IH-structure. A much better efficiency is obtained by using a multi-harmonic re-buncher or a double-drift buncher (DDB) [23]. However, a multi-harmonic buncher is difficult to implement due to the complicated adjustment of phases and amplitudes of the different harmonics. A DDB has the same bunching efficiency like a three-harmonic driven single buncher, i.e. about 67%. Therefore, a DDB has been chosen for the HITRAP setup, which is located 6.1 m in front of the IH-structure. It uses a 108.408 MHz cavity and a 216.816 MHz cavity separated by a 0.8 m drift. Both cavities are designed as coaxial quarter-wave resonators with four gaps each. The calculated shunt impedance of both cavities is larger than 60 MΩ/m and therefore RF generators with a power of 2 kW are sufficient to drive the cavities.

## 3.2. The radio frequency quadrupole structure

The design of the HITRAP 4-rod-RFQ is very similar to the 108 MHz structure of the high charge state injector LINAC at GSI [24]. However, the RF-power requirements are much more relaxed due to the lower A/q, which is <3 instead of 9. The low A/q allows a short structure of 143 cells with a lenght of 1.9 m and a maximum rod voltage of 70 kV. The mean aperture radius is about 4 mm which reduces the peak fields to safe values. The phase spread of ±40° of the ion bunches extracted from the IH-structure must be matched to the RFQ acceptance of ±10°. Thus in the matching section between IH-structure and RFQ, an existing two-gap spiral re-buncher will be used that has to be adapted to the right cell length. The beam at the exit of the RFQ has an emittance of 100 π mm mrad in both transverse directions.

The rod design had to be optimized with respect to the transverse angular spread of the ions at the RFQ exit. A small beam size and large angular spread are the result of strong quadrupole focusing. Ion beams with such properties would require lenses with large apertures close to the end of the rods. For sufficient transport to and injection efficiency into the cooler trap a minimum energy spread of the ions is required. To reduce the energy spread of ±7% from the RFQ a low energy de-buncher will be installed at the RFQ's exit. Hence a drift length of 200 mm towards the first focusing lens of the low energy beam line is necessary. Simulations have shown that the energy spread of the ions can be reduced to ±4% using a single-harmonic buncher. The transmission is about 90% in that case. With a saw tooth like RF-excitation of the buncher, a final energy spread of ±3% and full transmission could be reached.

## 3.3. The low energy beam line and the cooler trap

### 3.3.1. The low energy beam line

The main challenge for the low energy beam line is the transport of a beam with very large angular divergence and an energy spread of ±4 % (see also Tab. 1). Great care has to be taken when designing the ion optical elements in order to reduce geometrical as well as chromatic aberrations to a minimum. Usually electrostatic quadrupole lenses would be very well suited for beams with an energy of a few keV per charge [25], since the voltages at the electrodes can be kept low and the power supplies are cheap. However, due to the large angular spread quadrupoles are not advantageous in comparison to cylindrical einzel lenses. Furthermore, ion beams with a very low mass-to-charge ratio (< 3) are very vulnerable to stray magnetic fields and even to the magnetic field of the earth. Therefore, the distance between two focal points should be less than 1 m. For

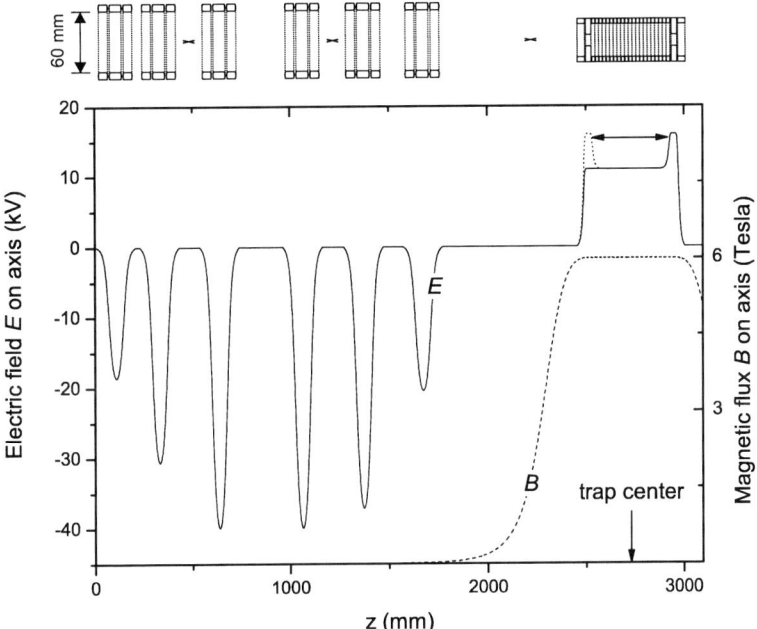

**FIGURE 2.** Plot of electric (full line) and magnetic field (dashed line) on the symmetry axis of the low-energy beam transfer line between RFQ ($z = 0$) and cooler trap (center at $z = 2750$ mm). The full line indicates the electric field when the ion bunch is injected into the trap and electrostatically retarded by 10 kV; the dotted line represents the case when the ion bunch is inside the trap and confined. The full line with two arrows indicates the trapped ion beam with an energy of 6 keV/u. The arrangement of einzel lenses and the trap electrode structure is shown in the upper part of the figure. Crosses mark focal points.

differential pumping at least two diaphragmas have to be foreseen, i.e. two focal points along the transfer line between RFQ and cooler trap. The presently best solution seem to be electrostatic einzel lenses in accelerating mode, relying on a spherically symmetric beam that is produced by the RFQ by shaping the RFQ rods accordingly.

An arrangement of six einzel lenses will generate the necessary two focal points and enable the matching of the beam into the magnetic field of the cooler trap. The electric and magnetic fields along the transfer line are plotted in Fig. 2. By the use of einzel lenses in the accelerating mode aberrations can be kept low since the lenses can be small and still the beam does not fill the lens too much.

## *3.3.2. The cooler trap*

The cooler trap serves to capture the highly charged ions in flight and to decrease its phase space volume and hence to cool the ion beam. A cylindrical Penning trap in a magnetic field of 6 Tesla is used. The trap will consist of 23 electrodes that can be used to shape the potential very flexibly and to create three harmonic trap regions for ions

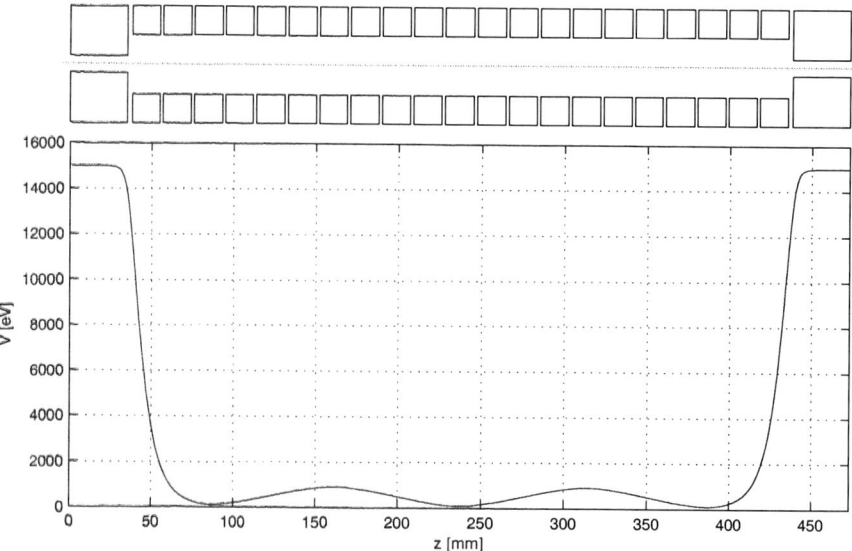

**FIGURE 3.** Schematics of the trap electrodes (top) and calculated potential on the axis of the cylindrical electrodes (bottom).

and two for electrons. The central electrode will be split into eight radial segments to enable different schemes of radial excitation and observation of the ion cloud. The two outermost electrodes have a reduced inner diameter for differential pumping. They are biased to about 18 kV in order to confine the ions as for instance $^{238}U^{92+}$ entering the trap with an energy of 6 keV/u. Figure 3 shows the design of the trap electrodes along with a possible potential slope on the symmetry axis of the cylindrical electrode stack.

The first step of the cooling process is the in-flight capture. For this, the ion bunch with a pulse length of about 1 $\mu$s will be injected into the trapping region and decelerated further from 6 keV/u to approximately 2 keV/u while the ion beam is kept radially small by the strong magnetic field. The incoming bunch is reflected at the downstream trapping electrode and trapped by switching the upstream trapping electrode from the initial potential of $\approx 11$ kV to about 18 kV just after injection of the ion bunch. The switching has to be performed within less than 400 ns.

During the second step the HCI interact with two electron clouds which are stored simultaneously. The electrons have a kinetic energy corresponding to 4 K since they emit synchrotron radiation during their storage in a high magnetic field and hence are in equilibrium with the surrounding. To avoid ion-electron recombination it is necessary to separate the ion and electron clouds as soon as their velocities get similar, i.e. for an ion energy below about 10 eV. Since dielectronic recombination is not possible for bare ions and the electron densities are still too low for three-body recombination processes, only radiative recombination must be taken into account. Calculations show that more than 90% of the bare uranium ions sent initially into the trap remain in the original charge state within the projected cooling time of a few seconds [26].

286

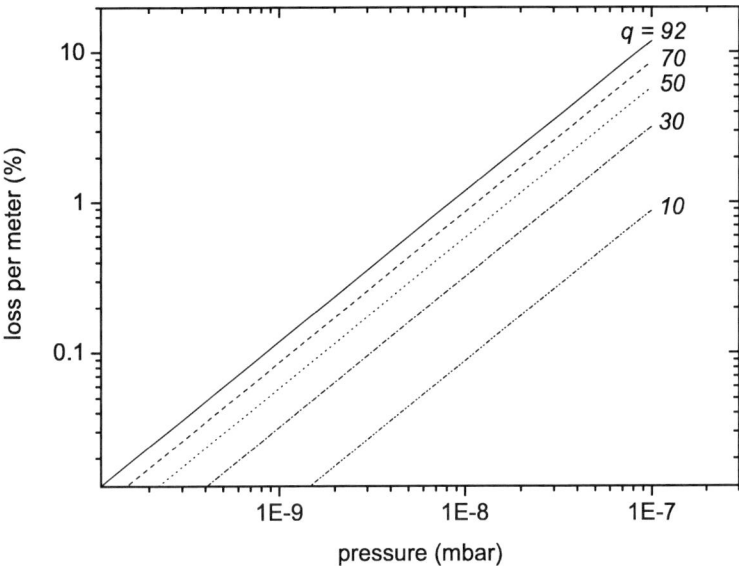

**FIGURE 4.** Loss per meter for highly-charged ions for the energy range between 10 eV and 25 keV versus pressure for different charge states

After separation of electron and ion clouds the ions will be cooled further using resistive cooling. The ion cloud oscillating in the Penning trap induces an image charge and hence an image current in the trap electrodes. If this signal is detected and fed into a resonant circuit kept at 4 K, the ion motion is damped [8].

Two modes of extracting the cooled ion cloud are foreseen: slow extraction and bunched extraction. If extracted in a bunch, the scheme is very similar to existing trap facilities where up to 100 million singly charged ions are extracted in a 15 $\mu$s pulse [27]. This mode is especially well suited such experiments where the ions need to be recaptured in a trap. The slow extraction mode will allow one to distribute the 100 000 ions over the 10 s that are available until the next decelerated ion bunch arrives. This mode is for example well suited for the ion-surface interaction experiments in order to avoid detector saturation.

### 3.3.3. Vacuum requirements

Due to the high charge states the cross sections for electron capture are rather high. Therefore, the residual gas pressure has to be kept as low as possible.

For heavy, HCI as for instance $U^{92+}$ no experimental data exists for ion energies

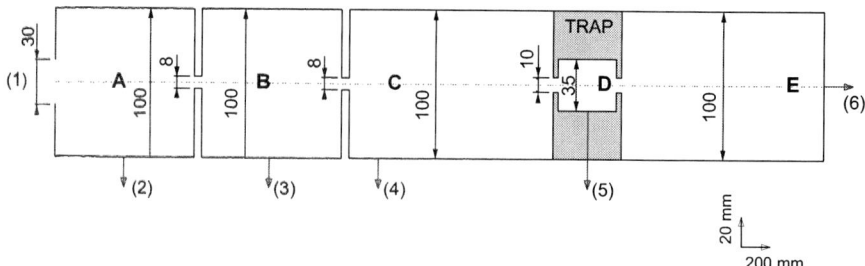

**FIGURE 5.** Sketch of the vacuum system for the low energy beam line between RFQ decelerator and cooler trap and for the cooler trap (all dimensions in mm). The letters name the sections. The arrows indicate positions where pumping takes place. The RFQ will be connected at position (1).

around 2 keV/u relevant for the low-energy beam lines of HITRAP. Therefore, two theoretical approaches have to be used instead: the classical over-barrier model [28] and an empirical formula by Schlachter [29] that is fitted to experimental data for slightly lower charge states. Both, model and formula, depend on the ionization potential of the residual gas atom and the charge state of the ion. For simplicity a constant ionization potential of 10 eV has been assumed, which is well below the one for common residual gas components like $O_2$, $H_2$ or $H_2O$ and hence yields rather pessimistic values. Then, the cross sections for single-electron capture from $U^{92+}$ into $U^{91+}$ are $3 \cdot 10^{-13}$ cm$^2$ for the classical over-barrier model and $5 \cdot 10^{-13}$ cm$^2$ for the Schlachter formula. Neither the over-barrier model nor the Schlachter formula depends on the energy of the ion. They are both regarded as valid for ion energies between 10 eV and 25 keV/u. The uncertainties are very hard to estimate for the Schlachter result, since it is in fact an extrapolation of experimental data. The cross sections calculated with the classical over-barrier model are normally valid within 20%. For ion energies below 10 eV, relevant for the last cooling stage in the trap, the semi-empirical model by Olsen and Salop is valid [30]. In this model the charge exchange cross section depends linearly on the relative velocities of the collision partners and gives for very low energy (equivalent to 4 K) slightly higher cross sections than the other two formulas.

Using this information it is possible to calculate the fraction of the ion beam that will undergo a charge exchange reaction and hence be lost for the experiment. The fractional loss per meter is calculated using the Schlachter formula as a function of the residual gas pressure and plotted in Fig. 4. Evidently, a residual gas pressure below $1 \cdot 10^{-9}$ mbar is required to keep the beam losses below 1% for a 10 m beam line. Since all estimates were done using the less favourable values for ionization energy and cross section a pressure below $1 \cdot 10^{-9}$ mbar should be sufficient for the transfer beam lines.

The vacuum requirements for the trap itself are of course much more stringent. Here, it should be possible to store heavy HCI at least for 10 seconds without losses due to vacuum conditions. Therefore, the pressure has to be well below $1 \cdot 10^{-13}$ mbar assuming above mentioned cross sections for electron capture. This is one of the reasons why the trap region will be kept at a temperature close to 4 K.

The vacuum system sketched in Fig. 5 is based on ultra-high vacuum technology. Turbo-molecular pumps will be used during the bake-out period. After bake-out the system will be pumped by standard ion getter pumps ($300\,\mathrm{l\,s^{-1}}$), indicated in the figure by the arrows numbered (2), (3), (4) and (6). The system has been analyzed using first-order principles like tube conductances at different temperatures [31]. An out-gassing rate of $10^{-12}\,\mathrm{mbar\cdot l\cdot s^{-1}\,cm^{-2}}$ has been assumed after bake-out for all surfaces outside the trap region. The pressure in the RFQ is assumed to be about $10^{-8}$ mbar (position (1) in Fig. 5). Assuming $H_2$ as main gas component one calculates a pressure of $5\cdot 10^{-9}$ mbar in section A and already $2\cdot 10^{-10}$ mbar in section B. In sections C and E the pressure is estimated to $1\cdot 10^{-11}$ mbar. The cryopumping effect of the trap has been accounted for by simulating a pump at position (5). The assmption of a pumping speed of $5\,\mathrm{l\cdot s^{-1}cm^{-2}}$ [32] yields a total pumping speed of about 12 000 l/s for the available surface at 4 K. A pressure of $< 10^{-13}$ mbar is obtained inside the trap with the installed diffusion barrier with an inner diameter of 10 mm. Even lower pressure can be envisaged if this diffusion barrier is opened only for the short period of time when the ion bunch is injected into the trap every 10 seconds.

# 4. STATUS

HITRAP is a project to produce and slow down highly charged ions up to bare uranium. The construction of the major parts of the decelerator has been started in the beginning of 2005. The superconducting magnet will be ordered mid of 2005. Many of the experimental setups presented in this paper are already prepared and are proceeding with tests using lower charged ions.

HITRAP is also an integral part of the low-energy antiproton and ion facility FLAIR. There the evaluation of the experimental program yielded very positive results for FLAIR and the technical design issues are being worked out.

# ACKNOWLEDGMENTS

We would like to thank those who contributed to the HITRAP technical design report. Many of the information presented here has been taken from this report. We thank also M. Kauschke and C. Welsch for valuable discussions. We acknowledge furthermore the support by the European Commission under contract number HPRI-CT-2001-50036 (HITRAP) and by the German Ministry for Education and Research (BMBF).

# REFERENCES

1. K. Blasche, and B. Franzke, "Status report on SIS-ESR," in *Proceedings of the 4th European Particle Accelerator Conference, London, 1994*, edited by V. Suller and Ch. Petit-Jean-Genaz, World Scientific, Singapore, 1994, p. 133.
2. J. W. McDonald, R. W. Bauer, and D. H. Schneider, *Rev. of Sci. Instr.*, **73**, 30–35 (2002).
3. B. Franzke, *Nucl. Instr. and Meth.*, **B 24**, 18–25 (1987).
4. H. Geissel, P. Armbruster, K.-H. Behr, A. Brünle, K. Burkard, M. Chen, H. Folger, B. Franczak, H. Keller, O. Klepper, B. Langenbeck, F. Nickel, E. Pfeng, M. Pfützner, E. Roeckl, K. Rykaczewski, I. Schall, D. Schardt, C. Scheidenberger, K.-H. Schmidt, A. Schröter, T. Schwab, K. Sümmerer, M. .Weber, G. Münzenberg, T. Brohm, H.-G. Clerc, M. Fauerbach, J.-J. Gaimard, A. Grewe,

E. Hanelt, B. Knödler, M. Steiner, B. Voss, J. Weckenmann, C. Ziegler, A. Magel, H. Wollnik, J. Dufour, Y. Fujita, D. Vieira, and B. Sherrill, *Nucl. Instr. and Meth.*, **B70**, 286 (1992).

5.  T. Beier, H. Häffner, N. Hermanspahn, S. Djekic, H.-J. Kluge, W. Quint, S. Stahl, T. Valenzuela, J. Verdú, and G. Werth, *Eur. Phys. J. A*, **15**, 41 – 44 (2002).
6.  G. Werth, H. Häffner, and W. Quint, *Adv. At. Mol. Opt. Phys.*, **48**, 191–217 (2002).
7.  L. S. Brown, and G. Gabrielse, *Rev. of Mod. Phys.*, **58**, 233–311 (1986).
8.  H. Häffner, T. Beier, S. Djekic, N. Hermanspahn, H.-J. Kluge, W. Quint, S. Stahl, J. Verdú, T. Valenzuela, and G. Werth, *Eur. Phys. J. D*, **22**, 163 – 182 (2003).
9.  V. A. Yerokhin, P. Indelicato, and V. Shabaev, *Phys. Rev. Lett.*, **89**, 143001 (2002).
10. J. Verdú, S. Djekic, S. Stahl, T. Valenzuela, M. Vogel, G. Werth, T. Beier, H.-J. Kluge, and W. Quint, *Phys. Rev. Lett.*, **92**, 093002 (2004).
11. P. J. Mohr, and B. N. Taylor, *Rev. Mod. Phys.*, **77**, 1–107 (2005).
12. I. Bergström, C. Calberg, T. Fritioff, G. Douysset, J. Schönfelder, and R. Schuch, *Nucl. Instr. and Meth.*, **A487**, 618–651 (2002).
13. G. Bollen, *Nucl. Phys. A*, **693**, 3–18 (2001).
14. M. P. Bradley, J. V. Porto, S. Rainville, J. K. Thompson, and D. E. Pritchard, *Phys. Rev. Lett.*, **83**, 4510 (1999).
15. J. Ullrich, R. Moshammer, R. Dörner, O. Jagutzki, V. Mergel, H. Schmidt-Böcking, and L. Spielberger, *J. Phys. B*, **30**, 2917–2974 (1997).
16. R. Dörner, V. Mergel, O. Jagutzki, L. Spielberger, J. Ullrich, R. Moshammer, and H. Schmidt-Böcking, *Phys. Rep.*, **330**, 95–192 (2000).
17. D. Fischer, B. Feuerstein, R. D. DuBois, R. Moshammer, J. R. C. López-Urrutia, I. Draganic, H. Lörch, A. N. Perumal, and J. Ullrich, *J. Phys. B*, **35**, 1369–1377 (2002).
18. H. Khemliche, T. Schlathölter, R. Hoekstra, R. Morgenstern, and S. Schippers, *Phys. Rev. Lett.*, **89**, 1219 (1998).
19. C. Laulhe, R. Hoekstra, S. Hoekstra, H. Khemliche, R. Morgenstern, A. Närmann, and T. Schlathölter, *Nucl. Instr. and Meth.*, **B157**, 304 – 308 (1999).
20. J.-P. Briand, S. Thuriez, G. Giardino, G. Borsoni, M. Froment, M. Eddrief, and C. Sébenne, *Phys. Rev. Lett.*, **77**, 1452 (1996).
21. C. Kitegi, U. Ratzinger, and S. Minaev, "The IH-cavity for HITRAP," in *Proceedings of the LINAC 2004, Lübeck, Germany*, GSI, Darmstadt, Germany, 2004, pp. 54–56, http://bel.gsi.de/linac2004/INDEX.HTML.
22. M. Steck, K. Beckert, P. Beller, B. Franczak, B. Franzke, and F. Nolden, "Improved performance of the heavy ion storage ring ESR," in *Proceedings of EPAC 2004*, European Physical Society Accelerator Group (EPS-AG), 2004, p. 1168, ISBN 92-9083-231-2 (Web version), ISBN 92-9083-232-0 (CD-ROM).
23. V. S. Pandit, P. R. Sarma, and R. K. Bhandari, "Modification of a double drift beam bunching system to get the efficiency of a six harmonic buncher," in *Proceeding of Cyclotrons and their Applications 2001, East Lansing, USA*, edited by F. Marti, AIP conference proceedings 600, 2001, pp. 452–454.
24. J. Klabunde, E. Malwitz, J. Friedrich, and A. Schempp, "Upgrade of the HLI-RFQ accelerator," in *Proceedings of the 1994 Int. Linac Conf.*, edited by K. Takata, Y. Yamazaki, and K. Nakahara, National Laboratory for High Energy Physics (KEK), Tsukuba, Japan, 1994, p. 704.
25. R. Baartman, *Nucl. Instr. and Meth.*, **B204**, 392–399 (2003).
26. J. Bernard, J. Alonso, T. Beier, M. Block, H.-J. K. S. Djekić, C. Kozhuharov, W. Quint, S. Stahl, T. Valenzuela, J. Verdú, M. Vogel, and G. Werth, *Nucl. Instr. and Meth.*, **A532**, 224–228 (2004).
27. F. Ames, G. Bollen, P. Delahaye, O. Forstner, G. Huber, O. Kester, K. Reisinger, and P. Schmidt, *Nucl. Instr. and Meth.*, **A538**, 17–32 (2005).
28. R. Mann, *Z. Phys.*, **D 3**, 85–90 (1986).
29. A. S. Schlachter, "Charge-Changing-Collisions," in *Proceedings of the 10th Int. Cyclotron Conf. (Michigan 1984)*, 1984, p. 563.
30. R. E. Olsen, and A. Salop, *Phys. Rev.*, **A 14**, 579 (1976).
31. A. Berman, *Vacuum Engineering calculations, formulas and solved excercises*, Academic Press Inc., 1992.
32. P. Redhead, "Ultrahigh and extreme high vacuum," in *Foundations of Vacuum Science and Technology*, edited by J. M. Lafferty, John Wiley & Sons, Inc., 1998, pp. 625–656.

# ANTIPROTON MANIPULATION

# Production of ultra-slow antiproton beams

Hiroyuki A. Torii*, N. Kuroda†, M. Shibata†, Y. Nagata*, D. Barna**,
M. Hori‡, J. Eades‡, A. Mohri†, K. Komaki* and Y. Yamazaki*,†

*Institute of Physics, University of Tokyo, 3-8-1 Komaba, Meguro-ku, Tokyo 153-8902, Japan
†Atomic Physics Lab., RIKEN, 2-1 Hirosawa, Wako-shi, Saitama 351-0198, Japan
**KFKI, Budapest, Hungary
‡CERN, Genève 23, CH-1211, Switzerland

**Abstract.**
We have recently succeeded in decelerating and confining millions of antiprotons, 50 times more
efficiently than conventional methods, in an electromagnetic trap. These antiprotons were cooled by
preloaded electron plasma to an energy below an electronvolt. They were then extracted out of the
magnetic field of 2.5 T and transported typically at 250 eV along a beamline, designed for efficient
transport at 10–1000 eV. This unique beam from our apparatus named MUSASHI opens up a new
field of atomic and nuclear physics probed by ultra-slow antiprotons.
In this paper, the whole experimental setup and procedure will be overviewed: deceleration, capture, cooling and extraction of antiprotons will be discussed in detail, including technical description
of diagnostic devices.

**Keywords:** antiproton, Antiproton Decelerator (AD), RFQD (Radio Frequency Quadrupole Decelerator), Multi-Ring electrode Trap (MRT), ultra-slow antiproton beam
**PACS:** 7.77.Ka, 39.10.+j, 36.10.-k, 52.27.Jt

## INTRODUCTION

Since its first discovery in 1955 at Bevatron in Lawrence Berkeley National Laboratory [1], antiproton has been considered as a suitable candidate (anti-)particle for test of
symmetry between matter and antimatter. With the advent of the Low-Energy Antiproton
Ring (LEAR) which operated from 1982 until 1996 at CERN (Organisation Européenne
pour la Recherche Nucléaire) in Geneva, low-energy antiproton physics started to flour-
ish. CPLEAR collaboration [2] used the provided 5 MeV antiproton beam to measure
precisely the parameters for CP violation between $K^0$ and $\overline{K^0}$. Following the first successful production of antihydrogen [3], reported in the last year of LEAR operation,
a plan to construct an alternative machine dedicated for studies using low-energy antiprotons came into reality as the Antiproton Decelerator (AD) [4] at CERN in the turn
of the century. Two of the experimental collaborations in the AD hall, ATHENA and
ATRAP, succeeded in synthesis of cold antihydrogen atoms out of trapped antiprotons
and positrons [5, 6], and are pursuing their goal of spectroscopy of the anti-atom. On
the other hand, our collaboration ASACUSA, as is represented by its name of origin
"Atomic Spectroscopy And Collisions Using Slow Antiprotons", is aiming at exploring
a wider field of atomic physics involving low-energy antiprotons [7]. The physics program includes spectroscopy of antiprotonic helium atoms which has developed so far as
to a precise determination of the antiproton mass [8], stopping-power measurements at
low-energies [9], and researches on antiprotonic collision dynamics such as ionization

CP793 Physics with Ultra Slow Antiproton Beams
edited by Yasunori Yamazaki and Michiharu Wada
© 2005 American Institute of Physics 0-7354-0282-5/05/$22.50

processes and antiprotonic atom formation [10, 11]. We are also preparing a cusp trap [12] and a Paul trap for production of antihydrogen as our near-future program.

Antiproton, as an exotic particle with infinite lifetime, is an interesting and ideal probe in atomic physics, as well as in nuclear physics. Having the same mass as the proton but with opposite charge, it acts as a "heavy electron" or as a "negetive nucleus". But in order to investigate atomic processes such as ionization and atomic capture, the antiproton as a probe particle needs to become available as a mono-energetic beam with a low energy comparable to the Rydberg energy. With the aim of decelerating and cooling the antiproton beams delivered from the AD facility at 5.3 MeV, the Trap Group of the ASACUSA collaboration prepared a sequential combination of a Radio-Frequency Quadrupole Decelerator (RFQD) and an electromagnetic trap in a strong magnetic field produced by a superconducting solenoid. The antiprotons were to be confined and cooled in the trap, before being extracted along a beamline specially designed to transport an ultra-low energy beam at 10–1000 eV.

## PRODUCTION AND DECELERATION

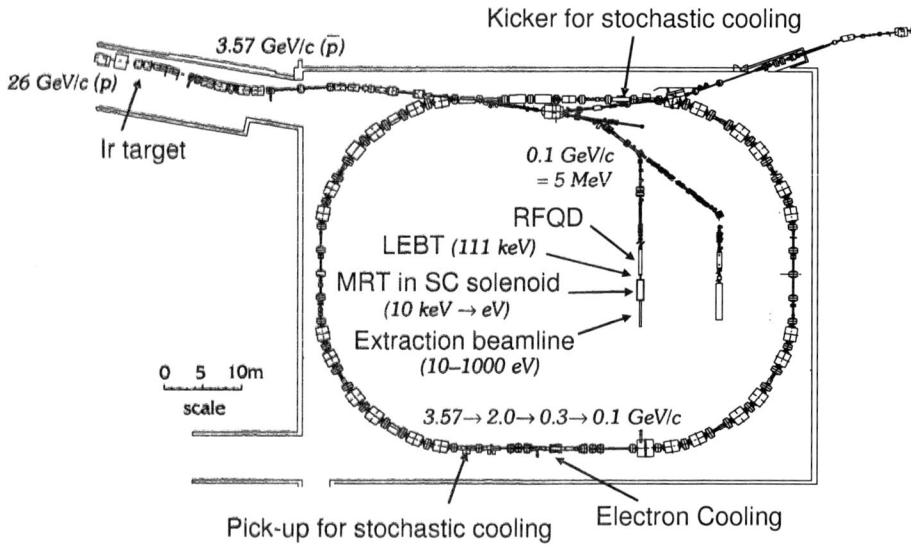

**FIGURE 1.** Layout of the Antiproton Decelerator (AD) at CERN.

Antiprotons ($\bar{p}$) are produced by protons colliding onto a fixed target above the threshold of 5.6 GeV kinetic energy in the laboratory frame. Research on antiproton is pursued also at Fermilab in the U.S., but if one wants a well-cooled low-energy beam, the Antiproton Decelerator (AD) is the only facility which provides such a beam available. Figure 1 shows a schematic of the AD ring. At CERN, proton beams accelerated to 26 GeV/c collide with an Ir target to produce antiprotons via the reaction $p + p \rightarrow p + p + p + \bar{p}$. A fraction of them, $5 \times 10^7$ in number, are collected by a strong magnetic horn at 3.6 GeV/$c$ and stored in the AD ring, which are then cooled via electron cooling [13] and stochastic cooling [14] techniques and decelerated down to a momentum of 100 MeV/$c$ or 5.3 MeV in kinetic energy. They are then extracted to experimental zones as a good-quality* pulsed beam consisting of typically $3 \times 10^7$ antiprotons in a bunch of 100–200 ns.

Thus produced antiproton beam, though already lower in energy by 3 orders of magnitude than at production, needs further deceleration for them to be captured in vacuo electrostatically. A conventional method using thick degrader foils inevitably caused considerable broadening of the energy spread, and most of the antiprotons either stopped inside the foil or were not decelerated enough to be confined in the prepared potential. At best the capture efficiency was 0.01%–0.1% [15, 16].

In order to decelerate antiprotons efficiently, the ASACUSA collaboration [7] developed a Radio Frequency Quadrupole Decelerator (RFQD) [17] in collaboration with the CERN PS group. RF wave was applied to the cavities in a similar way as in a normal RFQ LINAC, but with opposite phase, to decelerate (instead of accelerating) microbunches of the antiproton beam at 5.3 MeV down to 63 keV, with an efficiency of 30%. Since the RF cavities can be biased $\pm$ 60 keV, the output antiproton energy can be varied 10–120 keV.

# INJECTION

The output beam from the RFQD was injected into an eletromagnetic trap in the strong magnetic field of 2.5 T produced by a superconducting solenoid, via an Low-Energy Beam Transport line (LEBT), as shown in Fig. 2(a).

The beam was focused by a set of pulsed solenoids at LEBT and by the converging magnetic field of the superconducting solenoid to a diameter of 3–4 mm (see Fig. 2(b)). It was injected into the trap through a thin Mylar i.e. PET (polyethylene terephthalate) double-layered foil of 90 $\mu$g/cm$^2$ for each layer, used to isolate ultrahigh vacuum (UHV) of $10^{-10}$ Pa inside the trap from a relatively poor vacuum of $10^{-6}$–$10^{-7}$ Pa in the LEBT. This foil also served as a highly sensitive beam profile monitor for the antiproton beam. Silver was evaporatively plated onto the layers of the foil in a form of ten strips of 1 mm width and 25 nm thickness. Signals from each strips used as electrodes to detect emission of secondary electrons were amplified and read out, which allowed monitoring position and distribution of the penetrating antiproton beam for x and y direction for each layer (thus two-dimensional monitor). An example of the beam profile observed by this monitor is given in Fig. 2(c). For diagnosis of the beam transport at LEBT, we used yet another set of newly developed beam profile monitors consisting of thin wires of 10–20 $\mu$m diameter placed with a gap of 1 mm next to each other [18]. These beam profile monitors allowed for the first time non-destructive measurement of the antiproton beam at low energies, which was not possible with MWPCs (Multi-Wire Proportional Chambers), and they proved to be powerful tools in the delicate beam tuning at LEBT.

Taking into account the energy loss in the foil, we varied the bias voltage of the RFQD

---

* The emittance of the beam is $1\pi$ mm mrad and the momental bite is $\Delta p/p \sim 0.1\%$.

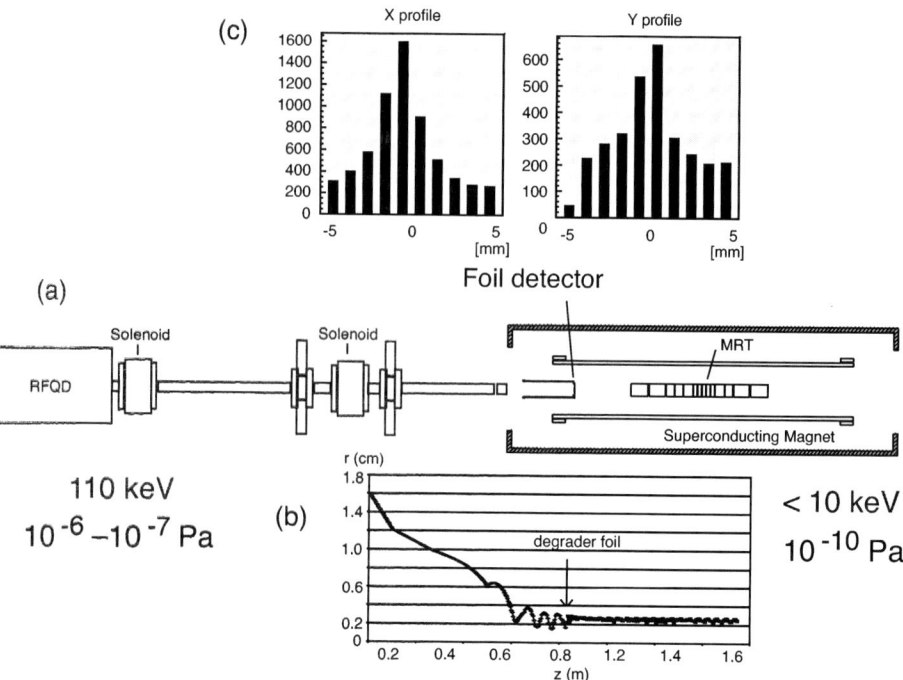

**FIGURE 2.** (a) Schematic of the LEBT line and the magnet. The antiproton beam out of the RFQD was focused into the foil in the 2.5 T magnetic field. Calculated beam diameter along the LEBT line (b) and an example of the beam profile monitored by the foil detector (c) are also shown.

so as to adjust the antiproton energy after penetration through the foil. In order for the antiprotons to be captured in a high-voltage potential of 10 kV as will be discussed later, their transverse energy should be less than 10 keV after the foil. We placed two Čerenkov detectors, one upstream and one downstream of the superconducting solenoid, in order to know the rough position of antiproton annihilation by detecting annihilation products such as fast charged pions. Figure 3 shows signals from the Čerenkov detectors, read out by a digital oscilloscope. When the antiproton beam was stopped at a closed gate valve #1 (see Fig. 3(a)), two peaks were observed, corresponding to the undecelerated fraction of the beam at 5.3 MeV and the decelerated fraction at 91.5 keV at later time. When the gate valve was opened (b), the second peak moved slightly to a later time, with the time difference due to time of flight, corresponding to annihilation at the position of the degrader foil, placed 40 cm downstream of the gate valve. This means that the antiproton beam has lost the energy and stopped inside the foil. This fact can be confirmed by the signal from the downstream Čerenkov detector (c), showing no peak of decelerated antiprotons reaching downstream. On the other hand, when the incident energy of the antiproton beam out of the RFQD was increased to 121.5 keV (d), little annihilation was observed in the foil, while a large broad peak was observed downstream (e), indicating that most of the antiprotons at 121.5 keV have penetrated through the foil.

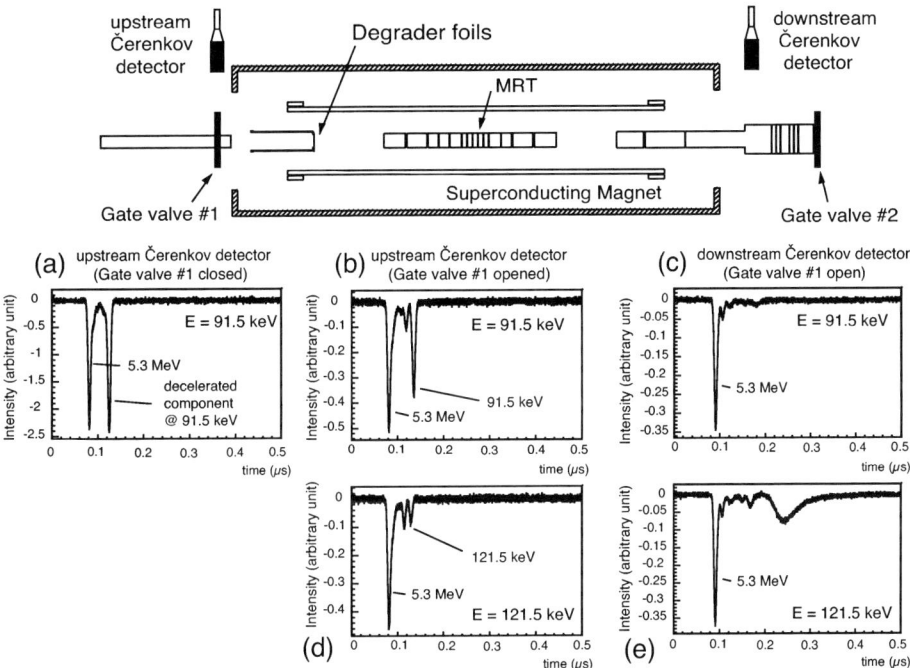

**FIGURE 3.** Signals from Čerenkov detectors at different conditions, shown together with a schematic of the apparatus. For details see the text.

We then applied a high voltage of 10 keV to the most downstream electrode (among the set of trap electrodes placed in the center of the superconducting magnet) to reflect back the antiproton beam. As shown in Fig. 4 for the case of 111.5 keV injection, the broad peak (A) observed at downstream became smaller (B) after application of the high voltage, because a major fraction of antiprotons bounced back to upstream. The broad peak still remaining corresponds to antiprotons with their energy greater than 10 keV after passage through the foil. By varying the voltage applied, we can measure transverse energy distribution after the foil, as the derivative of the reflection ratio as a function of the potential applied, as shown in Fig. 5. For each voltage, the area of the broad peak was compared between the cases with and without the high voltage, and their ratio was calculated. For example, at the voltage of 10 kV, from Fig. 4 we learn that among the antiprotons which have penetrated through the foil (A), 70% of them were reflected back by the potential toward upstream and disappeared from the downstream Čerenkov signal, while the rest 30% still travelled toward downstream (B). This means that, after the foil, 70% of the antiprotons had a transverse energy of 10 keV or less, and

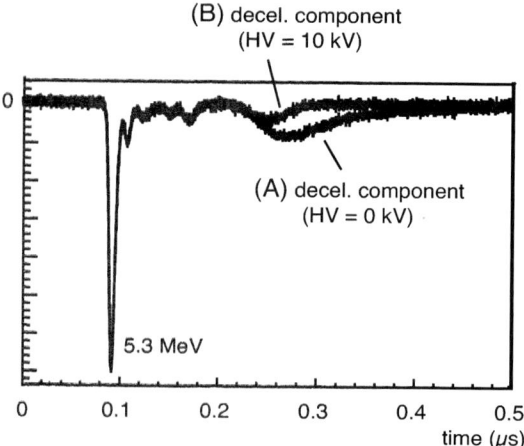

**FIGURE 4.** Signal from the downstream Čerenkov detector. When no high voltage was applied, the broad peak (A) corresponds to antiprotons with their broad energy variation around 10 keV, after energy degradation in the foil. When a high voltage of 10 kV was applied, the remaining broad peak (B) corresponds to antiprotons with their energy more than 10 keV which reached downstream, while the rest of the antiprotons were reflected by the high voltage and hence were not observed by the downstream Čerenkov detector.

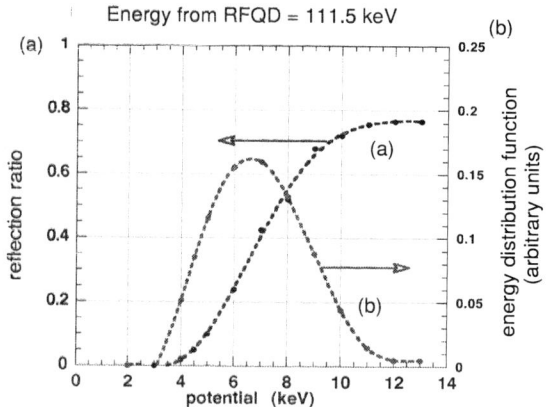

**FIGURE 5.** Reflection ratio (a) of the antiprotons by high voltage potential and deduced transverse energy distribution (b) of antiprotons after passage through the degrader foil, incident at 111.5 keV.

this number is plotted as graph (a) in Fig. 5. [†] By differentiating thus obtained curve, the

---

† The percentage discussed is normalized to the number of antiprotons which have penetrated through the foil. We must note that 30% of incident antiprotons stopped inside the foil under this condition.

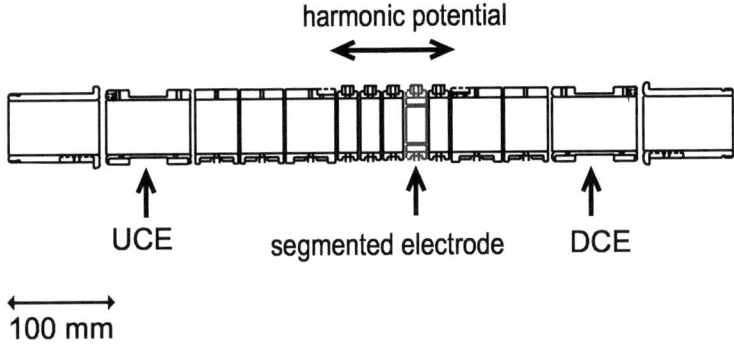

harmonic potential

UCE      segmented electrode      DCE

100 mm

**FIGURE 6.** A schematic cross section of the MRT along the trap axis.

energy distribution after the foil can be obtained and is plotted as the graph (b). **

## CONFINEMENT AND COOLING

Antiprotons were then captured and confined in an electromagnetic trap. We used a Multi-Ring Trap (MRT) [19] consisting of 14 cylindrical electrodes placed coaxially along the magnetic field line, as shown in Fig. 6. A favorable feature of the MRT compared with a normal Penning trap is that a harmonic electric potential can be prepared in a wider region near the trap axis by application of appropriate voltage on each electrode, which allows trapping of a much larger number of charged particles. A large trap volume can be prepared also by a trap with a well-type potential, but the MRT has a superior ability to trap particles far more stably.

Figure 7 shows sequential steps for antiproton capture, cooling and extraction. The pulse of incident antiprotons were reflected backward at the DCE (Downstream Catching Electrode) floated at $-10$ kV. By the time the pulse returned after its round trip of typically 500 ns back to the UCE (Upstream Catching Electrode), the trap was closed by a fast switch with a rise time of 200 ns which biased the UCE to $-10$ kV, confining the major part of the antiprotons. The antiprotons were then cooled by a plasma of typically $3 \times 10^8$ electrons preloaded in the harmonic potential. Antiprotons lost their energy by transferring it to electrons, while the heated electrons cooled by themselves by emission of synchrotron radiation in the magnetic field of 2.5 T with its time constant of about 1 s, until the antiprotons were trapped in the bottom of the harmonic potential of 50 V depth. We then opened one side of the potential for 550 ns to selectively release electrons: lighter and thus faster electrons escaped within this short period, while heavier and much slower antiprotons remained inside. This release of electron turned out to be

---

** We assumed here that the position distribution of antiprotons reaching downstream does not depend on the energy, so that the solid angle of the downstream Čerenkov detector for antiproton annihilation is constant.

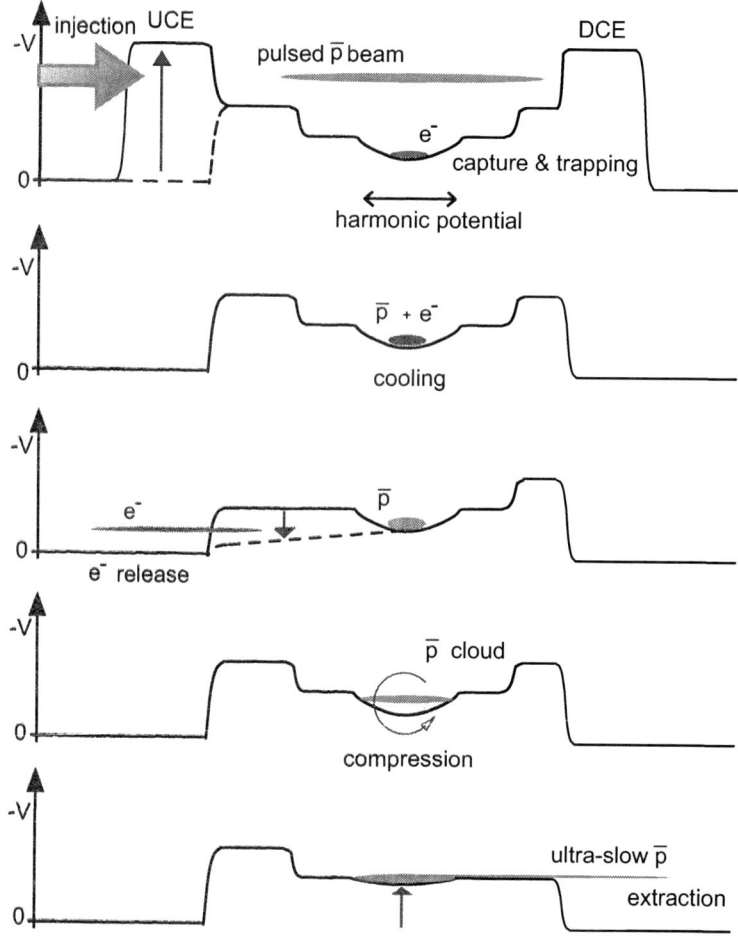

**FIGURE 7.** Sequential procedures of antiproton capture, cooling and extraction.

essential because our trial of antiproton extraction together with electrons never worked out [20].

The antiproton cloud were then given torque by a rotating electric field to be compressed radially. For this purpose, one of the electrodes was segmented azimuthally into four parts as can be seen in Fig. 6, and an RF voltage was applied to each segment with a phase difference of $\pi/2$ next to each other [21, 22].

As a diagnostic device of antiproton trapping, we prepared a set of track detectors to know the position and time of antiproton annihilation. Two scintillator bars of 2 m in length, with a rectangular cross section of 4(H) × 6(V) cm$^2$, were placed in the same plane and parallel to the trap axis, as shown in Fig. 8(a). Passage of charged particles such as pions, or electron-positron showers converted from gamma rays originating in

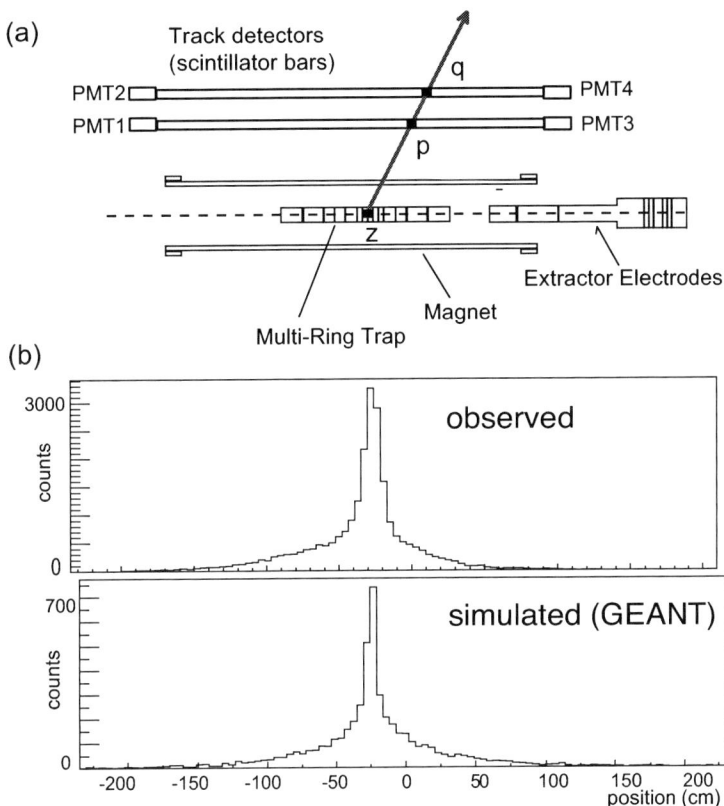

**FIGURE 8.** (a) Position of antiproton annihilation can be calculated from the trajectory of produced particles, detected by scintillator bars. (b) Observed position resolution agreed with our simulation using GEANT code.

the decay of neutral pions, was detected and the position of the hit points (shown as 'p' and 'q' in the figure) was calculated from the time-of-flight difference of the scintillation light, arriving at photomultiplier tubes (PMT) at both ends of the bars. Tracking back the reconstructed particle trajectory back onto the trap axis, antiproton annihilation ('z' in the figure) can be detected with a position resolution of 20 cm and with a detection efficiency of typically $\varepsilon = 5\%$, which agreed with our simulation [23] using GEANT trajectory calculation code [24] (see Fig. 8(b)) [‡].

Figure 9 shows detected annihilation counts as a function of time and position along the beam axis (i.e. the trap axis). Frequent annihilation was observed at early times

---

[‡] The detection efficiency is primarily determined by the solid angle of the detector and the multiplicity of pions. Relatively bad position resolution is due to scattering of electron-positron showers in the thick material of the magnet. The showers originate in gamma rays from the decay of neutral pions.

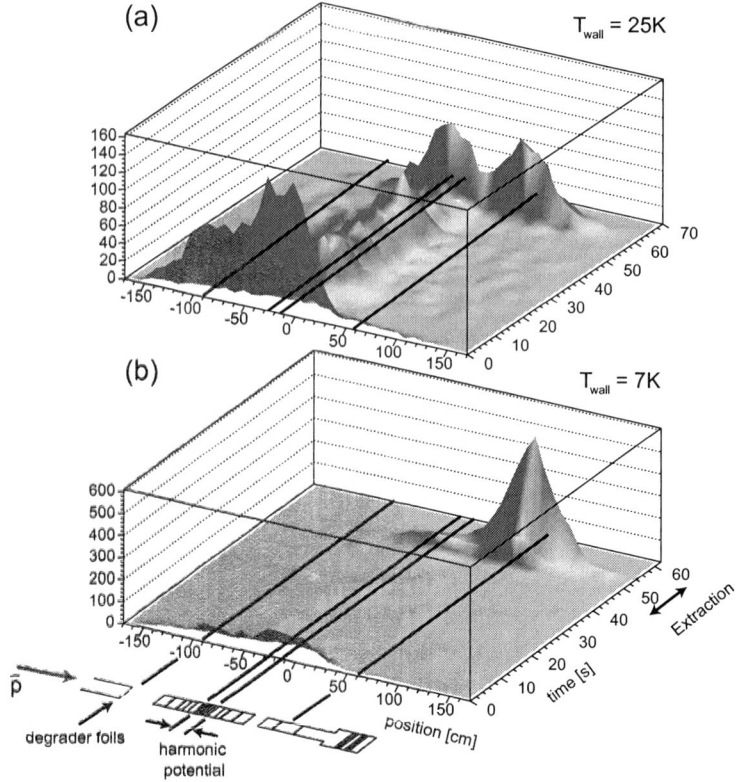

(a)                                                                    $T_{wall}$ = 25K

(b)                                                                    $T_{wall}$ = 7K

Extraction

time [s]

position [cm]

p̄

degrader foils

harmonic
potential

**FIGURE 9.** Annihilation counts observed as a function of time and position.

for a typical period of 10 s following antiproton injection at $t = 0$. They occurred at positions of the degrader foil and the trap. After the antiprotons have been cooled enough, there still remained constant annihilation in the trap center when the temperature of the bore which housed the trap electrodes was 25 K (top figure (a)). This was due to continuous antiproton annihilation against atomic nuclei of residual gases, mainly hydrogen. A standard "ultra-high" vacuum is not apparently good enough for stable confinement of antiprotons. On the other hand, no annihilation was observed during confinement at the bore temperature of 7 K (bottom figure (b)). At this cryogenic temperature, even the hydrogen gas froze out and an extremely high vacuum better than $10^{-10}$ Pa was achieved, preventing antiproton annihilation. Since the vacuum is very sensitive to the ambient temperature, stable cryogenic cooling was important in stable antiproton trapping. Antiprotons were then extracted and many annihilations were observed downstream of MRT.

Integrating the number of antiproton annihilations over all the positions, we obtain total number of antiprotons. Figure 10 shows cumulative count of antiproton annihilation as a function of time. The rapid increase at the last stage of the trapping cycle

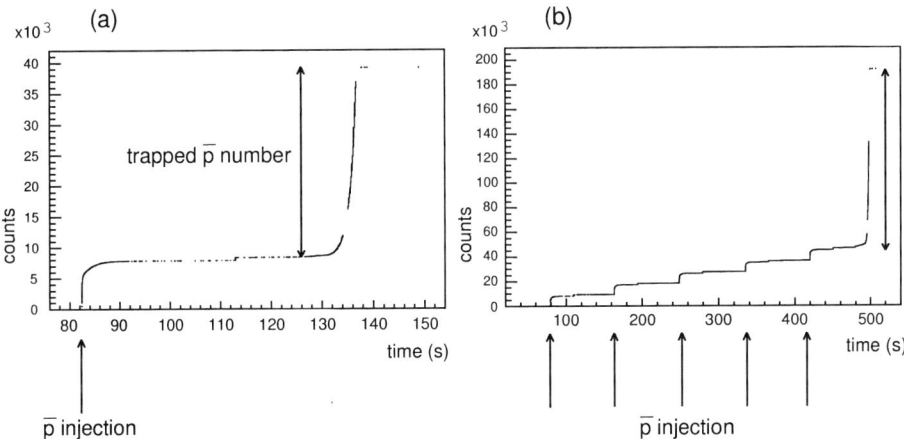

**FIGURE 10.** Cumulative number of antiproton annihilation as a function of elapsed time. The total number of antiprotons in the trap was (a) 1.2 million for a single AD shot and (b) 4.8 million for stacking of 5 AD shots.

corresponds to the antiprotons which were confined stably. Taking into account the detection efficiency $\varepsilon$ of antiproton annihilation by the track detectors, we concluded that $1.2 \times 10^6$ antiprotons were trapped stably until the end of our trap cycle of 1–5 minutes [20]. Typically 1 million antiprotons were trapped for each AD shot. We then accumulated antiprotons for several AD shots (see Fig. 10(b)). This technique of "stacking" also worked fine, and we trapped $4.8 \times 10^6$ antiprotons simultaneously for stacking of 5 AD shots, the largest number of antiprotons ever accumulated.

## EXTRACTION AND BEAM TRANSPORT

The antiprotons were then released from the trapping potential as it was gradually shallowed, and were extracted as an ultra-slow continuous beam of 10–500 eV. Since the antiprotons tend to expand in radial direction when they follow the strongly diverging magnetic field line out of the 2.5 T magnetic field, it was essential that the antiproton cloud be well compressed radially in the trap. This was achieved by application of rotating electric field [22]. Also important in efficient extraction was precise alignment. The magnetic field axis, electric trap axis, the beam transport axes: all these needed to be well aligned precisely.

The extraction beamline was designed to transport antiproton beams over a length of 3 m, at variable energies ranging from 10 to 1000 eV. The antiproton beams were refocused three times by sets of Einzel lenses at the position of apertures, as shown in Fig. 11. These variable apertures of diameter 4–10 mm allow differential pumping of 6 orders of magnitude along the beamline, which was necessary to keep the trap region at an extremely high vacuum better than $10^{-10}$ Pa so as to avoid antiproton annihilation, while the end of the beamline will be exposed to atomic or molecular gas jets of upto

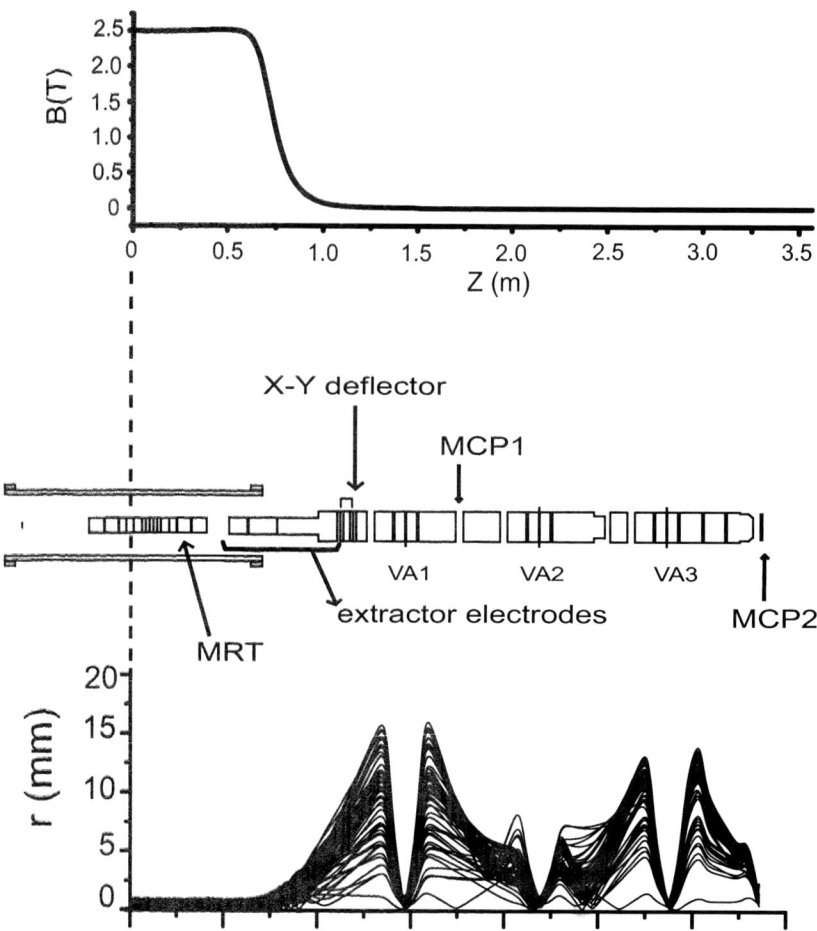

**FIGURE 11.** Schematic of the extraction beamline for ultra-slow antiproton beams. The beams are focused 3 times at positions of variable apertures, to comply with the requirement of effective differential pumping (bottom). Magnetic field strength is also shown in the top figure.

$10^{-4}$ Pa [25]

The MRT, the superconducting solenoid and the eV-beam transport line are jointly known as "MUSASHI", or the Monoenergetic Ultra-Slow Antiproton Source for High-precision Investigations. MUSASHI opens a new research field ranging from atomic physics to nuclear physics [10], including our near-future project of antihydrogen synthesis in a cusp trap [12]. As from nuclear physics point of view, low-energy antiproton is suited to give information on nuclear peripherals [10, 26, 27, 28], but it would be more interesting from atomic-physics point of view.

Especially, atomic formation and ionization processes by low-energy antiprotons can be studied under single collision conditions for the first time, and theoretical calculations [29] will now be tested experimentally. Taking advantage of the slow extraction i.e. continuous aspect of the beam, event-by-event data acquisition becomes possible, associated with each single antiproton. We are now preparing a supersonic gas-jet target for atomic collision experiments planned in the next years. The target is aimed to achieve a density of $3 \times 10^{13}$ cm$^{-3}$ with a gas-jet cross section of 5 mm $\times$ 1 cm, which will be crossed with the ultra-slow antiproton beams to produce antiprotonic atoms [30].

## SUMMARY

With the conbination of a RFQD and a MRT in the superconducting solenoid, we have successfully decelerated antiprotons from 5 MeV to less than 10 keV, confined them and cooled them down to sub-eV energies with preloaded electrons. Out of 30 million antiprotons delivered from the AD, 1.2 million antiprotons were trapped stably for a cycle of a few minutes, which was 50 times more efficient than the conventional method of using thick degrader foils for deceleration. Antiprotons were extracted at energies of 10–500 eV, and this continuous ultra-slow antiproton beam will be a powerful tool for the study of atomic collision dynamics and formation of antiprotonic atoms including antihydrogen, as well as nuclear physics.

## ACKNOWLEDGMENTS

We are grateful to the AD and PS stuff in the AB department of CERN for their tireless effort in providing us with the antiproton beam. We acknowledge Mr. W. Pirkl and his team for their engagement in the design and operation of the RFQD. This work was supported by the Grant-in-Aid for Creative Scientific Research (10P0101) of the Japanese Ministry of Education, Culture, Sports, Science and Technology (MonbuKagaku-shō), Special Research Projects for Basic Science of RIKEN, and the Hungarian National Science Foundation (OTKA T033079).

## REFERENCES

1. O. Chamberlain, E. Segrè, C. Wiegand, and T. Ypsilantis, Phys. Rev. **100,** 947 (1955).
2. CPLEAR Collabolation, A. Angelopoulos *et al.*, Phys. Lett. B **420,** 191 (1998).
3. G. Baur *et al.*, Phys. Lett. B **368,** 251 (1996).
4. CERN Antiproton Decelerator web page, http://psdoc.web.cern.ch/PSdoc/acc/ad/index.html
5. M. Amoretti *et al.*, Nature **419,** 456 (2002).
6. G. Gabrielse *et al.*, Phys. Rev. Lett. **89,** 213401 (2002).
7. ASACUSA collaboration web page, http://cern.ch/ASACUSA
8. M. Hori, J. Eades, R.S. Hayano, T. Ishikawa, W. Pirkl, E. Widmann, H. Yamaguchi, H.A. Torii, B. Juhász, D. Horváth, and T. Yamazaki, Phys. Rev. Lett. **91,** 123401 (2003).
9. S.P. Møller, A. Csete, T. Ichioka, H. Knudsen, U.I. Uggerhøj, and H.H. Andersen, Phys. Rev. Lett **88,** 193201 (2002).
10. Y. Yamazaki, Nucl. Instrum. Methods in Phys. Res. B **154,** 174 (1999).
11. ASACUSA collaboration, proposal submitted to CERN/SPSC, CERN/SPSC 97-19; CERN/SPSC 2000-04; CERN-SPSC 2005-002.

12. A. Mohri and Y. Yamazaki, Europhys. Lett. **63**, 207 (2003); A. Mohri, "Non-Neutral Plasma Confinement in a Cusp-Trap and Possible Application to Anti-Hydrogen Beam Generation", Contribution to Workshop on Physics with Ultra Slow Antiproton Beams, RIKEN, March 2005 (in this book). .

13. G.I. Budker, Proc. Int. Symp. On Electron and positron strage rings, Saclay, ed. by H. Zyngier and E. Cremieux-Alcan (PUF, Paris, 1967) p.II-1-1; S.P. Vytchanin, Sov. Phys. Dokl. **22**, 321 (1977); G.I. Budker and A.N. Skrinski, Sov. Phys. Usp. **21**, 277 (1978).

14. S. van der Meer, Stochastic Cooling and the Accumulation of Antiprotons, *Nobel lecture (1984)*; Stochastic cooling in the Antiproton Accumulator, IEEE Trans. Nucl. Sci. **NS28**, 1994 (1981) .

15. G. Gabrielse *et al.*, Phys. Lett. B **548**, 140 (2002).

16. M. Amoretti *et al.*, Nucl. Instrum. Methods in Phys. Res. A **518**, 679 (2004).

17. A. M. Lombardi, W. Pirkl, and Y. Bylinsky, in *Proceedings of the 2001 Particle Accelerator Conference, Chicago, 2001* (IEEE, Piscataway, NJ, 2001), pp. 585–587.

18. M. Hori, Nucl. Insturm. Methods. A **522**, 420 (2004).

19. A. Mohri, H. Higaki, H. Tanaka, Y. Yamazawa, M. Aoyagi, T. Yuyama, and T. Michishita: Jpn. J. Appl. Phys. **37**, 664 (1998).

20. N. Kuroda, H. A. Torii, K. Yoshiki Franzen, Z. Wang, S. Yoneda, M. Inoue, M. Hori, B. Juhász, D. Horváth, H. Higaki, A. Mohri, J. Eades, K. Komaki, and Y. Yamazaki, Phys. Rev. Lett. **94**, 023401 (2005).

21. H. Higaki, N. Kuroda, K. Yoshiki Franzen, Z. Wang, M. Hori, A. Mohri, K. Komaki, and Y. Yamazaki, Phys. Rev. E **70**, 026501 (2004).

22. N. Kuroda *et al.*, "Control of plasmas for production of ultraslow antiproton beams", Contribution to Workshop on Physics with Ultra Slow Antiproton Beams, RIKEN, March 2005 (in this book).

23. S. Yoneda, master thesis, Department of Physics, University of Tokyo, *in Japanese* (2003); D. Barna, private communication (2004).

24. GEANT3, CERN Program Library Long Writeup W5013.

25. K. Yoshiki Franzen, N. Kuroda, H.A. Torii, M. Hori, Z. Wang, H. Higaki, S. Yoneda, B. Juhász, D. Horváth, A. Mohri, K. Komaki, and Y. Yamazaki, Rev. Sci. Instrum. **74**, 3305 (2003).

26. A. Trzcińska *et al.* Neutron Density Distributions Deduced from Antiprotonic Atoms, Phys. Rev. Lett. **87**, 82501 (2001).

27. J. Jastrzebski *et al.* Nucl. Phys. A **588**, 405c (1993).

28. M. Wada and Y. Yamazaki, Nucl. Instr. Meth. B **214**, 196 (2004).

29. J. S. Cohen, Rep. Prog. Phys. **67**, 1769–1819 (2004).

30. V. L. Varentsov *et al.*, "ASACUSA Gas-Jet Target: Present Status And Future Development", Contribution to Workshop on Physics with Ultra Slow Antiproton Beams, RIKEN, March 2005 (in this book).

# Control of plasmas for production of ultraslow antiproton beams

N. Kuroda*, H.A. Torii†, M. Shibata*, Y. Nagata†, D. Barna**,
D. Horváth**, M. Hori‡, J. Eades§, A. Mohri*, K. Komaki† and
Y. Yamazaki*,†

*RIKEN, 2-1 Hirosawa, Wako-shi, Saitama, 351-0198, Japan
†Institute of Physics, University of Tokyo, 3-8-1 Komaba, Meguro-ku, Tokyo, 153-8902, Japan
**KFKI Research Institute of Particle and Nuclear Physics, H-1525 Budapest, Hungary
‡CERN, CH-1211 Genève, Switzerland
§Department of Physics, University of Tokyo, 7-3-1 Hongo, Bunkyo-ku, Tokyo, 113-0033, Japan

**Abstract.** To produce ultraslow antiproton beams, decelerated antiprotons were captured in an electro-magnetic trap and cooled by collisions with preloaded electrons. This electron cooling feature was nondestructively monitored by measurement of electrostatic oscillations of the confined electron plasma. The radial distribution of the plasma was controlled for efficient cooling and extraction by utilizing rotating wall field technique.

**Keywords:** antiproton, non-neutral plasma, multiring trap
**PACS:** 39.10.+j, 52.27.Jt, 52.35.Fp

## INTRODUCTION

The preparation of a large number of antiprotons at extremely low energy is an important step to the study of the initial process of antiprotonic atom ($\bar{p}A^+$) formation [1], to the high precision spectroscopy of $\bar{p}A^+$ [2], to the collision dynamics in the ionization of atoms by antiprotons [3], to the nuclear surface structure via antiprotonic atom formation and annihilation [4, 5], and to the synthesis and spectroscopy of antihydrogen atom[6]. Such exotic atoms can only be efficiently synthesized from component particles at eV and lower energies. The ASACUSA (Atomic Spectroscopy And Collisions Using Slow Antiprotons) collaboration prepared ultraslow antiproton source by the sequential combination of the CERN Antiproton Decelerator (AD), the radio frequency quadrupole decelerator (RFQD), and MUSASHI (Monoenergetic Ultra-Slow Antiproton Source for High-precision Investigation).

The AD at CERN cools and decelerates antiprotons with 2.7 GeV in kinetic energy to 5.3 MeV by stochastic and electron cooling techniques, and ejects them every 84 s with the pulse width of 90 ns, containing $2$–$3 \times 10^7$ particles. For further deceleration, we used the RFQD which reduced the 5.3 MeV antiproton beam energy around 110 keV. Some $5 \times 10^6$ antiprotons transmitted from the RFQD lost their energy, $\sim 100$ keV by passing through vacuum isolation foils, which is two 90 $\mu$g/cm$^2$ polyethtlenterephtalate (PET) foils located in front of the MUSASHI (see Fig. 1).

As shown in Fig.1, MUSASHI is composed of two parts, a multiring electrode trap (MRT) [7] housed in 2.5 T magnetic field by a superconducting solenoid and a transport

CP793 *Physics with Ultra Slow Antiproton Beams*
edited by Yasunori Yamazaki and Michiharu Wada
© 2005 American Institute of Physics 0-7354-0282-5/05/$22.50

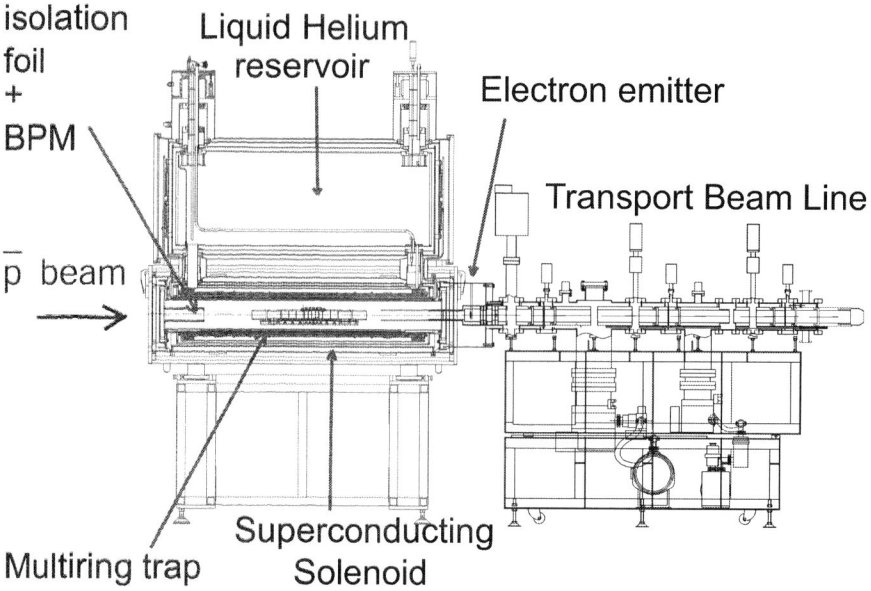

isolation foil + BPM

Liquid Helium reservoir

Electron emitter

Transport Beam Line

p̄ beam

Multiring trap

Superconducting Solenoid

**FIGURE 1.** A drawing of MUSASHI.

beam line for ultraslow antiproton beams [8]. Antiprotons were captured in the MRT, and then cooled via collisions with preloaded electrons [9]. About $1.2 \times 10^6$ among the captured antiprotons at the maximum could be stored in the MRT for periods of 10 min or more [10]. After the cooling, antiprotons were extracted to the field free region via the transport beam line.

In this experiment, we developed a nondestructive plasma diagnostic technique for the study of cooling process of energetic antiprotons by an electron plasma, and also developed the control technique of the radial distribution of plasma and/or charged particle cloud in the MRT for the efficient capture and extraction of antiprotons.

## EXPERIMENT

### Multiring electrode trap

The MRT consists of 14 ring electrodes as shown in Fig. 2. By applying a proper voltage (like as Fig. 3(a)) on each electrode, F1, FH2, FH1, BH1, S, BH2, and B1 (Fig.2), an electrostatic potential

$$\phi(\rho, z) = -V_0 \frac{\rho^2 - 2z^2}{2L^2 + R^2} + \delta, \tag{1}$$

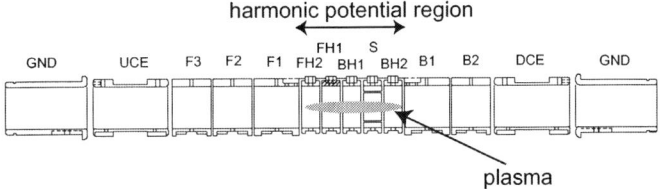

harmonic potential region

plasma

**FIGURE 2.** The schematic view of the MRT. The electrode "S" is segmented azimuthally.

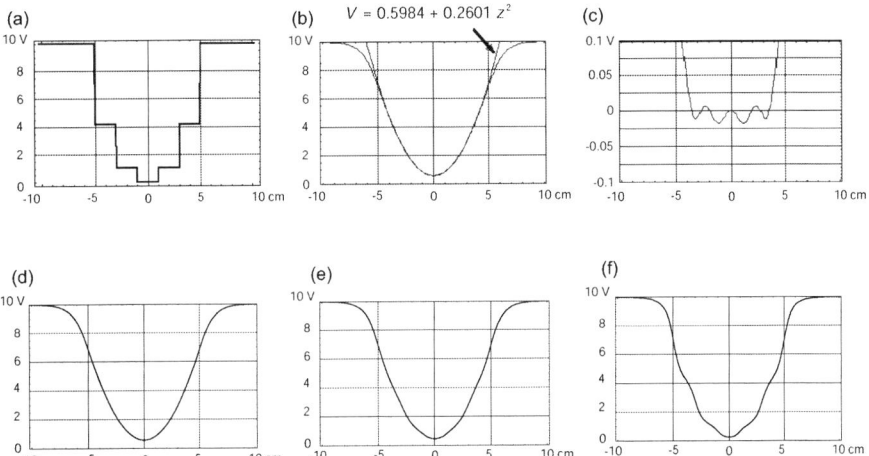

(a)

(b) $V = 0.5984 + 0.2601\, z^2$

(c)

(d)

(e)

(f)

**FIGURE 3.** (a) Applied voltages on each ring electrode. (b) The produced potential at $\rho = 0$ mm. Fitted quadratic function is also shown. (c) Difference between the produced potential and the quadratic function in the case of (b). (d) $\rho = 5$ mm, (e) $\rho = 10$ mm, (f) $\rho = 15$ mm.

is produced, which is described in the cylindrical coordinate, where the radius of the ring electrode is $R = 20$ mm, the axial length of the trap region is $2L = 125$ mm, the potential difference is $V_0 = \phi(0,L) - \phi(0,0)$ the offset from the surface potential on the center electrode "BH1" is $\delta$. Figure 3 (b) shows the resultant potential along the axis for $V_0 = 10$ V. The quadratic function is superimposed on the produced potential, which is called as harmonic potential. The difference between the produced potential along the axis and the quadratic function is shown in (c). The harmonic potential region is much larger than usual Penning traps. The calculated produced potential at $\rho = 5$ mm, 10 mm and 15 mm are shown in Fig.3 (d), (e), and (f), respectively. As is expected, the produced potential distorted at the larger $\rho$. During the following experiment, $V_0$ was chosen to 50 V.

One of the ring electrodes ("S" in Fig. 2) was azimuthally segmented into 4 sectors so that an rf field could be applied to compress the charged particle cloud in the MRT [11].

Electrons as a coolant were injected from the downstream side of the MRT. They were emitted from a field emission array cathode on a pneumatic type feedthrough, which was retracted during the extraction procedure of ultraslow beams. The typical number of stored electrons was $3 \times 10^8$.

## Plasma diagnosis

A well cooled charged particle cloud behaves as a nonneutral plasma in an electromagnetic trap. The shape of a plasma confined in a harmonic potential is spheroid under an equilibrium condition, a rigid rotor equilibrium [12]. For diagnosis of the trapped electron plasma, a white noise was applied to the plasma via one of the electrodes comprising harmonic potential region. The excited axially symmetric electrostatic oscillation, $(l,0)$ mode, indicating a number $l$ which corresponds to $l-1$ nodes of the charge distribution along the axial direction of the plasma and a number 0 for azimuthal direction can be observed via another ring electrode. The signal of the oscillation was monitored by a real time spectrum analyzer after amplification. These oscillation induced peaks in the power spectrum at the corresponding mode frequencies. The shift of these peaks with time was used to monitor the antiproton cooling process because these modes has temperature dependence as is discussed in following paragraphs.

The dispersion relation of axially symmetric electrostatic oscillation of cold spheroidal plasmas whose angular frequency $\omega_l$ is given by D.H.E. Dubin [13] as

$$\frac{\varepsilon_3}{\varepsilon_0} = 1 - \frac{\omega_p^2}{\omega_l^2} = \left( \frac{\alpha - \varepsilon_3/\varepsilon_0}{\alpha^2 - 1} \right)^2 \frac{P_l(\xi_1)Q_l'(\xi_2)}{P_l'(\xi_1)Q_l(\xi_2)}, \tag{2}$$

where $\varepsilon_3$ is the dielectric tensor element along the magnetic field, $\varepsilon_0$ is the vacuum dielectric constant, $a$ is the radius of the spheroid, $2b$ is the length of the plasma, $\alpha = b/a$ is the aspect ratio, and $\omega_p = \sqrt{n_e q^2/\varepsilon_0 m_e}$ is the angular frequency of the plasma oscillation. Here $n_e$ is the number density of the electrons, $q$ is the electron's charge, and $m_e$ is the mass. $P_l$ and $Q_l$ represent Legendre functions of the first and second kinds, while $P_l'$ and $Q_l'$ denote their derivatives where $\xi_1 = \alpha(\alpha^2 - \varepsilon_3/\varepsilon_0)^{-1/2}$ and $\xi_2 = \alpha(\alpha^2 - 1)^{-1/2}$. Even when the temperature of the plasma, $T_e$, is finite, but still the boundary of the plasma is regarded as sharp, Eq. 2 is extended to a finite temperature region as [14, 15],

$$\frac{\varepsilon_3}{\varepsilon_0} = 1 - \frac{\omega_p^2}{\omega_l^2} \left( 1 + \frac{3k_B T_e}{m_e} \frac{k^2}{\omega^2} \right) = \left( \frac{\alpha - \varepsilon_3/\varepsilon_0}{\alpha^2 - 1} \right)^2 \frac{P_l(\xi_1)Q_l'(\xi_2)}{P_l'(\xi_1)Q_l(\xi_2)}, \tag{3}$$

where $k = \pi(l-1)/2b$ is the wave number of the electrostatic wave.

The asterisks plotted in Fig. 4 show the observed frequencies of $(l,0)$ modes ($l = 1$–7). These observed frequencies were fitted by a calculated values (circles in the figure). Under an assumption of thermally stabilized plasma in the 50 V harmonic potential and the length of trapping region $L = 12.5$ cm for $3 \times 10^8$ electrons, the mode frequencies of the plasma was calculated as circles shown in Fig.4, which indicated the radius of the plasma was 5.0 mm.

When $\omega_l$ of the plasma is experimentally determined for more than two different $l$s, Eq. 2 gives the density and the aspect ratio. Table 1 shows the observed mode frequencies

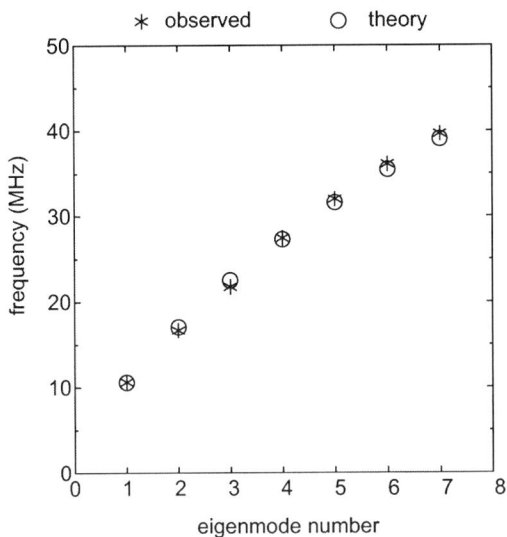

**FIGURE 4.** Observed and calculated axially symmetric frequencies up to 7th eigenmode.

**TABLE 1.** Observed frequencies and estimated aspect ratio and density of the electron plasma from two different frequencies. The size of the plasma, $a$ and $b$, is also shown, which is evaluated by using independently measured total number of the electron.

| $f_{(2,0)}$ [MHz] | $f_{(3,0)}$ [MHz] | $f_{(4,0)}$ [MHz] | aspect ratio | density [/cm$^3$] | a [mm] | b [mm] |
|---|---|---|---|---|---|---|
| 17.660 | 23.212 | – | 14 | $1.2 \times 10^8$ | 3.4 | 4.6 |
| 17.660 | – | 27.966 | 13 | $1.1 \times 10^8$ | 3.6 | 4.6 |
| – | 23.212 | 27.966 | 12 | $1.0 \times 10^8$ | 3.7 | 4.5 |

for $l = 2, 3$, and 4. From the relation between two different modes, the aspect ratio and the number density of the plasma was estimated. According to the total number of particles of the plasma which was measured independently, the real size of the plasma, $a$ and $b$, can be determined.

Figure 5 shows an example of the time evolution of the power spectrum by the antiproton beam injection where the resonance corresponding to the (1,0), (2,0), and (3,0) modes are clearly observed. The resonance frequency of the (1,0) mode was almost constant for the whole period monitored as is expected from Eq.3. The (2,0) and (3,0) modes increased after the antiproton injection, reached a maximum in a few seconds, and then slowly returned to the initial frequencies. It took about 30 s. Such a frequency variation is due to heat-up of the electron plasma by the antiproton injection followed by the cooling with synchrotron radiation.

These frequency shifts and time evolution were understood from a model considering a beam cooling by a plasma described in Ref. [16]; The energy transfer rate from the

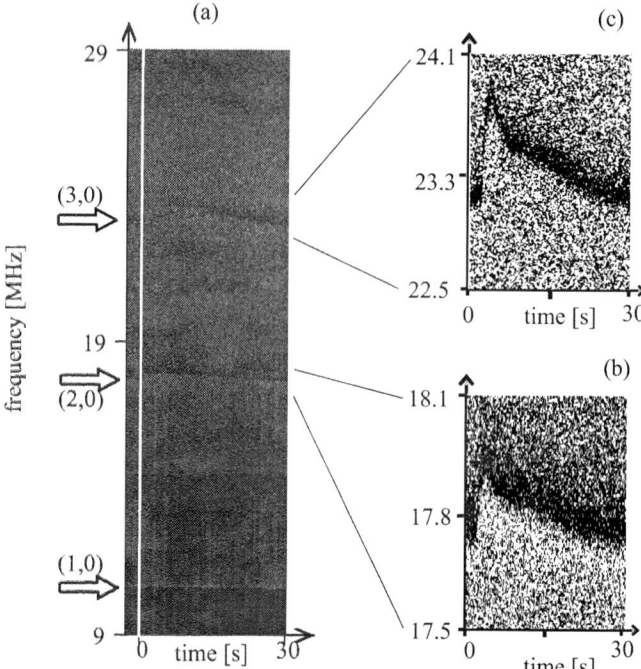

**FIGURE 5.** (a) Time evolution of the power versus frequency spectrum of the electron plasma after antiproton injection at $t = 0$. The three arrowed regions correspond to the (1,0), (2,0), and (3,0) modes, respectively; (b) and (c) show the (2,0) and (3,0) modes with expanded vertical scales.

antiproton beam to the electron plasma is given by [16],

$$\left\langle \frac{d\mathcal{E}}{dt} \right\rangle = A \frac{n_e q^4 \log \Lambda(T_e, E_{\bar{p}})}{4\pi \varepsilon_0^2 m_e v_{\bar{p}}(E_{\bar{p}})} F\left(\frac{v_{\bar{p}}}{v_e}\right), \tag{4}$$

where the velocities of electron and antiproton are $v_e$ and $v_{\bar{p}}$, respectively. The Coulomb logarithm $\log \Lambda$ is expressed as

$$\log \Lambda(T_e, E_{\bar{p}}) = \log\left(\frac{\lambda_D(T_e)}{L(T_e, E_{\bar{p}})}\right), \tag{5}$$

$$L(T_e, E_{\bar{p}}) = \frac{q^2}{4\pi\varepsilon_0 m_e} \frac{1}{(v_e v_{\bar{p}})^2}, \tag{6}$$

$$\lambda_D = \sqrt{\frac{\varepsilon_0 k_B T_e}{n_e q^2}}. \tag{7}$$

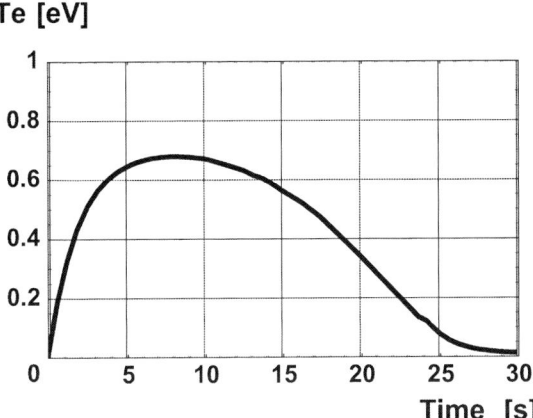

**Te [eV]**

**FIGURE 6.** Calculated time evolution of the electron plasma temperature.

Here we use the following functions defined as,

$$F(x) = \Phi(x) - \frac{2x}{\sqrt{x}}\left(1 + \frac{m_e}{m_{\bar{p}}}\right)\exp(-x^2), \tag{8}$$

where

$$\Phi(x) = \frac{2}{\sqrt{\pi}}\int_0^x \exp(-t^2)dt, \tag{9}$$

is the error function.

The antiproton beam emerging from the degrader foils is not a monochromatic beam. In this experimental condition, the diffusing time from the velocity perpendicular to the magnetic field $v_\perp$ to the parallel component $v_\parallel$ is much long compared with the deceleration time along the field. For $E_{\bar{p}} = 10$ keV, the diffusion time becomes c.a. $10^4$ s, while the deceleration time is the order of 10 s. Therefore we estimate the energy deposit of the antiproton beam to the electron plasma by considering only the energy distribution along the magnetic field axis which experimentally obtained [17]. For simplicity, interactions among antiprotons are also neglected. Considering the experimentally obtained synchrotron radiation cooling time, $\tau_c[\text{s}] \sim 8.6/B[\text{T}]^2$, the energy transfer between the beam to the plasma is found by solving the following simultaneous equation with respect to time $t$:

$$\frac{dE_{\bar{p}}(t)}{dt} = -\left\langle\frac{d\mathcal{E}}{dt}\right\rangle, \tag{10}$$

$$\frac{dT_e(t)}{dt} = \frac{2}{3k_B}\left(\frac{N_{\bar{p}}}{N_e}\right)\left\langle\frac{d\mathcal{E}}{dt}\right\rangle - \frac{T_e(t)}{\tau_c}, \tag{11}$$

with initial value of $E_{\bar{p}}(0)$ and $T_e(0)$.

313

The temperature of the electron plasma was evaluated from the frequency shift. The (2,0) and (3,0) mode frequency shift after the antiproton injection were around 0.6 eV and 0.8 eV at the maximum, respectively. Figure 6 shows a calculated time evolution of the electron plasma temperature, which shows the estimated maximum temperature rise was about 0.7 eV. The shift came back to the initial temperature within 30 s. These features by numerically solving the simultaneous Eqs. 10 and 11 agreed with experimental results shown in Fig. 5. Here the coupling constant was chosen to 0.073.

## Radial (de)compression

It is known that a rotating rf field can give a torque for confined plasma which is in a rigid rotor equilibrium, and so the radial distribution of the plasma is controlled [18, 19, 20]. In this experiment, the electron plasma was successfully compressed radially by rotating electric dipole field whose frequency and amplitude were around 1–10 MHz and 0.8 $V_{pp}$. We found that the counter-rotating electric dipole fields rotating inversely against the direction of rigid rotation of the plasma expanded the plasma radially. The parameters of the counter-rotating field were the same as for compression except for the rotation direction. After the expansion, the radius of the plasma was estimated by using Eqs. 2 and 3 with two different electrostatic mode frequencies as described in the above section. By applying the counter-rotating field, the initially formed electron plasma has the radius of 0.7 mm was expanded to 1.7 mm for 15 s counter-rotation, 2.1 mm for 30 s, 3.4 mm for 60 s, and 4.9 mm for 120 s, respectively, as summarized in Tab. 2.

Filled circles in Fig.7 show the number of confined antiprotons as a function of electron plasma radius. This result shows that the injected antiproton beam was broader than the initial electron plasma, that is, without any radial distribution control a considerable amount of injected antiprotons which were not cooled and trapped in the trapping region were lost. The other result by a beam profile monitor located on the entrance of the MRT ("BPM" in Fig. 1) also confirmed that the injected antiproton beams from the RFQD had a larger radius (Fig. 8). From the confined number of antiprotons against radii of plasmas, we roughly estimated the injected antiproton beam profile. Figure 9 shows an example of contained antiproton number within each shell under the assumption that the radial distribution of cooled antiprotons by collisions with the electrons was same as the electron plasma radius, which agrees with the result of the beam profile monitor.

Triangle symbols in Fig. 7 show the extracted number of antiprotons after the electron cooling after the electron release from the trap region[10]. When the radius of the

TABLE 2. Observed frequencies for different radii of plasmas controlled by rotating electric fields.

| $f_{(2,0)}$ [MHz] | $f_{(3,0)}$ [MHz] | aspect ratio | radius [mm] | half length [mm] | density [/cm$^3$] |
|---|---|---|---|---|---|
| 17.669 | 24.034 | $\sim 89$ | $\sim 0.7$ | 62 | $\sim 2.7 \times 10^9$ |
| 17.882 | 23.949 | 30 | 1.7 | 51 | $4.2 \times 10^8$ |
| 17.760 | 23.689 | 24 | 2.1 | 50 | $3.0 \times 10^8$ |
| 17.660 | 23.212 | 14 | 3.4 | 49 | $1.2 \times 10^8$ |
| 17.561 | 22.756 | 9 | 4.9 | 45 | $6.5 \times 10^7$ |

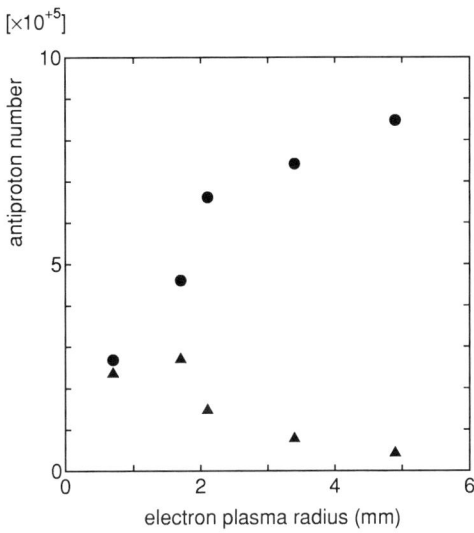

**FIGURE 7.** Confined antiproton number against the radii of the electron plasmas shown as filled circles. Also extracted number of antiprotons is shown as filled triangles.

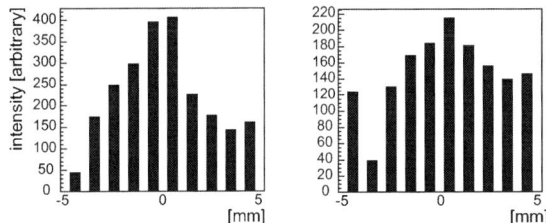

**FIGURE 8.** An example of the incident antiproton beam profile at the entrance of the MRT.

electron plasma was 1 mm, most of confined antiprotons were extracted as we expected from our beam line design, that is, antiprotons within 1 mm from the axis of the beam line were completely extracted [8]. But for larger radii of plasmas, the extraction numbers of antiprotons were decreased. For example, from the result shown in Fig. 7, smallest size electron plasma whose radius was about 0.7 mm confined and cooled only $2.7 \times 10^5$ antiprotons, but most of the cooled antiprotons, that is, $2.3 \times 10^5$, were extracted out of the field free region. On the other hand, about 10 mm diameter electron plasma cooled more than $8.0 \times 10^5$ antiprotons, but only less than $5 \times 10^4$ antiprotons were extracted.

To increase the extraction number of the antiproton, the radial compression technique of antiproton cloud was developed. Before the antiproton cloud compression, electrons were removed from the trapping region by opening and closing the trapping potential well for 550ns three times. Then the antiproton cloud was successfully compressed

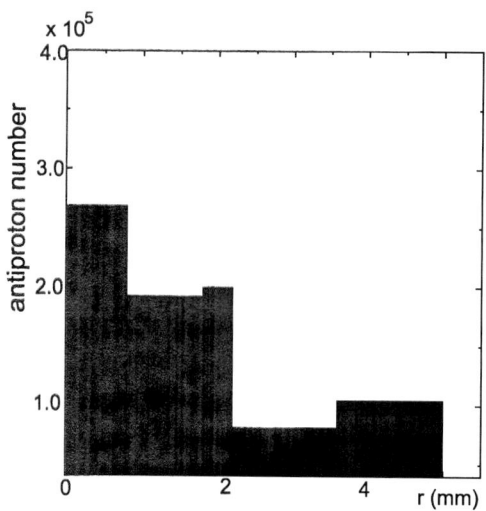

**FIGURE 9.** An example of the profile of an injected antiproton beam.

radially and extracted out of the field free region.

## SUMMARY

The antiprotons passing through the isolation foils were cooled and confined in the harmonic potential provided by the MRT. The cooling process was nondestructively observed by monitoring electrostatic mode frequencies of the electron plasma. This feature was understood by the model considering the beam cooling by the electron plasma and the synchrotron radiation cooling of the electrons in the 2.5 T magnetic field.

The radial distribution of the electron plasma was controlled by utilizing the rotating electric field technique. The injected antiproton beams with large radius were cooled by the expanded electron plasma decompressed by the counter-rotating field against the rigid rotation direction of the electron plasma.

## ACKNOWLEDGMENTS

This work was supported by the Grant-in-Aid for Creative Scientific Research (10P0101) of the Japanese Ministry of Education, Culture, Sports, Science and Technology (MonbuKagaku-shō), Special Research Projects for Basic Science of RIKEN, and the Hungarian National Science Foundation (OTKA T033079).

# REFERENCES

1. Y. Yamazaki, *Nucl. Instrum. Methods Phys. Res. B*, **154**, 174 (1999).
2. M. Hori, J. Eades, R. Hayano, T. Ishikawa, J. Sakaguchi, E. Widmann, H. Yamaguchi, H. Torii, B. Júhasz, D. Horváth, and T. Yamazaki, *Phys. Rev. Lett.*, **87**, 093401 (2001).
3. H. Knudsen, U. Mikkelsen, K. Paludan, K. Kirsebom, S. P. Møller, E. Uggerhøj, J. Slevin, M. Charlton, and E. Morenzoni, *Phys. Rev. Lett.*, **74**, 4627 (1995).
4. A. Trzińska, J. Jastrzębski, P. Lubiński, F. Hartmann, R. Schmidt, T. von Egidy, and B.Kłos, *Phys. Rev. Lett.*, **87**, 082501 (2001).
5. M. Wada, Y. Ishida, T. Nakamura, Y. Yamazaki, T. Kambara, H. O. ad Y. Kanai, T. Kojima, Y. Nakai, N. Oshima, A. Yoshida, T. Kubo, Y. Matsuo, Y. Fukuyama, K. Okada, T. Sonoda, S. Ohtani, K. Noda, H. Kawakami, and I. Katayama, *Nucl. Instrum. Methods Phys. Res. B*, **204**, 570 (2003).
6. A. Mohri, and Y. Yamazaki, *EuroPhys. Lett.*, **63**, 207 (2003).
7. A. Mohri, H. Higaki, H. Tanaka, Y. Yamazawa, M. Aoyagi, T. Yuyama, and T. Michishita, *Jpn. J. Appl. Phys.*, **37**, 664 (1998).
8. K. Yoshiki Franzen, N. Kuroda, H.A. Torii, M. Hori, Z. Wang, H. Higaki, S. Yoneda, B. Juhász, D. Horváth, A. Mohri, K. Komaki, and Y. Yamazaki, *Rev. Sci. Instrum.*, **74**, 3305 (2003).
9. G. Gabrielse, X. Fei, L. Orozco, R. Tjoelker, J. Haas, H. Kaliowsky, T. Trainor, and W. Kells, *Phys. Rev. Lett.*, **63**, 1360 (1989).
10. N. Kuroda, H.A. Torii, K. Yoshiki Franzen, Z. Wang, S. Yoneda, M. Inoue, M. Hori, B. Júhasz, D. Horváth, H. Higaki, A. Mohri, J. Eades, K. Komaki, and Y. Yamazaki, *Phys. Rev. Lett.*, **94**, 023401 (2005).
11. H. Higaki, N. Kuroda, K. Yoshiki Franzen, Z. Wang, M. Hori, A. Mohri, K. Komaki, and Y. Yamazaki, *Phys. Rev. E*, **70**, 026501 (2004).
12. R. C. Davidson, *PHYSICS OF NONNEUTRAL PLASMAS*, Imperial College Press and World Scientific, London, 2001.
13. D. Dubin, *Phys. Rev. Lett.*, **66**, 2076 (1991).
14. M. Tinkle, R. Greaves, C. Surko, R. Spencer, and G. Mason, *Phys. Rev. Lett.*, **72**, 352 (1994).
15. H. Higaki, N. Kuroda, T. Ichioka, K. Yoshiki Franzen, Z. Wang, K. Komaki, and Y. Yamazaki, *Phys. Rev. E*, **65**, 046410 (2002).
16. D. Sivukhin, "COULOMB COLLISIONS IN A FULLY IONIZED PLASMA," in *Review of Plasma Physics*, edited by A. M. Leontovich, Consultants Bureau, New York, 1966, vol. 4.
17. H. Torii, N. Kuroda, M. Shibata, Y. Nagata, D. Barna, M. Hori, J. Eades, A. Mohri, K. Komaki, and Y. Yamazaki, "Production of ultra-slow antiproton beams," in *Workshop on Physics with Ultra Slow Antiproton Beams*, RIKEN, AIP, 2005 (in press).
18. X.-P. Huang, F. Anderegg, E. Hollmann, C. Driscoll, and T. O'Neil, *Phys. Rev. Lett.*, **78**, 875 (1997).
19. F. Anderegg, E. Hollmann, and C. Driscoll, *Phys. Rev. Lett.*, **81**, 4875 (1998).
20. T. Ichioka, *Development of intense beam of ultracold antiprotons*, Ph.D. thesis, University of Tokyo (2001).

# The Anticyclotron Project [1]

Dezső Horváth

*KFKI Research Institute for Particle and Nuclear Physics, H–1525 Budapest, Hungary
and Institute of Nuclear Research (ATOMKI), Debrecen, Hungary*

**Abstract.** A progress report is presented on the development of the *anticyclotron* — deceleration of antiprotons and negative muons via collisions in a low–pressure gas or thin foils during revolutions in a cyclotron field. Beam tests performed at CERN and PSI are reported and present applications outlined.

**Keywords:** anticyclotron, antiproton, cyclotron trap, extraction, trapping
PACS: 29.20.Hm, 29.27.Ac, 36.10.Gv

## HISTORICAL INTRODUCTION

There are several interesting experiments in basic physics, which should become feasible with the availability of a great number of slow, *exotic*, negatively charged particles, such as direct experimental studies of the fundamental symmetry between matter and antimatter [1] and other high–precision studies in low-energy atomic, nuclear and particle physics. All of these particles ($\mu^-$, $\pi^-$, $K^-$, $\bar{p}$) are produced with relatively high energies and have to be decelerated for the experimental studies. Usually, this is done by letting them pass through degrader materials and then stopping them in a target which has to be thick due to particle straggling in the degrader. For spectroscopic measurements, however, one has to use thin targets in order to minimize self–absorption and also to exclude interatomic collisions, the latter being the stronger condition.

The antiproton is perhaps the most interesting of these particles, mainly owing to its stability. At present the only available source of low-energy antiprotons is the Antiproton Decelerator (AD) [8] at CERN where the antiprotons, produced at a few GeV kinetic energy by inelastic scattering of high energy protons, undergo a many–step cooling — deceleration — cooling procedure. The AD serves as a replacement for LEAR, the former Low Energy Antiproton Ring of CERN which was decommissioned in 1996 after more than 10 years of work. The importance of having cold antiprotons cannot be overestimated; its physics possibilities are thoroughly discussed at all conferences and workshops related to antiproton physics.

In its *slow extraction* regime (with spill lengths from 400 $\mu$s to about one hour) LEAR could reliably deliver $3 \times 10^9$ antiprotons per filling with a kinetic energy of 5 MeV, whereas in fast extraction 50 – 400 ns long bunches of $> 10^8$ $\bar{p}$ [3]. The AD gives $4 \times 10^7$ 5 MeV antiprotons in 200 ns bunches with a repetition period of 100 s. However,

[1] Invited paper presented at *Workshop on Physics with Ultra Slow Antiproton Beams, RIKEN, Wako, Japan, 14-16 March 2005*

the need for cold antiprotons requires further deceleration stages as ion traps can accept antiprotons only up to a few tens of keV for subsequent cooling [4] down to energies corresponding to cryogenic temperatures.

Trapping and cooling of antiprotons is now considered to be routine. Thus the problem of decelerating antiprotons from GeV to $\mu$eV energies is solved, except from the interval between a few MeV and a few tens of keV. Numerous deceleration schemes have been proposed to bridge this gap since as early as the beginning of LEAR. The PBARMASS (PS189) experiment at LEAR [5] was originally proposed in 1981 to be performed using the planned ELENA (Extra Low ENergy Antiproton) storage ring. In 1989 this was changed to a Radiofrequency Quadrupole Decelerator (RFQD) [6] designed to accept 61 MeV/c antiprotons from LEAR. The anticyclotron (AC) project was proposed in 1989 [7] for an antiproton gravity experiment. ELENA was never built at LEAR. The RFQD and the AC were installed and tested at LEAR and both tests failed due to the not satisfactory specifics of the injected beam: below 100 MeV/c LEAR could not deliver antiprotons with the required (and previously published [3]) specifications.

Now history repeats itself: our ASACUSA collaboration installed an RFQ post-decelerator at the Antiproton Decelerator (AD) which works at a high efficiency [9] and the ELENA project has again emerged [10] as a possible way of obtaining keV antiprotons at the AD. My talk aims at reviving some interest in the anticyclotron as well.

# THE ANTICYCLOTRON

The anticyclotron is a small superconducting cyclotron with no RF field, operating in an inverse way: tangential injection at a high radius and — after deceleration in a low pressure gas or in a thin degrader foil — axial extraction (Fig. 1). It was used to study $\bar{p}$ X-ray cascades in low-pressure gases (*cyclotron trap* [11, 12]). The anticyclotron was proposed to serve as a basic apparatus to provide an ultra-low energy antiproton beam [2, 7] with a predicted transmission efficiency $\varepsilon(7–10\ \text{keV})$ up to 20% with 0.3 mbar $H_2$ as moderator.

## The anticyclotron with moderator gas

In 1990–91 the performance of the anticyclotron was tested at LEAR by the P118 Collaboration [13, 14, 15] in the following operating conditions: 72 MeV/c antiprotons of 100 ns long bursts were injected at an initial radius of $R = 109$ mm, $\Delta Z = 10$ mm off the axis of the magnetic field, through a Be window and a thin scintillator, into the target chamber. A properly timed high-voltage pulse applied on the inflector electrodes (Fig. 2) shifted the antiproton trajectories to the median plane where — having avoided hitting the injection structure — they slowly decelerated in the low-pressure ($<$ mbar) gas. Before atomic capture, a suitable high-voltage pulse was to be applied on the extraction electrodes (two Ø100 mm rings covering the bore holes) to push the $\bar{p}$ swarm out of the target chamber, through a thin carbon foil in one of the axial bore holes of the magnet

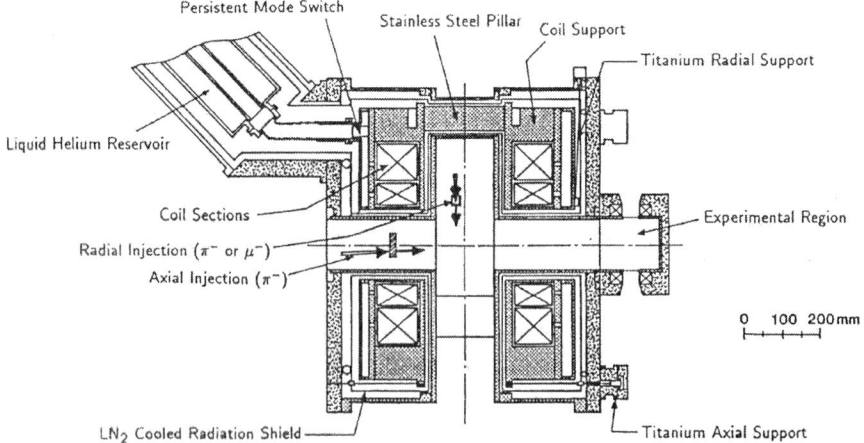

**FIGURE 1.** The cross section of Anticyclotron I. The fast particles (muons, pions or antiprotons) to be moderated can be injected radially or axially. Axial injection is for pions which then decay to muons.

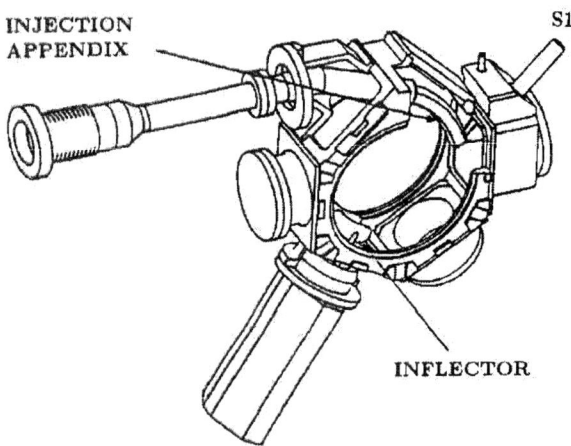

**FIGURE 2.** The vacuum chamber of Anticyclotron I. The fast antiprotons are injected radially via the injection appendix and initially degraded by the beam monitor scintillator S1. For the antiproton tests at LEAR a pair of inflector electrodes were installed to shift the particle trajectories away from the injector appendix and low-pressure helium gas served as moderator.

(located on either side of the median plane), into a cylindrical trap. In optimal conditions, conversion efficiencies (from 2 MeV to keV energies) up to $10 - 20\%$ were expected.

Each antiproton annihilation produces several fast, charged particles and the vacuum chamber was lined on the outside with plastic scintillators to detect them. The stopping

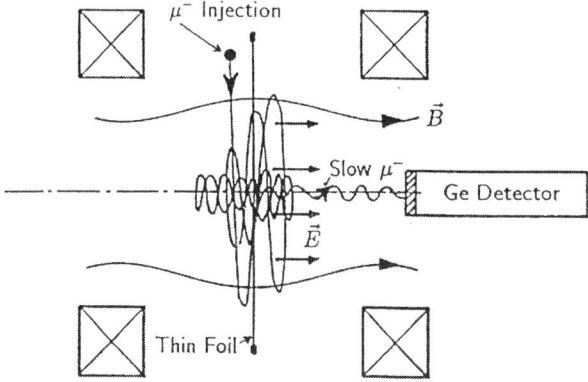

**FIGURE 3.** The setup of our $\mu^-$ tests: Cross-section of the target chamber with the foil in the median plane, the Ge(Li) detector in the bore hole and a simulated $\mu^-$ trajectory [17].

of antiprotons in the centre was demonstrated by the presence of a broad delayed peak (*gas peak*) in the spectrum of the time delay between the S1 beam counter and the annihilation signal.

In the bore hole opposite to that containing the trap three X–ray sensitive CCD detectors were installed at $Z = 100$ mm from the median plane in a vertical arrangement. A 12.5 $\mu$m Kapton window separated the CCD vacuum from the anticyclotron chamber. During these beam tests, in 1990–91, we measured the annihilation time spectra and deduced from them the stopping efficiency and the deceleration times at various pressures and with several gases [16], studied the antiproton trajectories with a radially movable scintillator and observed $\bar{p}p$ Balmer X-rays with CCD's.

Unfortunately, the antiproton tests could not be completed owing to the unexpectedly poor characteristics of the 72 MeV/c $\bar{p}$ beam; in spite of the estimations and studies published in [3], LEAR never could deliver a good quality $\bar{p}$ beam with momenta below 100 MeV/c.

## The anticyclotron with foil moderator

In the middle of 1991 the anticyclotron was returned to PSI to serve as a basic device to produce a low energy (few keV) muon beam. The characteristics of the muon beam makes this endeavor much more problematic than that for antiprotons [17]:

- The emittance of the muon beam is almost two orders of magnitude higher so the relative acceptance of the system will be much lower.
- The momentum resolution is in the percent region, an order of magnitude worse than that for the LEAR beams.

321

- The lifetime of the muons (2.2 $\mu$sec) requires a much higher gas pressure which would immediately lead to discharges during extraction and serious problems of vacuum separation between the anticyclotron and the beam line of the extracted slow muons.

Therefore a new idea was proposed: to use a thin foil ($\sim \mu$g/cm$^2$), or a set of them, in, or parallel to, the median plane of the anticyclotron instead of gas [17]. Computer simulations have shown that for a radial injection of $8 \times 10^6$ $\mu^-$/s of 40 MeV/c momentum should yield $2.4 \times 10^4$ slow (5 – 25 keV) muons/s with a $5 \times 5$ cm$^2$ spot size. These data were calculated for the high–intensity $\pi$E5 channel of PSI assuming a 1 mA proton current [18].

## Muon extraction tests

For our first test run, in November 1991, in the $\pi$E1 channel at PSI, a 240 $\mu$g/cm$^2$ Mylar foil was installed in the target chamber of the anticyclotron (Fig. 3). We had $\approx 1500$ $\mu^-$/s (per 100 $\mu$A·s proton current) entering the anticyclotron, accompanied by three orders of magnitude more electrons. The muons and the electrons were separated using time-of-flight and by the energy deposited in the scintillators monitoring the beam. The beam entering the anticyclotron at a tilting angle of 80 mrad to the median plane was accepted within radii 115 – 135 mm and impact parameter (axial shift) values 12.5 – 22.5 mm as determined by the monitor scintillator.

The computer simulations predicted that within these conditions $\approx 20\%$ of the muons will stop in the foil. A planar GeLi detector in the bore hole of the anticyclotron detected the carbon and oxygen X-rays produced by the muons stopped in the foil (Fig. 3). The results of this first test were very encouraging: the efficiency of muon injection into the anticyclotron was 15%, and about 17% of the injected $\mu^-$ stopped in the foil.

Extraction tests of muons started in June 1992 [17, 18, 19, 20]. An 80 $\mu$g/cm$^2$ thick, $\emptyset$90 mm Ti layer was evaporated on a 240 $\mu$g/cm$^2$ Mylar foil. The foil was installed in the median plane of the target chamber with the Ti spot away from the GeLi detector. For extraction, a static high voltage was applied to the metal layer of the foil. The extracted muons were stopped in the 300 $\mu$m thick beryllium window of the GeLi detector, placed in the bore hole behind the maximum of the magnetic field, and were identified by the 33 keV $\mu$Be$(2P - 1S)$ line. 1% of the muons slowed down in the foil were extracted with no electric field, while applying a –9.5 kV voltage to the metal-deposited foil yielded a 2% extraction (the simulations predicted 2.3%). The self–extraction of muons with no field on was also predicted by the simulations: the confining force of the magnetic field diminishes with decreasing radius and velocity of the particles whereas large-angle scattering in the foil increases.

## Slow extracted muon beam

Based on the experience with Cyclotron Trap I a new device was designed and built at PSI (Fig. 4). As the fixed bore hole of the old one largely restricted the geometry for

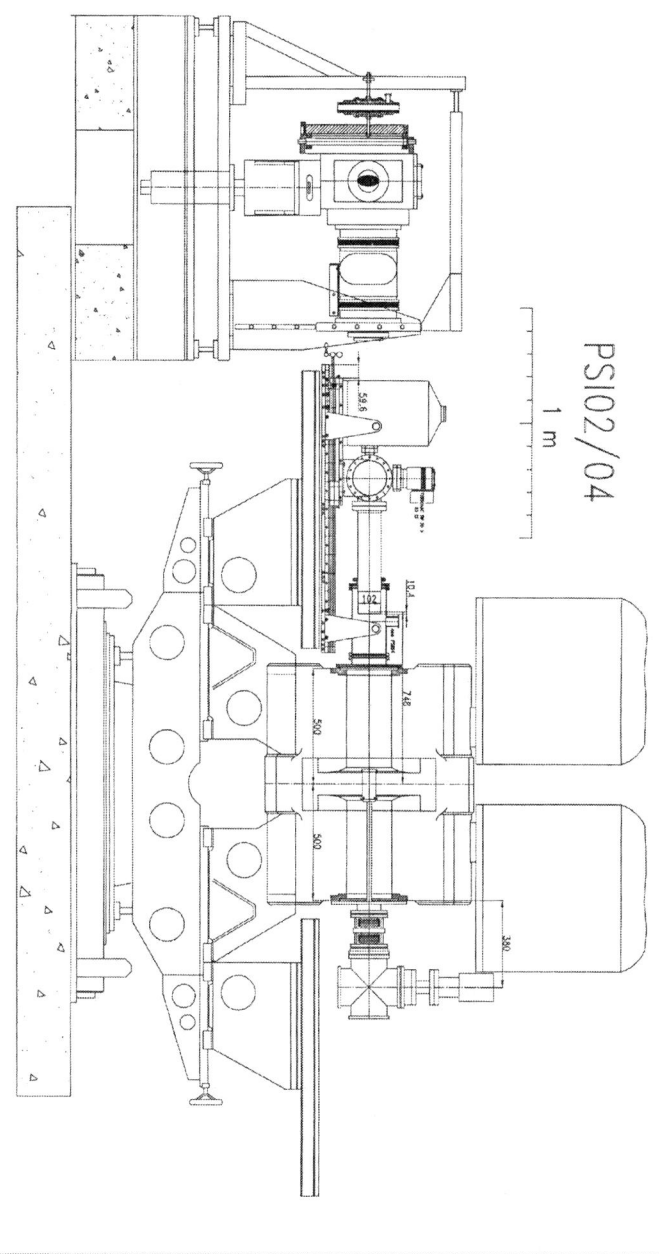

**FIGURE 4.** The design of Anticyclotron II. The two coils can be moved apart to make space for the experimental setup.

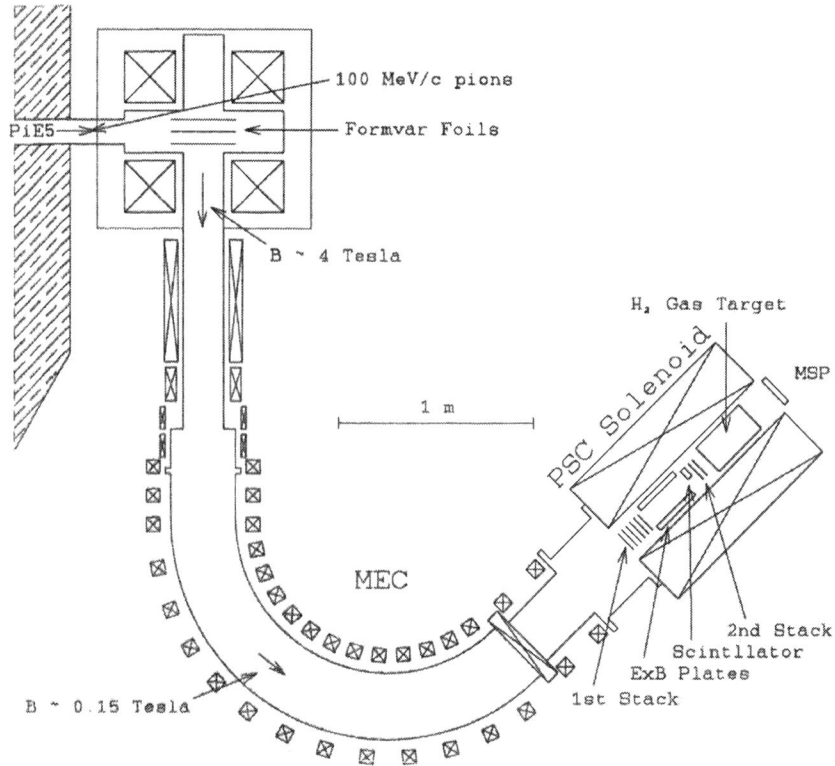

**FIGURE 5.** Slow extracted muon beam line at PSI [22]. The axially injected negative pions decay to fast muons which get trapped in the $B = 4$ T field of Anticyclotron II and slow down in a stack of thin foils from MeV to keV energies. The 20 keV muons are extracted with an electric field into the curved muon extraction channel (MEC) which helps cleaning and monochromatizing the slow muon beam. The slow muons are stopped in a hydrogen gas target placed in a superconducting solenoid (PSC).

experiments, Cyclotron Trap II consists of two separate superconducting coils in order to liberate the space for various experiments between them. Its field is also higher: up to 4 Tesla. It is used as a low-pressure hydrogen target for pionic hydrogen experiments [21] and also as a source of low-energy muons for extraction (Anticyclotron II) [22].

Recently an extraction beam line was constructed [22] at PSI based on the new anticyclotron (Figs. 5 and 6). Negative pions of 100 MeV/$c$ momentum are injected axially in the anticyclotron where they decay in flight to muons. The muons get trapped in the cyclotron field and slow down in a series of collisions in a stack of thin foils from MeV to keV energies. An axial electric field extracts $\sim 20$ keV muons from the $B = 4$ T magnetic field of the trap into the curved muon extraction channel (MEC). The toroidal $B = 0.15$ T field of the MEC helps to eliminate the electrons and fast muons mixed with the slow muon beam.

**FIGURE 6.** Slow extracted muon beam line at PSI [22].

At $2 \times 10^8$/s 100 MeV/$c$ $\pi^-$ injected in the anticyclotron, $10^4$/s $\mu^-$ arrived to the Be foil before the hydrogen target with kinetic energies $E_\mu = 10 - 50$ keV and diameter about 3 cm. 100 muon stops/s were measured in the hydrogen target, but that was limited by the laser experiment, not the muon extraction system.

## Application for antiprotons

If one wanted to apply the foil method to antiprotons, it would correspond to the following conditions [17]: a plastic foil of a few $\mu$g/cm$^2$ thickness (crossed twice in a betatron oscillation, 100 ns) has a similar slowing-down efficiency as hydrogen gas of a few mbar pressure (traversing about 60 cm of it in every 50 ns). The anticyclotron then could be completely evacuated (at present the limit is $2 \times 10^{-6}$ mbar) which should make the problems of active inflector, vacuum separation and discharges during extraction nonexistent or easy to solve. Using the foil method should also leave us the freedom to choose an optimal thickness, not limited by other conditions.

Efficient moderation of antiprotons needs extremely thin foils. First of all, in order to have an acceptable stopping efficiency, the foil moderator has to simulate the gas as much as possible, i.e. one may need an extended array of parallel, very thin foils. Note, that the extended foil structure ensures that the time structure of the injected antiproton bunch is conserved with some spread from straggling; this is important for most applications. By extrapolating the $\mu^-$ results we expect to be able to convert up to 5% of the 5 MeV $\bar{p}$ beam into an extracted beam of a few keV energy. The efficiency and characteristics of an antiproton anticyclotron are expected to be comparable to those achieved by the RFQ decelerator, so it can be considered by the antihydrogen experiments as an alternative solution.

## CONCLUSION

The development of the anticyclotron method has been proceeding along several lines at PSI. The low–energy $\mu^-$ beam is developed as part of the muonic hydrogen Lamb shift experiment, whereas the new cyclotron trap is used as a low-pressure gas target for the pionic hydrogen studies. This method is proposed for consideration by the antihydrogen experiments at the Antiproton Decelerator at CERN.

## ACKNOWLEDGMENTS

This talk could and should have been presented by any one of the following colleagues: Leopold Simons (PSI), Franz Kottmann (ETHZ), John Eades (CERN) and Detlev Gotta (Jülich). Instead, they asked me to do that and supplied all necessary information for which I am grateful. The present work was supported by the Hungarian National Research Foundation (Contract OTKA T046095). My participation at the conference was made possible by the financial support of the organizers.

## REFERENCES

1. M. Charlton, J. Eades, D. Horváth, R. J. Hughes, C. Zimmermann: Physics Reports **241**, 65 (1994).
2. J. Eades *et al.*: *Experiments with very low energy antiprotons and a possible new method for measuring the gravitational mass of the antiproton*, CERN/PSCC/89–30, PSCC/I77, 1989.
3. P. Lefevre, In Physics at LEAR with low energy antiprotons, Proc. 4th LEAR Workshop, Villars-sur-Ollon, 1987, p. 19;
   M. Chanel, In LEAP'90, Proc. First Biennial Conf. on Low Energy Antiproton Physics, Stockholm, 1990, P. Carlson *et al.*, eds., World Scientific, p. 470 (1991);
   P. Lefevre *et al.*, *Production of MeV antiprotons*, Antihydrogen Workshop, Munich, 1992; Hyperfine Interactions, **76** (1993).
4. G. Gabrielse *et al.*: Phys. Rev. Lett. 65, 1317 (1990).
5. G. Audi *et al.*: Proposal Nr. CERN-PSCC-81-84, 1981.
6. PS189 Status Report, CERN-PSCC-89-26, 1989.
7. J. Eades, L. M. Simons: Nucl. Instr. Meth. **A278**, 368 (1989).
8. The Antiproton Decelerator of CERN: http://www.cern.ch/PSdoc/acc/ad/index.html.
9. A.M. Lombardi, W. Pirkl, Y. Bylinsky: *First Operating Experience with the CERN Decelerating RFQ for Antiprotons*, Report CERN/PS 2001-064.
10. P. Beloshitskii *et al.*: *Proposal for an Extra Low ENergy Antiproton ring: ELENA*, SPSC-2005-011, CERN, 2005.
11. L. M. Simons: Physica Scripta **T22**, 90 (1988).
12. K. Heitlinger *et al.*: Z. Phys. A **342**, 359 (1992).
13. E. C. Aschenauer *et al.*: *Report to the SPSL Committee on the beam tests made by the P118 collaboration at LEAR*, CERN/SPSLC/92/28, SPSLC/M494, 20 May 1992.
14. E. C. Aschenauer *et al.*: Materials Science Forum **105–110**, 1837 (1992).
15. E. C. Aschenauer *et al.*: Yadern. Fiz. **55** (1992) 1555; Sov. J. Nucl. Phys. **55**, 856 (1992).

16. D. Horváth *et al*.: In: *Exotic Atoms, Molecules and their Interactions*, Proc. 6th Workshop of the International School of Physics of Exotic Atoms (Erice, Sicily, 1994), C. Rizzo and E. Zavattini, eds., INFN Press, Trieste, 1994, p. 151.
17. P. De Cecco *et al*.: Hyperfine Interactions, **76**, 275 (1993).
18. L. M. Simons, F. Kottmann: In: *Muonic Atoms and Molecules*, L. A. Schaller, C. Petitjean, eds., Birkhäuser, 1993, p. 307.
19. D. Horvath *et al*.: Nucl. Instr. Meth. B **85**, 736 (1994).
20. P. De Cecco *et al*.: Nucl. Instr. Meth. A **394**, 287 (1997).
21. D. Gotta: Prog. Part. Nucl. Phys. **52**, 133 (2004).
22. F. Kottmann et al.: Hyperfine Interactions, **138**, 55 (2001).

# ASACUSA Gas-Jet Target: Present Status And Future Development

V.L. Varentsov[2,3], N. Kuroda[2], Y. Nagata[1,2], H. A. Torii[1], M. Shibata[2], Y. Yamazaki[1,2]

[1]Institute of Physics, University of Tokyo, Komaba, Meguro-ku,Tokyo 153-8902, Japan
[2]Atomic Physics Laboratory, RIKEN, Wako, Saitama 351-0198, Japan
[3]V.G. Khlopin Radium Institute, 2nd Murinskiy Ave. 28, 194021, St. Petersburg, Russia

**Abstract.** A supersonic gas-jet target apparatus that have been prepared to study elementary processes of antiprotonic atoms formation using monoenergetic ultra-slow antiproton beams is described. We investigated an operation of this target with cryogenically cooled nozzle by both gas dynamic simulations and supersonic jet measurements. In result, the helium target density of $2 \cdot 10^{12}$ atoms/cm$^3$ has been obtained.
For considerable increasing of the target density, a qualitative modification of the present target setup is suggested. The goal can be achieved by the use of pulsed high-pressure supersonic gas jet that operates in accordance with the pulsed mode of the MUSASHI penning trap. For this purpose an additional stage of differential pumping with a skimmer will be set into the present target setup. To avoid the clusters in the gas-jet target, a sonic nozzle equipped with a solenoid driven pulsed gas valve will be used at room or higher temperatures. The operation of this future version of the gas-jet target apparatus has been studied by means of detailed computer simulations. Results of these calculations for helium, which show the possibility of pulsed gas target production of $3 \cdot 10^{13}$ atoms/cm$^3$ density are presented also.

**Keywords:** Supersonic gas jet, nozzle, internal target, skimmer, computer simulation.
**PACS:** 39.10.+j; 47.40.Ki

## INTRODUCTION

Antiproton, the antiparticle counterpart of the proton is an exotic but interesting probe for atomic physic. Now that antiprotons can be prepared as a monoenergetic beam at electronvolt energies from the ASACUSA-MUSASHI apparatus [1], the next natural step is to study collision dynamics including ionization of atoms by antiproton impact and capture of an antiproton to form an antiprotonic atom. Formation processes of antiprotonic atoms can be studied by employing collisions between an ultra-slow antiproton beam and a thin gas target $A$ under single collision conditions:

$$\bar{p} + A \rightarrow \bar{p}A^+ + e^-$$

at an energy comparable to the ionization potential of the target atom. Atomic formation and ionization processes will be detected by taking coincidence between signals of antiproton annihilation and the electron released from the target atom.

Since the number of antiprotons available as ultra-slow beams is very much limited, the reaction probability must be maximized in order to make best use of the antiprotons, and therefore a maximum possible density of atomic gas jets need to be prepared to be crossed with the antiprotonic beams. The number of antiprotonic atoms formed is given by the formula

$$N_{\bar{p}A+} = \sigma\, n_A\, L\, N_{\bar{p}}\,,$$

where $N_{\bar{p}A}$ is the number of the antiprotonic atoms, $\sigma$ is the formation cross section, $n_A|$is the number density of the atomic target, $L|$ is the interaction length, and $N_{\bar{p}}|$is the number of antiprotons.

Atomic formation cross sections are naturally of the order of $10^{-16}$ cm$^2$, as was confirmed by a few theoretical calculations [2]. Suppose a beam with $10^5$ antiprotons is available every shot per several minutes [1], we will obtain $10^2$ antiprotonic atoms for a gas density of $10^{13}$ atoms/cm$^3$ at an interaction length of 1 cm. This number is almost at the lower limit of detection in order to assure enough statistical significance for the formation events to be distinguished from background events. At our early stage of development, several ideas for realization of an atomic-beam target were considered [3]. Usage of effusive gas out of micro-capillary array was also considered, but was rejected in favor of supersonic gas jets with a cryo-cold head for two main reasons: 1) the gas flow should be well-collimated and 2) the background vacuum in a collision chamber should be kept at a level below $10^{-6}$ Torr to allow operation of multichannel plate (MCP) for particle detection and to ensure an ultra-high vacuum of $10^{-12}$ Torr in the antiproton trap 3 m upstream by differential pumping with three apertures in the antiproton beamline [4].

To begin with, we will prepare rare gas atomic targets starting with the simplest helium atom, before we shall consider atomic and molecular hydrogen targets, which are though simple from theoretical point of view, confront experimental difficulties including special safety precautions needed to be taken.

## EXPERIMENTAL SETUP

A schematic figure of the present ASACUSA gas-jet target setup is presented in Fig. 1. It has five differentially pumped vacuum chambers, which are evacuated by turbo molecular pumps (TMP). The first chamber, where the supersonic jet expands into vacuum, is evacuated by a large TMP (Pfeiffer TPH2101P), and backed by a rotary vane pump (Alcatel, 63 m$^3$/h). An effective pumping speed of this chamber consists of 1076 l/s for helium, when a gas flow rate through the nozzle is 6.6 mbar l/s. Through the conical skimmer the expansion chamber communicates with the next chamber that is pumped by two TMP (Pfeiffer TMH261), each having a nominal pumping speed for helium of 220 l/s, backed by one rotary vane pump (Alcatel 33 m$^3$/h). The third chamber, where the ultra slow antiproton beam crosses the target-beam and a detector system is placed, provides an additional stage of differential pumping. This collision chamber, communicating with the second chamber by means of a collimator, is pumped by TMP (Leybold TURBOVAC600), and backed by a rotary vane pump (Alcatel, 15 m$^3$/h). The last two differentially pumping stages are provided by target-beam dump chambers, which are communicate with each other and the collision chamber by means of short rectangular tubes which are used as diaphragms. The first of these chambers is pumped by TMP (Leybold TURBOVAC600), backed by a rotary vane pump (Alcatel, 15 m³/h). The other one is evacuated by a large TMP (Pfeiffer TPH2101P), backed by a rotary vane pump (Alcatel, 33 m³/h). The effective pumping speeds and background gas pressures under typical gas-jet operation conditions are shown in the Fig. 1.

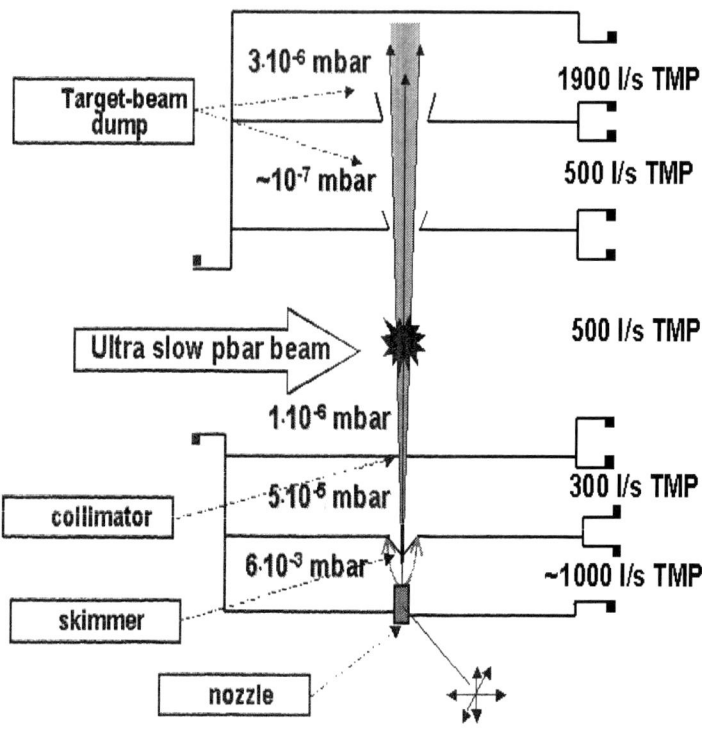

**FIGURE 1.** Schematic figure of the present gas-jet target setup. Five differentially pumped vacuum chambers are evacuated by turbo molecular pumps (TMP). The effective pumping speeds and background gas pressures under typical helium gas-jet operation conditions are shown.

The supersonic jet is produced by gas expansion into vacuum through the nozzle (in our case it is a hole of 0.1 mm in diameter in a gas cell wall). The gas cell with the nozzle is mounted on a cryogenically cooled head that allows decreasing a gas stagnation temperature (in the gas cell) down to 30 K. The cryogenic head, in its turn, is mounted on a movable base that can be displaced under vacuum along the three perpendicular x-y-z axes by means of two bellows. This x-y-z movement of the nozzle allows getting an alignment of the supersonic gas jet relatively to the fixed skimmer-collimator axis and regulating the nozzle-skimmer distance. We used it also for the jet profiles measurements, which are described in the next section.

The skimmer cuts out the small axial part of the supersonic jet with the forming of the target-beam. In order to minimize the gas flow perturbations at the skimmer entrance, a high quality and precision commercial one (Beam Dynamic Inc., model 10.2) is used. It has a parabolic profile, ultra-thin walls and ultra-sharp orifice edges. The skimmer is mounted on a removable supporting flange and has an entrance orifice diameter of 0.6 mm.

For final target-beam formation, a thin rectangular-shape collimator is used. The collimator has 2.2x4.4 mm aperture and is placed at 35 mm distance from the skimmer entrance.

The shape and sizes of the target-beam in the collision chamber at 50 mm distance from the collimator, where it crosses the antiproton beam, is determined by the described skimmer-collimator geometry. So, the target has 10 mm length at the antiproton beam direction and 5 mm length at the perpendicular one. It should be noticed that such rectangular target-beam geometry has an advantage over a circular beam of 10 mm diameter. Target thickness (it is a product of target density and length in the beam direction) for both cases is near the same. But, taking into account that a 5 mm target size at the perpendicular direction is enough to overlap the antiproton beam, the described rectangular target-beam shape allows improvement of vacuum in the collision chamber due to the decreasing of the gas load into the collision chamber and increasing, at the same time, the operation efficiency of the target-beam dump system.

The distances from the antiproton beam axis to entrances into the 1$^{st}$ and 2$^{nd}$ target-beam dump chambers are 50 mm and 100mm, correspondingly. The diaphragm of the 1$^{st}$ dump chamber has 23.3x11.7 mm aperture with a tube length at the target-beam direction of 15 mm, the 2$^{nd}$ one has 31.7x15.9 mm aperture with the tube length of 25 mm. The length of the 2$^{nd}$ dump chamber at the target-beam direction is 80 mm.

## SUPERSONIC JET MEASUREMENTS AND SIMULATIONS

The operation of the described supersonic gas-jet target apparatus has been investigated in supersonic jet measurements and gas dynamic simulations. The supersonic jet profiles were measured with an impact pressure probe (often called a Pitot tube), which design is presented in Fig.2.

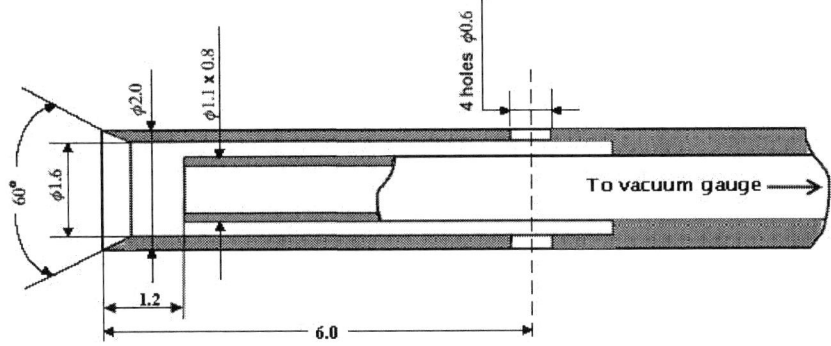

**FIGURE 2.** Schematic of the Pitot tube design.

The Pitot tubes of this design (Fig.2) have been used first in old experiments at LNPI [5] (Russia), then at GSI [6], LMU-Munich [7] (Germany) and NSCL-MSU [8] (USA). Such a Pitot tube consists of two coaxial combined tubes with 4 small holes drilled in the outer tube wall (see Fig.2). The Pitot tube of this design has an advantage over a simple Pitot tubes because it is insensitive to an angle between Pitot tube axis and an impact gas flow direction. It provides for measurements with accuracy better than 1% up to the angle of 20 degree even at high Mach numbers in the jet [9]. It is interesting to note that a simple tube of 1.02 mm outer diameter has been used in [10] to determine the gas flow angle to the axis in the supersonic nozzle and a low-density nitrogen supersonic jet. In our jet profiles measurements the skimmer was replaced by the Pitot tube (mounted on the holding flange) exactly at the position of the skimmer entrance orifice. In order to determine an optimum stagnation condition (temperature $T_0$ and pressure $P_0$) and a corresponding nozzle-skimmer distance, at which the greatest possible target density is achieved, measurements of the total gas flow rates through the collimator have been carried out also.

Potentialities of the present gas-jet target setup have been also investigated in computer simulations with the VARJET gas dynamic code based on a solution of full time-dependent system of Navier-Stokes equations. This code is described in details in [11], where results of computer experiments for generation of various internal molecular and atomic beam targets from gases and nonvolatile substances are presented as well.

Some results of our Pitot pressure measurements and corresponding calculations for the nozzle temperature $T_0 = 300$ K and stagnation pressure $P_0 = 2$ bar are shown in the Fig. 3. Large disagreement between measurement and calculation at 2 mm distance from the nozzle in Fig.3a may be explained by proximity of the Pitot tube of 2 mm diameter to the nozzle (e.g. the Pitot tube can change local pumping conditions due to its relatively big size). There is some disagreement also between measurements and calculations in description of a radial supersonic jet shape (Fig.3b). The reason has to do with a design defect of our present manipulator for the nozzle movement in x-y directions (perpendicular to the skimmer-collimator axis), because the bellows, which are used to enable the nozzle movement in vacuum, have a resistance force for the x-y movement. It is apparent from an asymmetry of the experimental jet profile in the Fig.3b, because it is obvious that the gas flow from a circular nozzle should be cylindrically symmetric.

In result of investigation of the present target setup operation we have found that the maximum helium target density is achieved when the stagnation pressure $P_0 = 1.4$ bar at the nozzle temperature $T_0 = 150$ K. Figure 4 shows results of measurements of the helium target density as a function of the nozzle-skimmer distance.

The target density $n$ (in atoms/cm$^3$) can be obtained from measurements of total gas flow $\Phi_{tot} = \Phi_3 + \Phi_4 + \Phi_5$ (in mbar·l/s) through the collimator absorbed by the pump of collision chamber $\Phi_3$ (when it is disconnected by flange from the pbar beam line) and $\Phi_4$ and $\Phi_5$ evacuated by pumps of two target-beam dump chambers. The gas flow $\Phi_i$ is equal to a product of the pressure increase $\Delta P_i$ in the vacuum chamber in response to the jet flow from the nozzle (measured by vacuum gauge) and known pumping speed $S_i$ (in l/s): $\Phi_i = \Delta P_i \cdot S_i$. Notice, that at room temperature 1 mbar·l/s = $2.4 \cdot 10^{19}$ atoms/s.

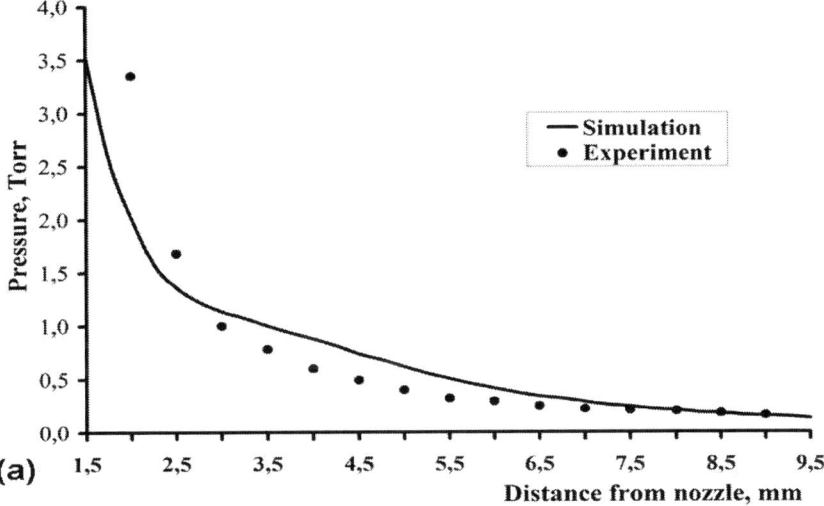

(a)

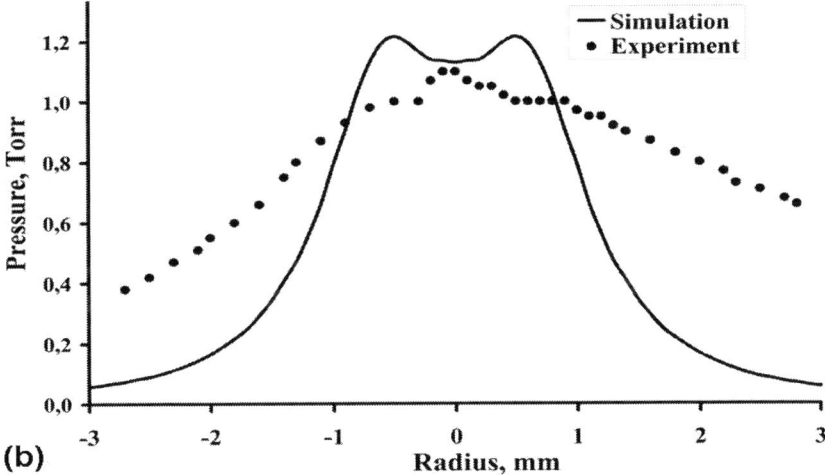

(b)

**FIGURE 3.** The Pitot pressure profiles of the helium supersonic jet for the nozzle temperature $T_0 = 300$ K and stagnation pressure $P_0 = 2$ bar. (a) The pressure along the axis. (b) The radial pressure distribution at 3 mm downstream the nozzle.

To determine the target density, one can use the following equation $n = \Phi_{tot} / (A_b V_b)$, where $A_b$ is area of a target-beam cross section (in cm$^2$) and $V_b$ is value of an atomic beam velocity (in cm/s). The target-beam area $A_b$ is defined only by the skimmer-collimator geometry and in our case it is about 0.5 cm$^2$. The beam velocity

$V_b$ is equal to the gas jet velocity $V_{jet}$ at the skimmer entrance. The $V_{jet}$ for big Mach numbers can be estimated in the isentropic approximation using the following equation:

$$V_{jet} = \sqrt{\frac{2\gamma k T_0}{m(\gamma - 1)}} \;,$$

where $\gamma$ is the ratio of specific heats, $k$ is Boltzmann's constant and $m$ is atomic mass.

The $V_{jet}$ can be obtained with a good accuracy for any Mach number from our VARJET simulations as well. E.g. the described method of the target density measurement has been used in works [12-14] as well.

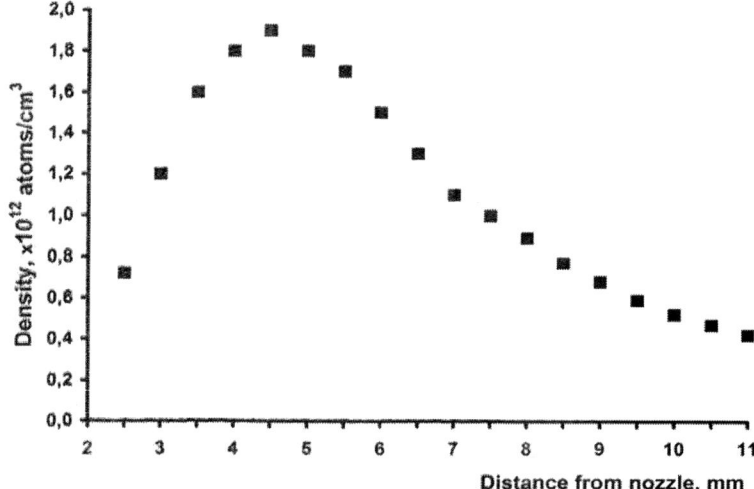

**FIGURE 4.** The helium target density as a function of the nozzle-skimmer distance for the nozzle temperature $T_0 = 150$ K and the stagnation pressure $P_0 = 1.4$ bar.

It can be seen that a target density curve has a maximum when the skimmer is placed at 4-5 mm distance from the nozzle. It is a matter of common knowledge that for an optimum skimmer position, a Knudsen number determined as a ratio of a free pass length to the skimmer orifice diameter should be about one. According to our simulation for $T_0 = 150$ K and $P_0 = 1.4$ bar, the helium density on the axis of supersonic jet at 4 mm downstream the nozzle is $5.7 \cdot 10^{15}$ atoms/cm$^3$. So, the Knudsen number, in this instance, for the skimmer diameter of 0.6 mm is about 1.1.

Some results of computer simulations for the present gas-jet target setup are listed in the Table 1.

**TABLE1.** Main characteristics of He supersonic gas-jet target for the present target setup

| | |
|---|---|
| Nozzle diameter | 0.1 mm |
| Skimmer diameter | 0.6 mm |
| Collimator | 2.2x4.4 mm |
| Collimator-skimmer distance | 35 mm |
| Nozzle-skimmer distance | 4 mm |
| Stagnation temperature | 150 K |
| Stagnation pressure | 1.4 bar |
| Gas flow rate through the nozzle | 6.6 mbar·l/s |
| Beam temperature | 13.1 K |
| Mach number | 5.9 |
| Beam velocity | 1190 m/s |
| Target cross section | 5 x 10 mm |
| **Target-beam density** | $1.8 \cdot 10^{12}$ atoms/cm$^3$ |

As evident from Tab. 1, the calculated helium target density is in a good agreement with the experimental one.

It should be pointed out that, in principle, the decreasing of stagnation temperature at constant value of the total gas flow rate through the nozzle makes possible higher Mach numbers in the jet and thus higher gas target densities. The reason is that gas viscosity effects decrease with temperature decreasing. So, our simulation for 6.6 mbar·l/s the helium flow rate through the nozzle of 0.1 mm throat diameter and stagnation temperature $T_0 = 30$ K revealed that the target density may be as much as $1 \cdot 10^{13}$ atoms/cm$^3$ even for the present gas-jet target apparatus. Unfortunately, due to some design defect of the present setup, a tilting of the cryogenic head (with the nozzle) about skimmer-collimator axis takes place on cooling, causing the target density to decrease.

## ASACUSA TARGET FUTURE DEVELOPMENT

For considerable increasing of the ASACUSA gas-jet target density, a qualitative modification of the present target setup is scheduled for future experiments with ultra slow antiprotons. The goal can be achieved by the use of pulsed high-pressure supersonic gas jet that operates in accordance with the pulsed mode of the MUSASHI penning trap [15]. For this purpose an additional stage of differential pumping with a skimmer will be set into the present target setup. To avoid the clusters in gas-jet targets of different gases, the nozzle will operate at room or higher temperatures.

Typical ultra slow antiproton beam pulse duration is about 10 s with a repetition rate of 0.01 Hz (one antiproton shot per 100 s). So, the using of pulsed mode of supersonic gas jet from a converging sonic nozzle equipped with a solenoid driven pulsed gas valve will allow using much higher stagnation pressures keeping the same level of time-average gas consumption. The operation of the pulsed nozzles for producing molecular and cluster beams have been described elsewhere [16-21].

The using of higher stagnation pressures provides a way of achieving larger Mach numbers in the jet and, as a result, obtaining higher target densities.

Schematic figure of a future gas-jet target setup is presented in Fig. 5. The distance between skimmers is 25 mm. The higher stagnation pressure $P_0$ in the nozzle leads to increase of background gas pressure in the nozzle exhaust chamber, so that this background pressure during the helium jet-pulse from the nozzle of 0.1 mm throat diameter at $P_0$ = 23 bar will be about 0.1 mbar. That's why we will need to use a Roots pump of about 800 l/s (see Fig.5) instead of presently used turbo molecular pump (Pfeiffer TMH261).

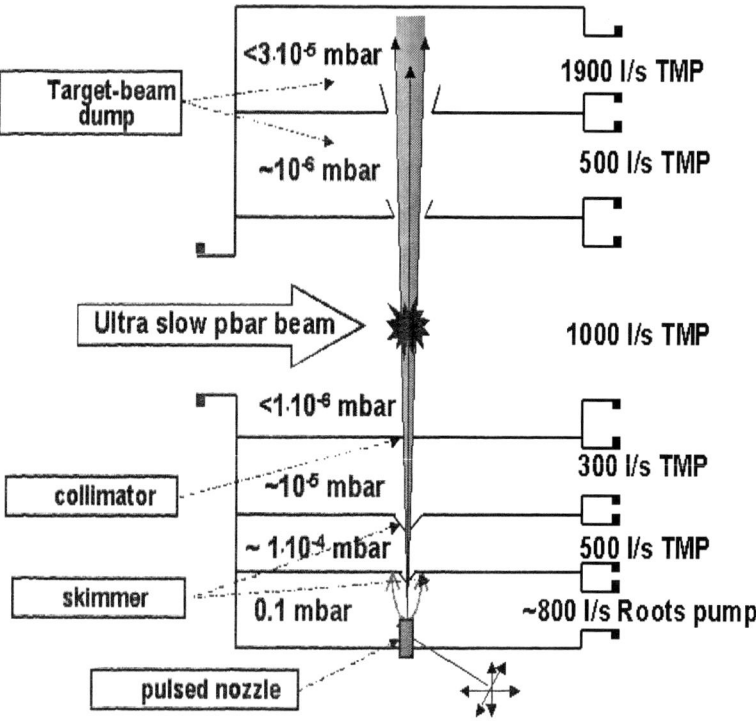

**FIGURE 5.** Schematic figure of a future gas-jet target setup. The sonic nozzle is equipped with a solenoid driven pulsed gas valve. Given pressure values show typical vacuum conditions in the system during the helium jet-pulse.

The operation of the described future version of the ASACUSA gas-jet target apparatus has been studied by means of detailed time-dependent simulations with VARJET code [11]. Among other things the simulations revealed that a steady flow of the He supersonic jet is attained within 1 ms after the pulsed gas valve opening. So, to ensure that the gas target density remain constant throughout an antiproton beam shot, all one only has to do is to open the gas valve 1 ms before the first antiproton from MUSHASHI penning trap will reach the target position in the collision chamber. By way of illustration, Fig. 6 shows calculated time profile of the axial He jet velocity at 11 mm downstream the nozzle.

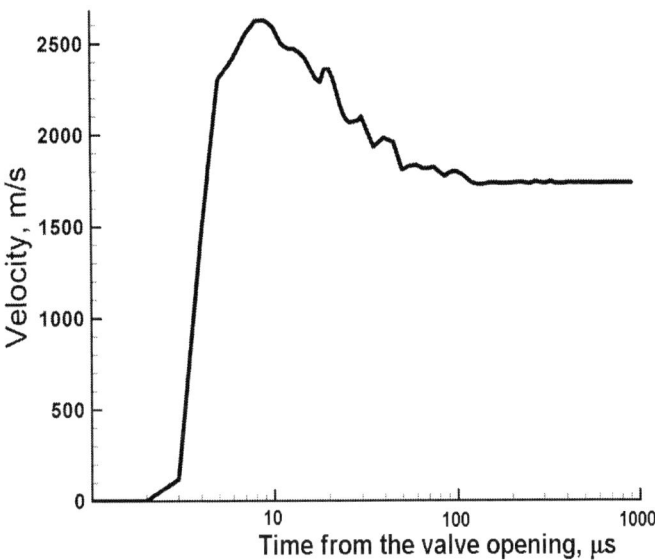

**FIGURE 6.** Calculated time profile of the axial He jet velocity at 11 mm downstream the nozzle for $P_0 = 22.6$ bar, $T_0 = 300$ K. The nozzle throat diameter is 0.1 mm.

Some results of these calculations for pulsed He gas-jet target are presented in Figs. 7 and 8. Figure 7 shows the static pressure, density, Mach number and static temperature distributions along the axis of He pulsed jet for $P_0 = 22.6$ bar, $T_0 = 300$ K. The converging sonic nozzle throat diameter is 0.1 mm. The helium density on the jet axis at 11mm downstream the nozzle is $5.1 \cdot 10^{15}$ atoms/cm$^3$ and this place very suits for the position of the skimmer with 0.6 mm orifice, because the Knudsen number in this case is about 1.2. It is interesting to note that at 11 mm distance from the nozzle the pressure on the jet axis (Fig. 7(a)) is 200 times less than background gas pressure. Whereas the gas density here (Fig. 7(b)) is twice as large as the background gas density outside the supersonic jet. One can see also from this figure that due to the supersonic gas expansion into vacuum the Mach number (Fig. 7(c)) in the jet is increased up to 36 and static temperature (Fig. 7(d)) drops below 1 K.

337

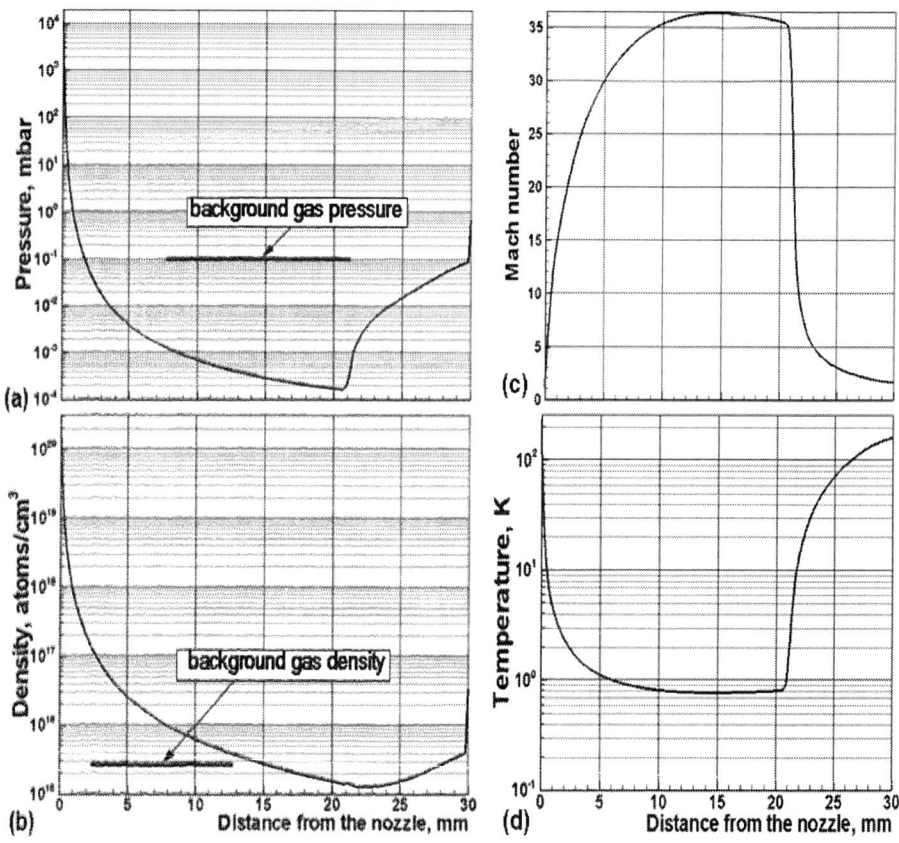

**FIGURE 7.** Calculated static pressure (a), density (b), Mach number (c) and static temperature (d) distributions along the axis for the same pulsed jet as that of Fig. 6.

Figure 8 shows radial profiles of the static pressure, density, static temperature and gas flow velocity at 11 mm downstream the nozzle. A strong barrel shock wave structure of the supersonic jet that is clear defined in the Figs.7 and 8 shields an axial jet region against background gas penetration into the gas-jet target and it is reasonable also to say that the nozzle-skimmer distance of about 11 mm looks as an optimum one.

Main design and operation parameters of the future gas-jet target apparatus for the case of the pulsed He supersonic jet are listed in the Table 2.

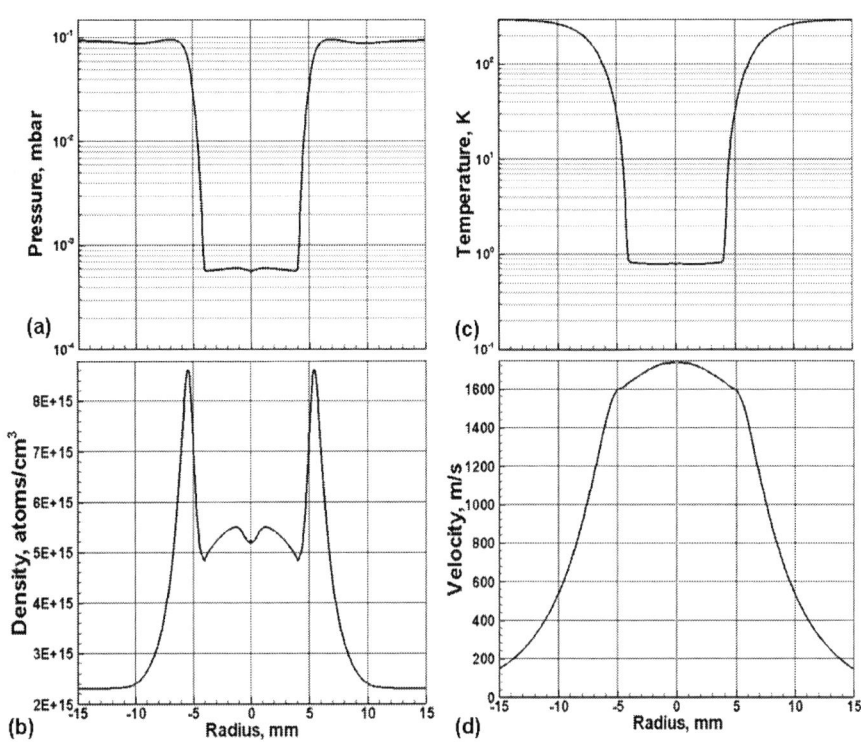

**FIGURE 8.** Calculated radial profiles of the static pressure (a), density (b), static temperature (c) and gas flow velocity (d) at 11 mm downstream the nozzle for the same pulsed jet as that of Figs. 6 and 7.

**TABLE 2.** Main design and operation parameters of the future gas-jet target apparatus for the case of pulsed He supersonic jet

| | |
|---|---|
| Nozzle diameter | 0.1 mm |
| 1st skimmer diameter | 0.6 mm |
| 2nd skimmer diameter | 2.2 mm |
| Nozzle-skimmer distance | 11 mm |
| Skimmer-skimmer distance | 25 mm |
| Skimmer-collimator distance | 35 mm |
| Stagnation temperature | 300 K |
| Stagnation pressure | 22.6 bar |
| Gas flow rate though the nozzle during pulse | 78.5 mbar·l/s |
| Beam temperature | 0.78 K |
| Mach number | 35.8 |
| Beam velocity | 1733 m/s |
| Target cross section | 4 x 8 mm |
| **Target-beam density** | $3.0 \cdot 10^{13}$ atoms/cm$^3$ |

339

As seen from Table 2, the helium gas target of $3 \cdot 10^{13}$ atoms/cm$^3$ density will be accessible for atomic collision experiments with ultra slow antiproton beams. There can be no doubt that targets of other gases with similar densities will be possible here as well.

## ACKNOWLEDGMENTS

We acknowledge Dr. Z. Wang and Mr. S. Yoneda who started designing the initial version of our gas-jet target and the collision chamber, for their work at early stage of development.

## REFERENCES

1. H.A. Torii, "Production of ultra-slow antiproton beams", Contribution to Workshop on Physics with Ultra Slow Antiproton Beams, RIKEN, March 2005 (in this book).
2. J. S. Cohen, Rep. Prog. Phys., **67**, 1769 (2004).
3. Z. Wang, "Development of an experimental apparatus for collisions between ultra-slow antiprotons and atoms/molecules" (internal report within the collaboration), January 2002.
4. K. Yoshiki Franzen, et al., "Transport beam line for ultraslow monoenergetic antiprotons", Rev. Sci. Instrum., **74**, 3305 (2003).
5. V. L. Varentsov, *Ph.D. Dissertation*, Leningrad Nuclear Physics Institute, 1986.
6. H. Reich, W. Bourgeois, B. Franzke, A. Kritzer, V. L. Varentsov, Nucl. Phys. **A 626**, 417 (1997).
7. O. Engels, et al., Hyperfine Interactions, **132**, 505(2001).
8. P. A. Lofy, *Ph.D. Dissertation*, Michigan State University, 2003.
9. A. N. Petunin, Uchenie zapiski TsAGI, vol. 7, 112 (1976).
10. I. D. Boyd, P. F. Penko, D. L. Meissner, K. J. DeWitt, AIAA Jornal, **30**, 2453 (1992).
11. V. L. Varentsov, A. A. Inatiev, Nucl. Instr. And Meth., **A 413**, 447 (1998)
12. R. Burgei, et al., Nucl. Instr. And Meth., **204**, 53 (1982).
13. D. Anghinolfi, et al., Nucl. Instr. And Meth., **A 396**,23 (1997).
14. D. Allspach, et al., Nucl. Instr. And Meth., **A 410**,195 (1998).
15. N. Kuroda, et al., Phys. Rev. Lett., **94**, 023401 (2005).
16 O. F. Hagena, W. Obert,, J. Chem. Phys., **56**, 1793 (1972).
17 L. Abad, D. Bermejo, V.J. Herrero, J. Santos, Rev. Sci. Instrum, **66**, 3826 (1995).
18. Smith, R. A., T. Ditmire, J. W.G. Tisch, Rev. Sci. Instrum, **69**, 3799 (1998).
19. J. Libuda, I. Meusel, J. Hartmann, H.-J. Freund, Rev. Sci. Instrum, **71**, 4395 (2000).
20. S. Semushin, V. Malka, Rev. Sci. Instrum, **72**, 2961 (2001).
21. V. N. Popok, S. V. Prasalovich, M. Samuelsson, E. E. B. Campbell, Rev. Sci. Instrum, **71**, 4395 (2000).

# A compact setup of fast pnCCDs for exotic atom measurements

H.Gorke*, W.Erven*, D.Gotta†, R.Hartmann**, L.Strüder‡ and L.Simons§

*Zentralinstitut für Elektronik des Forschungszentrum Jülich (ZEL), Germany
†Institut für Kernphysik (IKP), Forschungzentrum Jülich, Germany
**PNSensor GmbH, München Germany
‡MPI für extraterrestische Physik, Garching Germany
§Paul-Scherrer-Institut, Villigen Switzerland

**Abstract.** X-ray measurements at particle accelerators suffer from a high beam-induced background. For low-energy X-rays CCDs can solve this problem due to their pixel structure, which allows to reduce the background by a cluster analysis. To be able to measure at reasonably high count rates, fast read-out capable CCDs, so called pnCCDs, are used, which have been developed at the semiconductor laboratory of the Max Planck Institute (MPI) for the XMM satellite mission. Recently an improved version of pnCCDs was developed. These chips will be available with different geometries, pixel size and allow a frame-store read-out. At the ZEL (Central Laboratory for Electronics) of the reseach center Jülich a dedicated electronics for fast read-out was developed in collaboration with the MPI.

**Keywords:** PN-CCD, X-ray detectors, Data acquisition, exotic atoms
**PACS:** 07.85.Fv; 07.05.Hd; 36.10.-k

## INTRODUCTION

The aim was to obtain a compact and flexible detector setup with an easily manageable user interface providing computer controlling of all relevant detector parameters [1]. The working principle and performance of pnCCDs setup will be demonstrated from first measurements at the high intensity pion beam at the Paul-Scherrer-Institut (PSI).

## PNCCD AND READ-OUT CHIP CAMEX

The pnCCDs are a development of the semiconductor laboratory of the Max Planck Institut für extraterrestische Physik. These devices are fully depleted with a sensitive thickness of $450\mu m$ and are backside illuminated which allow an ultra-thin entrance window [2]. The full depletion yields their intrinsic quantum efficiency of more than 90% in the energy range between 300 eV and 10 KeV. The pixel matrix is formed with p- and n-structures on the front side building up lines (p) and channels (n). Every channel ends at its own readout anode with an on-chip amplifier. In this way the detector capacitance is reduced to $\leq 30fF$ which leads to an outstanding energy resolution and allows a parallel read-out of all pixel in one line minimising the read-out time.

The on-chip amplifiers are directly connected to the read-out CAMEX. The CAMEX chip is build up with 132 or 128 parallel working correlated double sampling amplifier

CP793 *Physics with Ultra Slow Antiproton Beams*
edited by Yasunori Yamazaki and Michiharu Wada
© 2005 American Institute of Physics 0-7354-0282-5/05/$22.50

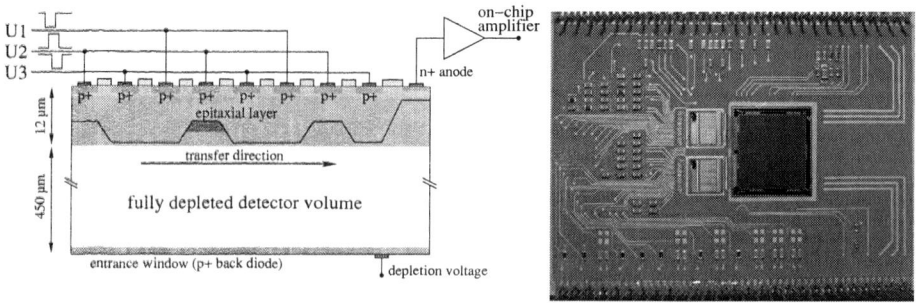

**FIGURE 1.** Left - cut through a n-channel of a pnCCD. Right - a $50\mu$m pixel size pnCCD with two CAMEX read-out chips mounted on a ceramic support.

channels. Read-out is done serially during analog signal processing with up to 10 MHz over one or two read-out nodes. The chip offers programmable gains and freely programmable correlated double samplings. Electronic noise less than 3 electrons (rms) is achieved. The parameters of analog signal processing and read-out are controlled and performed by digital control-pulses with have to been synchronised with the shift-pulse of the CCD device. CCD and CAMEX are mounted on a ceramics support containing filters for power connections and cooling contact.

## PNCCD DETECTOR SYSTEM

The condition for a compact system setup is a small and flexible setup, which allows a fast and easy installation at different locations. The setup consists of a detector cryo-

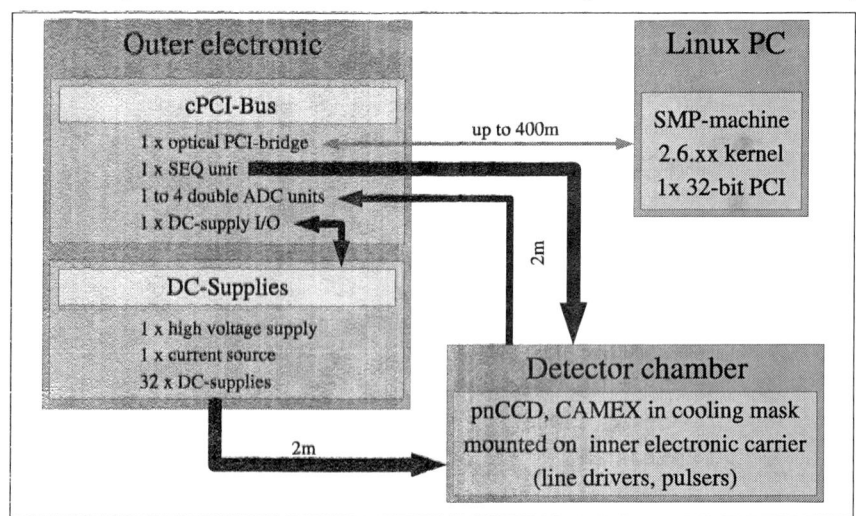

**FIGURE 2.** Setup of the pnCCD detector system

342

stat (chamber), an outer electronic which has to be located close to the chamber (less than 2m) and a DAQ system for measurements at accelerators in larger distance from the chamber. The chamber has to combine the specific geometric conditions of the experiment and the requirements of the detector like cooling and close-to-chip electronics. The outer electronics supports the electronic modules located at the detector chamber. To use the same electronics for different detectors and setups, it is made configurable and controllable form a farther away PC.

The setup which was developed at ZEL in collaboration with the MPI is realised by an electronic crate in which the 32-bit, 33MHz PCI-bus of a LINUX workstation is extended via a fiber optical bridge in the outer electronic. This allows to place the support electronics near the detector chamber having remotely full software control.

## Outer Electronics

The outer electronics is realised as one crate containing all DC-powersupplies and as PCI devices a I/O-unit, the sequencer and the ADCs. The DC-powersupplies are fully controlled and parametrised through the cPCI I/O-unit by software. The sequencer device is freely programmable with a time resolution of 10 ns and 64 LVDS-output lines. The program code is generated by a special software on the LINUX workstation and can be uploaded at any time to the processor realised in the FPGA (Field-Programmable Gate Array). The sequencer unit generates all digital control pulses for the whole system

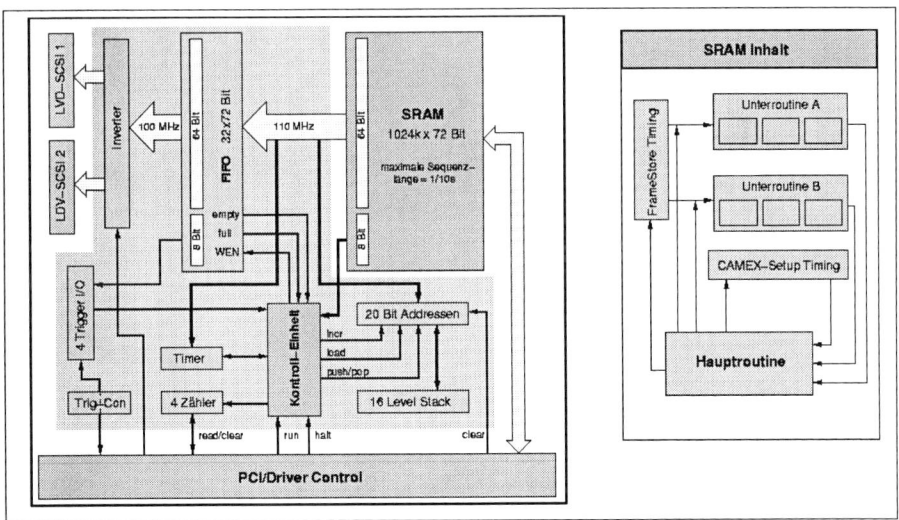

**FIGURE 3.** Block diagram of the state machine in the FPGA of the cPCI sequencer device with an example scheme of a program loaded into the SRAM.

such as the shift pulses for the CCD, the timing for the CAMEX chip and the read-out clock for the ADCs. A running program can be switched to different modi (such as start/stop ADC clock, start/stop CCD running as frame-store device) by software triggers (device driver program) or by hardware triggers. As an example the ADC device starts its read-out clock by a hardware trigger, this avoids race conditions between start of DAQ and read-out timing.

One ADC device holds two 80 MHz 14-bit flash ADCs controlled and read-out by the PCI-bus. As show in Figure 4 the CCD read-out requires up to eight ADCs. The four ADC devices generate up to 140 MByte data per second which overloads the connected PCI-bus. Even for less ADC devices such a dataflow is not suitable to be stored for longer measuring periods. For that data reduction by using a threshold is needed. To

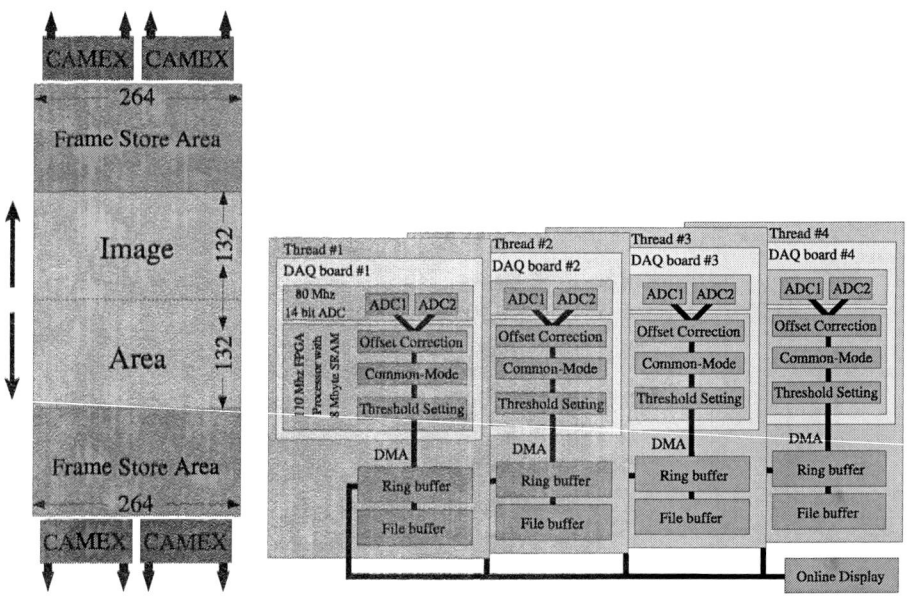

**FIGURE 4.** Configuration of the DAQ for double-sided read out pnCCD throw eight read out nodes. The configuration allows to operate a 264 × 264 CCD at 1000 frames/s generating a constant data flow of 140 MBytes/s.

use thresholds in the CCD read-out, the data have to be corrected for offset and common mode. As long as the dataflow does not extend the PCI-bus speed, data reduction could be done on-line by software. But for a setup covering all possibilities a hardware data reduction is needed. The offset and common mode correction and the data reduction for two ADCs is done with a FPGA processor on the ADC device. As seen later a further reduction with cluster analysis is sometimes desirable. Such a feature is under development.

The DAQ for the system has to take into account, that it has to work with up to four parallel ADC devices at the same time. Because of this the DAQ software has to be threaded. For the shown configuration at least five threads are needed. For every ADC device exists one thread, which handles the IRQs from the device, stores the data from the ring buffer in files and passes the data to an on-line display the fifth thread.

## MEASUREMENT AT PSI

At the high intensity pion beam of the Paul-Scherrer-Institut a pnCCD system was setup for measurements with pionic atoms in order oto optimise injection into the cyclotron trap [3]. For this measurement, a $1\,cm^2$ non frame-storage device was used, which is based on the XMM CCDs having pixels of $150 \times 150m^2$ and a thickness of $300\mu m$.

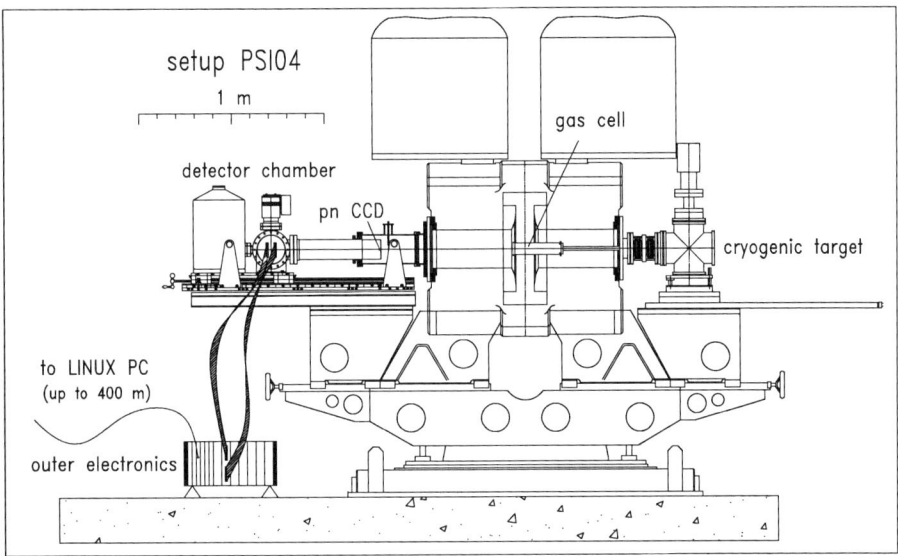

**FIGURE 5.**  Setup for the measurement of pionic-atom X-rays at PSI.

The detector chamber was placed at one bore hole from the cyclotron trap. As target a gas filled cell was used. The gas could be cooled down to incease the density of the target because only thin windows can be used for the low-energy Xrays. The pnCCD was placed 700 mm from the center of the gas cell and was operated at full beam intensity. The CCD was read-out continuously every 2 ms. In this mode the CCD has only a relative position resolution. Which was sufficient for this measurement because this still allows a cluster analysis. The cluster analysis is needed because for the beam-induced background. The improvement of the spectra by doing a cluster analysis is shown in Figure.6. Data reduction and analysis of the 500 frames/s in this setup was done all by software.

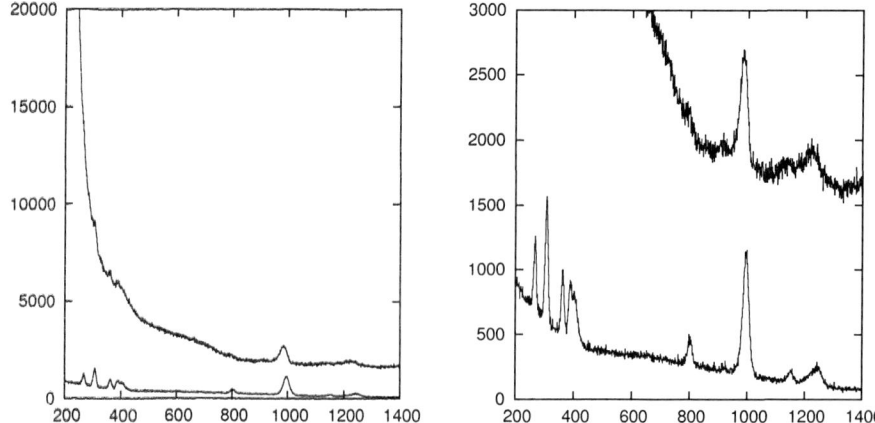

**FIGURE 6.** On-line spectrum of $\pi^4$He at 10 bar measured at PSI. The spectrum is holding the data of $5.3 \times 10^6$ frames, taken in $\sim$3 hour. The upper one represents offset-, common-mode corrected data above threshold. The lower spectrum shows the result after cluster analysis. The right side shows the same spectrum in a different scale. The lines start with from left Al (1.4 keV), $\pi^4$He(n-2) from 2 to 3 keV, at 800 ADU the $\mu^4$He(2-1) (8 keV) is to seen. The lines from 1000 ADU belong to $\pi^4$He(n-1)transitions starting at 10.6 keV

# REFERENCES

1. H. Gorke, thesis, Universität zu Köln.
2. N. Meidinger, R. Hartmann, L. Strüder et al., *Nucl. Instr. and Meth.* **A 512** (2003) 341.
3. http://pihydrogen.web.psi.ch.

# Intense Source of Slow Positrons

P. Perez and A. Rosowsky*

*DSM/Dapnia/SPP, CEA/Saclay, F-91191 GIF-SUR-YVETTE

**Abstract.** We describe a novel design for an intense source of slow positrons based on pair production with a beam of electrons from a 10 MeV accelerator hitting a thin target at a low incidence angle. The positrons are collected with a set of coils adapted to the large production angle. The collection system is designed to inject the positrons in a Greaves-Surko trap. Such a source could be the basis for a series of experiments in fundamental and applied research and would also be a prototype source for industrial applications which concern the field of defect characterization in the nanometer scale.

**Keywords:** positron positronium accelerator/linac defect characterization matter antimatter symmetry antigravity
**PACS:** 41.75.Fr

An apparatus is designed to produce and collect MeV positrons which could subsequently be stored and/or cooled in a Greaves-Surko trap [1] at a rate of order $10^9$ per second. In contrast, radioisotope sources currently in use produce fluxes $\leq 2 \; 10^7 s^{-1}$. Its size is also much smaller than that of a large linac or of a nuclear reactor but could be used as a small facility for interdisciplinary experimental studies with positrons.

Such a source would allow the production of positronium (Ps), hydrogen and anti-hydrogen in a symmetric way in the same experimental conditions through the set of reactions: $e^+ + e^- + e^\pm \rightarrow Ps + e^\pm$, followed by $Ps + p \rightarrow H + e^+$ or its antimatter counterpart $Ps + \bar{p} \rightarrow \bar{H} + e^-$ and the production of ions via $H + Ps \rightarrow H^- + e^+$ or $\bar{H} + Ps \rightarrow \bar{H}^+ + e^-$ [3].

This setup uses $e^+ e^-$ pair creation from the interaction of a flat electron beam with a 50 microns tungsten foil. The angle between the beam plane and the foil is very small, approximatively 3 degrees. The beam comes from a 10 MeV electron accelerator with an average intensity of 2.3 mA. The positron kinetic energy spectrum is peaked at 1.2 MeV and extends to 8 MeV.

The pair production cross section increases with energy. However the development of such a setup for university or industrial applications limits the beam energy to 10 MeV because a higher energy would start to activate the environment (legal limit).

The first step in the positron capture by the trap is the moderation process. This process involves the slowing down of the positrons, the creation of meta-stable states with collective charge oscillations in the moderator and its re-emission at $\approx 1.5$ eV. The moderation efficiency decreases with the incident positron kinetic energy and is negligible at a few MeV. Therefore a magnetic collector was designed to separate the positrons from the electrons while preserving the positrons with a kinetic energy below 1 MeV and is described in detail in [2].

The setup allows to collect more than $5 \; 10^{11} s^{-1}$ positrons of less than 600 keV in a 2 cm radius aperture at 2 m from the target. The flux of electrons and photons reaching

CP793 *Physics with Ultra Slow Antiproton Beams*
edited by Yasunori Yamazaki and Michiharu Wada
© 2005 American Institute of Physics 0-7354-0282-5/05/$22.50

this aperture is of the order of 50 W.

# REFERENCES

1. T. J. Murphy and C. M. Surko, *Phys. Rev.*, **A 46** 5696 (1992); C. M. Surko, S. J. Gilbert and R. G. Greaves, *Non-Neutral Plasma Phys. III*, edited by J. J. Bollinger, R. L. Spencer and R. C. Davidson, American Institute of Physics, New York, 1999, pp. 3–12; http://physics.ucsd.edu/research/surkogroup/positron/buffergas.html
2. P. Perez and A. Rosowsky, *Nucl. Intr. Meth.*, **A 532** 523 (2004).
3. P. Perez and A. Rosowsky, these proceedings; P. Perez and A. Rosowsky, *Nucl. Intr. Meth.*, **A** (2005) in press.

# NON-NEUTRAL PLASMA

# On the Possibility of Non-Neutral Antiproton Plasmas and Antiproton-Positron Plasmas

H. Higaki

*Plasma Research Center, University of Tsukuba*
*1-1-1, Tennoudai, Tsukuba, Ibaraki, Japan 305-8577*

**Abstract.** Progresses in accumulating a large number of low energy antiprotons with Antiproton Decelerator (AD), Radio Frequency Quadrupole Decelerator (RFQD), and a multiring trap in Atomic Spectroscopy And Collisions Using Slow Antiprotons (ASACUSA) enables the confinement of more than $10^6$ antiprotons. Confinement of a larger number of antiprotons in the trap will result in a non-neutral antiproton plasma. This is also favorable for the effective production of low energy antiproton beams. Possibility of an antiproton-positron plasma is also considered in a magnetic mirror field.

**Keywords:** antiproton plasma, antiproton-positron plasma, magnetic mirror field
**PACS : 52.27.Jt, 52.35.Fp, 52.55.Jd**

## INTRODUCTION

A large number of low energy antiprotons ($<10^5$) have been confined in particle traps at Antiproton Decelerator (AD) in CERN [1,2]. However, plasma oscillations of a non-neutral antiproton plasma have not been observed. It is thought that the size of the antiproton cloud was smaller than its Debye length. Progresses in accumulating a larger number of low energy antiprotons with AD, Radio Frequency Quadrupole Decelerator (RFQD), and a multiring trap in Atomic Spectroscopy And Collisions Using Slow Antiprotons (ASACUSA) enabled the confinement of a larger number of antiprotons ($>10^6$) [3]. Further improvement in stacking antiproton pulses will lead to the observation of the plasma oscillations of a non-neutral antiproton plasma. The large amount of antiprotons is also favorable for the effective production of low energy antiproton beams to be used in atomic collision experiments.

The possibility of producing antiproton-positron plasmas is also considered with a magnetic mirror configuration. Although antihydrogen atoms were created by mixing a positron plasma and antiproton cloud [2,4], it was a beam-plasma system. The basic procedure here is to accumulate both antiprotons and positrons in a nested Penning trap in a magnetic mirror field. It is assumed that more positrons are trapped than antiprotons. Adiabatic expansion of particles along the magnetic field by reducing the trapping potential results in an anisotropic energy distribution. Then, the positron plasma can be trapped with the magnetic mirror field and antiprotons can be trapped simultaneously with the magnetic field and the space potential of the positron plasma. A small power positron cyclotron resonance heating may enhance the confinement time of the antiproton-positron plasma.

CP793 *Physics with Ultra Slow Antiproton Beams*
edited by Yasunori Yamazaki and Michiharu Wada
© 2005 American Institute of Physics 0-7354-0282-5/05/$22.50

# ANTIPROTON PLASMAS

When a large number of charged particles are confined in a Penning trap, they behave as a non-neutral plasma. For examples, $4.3 \times 10^4$ electrons confined in a Penning trap in a cryogenic environment were characterized with its electrostatic electron plasma oscillations [5]. Unfortunately, such a small number of electrons are not enough for the effective cooling of high energy antiprotons provided through RFQD. A larger number of electrons ($>10^8$) with aspect ratio much larger than unity behaves as a prolate spheroidal plasma even if its temperature is larger than 1eV [6]. This spheroidal non-neutral electron plasma has been used to cool high energy protons [6] and antiprotons [3]. Since the frequencies of the electrostatic oscillations depend on the electron plasma temperature, non-destructive measurement of the frequencies works as a real time monitor of the electron cooling process.

There are also some examples of non-neutral ion plasmas. When a laser cooling technique is available, as in $Be^+$ ions [7], strongly coupled plasmas can be realized with the ion numbers less than $10^5$. Otherwise, a larger number of ions are required in general to form ion plasmas of Li [8] and Mg [9], since the plasma temperature is not low enough.

As far as antiparticles are concerned, positron plasmas have been already realized with more than $10^7$ positrons [10]. They were used for antihydrogen synthesis [2], low energy positron collision experiments [11], and so on. Positron plasmas are to be used for the cooling of highly charged ions [12]. However, there has been no report on the observation of plasma oscillations for antiproton plasmas. This might be because the temperature of antiprotons were not low enough that the estimated Debye length was larger than the cloud size, or the sensitivity of the detectors were not enough to detect the electrostatic oscillations of antiprotons.

Some parameters for above examples are listed in TABLE 1. Here, N, n, T, r, z, and $\lambda_D$, denote the particle number, density, temperature, radius, half axial length, and estimated Debye length, respectively.

TABLE 1. Various non-neutral plasmas and clouds.

| Species | N | n [cm$^{-3}$] | T | r [mm] | z [mm] | $\lambda_D$ [mm] |
|---|---|---|---|---|---|---|
| Electron [5] | $4.3 \times 10^4$ | $4.5 \times 10^7$ | ~ 4 K | 3.83 | 0.015 | 0.020 |
| Electron [6] | $1 \times 10^8$ | $3.5 \times 10^7$ | < 1eV | 4.6 | 30 | 1.3 |
| Positron [10] | $2 \times 10^7$ | $1 \times 10^6$ | 0.025eV | 15 | 20 | 1 |
| Be$^+$ [7] | $6 \times 10^4$ | $2.7 \times 10^9$ | 20mK | 0.12 | 0.38 | 0.0002 |
| Li+ [8] | $6.7 \times 10^9$ | $7.3 \times 10^6$ | 1.2eV | 16 | 580 | 3 |
| Mg+ [9] | $10^9$ | $3 \times 10^7$ | 0.05eV | 10 | 50 | 0.3 |
| Proton [13] | $1 \times 10^6$ | $1.2 \times 10^7$ | ~ 5eV | 1 | 20 | 5 |
| Antiproton [3] | $1 \times 10^7$ | $1.4 \times 10^8$ | 1eV ? | 1 | 16 | 0.6 ? |

As seen from above examples, one possibility to obtain antiproton plasmas is accumulating a large amount of low energy antiprotons in the trap. For the purpose of estimating the behavior of an antiproton cloud in the multiring trap, properties of ~ $10^6$ protons were investigated in advance [13]. The trap is composed of 12 ring electrodes aligned along the uniform magnetic field [3,6]. A unique feature of the trap is the ability to confine a large number of electrons ($> 10^8$) in a large volume with a harmonic potential, so that a large number of high energy antiprotons ($> 10^6$) can be cooled effectively. Since the production of low energy (< 1keV) antiproton beams in field free region is one of the main purposes of the multiring trap, the accumulated low energy antiprotons have to be extracted from the strong magnetic field through the transport beam line [14]. The crucial condition for the effective transport of low energy antiprotons is that the antiproton cloud (or plasma) should have small radial extent inside the trap. Thus, the radius of the antiproton cloud has to be controlled by the sideband cooling [5] or rotating electric field [9], depending on if it is a cloud or plasma. In case of $10^6$ protons in the multiring trap, the temperature was unfortunately too high to be a plasma. Therefore, the sideband cooling was effective for the radial compression of proton clouds [13]. Only the center of mass motion was observed before and after the radial compression of the proton cloud, as shown in FIG.1.

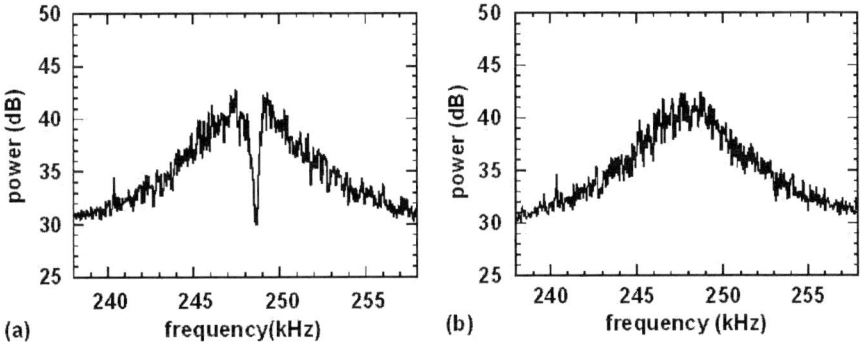

**FIGURE 1.** Tank circuit signals (a) with and (b) without $10^6$ protons in the multiring trap before radial compression. After radial compression, the dip appeared in (a) shifts lower by 4 kHz.

What happened with antiprotons provided through RFQD was that $10^6$ low energy antiprotons were confined routinely in the multiring trap. In this case, electrons can cool antiprotons less than 1 eV before electrons are discarded for the radial compression of antiprotons. Furthermore, some AD pulses are stacked to obtain a larger amount of antiprotons. Any improvements in the cooling efficiency in AD, RFQD, or multiring trap, and further progresses in stacking antiproton pulses from RFQD into the multiring trap will increase the number and volume of low energy antiprotons in the trap. As far as the radial compression takes much longer time compared with the AD cycle, stacking as many antiproton pulses as possible will reduce the waste of AD pulses. Currently, $10^7$ slow antiprotons in the trap are foreseen. Depending on its temperature, the antiproton cloud can behave as a non-neutral plasma. For example, the temperature of 1eV after radial compression with the density of $1.4 \times 10^8$ cm$^{-3}$ ($r_p \sim$ 1mm, $z_p \sim$ 16mm), results in the Debye length of about 0.6mm. This antiproton cloud should behave as a non-neutral plasma at least in the axial direction. Small amount of remnant electrons may help cooling antiprotons. Then, the electrostatic oscillation of the antiproton plasma should be observed near 400kHz with the normal experimental parameters of the multiring trap. This frequency may be shifted due to the remaining electrons. Detecting continuously the second harmonic axial oscillation of antiproton plasma will be a powerful tool for the nondestructive measurement inside the multiring trap.

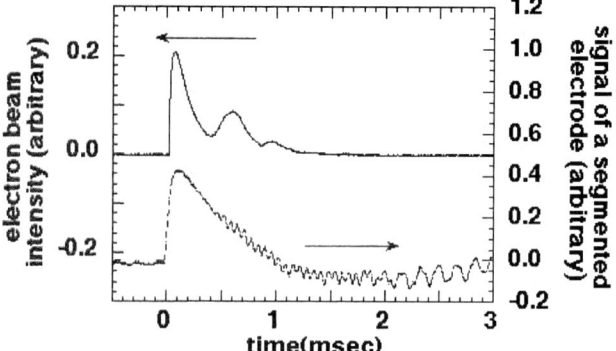

**FIGURE 2.** The electron beam intensity (solid line) fluctuates with diocotron oscillations (dashed line) when electrons are extracted with the time scale of a few msec.

Once antiproton plasmas are obtained, another problem may arise during their extraction. In the course of test experiments in the multiring trap with electron plasmas, depending on the duration of extraction, diocotron oscillations were observed and the extracted beam intensity fluctuated a lot. Shown in FIG.2 are the extracted electron beam intensity (solid line) and the signal obtained by an azimuthally segmented electrode (dashed line). It is seen that the later includes the fluctuations around10 kHz corresponding to diocotron oscillations. This implies that proper manipulations of antiproton plasmas will be also necessary for the effective production of low energy antiproton beams to be used in atomic collision experiments.

## ANTIPROTON-POSITRON PLASMAS

As mentioned in the introduction, the simultaneous confinement of both antiprotons and positrons is considered with a magnetic mirror configuration. Since the plasma confinement in a magnetic mirror has been investigated more than decades for thermonuclear fusion, characteristics of the magnetic mirror confinement is presented before the confinement of a dilute antiproton-positron plasma is considered.

### Brief Review of Magnetic Mirror Confinement

Principles of a single particle confinement in a magnetic mirror with the mirror ratio $R = B_1/B_0$ can be found in any textbooks on plasma physics [15]. Here, $B_0$ and $B_1$ denote the field strength at the center of the magnetic mirror (weakest) and at the mirror throats (strongest), respectively. The magnetic moment $\mu$ of a charged particle (charge:e, mass:m) in a magnetic field B is given by

$$\mu = IS = \frac{e\omega_c}{2\pi}\pi\rho^2 = \frac{mv_\perp^2}{2B} \tag{1}$$

354

with the cyclotron frequency $\omega_c$, Larmor radius $\rho$, and the perpendicular velocity of the particle $v_\perp$. This magnetic moment is the adiabatic invariant under the condition that the magnetic field changes slowly, which, in this case, is expressed by

$$\omega_c \ll \frac{L}{v_{//}}. \tag{2}$$

Here, L denotes the scale length of the magnetic field gradient. The conservation of energy and magnetic moment in the magnetic mirror field gives the following equations.

$$\varepsilon = m\frac{v_{//0}^2 + v_{\perp 0}^2}{2} = m\frac{v_0^2}{2} = m\frac{v_{//1}^2 + v_{\perp 1}^2}{2} \tag{3}$$

$$\mu = \frac{mv_{\perp 0}^2}{2B_0} = \frac{mv_{\perp 1}^2}{2B_1}$$

The axial confinement is provided if $v_{//1}^2 = 0$. Thus, the loss cone angle $\theta_0$ is determined with the mirror ratio by the following equation.

$$\frac{1}{R} = \frac{B_0}{B_1} = \frac{v_{\perp 0}^2}{v_{\perp 1}^2} = \frac{v_{\perp 0}^2}{v_0^2} \equiv sin^2\theta_0 \tag{4}$$

It means that a particle is confined if the pitch angle $\theta$ at the center is larger than $\theta_0$ and is lost if $\theta < \theta_0$. The gray region in FIG.3 is called loss cone. As an example, R=5 gives $\theta_0 \approx 27°$.

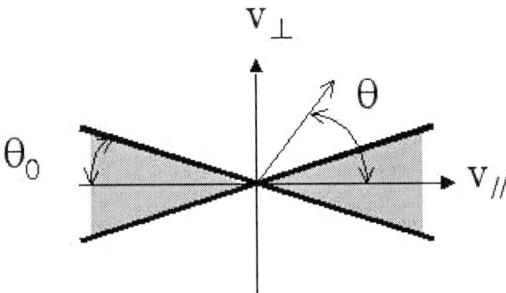

**FIGURE 3.** A single particle in a magnetic mirror. A particle with a pitch angle $\theta < \theta_0$ is not confined with the mirror ratio R = $1/sin^2\theta_0$.

When a plasma composed of protons and electrons is confined in a simple magnetic mirror, the loss cone is modified due to its self field potential $\phi(z)$. The conservation of energy and magnetic moment are given by the following equations.

$$\varepsilon = m\frac{v_{//}^2 + v_\perp^2}{2} + e\,\phi(z) = const. \tag{5}$$

$$\mu = \frac{mv_\perp^2}{2B(z)} = const.$$

The condition for the axial confinement inside the magnetic mirror field is given by the following equation.

$$\frac{mv_{\perp1}^2}{2} = \varepsilon - \mu B_1 - e\phi_1$$

$$= m\frac{v_0^2}{2} - \frac{B_1}{B_0}\frac{mv_{\perp0}^2}{2} + e(\phi_0 - \phi_1) \tag{6}$$

$$= \frac{mv_{//0}^2}{2} - (R-1)\frac{mv_{\perp0}^2}{2} + e\phi_c < 0$$

Therefore, the loss cone boundary is modified with the presence of the confinement potential $\phi_c = \phi_0 - \phi_1$ as follows.

$$sin^2\theta = \frac{1}{R}(1 + \frac{2e\,\phi_c}{mv_0^2}) \tag{7}$$

The schematics of the boundaries for protons and electrons are shown in FIG.4 for $\phi_c > 0$. In a simple magnetic mirror field, the ion confinement time $\tau_i \sim 1/v_{ii}$ and the electron confinement time $\tau_e \sim 1/v_{ee}$ are denoted by the ion-ion collision frequency $v_{ii}$ and electron-electron collision frequency $v_{ee}$. Since $v_{ii}$ is much less than $v_{ee}$ in general, the electrons are scattered into the loss cone more rapidly than ions. Thus, the plasma becomes ion rich and $\phi_c$ becomes positive. This positive $\phi_c$ tends to confine electrons and expel ions as shown in FIG.4. As a result, the loss rates of ions and electrons balance at a certain point. This feature is called ambipolar diffusion. A mirror confined plasma observed in a large scale thermal fusion device [16] has a positive confinement potential $\phi_c \sim 100$ V with the plasma parameters listed in TABLE 2.

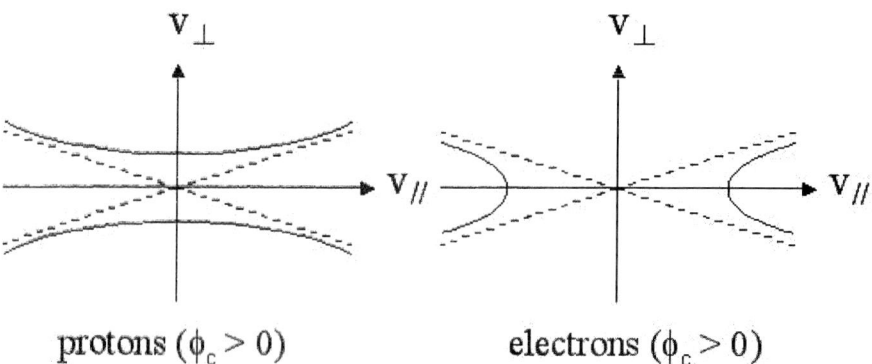

$V_\perp$         $V_\perp$

$V_{//}$         $V_{//}$

protons ($\phi_c > 0$)       electrons ($\phi_c > 0$)

**FIGURE 4.** The schematics of loss cone distributions modified with the self filed potential $\phi_c > 0$, which are usually observed in a mirror confined plasma for thermonuclear fusion.

The confinement time in this case is expressed by [17]

$$\tau_M = 0.78 \, \tau_i \, \ln\left[\frac{R}{1 + e\,\phi_c/T_i}\right] \quad with \quad \tau_i = \left(\frac{\pi}{2}\right)^{3/2}\frac{m^{1/2}k_B T_i^{3/2}}{ne^4 \ln\Lambda} \tag{8}$$

and is in the order of 10 msec. It should be noted in this example that the plasma is almost neutral. Excess protons are less than $10^{-4}$ of the plasma density. If the densities of both electrons and protons are reduced by four orders of magnitude, the confinement time seems to be much longer ($\sim$100 sec). However, the self field potential $\phi_c$ is expected to be much smaller (~10 mV). Furthermore, protons and electrons are provided continuously by feeding hydrogen gas into the plasma. Obviously, these conditions are not fulfilled for low energy antiprotons and positrons. Some modifications are necessary for the simultaneous confinement of antiprotons and positrons.

**TABLE 2.** A hydrogen plasma in a magnetic mirror field

| Plasma parameters | |
|---|---|
| Plasma density | $n \sim 2 \times 10^{12}$ cm$^{-3}$ |
| Parallel Proton temperature | $T_{p//} \sim 400$ eV |
| Perpendicular Proton temperature | $T_{p\perp} \sim 4$ keV |
| Electron temperature | $T_e \sim 80$ eV |

## Simultaneous Confinement of Antiprotons and Positrons in a Magnetic Mirror Field

Here, the simultaneous confinement of antiprotons and positrons are considered in a simple magnetic mirror field. Compared with the previous example, the number of antiprotons and positrons are limited and their densities will be inevitably much smaller. Since the number of antiprotons is much more limited in experiments, it is assumed that there are much more positrons ($10^8$) than antiprotons ($10^7$) and that the density of positron is in the order of $10^6$ to $10^8$ cm$^{-3}$.

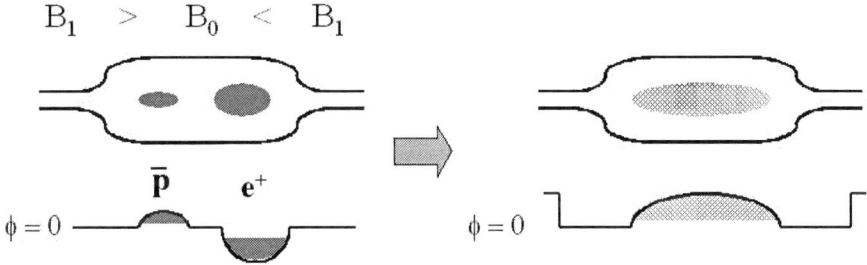

**FIGURE 5.** Antiprotons and positrons confined with a nested Penning trap inside a magnetic mirror field is mixed by the adiabatic expansion of both particles in axial direction.

The basic procedure for the simultaneous confinement is as follows. At first the low energy antiprotons and positrons are confined by a nested Penning trap inside a magnetic mirror field (FIG.5). Then, both particles are expanded adiabatically along the magnetic field by reducing the confinement potential. The adiabatic expansion results in anisotropic velocity distribution [18]. The perpendicular energy becomes larger than parallel energy, which enhances the particle confinement in the magnetic mirror field at the beginning. Since the excess positrons provide the confinement potential for antiprotons, both particles will be trapped simultaneously inside the magnetic mirror field. The confinement potential $\phi_c$ of 1V will be necessary for $10^7$ antiprotons.

The loss cone distribution will be modified as schematically shown in FIG.6 because there are much more positrons. In this case, the confinement time of low energy antiprotons is determined by the confinement time of positrons. This is because antiprotons are confined by the potential $\phi_c$ produced by positrons. As in the previous example, the confinement time of positrons in the magnetic mirror field is denoted by the collision frequency between positrons. When low energy positrons are confined in a strong magnetic mirror field, the collision frequency can be estimated [19]. A mirror ratio of 5 with $B_0$=1T and $B_1$=5T is a plausible example. Shown in FIG.7 is the calculated collision frequency for three different positron densities ranging from $10^6$ to $10^8$ cm$^{-3}$. It is seen that the lower density has the lower collision frequency, which leads to the longer confinement time. With the positron density of $10^6$ cm$^{-3}$ and the temperature of 1eV, the confinement time in a simple magnetic mirror field is expected to be less than 0.1 second. The strong magnetic field enhances the cyclotron radiation, which eventually moves positrons into the loss cone. Since the theoretical cyclotron radiation cooling time is about a few second for $B_0$=1T, its effect is not so serious in this case. It is thought that most of the characteristics of the antiproton-positron plasma can be investigated with the confinement time of 0.1sec.

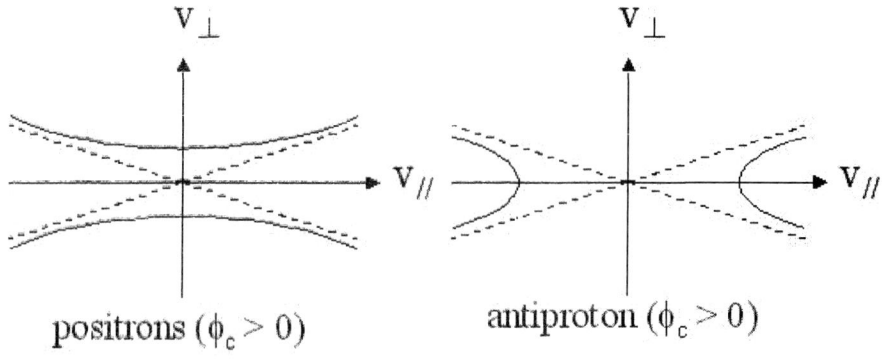

**FIGURE 6.** The schematics of loss cone distributions modified with the self filed potential $\phi_c$ >0, when much more positrons are confined than antiprotons.

If necessary, another technique should be employed to achieve a longer confinement time of the antiproton-positron plasma. One possibility is to apply positive electrostatic field outside the magnetic mirror as shown in FIG.4. This potential reflects back escaping positrons and the confinement time of the positrons can be much longer than the cyclotron radiation cooling time. However, this is not enough for the longer confinement of both particles. Since the confinement potential $\phi_c$ becomes the order of positron temperature [20], the low energy positrons (<0.1 eV) after the long confinement time results in the small $\phi_c$ for antiprotons. Therefore, the cyclotron resonance heating for positrons inside the magnetic mirror field by a small power microwave will be necessary to keep the positron temperature of 1eV and confinement potential $\phi_c$ of 1V.

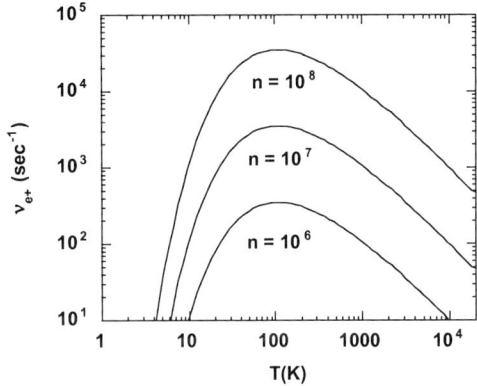

**FIGURE 7.** Calculated collision frequencies between positrons in a strong magnetic field (B=1T) for three different densities.

The same procedure is expected to work more effectively for the production of electron-positron plasma. In this case, equal densities of electrons and positrons can be confined simultaneously with the cyclotron resonance heating for both particles. Injection of moderated low energy positrons into the magnetic mirror field with the application of cyclotron resonance heating at the minimum B field may enhance the accumulation rate of low energy positrons in the trap without a buffer gas which was necessary to obtain $10^8$ low energy positrons in a trap so far.

## SUMMARY

Non-neutral antiproton plasmas will be realized near future. Proper manipulation of antiproton plasmas will be important for the effective production of low energy antiproton beams. Nondestructive measurement of the antiproton plasma with its electrostatic oscillations should be used as a real time monitor inside a trap. A magnetic mirror confinement may be used for the production of antiproton-positron plasmas and electron-positron plasmas.

## ACKNOWLEDGMENTS

The author acknowledges the ASACUSA trap group members and those involved in AD, RFQD operation.

# REFERENCES

1. G. Gabrielse, X. Fei, L. A. Orozco, S. L. Rolston, R. L. Tjoelker, T. A. Trainor, J. Haas, H. Kalinowsky, and W. Kells, *Phys. Rev. Lett.* **63**, 1360 (1989)
2. M. Amoretti *et al.*, *Nature* (London) **419**, 456 (2002).
3. N. Kuroda, H. A. Torii, K. Yoshiki Franzen, Z. Wang, S. Yoneda, M. Inoue, M. Hori, B. Juhasz, D. Horvath, H. Higaki, A. Mohri, J. Eades, K. Komaki, and Y. Yamazaki, *Phys. Rev. Lett.* **94** 023401 (2005).
4. G. Gabrielse *et al.*, *Phys. Rev. Lett.* **89**, 213401 (2002).
5. C. S. Weimer, J. J. Bollinger, F. L. Moore, and D. J. Wineland, *Phys. Rev. A* **49**, 3842 (1994).
6. H. Higaki, N. Kuroda, T. Ichioka, K. Yoshiki Franzen, Z. Wang, K. Komaki, Y. Yamazaki, M. Hori, N. Oshima, and A. Mohri, *Phys. Rev. E* **65** 046410 (2002)
7. T. B. Michell, J. J. Bollinger, X. -P. Huang, and W. M. Itano, *Opt. Express* **2** (1998) 314.
8. G. Dimonte, *Phys. Rev. Lett.* **46**, 26 (1981)
9. X. –P. Huang, F. Anderegg, E. M. Hollmann, C. F. Driscoll, And T. M. O'Neil, *Phys. Rev. Lett.* **78**, 875 (1997).
10. R. G. Greaves and C. M. Surko, *Phys. Rev. Lett.* **75** 3846 (1995).
11. K. Iwata, R. G. Greaves, T. J. Murphy, M. D. Tinkle, and C. M. Surko, *Phys. Rev. A* **51** 473 (1995).
12. N. Oshima, T. M. Kojima, M. Niigaki, A. Mohri, K. Komaki, and Y. Yamazki, *Phys. Rev. Lett.* **93**, 195001 (2004).
13. H. Higaki, N. Kuroda, K. Yoshiki Franzen, Z. Wang, M. Hori, A. Mohri, K. Komaki, and Y. Yamazaki, *Phys. Rev. E* **70** 026501 (2004)
14. K. Yoshiki Franzen, N. Kuroda, H. A. Torii, M. Hori, Z. Wang, H. Higaki, S. Yoneda, B. Juhasz, D. Horvath, A. Mohri, K. Komaki, and Y. Yamazaki, *Rev. Sci. Instrum.* **74** 3305 (2003).
15. F. F. Chen, Introduction to *Plasma Physics and Controlled Fusion*, New York: Plenum, 1984.
16. H. Higaki, M. Ichimura, K. Horinouchi, K. Nakagome, S. Kakimoto, Y. Yamaguchi, K. Ide, D. Inoue, H. Nagai, M. Yoshikawa, Y. Nakashima, and T. Cho, *Rev. Sci. Instrum.* **75** 4085 (2004).
17. D. V. Sivukhim, "Coulomb collisions in a fully ionized plasma," in *Reviews of Plasma Physics*, **4** edited by M. A. Leontovich, New York: Consultants Bureau, 1966, pp. 93-241.
18. A. W. Hyatt, C. F. Driscoll, and J. H. Malmberg, *Phys. Rev. Lett.* **59** (1987) 2975
19. B. R. Beck, J. Fajans, and J. H. Malmberg, *Phys. Rev. Lett.* **68** (1992) 317.
20. J. Fajans, *Phys. Plasmas* **10** 1209 (2003).

# Do Electrons and Ions Coexist in an EBIT?

A. Lapierre[1], J. R. Crespo López-Urrutia[1], J. Braun[1], G. Brenner[1]. H. Bruhns[1], D. Fischer[1], A. J. González Martínez[1], V. Mironov[1], C. J. Osborne[1], G. Sikler[1], R. Soria Orts[1], H. Tawara[1], Y. Yamazaki[2], J. Ullrich[1]

[1] *Max-Planck-Institut für Kernphysik, Saupfercheckweg 1, D-69117 Heidelberg, Germany*
[2] *RIKEN, Hirosawa 2-1, Wako-shi, Saitama 351-01, Japan*

**Abstract.** Detailed analysis of an extended exponential tail observed in the decay curve of the $^2P_{3/2}$-$^2P_{1/2}$ magnetic dipole (M1) transition in boron-like $Ar^{13+}$ provides evidence that electrons and ions might coexist in the same spatial region in the Heidelberg Electron Beam Ion Trap (HD-EBIT). On this basis, new trapping–cooling–recombination schemes for positron-antiproton plasmas are envisioned, integrated in a magnetic bottle configuration that should be able to trap the subsequently formed recombined cold antihydrogen. Moreover, the EBIT configuration, providing excellent spectroscopic access to the trapping region via seven view ports is shown to be well suited for performing precision spectroscopy of antiprotonic ions. Those might be generated either by recombination of antiprotons with neutral gas atoms or through radiative recombination and state-selective dielectronic-recombination-like processes with highly charged ions produced and stored in the EBIT simultaneously in a nested trap configuration.

## INTRODUCTION

Coexistence of antiprotons ($\bar{p}$) and positrons ($e^+$) in the same volume of space is one essential prerequisite for antihydrogen ($\bar{H}$) formation in a trap. In present machines this is approximated by first confining $\bar{p}$ and $e^+$ in separate but nested Penning traps in a homogeneous magnetic field environment and then forcing the positrons to repeatedly pass through the electron-cooled $\bar{p}$ cloud at some velocity (hot or cold mixing) [1, 2]. Recombination and $\bar{H}$ formation has been successfully observed either by imaging $\bar{H}$ annihilation products at the walls of the trap [2] or by detecting $\bar{p}$ produced via field-ionization of $\bar{H}$ transported through a region of a strong electric field [1]. In all present scenarios, $\bar{H}$ is not trapped and its existence is only demonstrated by destroying it again. Thus, substantial efforts are undertaken to combine such recombination schemes with magnetic gradient field geometries that could confine cold $\bar{H}$ after its formation which will be the next decisive step towards high-precision spectroscopy, the ultimate goal of such experiments. Several recombination and trapping scenarios have been and are still being discussed in order to reach that goal, among them, sophisticated "cusp-shaped" traps in an anti-Helmholtz configuration [3], laser-induced schemes in $\bar{p}$-$e^+$ plasmas and laser-induced recombination involving positronium, which was successfully demonstrated recently [4].

In this paper we provide evidence that electrons and $Ar^{13+}$ ions have been confined simultaneously in the same volume of space inside the Heidelberg Electron Beam Ion

CP793 *Physics with Ultra Slow Antiproton Beams*
edited by Yasunori Yamazaki and Michiharu Wada
© 2005 American Institute of Physics 0-7354-0282-5/05/$22.50

Trap (HD-EBIT). As schematically depicted in the upper part of Figure 1, low-energy electrons might be trapped either by the positive ion space charge potential (ISCP) created by the confined ions or by the axial magnetic bottle potential (dotted line in the upper part of Figure 1) close to the trapping volume of the ions, which are confined axially by electrostatic potentials applied to the drift tubes of the trap (full line in the upper part of Figure 1) as well as radially by the strong magnetic field.

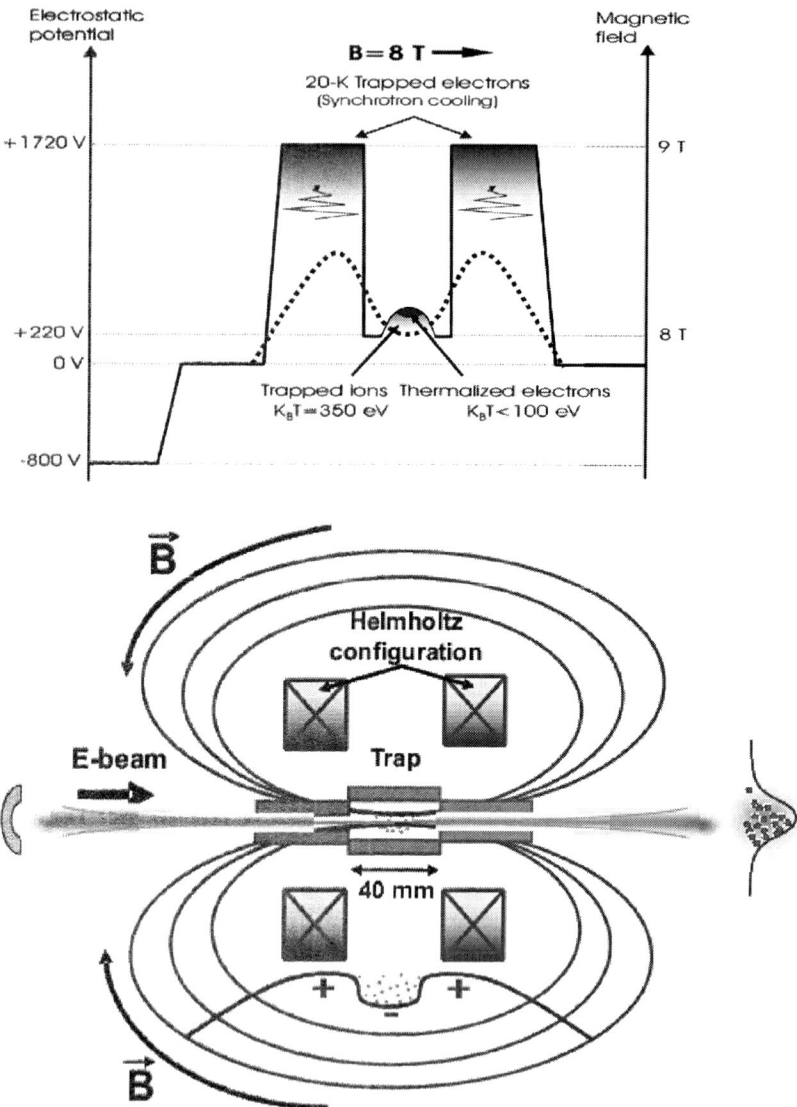

**FIGURE 1.** Schematic layout of the HD-EBIT (lower part) with superconducting coils, the electron beam, drift tubes with applied electrostatic potentials (full line, upper part) and magnetic bottle configuration (dotted line, upper part).

# LIFETIME MEASUREMENTS AT THE HD-EBIT

The HD-EBIT along with the on-going physics programme has been described in detail elsewhere [5]. In short, a high-energy electron beam of up to $E_e$ = 100 keV at currents as large as $I_e$ = 500 mA is collinearly injected along the Helmholtz-type magnetic field reaching 9 T at the centre of the trap. The B-field strongly compresses the electron beam in the radial direction to a diameter of less than 100 $\mu$ m thus achieving current densities close to $10^4$ A/cm$^2$. Ions, efficiently produced by sequential electron-impact ionisation of atoms injected into the trap via a collimated gas jet are radially trapped by the electron beam space charge potential (ESCP), which can reach several hundreds of volts. Longitudinally they are confined by applying appropriate potentials to the various drift tubes (nine in the present set-up). Two experimental view ports, six in next generation machines, provide optimum access for installation of various spectrometers covering essentially all wavelengths of interest. Under present conditions, ions in any charge-state up to helium-like $Hg^{78+}$ can be created, trapped, cooled via evaporative cooling and investigated inside the trap as well as extracted out of the trap in order to perform collision experiments with atoms, molecules, clusters, and surfaces. In the experiment to be discussed in this contribution, the electron beam current was typically set to 100 mA, the electron beam energy was about 700 eV (below threshold for ionization of $Ar^{13+}$) and the magnetic field was 8 T in the centre of the trap.

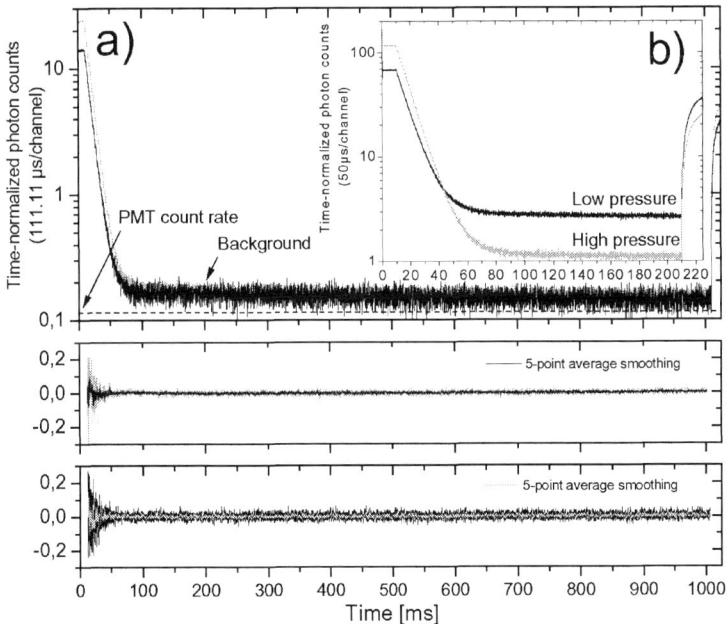

**FIGURE 2.** Decay curves of the $^2P_{3/2}$-$^2P_{1/2}$ metastable M1 transition in boron-like $Ar^{13+}$ taken for an electron beam switch-off time of a) 1000 ms and b) 200 ms for two different injection gas pressures (see text). Lower part: residuals after fitting to a linear combination of two exponential functions.

As described in detail in Ref. [6], the goal of the present experiment was to measure the lifetime of the $^2P_{3/2}$ metastable level of $Ar^{13+}$. For hat purpose, the electron beam was periodically switched off for time intervals of up to one second and the photons of the M1 transitions, filtered by an interference window, were detected by a photomultiplier tube (PMT) (*ca.* 34 counts/s dark rate). In a previous experiment the energy of the M1 transition was measured to be 441.2559(1) nm with a $5 \cdot 10^{-7}$ accuracy providing the most sensitive test of five-electron quantum electro dynamic (QED) contributions whose overall effect is estimated to be 0.96 nm [7].

Typical lifetime curves taken at two different Ar gas injection pressures are shown in Figure 2. In addition to the expected fast decay of the metastable level a slowly decreasing background with an apparent lifetime between one to two seconds was observed by fitting the observed decay curve to a linear combination of two exponentials. As demonstrated and discussed in detail in Ref. [6] and as becomes obvious from the residuals from such a fit shown in the lower part of Figure 2, this background did not influence our lifetime measurement beyond the experimental error bar stated in Ref. [6] which is dominated by other sources of uncertainty.

## BACKGROUND LEVEL AND TRAPPED ELECTRONS

Essentially four mechanisms might be responsible for the observed background: (i) ion-ion collisional excitation, (ii) charge-exchange collisions or (iii) cascade repopulation, both feeding the metastable level from upper excited states, and (iv) electron- impact excitation of the metastable level by low-energetic Penning electrons. Excitation by ion-ion collisions (i) or feeding of the metastable level through ion-atom charge-exchange encounters (ii), can both be ruled out since the slowly decaying background level was found to be lower for a higher injection gas pressure as shown in Figure 2. Moreover, the ion temperature estimated from the measured Doppler broadening of the spectral line to be about 350 eV was too low for any highly charged ion to come close enough for efficient ion-impact excitation of the level, *i.e.,* for reaction (i) to occur. Feeding of the metastable level through charge-exchange collisions of $Ar^{14+}$ with rest gas atoms, reaction (ii), can additionally be ruled out since the $Ar^{14+}$ abundance in the trap was negligibly small by keeping the electron beam energy safely below the ionization threshold of $Ar^{13+}$. Cascade repopulation, process (iii), can be discarded since the decay times of all known higher lying levels that could feed the metastable state are less than a few µs.

Thus, we conclude that reaction (iv), namely electron-impact excitation of the 2.8 eV M1 transition by low-energy electrons ($E_e < 100$ eV), trapped in the same volume of space as the ions for even long time periods after the electron beam was switched off, are the reason for the slowly decaying background contribution. In order to substantiate this result we have investigated the behaviour of the background level as a function of several parameters. In Figure 3, for example, we explore the background photon count rate for various "left-over" electron beam current in "switched-off" phases. In other words, the electron beam was not completely turned off, but reduced from the "switched-on" current of 100 mA to values of 40 mA or less, as indicated in Figure 3. For "switched-off" electron beam currents above 20 mA we observed the expected behaviour, namely a background level that is larger than the one for 0 mA and which increases with increasing "switched off" current due to the ongoing current-depended continuous electron impact excitation of the metastable level by the left-over electron beam. Most surprisingly, however, further reduction of the electron beam to currents of 10 mA or less, decreased the background level substantially by nearly one order of magnitude for 1 mA below the background level observed when the electron beam was completely turned off.

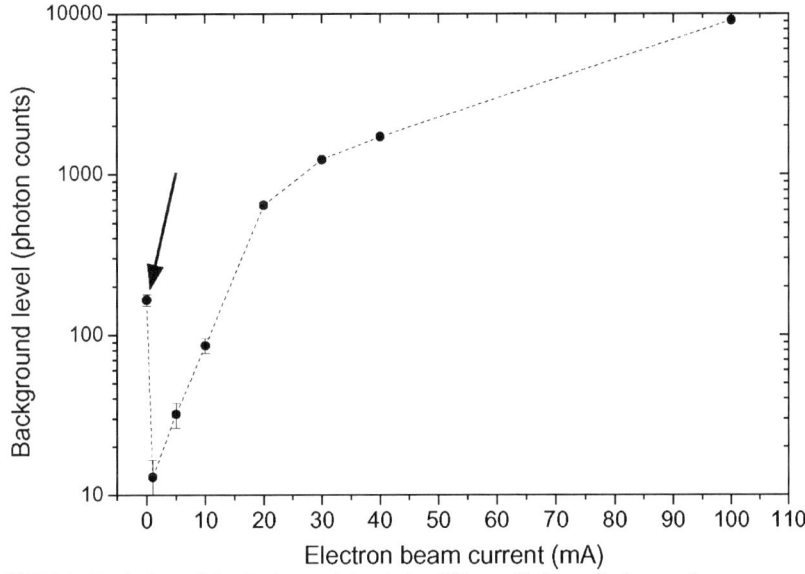

**FIGURE 3.** Evolution of the background level for different "left-over" electron beam currents in "switched-off" position (see text). The background level observed at 10 mA and less was lower than that observed at 0 mA, *i.e.*, without electron beam current (see arrow).

At first glance, one might interpret this observation to be a result of a depletion of the $Ar^{13+}$ population by electron impact ionization at lower currents due to an effective simultaneous enhancement of the electron beam energy above the $Ar^{13+}$ ionization threshold. Decreasing the current of the electron beam decreases its space charge potential (ESPC), and as a result, increases the beam energy, which can then be sufficient to ionize $Ar^{13+}$ ions. For instance, at a beam current of 1 mA, the beam energy is increased to about 1 KeV. However, simulation calculations of electron capture, electron-impact excitation and -ionization have shown that for beam currents lower than about 20 mA, depletion by electron impact ionization is insignificant because of its squared dependence on the electron beam current.

Therefore, we are led to interpret the observed suppression of the background to be a consequence of a reduction of the number of simultaneously trapped low-energy electrons being pushed away by the space charge of the left-over weak electron beam. This conclusion is further supported by inspecting the decay time of the slowly decreasing background part, extracted from the two-exponential fit. It is found to be considerably longer for "switched-off" currents between a few to 20 mA compared to zero. In most cases indeed, it is compatible with an infinite lifetime within the error bars of the fit, providing additional evidence that excitation by slow trapped electrons is getting less efficient since less such electrons are around if the electron beam is not completely switched off. Above a "switched-off" current of 20 mA, the whole decay curve cannot be fitted any more by a sum of two exponential functions which is consistent with the expectation, that now, in addition to the possible excitation by slow electrons, direct excitation by the left-over electron beam plays an increasing role with increasing "switched-off" current. Indeed, a three-exponential fit delivers increasing decay times of the background rate with higher left-over current, again becoming consistent with infinity for large currents due to a continuous ongoing excitation once a new equilibrium is reached.

Furthermore, we have systematically investigated the behaviour of the background level as a function of the injection gas pressure as indicated in Figure 2 and observed a decreasing background photon count rate with increasing pressure. For higher injection gas pressure a larger numbers of ions is created and confined in the trap as is clearly seen from the increased photon rate at high pressure when the electron beam is switched on. An increased number of stored ions enhances the ISCP and, thus, the potential well depth and the temperature of trapped low-energy electrons (see section four). Since the excitation cross section decreases with increasing electron velocity the background level might indeed expected to be lower in agreement with the observations.

In summary, we conclude that low-energy Penning electrons simultaneously trapped in the same volume in space as the ions are intimately related to the observed slowly decaying background.

## LOW-ENERGY ELECTRON TRAPPING

As illustrated in Figure 1, low-energy electrons which are in general always present in the apparatus could be trapped near the location of the ion cloud after the electron beam was switched off by essentially two mechanisms:

First, they might be confined axially by the magnetic bottle potential at the borders of our 40 mm trap. Due to the specific geometrical arrangement of the Helmholtz coils in our EBIT, an 8 T magnetic field at the centre of the trap is increased by about 5% to 8.4 T at an axial distance of 26 mm away from the trap centre with a maximum gradient at $\pm 14$ mm off centre. Accidentally this nearly coincides with the edge of the closest drift tubes and, thus with the axial border of the ion trap. Hence, it can be estimated, neglecting plasma and collision effects (see Ref. [8]), that electrons that are born in the middle of the trap with a pitch angle larger than about $\alpha > 77°$ are reflected before they could fall into the deep positive potential wells on both sides of the centre drift tubes, which axially confine the ion cloud.

Second, they might be trapped by the ISCP, a scenario seemingly supported for three reasons: (i), even a weak residual electron beam with a current of 1 mA providing a negative ESCP of only about -5 V at the edge of the electron beam radius, about half of which is compensated by the ISCP to an effective value of not more than

-3 V, is able to efficiently push away the low-energy electrons from the trapping region. (ii), as shown in Figure 4, increasing the trapping potential for the ions (see inset for the geometry) and, thus the number of trapped ions increased the background level indicating that the ISCP is directly correlated with the number of electrons overlapping with the ion cloud. (iii), whereas the lifetime of the metastable level was found to be independent of the drift tube potential for values higher than about +300 V, the background decay time decreased continuously with increasing the ion trapping potentials up to values as large as 2.7 kV. Thus, the background decay time seems to essentially reflect electron losses from the trap as a function of the ion trapping potential, which are expected to become larger for tighter confinement for the ions, *i.e.*, higher trapping potentials for the electrons outside the ion cloud.

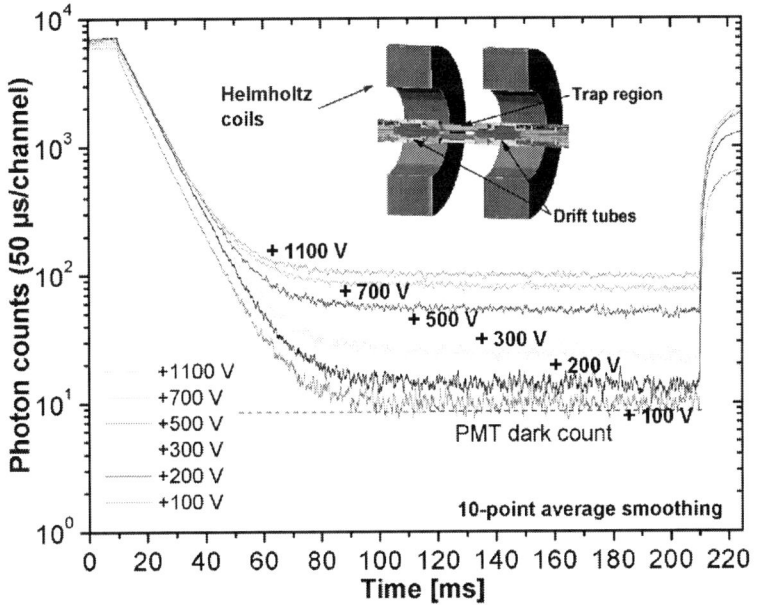

**FIGURE 4.** Increase of the slowly decaying background level as a function of the drift tube trapping potential as indicated in the Figure.

Whereas we cannot come to a final conclusion about the question on the dominant reason for trapping low-energy electrons on the basis of the present data, confinement of electrons by the positive space charge of the trapped ion cloud (ISCP) seems to be the more likely scenario and was further investigated by the following procedure. The sum space charge potential of both the electron beam and the trapped ion cloud was obtained by measuring the energy position of the maximum fluorescent signal for a given electron beam current and a given potential difference between the cathode and the whole drift tube assembly (the position of this maximum corresponds to an electron beam energy having the ionization potential of $Ar^{13+}$). Estimating theoretically the ESCP, the ISCP alone is found to be about +250 V, which is enough to trap electrons of sufficiently low energy. As described in detail in Ref. [6], the temperature of the low-energy trapped electron $k_B T_e$ is estimated to be $k_B T_e = 25-125\ eV$ using $\omega = qV/k_B T_e$ ( $q$ : charge of the trapped particles, $V$ trapping potential, $k_B$ : Boltzmann constant), where the dimensionless ratio $\omega$ is ranging from 2 to 10.

For an average electron temperature of 75 eV and from the measured background count rate we estimate the density of trapped electrons to be $5 \cdot 10^7$ cm$^{-3}$ assuming a Maxwellian-averaged rate coefficient for electron-impact excitation of the metastable level. The ion density is calculated approximately to be $3 \cdot 10^9$ cm$^{-3}$ from the measured total number of trapped ions extracted in an independent experiment, and taking the known length of the trap of 40 mm as well as an ion cloud of about 0.7 mm in diameter measured previously by an imaging optical spectrometer and by imaging the cloud on a CCD camera. The total numbers of electrons and ions are then estimated to be $N_e = 8 \cdot 10^5$ and $N_i = 5 \cdot 10^7$, respectively. Also, the equilibrium temperature of the ions under continuous electron bombardment at $I_e = 100$ mA for a 2.5 kV drift tube trapping potential is deduced from the measured Doppler width of the spectral line to be about $k_B T_i = 350 eV$.

As indicated in Figure 1, low-energy electrons outside the ion cloud rapidly cool down by emission of synchrotron radiation with a cooling time of 56 msec in an 8 T magnetic field. Thus, one might expect that after a short while, electrons will no longer have sufficient energy to excite the trapped ions, and consequently, the shape of

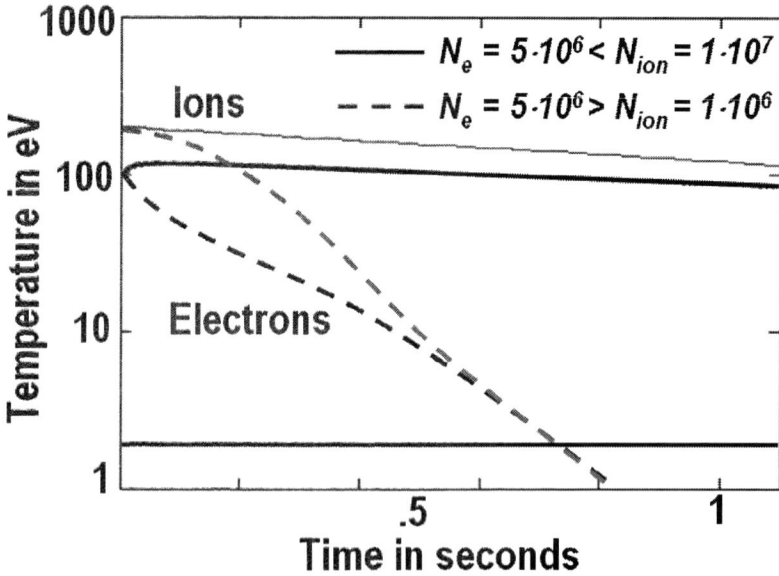

**FIGURE 5.** Temperature of electrons (dark lines) and ions (grey lines) as a function of time for two different scenarios (dashed and full lines, respectively) of relative numbers of trapped particles. In the present experiment, the ion-to-electron number density ratio is about 60.

the exponential background decay curve should abruptly change. Inspection of the decay behaviour by fitting a two-exponential function to 100 times the lifetime did not show any effect as can be seen from the residuals in Figure 2. Calculations presented in Figure 5, using the Spitzer equation [9] with different ratios between the numbers of trapped electrons and ions, show that for a number of ions (dark full line) exceeding that of electrons (grey full line) by only a factor of two (a factor of 60 in the present experiment) the trapped ions act as effective heat sink for the electrons. Thus, electrons and ions are nearly in thermal equilibrium, and they cool down only very slowly such that the metastable level can be excited for seconds and more, as observed.

## PROSPECTS AND CONCLUSION

In the future, several tests will be performed in order to substantiate our present findings and to clarify what might be the main mechanism for trapping low-energy electrons in the volume of space where the ions are confined. In a first experiment one could look for high-energy radiation from radiative recombination of the trapped electrons with the ions after turning off the electron beam even though the recombination rate is so small that estimates for the above numbers of electrons and ions in the trap yield an observable x-ray rate of only $10^{-4}$ per second at an assumed electron temperature of 50 eV. Since the recombination rate depends on the square root of the temperature, it should be enhanced for the decreased temperatures of the plasma after longer trapping times and might then become observable. Second, we are presently preparing a YAG pumped dye laser for laser spectroscopy of trapped ions which could be used for stimulating radiative recombination into high-lying levels of the ions as it has been observed in cooler sections of storage rings [10] and proposed as a promising way to enhance radiative recombination in an $\overline{H} - e^+$ plasma. Third, with one of our newly designed traps it is straightforward to extract the trapped electrons and measure their number as a simple proof of their existence. Fourth, it will be very instructive to image the ion cloud via the M1 transition radiation by opening the entrance slits of our optical grating spectrometer to get an insight into the steady-state situation of the trapping region with the electron beam on. By switching off the electron beam, this will provide detailed information on the development in space of the ion cloud excited by the low-energy trapped electrons and, thus, give us direct information on their location and expansion in the trap along with the evolution in time of the electron and ion clouds when they cool down.

To conclude, we have provided the first evidence that low-temperature electrons might coexist with highly charged ions in an EBIT configuration. The available extended data sets indicate that the positive space charge potential of the trapped ion cloud might be mainly responsible for the confinement of low-energy electrons whereas an additional effect by the magnetic bottle potential reflecting electrons axially at the border of the ion trap cannot be excluded and is certainly important.

Our findings are considered to be of significant importance for future schemes to efficiently produce, trap, and cool $\overline{H}$ in one region of space in a magnetic bottle configuration as planned e.g. in the next generation experiments within the ALPHA collaboration. In addition to the trapping of positrons by the ISCP of the antiprotons that seems be feasible according to our observations, axial drift tube potentials for the antiprotons might be dimensioned such that the $\bar{p}$ trapping electrostatic potential rises not earlier than the magnetic field, slightly different from our present EBIT configuration, such that the magnetic bottle effect is exploited more efficiently. If our present results are correct, $e^+$ and $\bar{p}$ would coexist in the same spatial region, the $e^+$ being confined by the combined effect of the space charge of the $\bar{p}$ and of the magnetic bottle, the latter possibly allowing the number of stored $e^+$ to exceed the one of $\bar{p}$. Then, synchrotron cooling could be quite effective as is illustrated in Figure 5 (dashed lines with $N_e > N_i$) and relatively short cooling times of a few seconds are achieved along with a considerably enhanced recombination efficiency at low positron temperatures. In addition, as will be tested in our present trap, a strong laser could be used to stimulate recombination into high-lying states, which then in turn can be readily de-excited with further lasers, including a $Ly_\alpha$ laser demonstrated recently [11]. Through the many viewports in our geometry, fluorescence radiation can be easily detected with reasonable solid angles.

Finally, we would like to point out that an EBIT configuration along with its viewports giving direct access to the centre of the trap shall be exploited in the future for precision spectroscopy of antiprotonic ions. Trapping low-energy $\bar{p}$ in the central drift tube and injecting any kind of atoms through an atomic jet (already installed in our present machines) should effectively generate antiprotonic ions ($\bar{p}$-capture) which are efficiently trapped in the same region of space as the antiprotons. Emitted radiation after de-excitation of the bound $\bar{p}$ can be straight-forwardly detected through the viewports by large solid angle germanium detectors or by high-resolution x-ray crystal spectrometers. Envisioning cycle times of 10 seconds for loading of the trap with up to $10^7$ $\bar{p}$ per shot as anticipated at the future Facility for Low-energy Antiproton and Ion Research (FLAIR) [12] at GSI might allow us to perform precision spectroscopy with effective signal rates of up to a few per second.

Moreover, we would like to emphasize, that new EBITs are presently under construction for charge breeding of radioactive ions to be generated at TRIUMF in Vancouver or at FAIR at GSI. Having trapped highly charged ions and $\bar{p}$ in a nested trap configuration, the ions in the central trapping region and the $\bar{p}$ at some neighbouring off-centre drift tube, would provide the unique possibility to let the $\bar{p}$ repeatedly collide with the highly charged radioactive heavy ions oscillating axially at well-controlled energies in the extended trapping region. $\bar{p} + A^{q+} \rightarrow A^{(q-1)+**}$ resonant recombination to doubly excited combined electronic and antiprotonic states might provide a unique possibility to form and investigate well-defined antiprotonic states with only few bound electrons around as it has been recently demonstrated for pure doubly excited electronic states of He-, Li-, Be-, and B-like ions formed in our EBIT through dielectronic recombination [13].

Presently, we are designing a modified $\bar{p}$-EBIT and anticipate to start with tests by the end of the year, among them the above mentioned ones using ions and electrons to prove whether or not clouds of opposite charge can coexist in an EBIT configuration.

# REFERENCES

1. G. Gabrielse, N. S. Bowden, P. Oxley, A. Speck, C. H. Storry, J. N. Tan, M. Wessels, D. Grzonka, W. Oelert, G. Schepers, T. Sefzick, J. Walz, H. Pittner, T. W. Hänsch, E. A. Hessels, and (ATRAP Collaboration), Phys. Rev. Lett. **89**, 213401 (2002)

2. M. Amoretti *et al.*, Nature (London) **419**, 456 (2002); M. C. Fujiwara *et al.* (ATHENA Collaboration), Phys. Rev. Lett. **92**, 065005 (2004).

3. A. Mohri and Y. Yamazaki, Europhys. Lett. **63**, 207 (2003).

4. C. H. Storry, A. Speck, D. L. Sage, N. Guise, G. Gabrielse, D. Grzonka, W. Oelert, G. Schepers, T. Sefzick, H. Pittner, M. Herrmann, J. Walz, T. W. Hänsch, D. Comeau, E. A. Hessels, and (ATRAP Collaboration), Phys. Rev. Lett. **93**, 263401 (2004).

5. J. R. Crespo López-Urrutia, J. Braun, G. Brenner, H. Bruhns, C. Dimopoulou, I. N. Draganic, D. Fischer, A. J. González Martínez, A. Lapierre, V. Mironov, R. Moshammer, R. Soria Orts, H. Tawara, M. Trinczek, and J. Ullrich J. Phys: Conf. Series **2** 42–51 (2004); J. R. Crespo López-Urrutia, B. Bapat, I. Draganic, B. Feuerstein, D. Fischer, H. Lörch, R. Moshammer, J. Ullrich, R. D. DuBois, and Y. Zou, Hyperfine Interactions **146-147**, 109 (2003); J. R. Crespo López-Urrutia, *et al.*, Phys. Scr. **T 80** (1999).

6. A. Lapierre, U. Jentschura, J. R. Crespo López-Urrutia, J. Braun, G. Brenner, H. Bruhns, D. Fischer, A. J. González Martínez, V. Mironov, C. J. Osborne, G. Sikler, R. Soria Orts, H. Tawara,. Ch. Keitel, I. I. Tupitsyn, W. R. Johnson, J. Ullrich, A. Volotka, Phys. Rev. Lett. (submitted)

7. I. Draganic, J. R. Crespo López-Urrutia, R. DuBois, S. Fritzsche, V. M. Shabaev, R. S. Orts, I. I. Tupitsyn, Y. Zou, and J. Ullrich, Phys. Rev. Lett. **91**, 183001 (2003).

8. J. Fajans, Physics of Plasmas, **10** 1209 (2003)

9. L. Spitzer, Physics of Fully Ionized Gases, Interscience, New York 1956

10. U. Schramm, J. Berger, M. Grieser, D. Habs, E. Jäschke, G. Kilgus, D. Schwalm, A. Wolf, R. Neumann, and R. Schuch Phys. Rev. Lett. **67**, 22-25 (1991); T. Schüssler, U. Schramm, T. Rüter, C. Broude, M. Grieser, D. Habs, D. Schwalm, and A. Wolf, Phys. Rev. Lett. **75**, 802-805 (1995).

11. K.S.E. Eikema, J. Walz, and T.W. Hänsch, Phys. Rev. Lett. **83**, 3828 (1999).

12. FLAIR Technical Design Report http://www-linux.gsi.de/~flair/.

13. A.J. González Martínez, A. Artemyev, J. Braun, G. Brenner, H. Bruhns, J.R. Crespo López-Urrutia, A. Lapierre, V. Mironov, J. Scofield, R. Soria Orts, H. Tawara, M. Trinczek, I. Tupytsin, and J. Ullrich, Phys. Rev. Lett. (accepted)

# Confinement of Toroidal Non-neutral Plasma in Proto-RT

## H. Saitoh*, Z. Yoshida* and S. Watanabe*

*Graduate School of Frontier Sciences, University of Tokyo,
5-1-5 Kashiwanoha, Kashiwa, Chiba 277-8583, Japan

**Abstract.** In contrast to linear configurations for non-neutral plasmas, toroidal devices allow us to trap charged particles without the use of a plugging electric field. Thus it has a potential ability to confine high-energy particles or to simultaneously trap multiple particles with different charges. In spite of the relatively long history of the study in pure toroidal magnetic field devices, toroidal non-neutral plasmas are attracting renewed interest with the use of magnetic surface configurations [1, 2]. Possible applications of toroidal traps for non-neutral plasmas are formation of matter-antimatter plasmas [2], investigation on the fundamental properties of exotic plasmas including pair (equal mass) plasmas, and experimental test on the equilibrium and stability of flowing plasmas [3, 4, 5]. As an initial test on non-neutral plasmas in the toroidal magnetic-surface geometry, formation and confinement properties of pure electron plasma have been investigated at Prototype-Ring Trap (Proto-RT) device with a dipole magnetic field [1, 6, 7]. Electrons can be injected by using chaotic orbit near a magnetic null line generated by the combination of dipole and vertical magnetic fields [6]. The confinement time of electrons is limited due to the effects of collisions with remaining neutral gas, and electrons of $\sim 10^{12}$ are trapped for $\sim 0.5$ s in the typical magnetic field strength of 100 G and back pressure of $4 \times 10^{-7}$ Torr in Proto-RT. Although the present experiment was carried out on the single-component plasma, the result shows that a stable confinement configuration has been realized for toroidal non-neutral plasmas by using the magnetic surface configuration. Together with the experiment on the toroidal pure electron plasma in Proto-RT, preliminary prospects for the injection and trap of anti-protons and positrons in the toroidal magnetic surface configuration, and creation of multi-component plasmas will be described.

**Keywords:** Non-neutral plasma, electron plasma, toroidal geometry, magnetic surface configuration, antimatter plasmas
**PACS:** 52.27.Jt, 52.27.Aj, 52.35.Fp

# INTRODUCTION

## Confinement of non-neutral plasmas in a toroidal geometry

In recent years, it has been recognized that toroidal geometries might be suitable for the various kinds of non-neutral plasmas [1, 2]. Although excellent confinement properties of single-component plasmas are realized in Malmberg-Penning traps [8], and also the partial overlap of negatively and positively charged particles is successfully demonstrated in a "nested" Penning trap [9, 10], it is not straightforward to simultaneously confine both charges with satisfying

$$\lambda_D \ll L, \tag{1}$$

which is an essential criteria for the observation of plasma phenomena. Here $\lambda_D$ is the Debye length of each component of particles defined as $\lambda_D \equiv (\varepsilon_0 T/ne^2)^{1/2}$, where $\varepsilon_0$

CP793 *Physics with Ultra Slow Antiproton Beams*
edited by Yasunori Yamazaki and Michiharu Wada
© 2005 American Institute of Physics 0-7354-0282-5/05/$22.50

is the dielectric constant of vacuum, $T$ is the temperature, $n$ is the number density, $e$ is the charge of particles, and $L$ is the scale size of the plasma. In contrast to linear traps for non-neutral plasmas, toroidal devices can produce closed magnetic field lines inside the device geometry. Thus it has a potential ability to confine charged particles at any degree of non-neutrality, or to trap high-energy beam particles and multiple species of different charges, without the use of plugging electric field. Mirror, cusp [11], and hybrid Penning-Paul [12] devices can also trap multi-component plasmas, but the toroidal geometry is one of the candidates for the simultaneous confinement of both species of particles in a pure magnetic field configuration.

The realization of the stable confinement of multi-component non-neutral plasma will make possible various experiments in the fields of atomic and plasma physics. One of the possible applications of multi-component plasma traps is the formation of matter-antimatter plasmas [1, 2]. These experiments include the creation of antiproton or positron plasmas and their mixtures, such as antihydrogen plasmas. Because the "overlap" scale length of both species are not limited to $\sim \lambda_D$ caused by the electrostatic forces, the pure magnetic field confinement in the toroidal geometry is potentially useful for the efficient creation of antihydrogen atoms. A positron-electron plasma is a so-called pair plasma with equal-mass particles, and several unique properties on wave propagations are theoretically predicted [13]. A positron-electron plasma is also expected to play important roles in many astrophysical phenomena, although little is known about the experimental issue in laboratory plasmas.

As well as the investigation on the properties of such kinds of exotic plasmas, fundamental experimental test on the equilibrium and stability of a flowing plasma (double Beltrami state) [3, 4, 5] is also possible in a toroidal geometry. When the velocity of plasma flow is comparable to the Alfvén speed, the plasma pressure can be balanced by the dynamic pressure of the flow, and it will open a new path toward the advanced fusion concept. The required flow can be induced by the drift motion of a non-neutralized plasma in the toroidal geometry. Thus the creation of a toroidal multi-component non-neutral plasma will form a basis for many scientific applications.

## Pure toroidal magnetic field configuration

Studies on toroidal non-neutral plasmas have been carried out in a pure toroidal magnetic field [14, 15, 16, 17, 18, 19, 20, 21, 22, 23, 24]. An interesting equilibrium state was revealed in these experiments, one is in marked contrast to the confinement of a neutral plasma, which requires rotational transforms of magnetic field lines in order to avoid the drift loss of particles. In the drift orbit approximation, the total energy of a charged particle in a toroidal magnetic field is

$$E = q\phi + \frac{\mu_0 B_0 R_0}{R} + \frac{L_z^2}{2mR} + \frac{m(\nabla\phi)^2}{2B^2}, \qquad (2)$$

where $q$ is charge, $\phi$ is the electrostatic potential, $R_0$ is the major radius, $B_0$ is the magnetic field strength at $R = R_0$, $\mu_0$ is the space permeability, $L_z$ is the angular momentum of the particle around $z$ axis, and $m$ is the particle mass. Because the total energy of each particle is conserved, surfaces defined by $E = const.$ constrain the positions of the guiding center of the particles. When there is no electric field (neutral plasmas), Eq. (2) becomes

$$E = \frac{\mu_0 B_0 R_0}{R} + \frac{L_z^2}{2mR} \propto 1/R, \tag{3}$$

and the guiding center motions are on cylinders, which inevitably intersects the boundary wall of a containing chamber. In non-neutral plasmas, however, due to the strong radial electric field caused by the space and image charges, $E = const.$ surfaces can take closed circular shape in the poloidal cross section of the torus. The particles might also take closed orbits in a pure toroidal field undergoing the poloidal $\mathbf{E} \times \mathbf{B}$ rotation [15] according to the trap configuration. In a conducting toroidal vessel, the particles can take closed orbits due to the $\mathbf{E} \times \mathbf{B}$ rotation, which is inward-shifted from the equipotential contours because of the other first order drifts. By substituting $\mathbf{E} \times \mathbf{B}$ drift speed $\mathbf{v} = -\nabla \phi \times \mathbf{B}/B^2$, the steady state continuity equation $\nabla \cdot (n\mathbf{v}) = 0$ becomes

$$\left[ \nabla \phi \times \nabla \left( \frac{n}{B^2} \right) \right] \cdot \mathbf{B} = 0. \tag{4}$$

When electric and magnetic fields and plasmas are axisymmetric, Eq. (4) implies that

$$\frac{n}{B^2} = f(\phi). \tag{5}$$

By using Eq. (5), one can solve Poisson's equation

$$\nabla^2 \phi = -\frac{nq}{\varepsilon_0} \tag{6}$$

in order to obtain equilibrium potential and density profiles of toroidal non-neutral plasmas.

The early experiments of toroidal electron plasmas were conducted for the storage or acceleration of heavy ions [16] or creation of high-current relativistic electron beams [18]. In a series of experiments in low-aspect-ratio torus devices [21, 22], toroidal effects, electron injection methods, and other collective behaviors of electron plasma were studied. Existence of the above mentioned equilibrium of electron plasma in pure toroidal field was demonstrated in these experiments. An electrostatic force balance equilibrium model, including the effects of the space and image charges, was also studied [20]. Toroidal electron plasmas feel repulsive "hoop force" caused by the self electric field and also by the diamagnetic effects, and the confinement can be achieved by the external radial electric fields due to the image charges on the vessel wall or electrodes.

## Toroidal magnetic surface configuration

Magnetic surface configurations are most efficient trap devices for plasmas and are intensively studied in the field of high temperature fusion plasma research. In a simple torus device with a pure toroidal magnetic field $B_\phi$, the inhomogeneous field strength

($\propto 1/R$) induces the vertical charge separation, which lead to the outward $\mathbf{E} \times \mathbf{B}$ loss of the both particles. In order to avoid such rapid loss of plasmas, a poloidal field plays essential role in order to form closed magnetic surfaces. Poloidal fields are generated by means of an internal plasma current (tokamaks), an external coil current (stellarators), or an internal coil current (internal conductor and multi-pole devices).

Magnetic surfaces are defined as $\psi(\mathbf{r}) = const.$ planes, where magnetic field lines lie on $\psi(\mathbf{r}) = const.$ and thus

$$\nabla \psi(\mathbf{r}) \cdot \mathbf{B} = 0 \qquad (7)$$

is satisfied. When the magnetic field configuration is axisymmetric, the analytic representation of the magnetic surfaces is

$$\psi = r A_\theta(r, z) = const. \qquad (8)$$

and a poloidal field is given by

$$B_r = -\frac{1}{r}\frac{\partial \psi}{\partial z} \text{ and } B_z = \frac{1}{r}\frac{\partial \psi}{\partial r}, \qquad (9)$$

which satisfies Eq. (7).

The Lagrangian equation of the motion of a charged particle in a magnetic field $\mathbf{B}$ and an electric field $\mathbf{E}$ is given by

$$\frac{d}{dt}\left(\frac{\partial L}{\partial \dot{q}_i}\right) - \frac{\partial L}{\partial q_i} = 0 , \qquad (10)$$

where

$$L = \frac{mv^2}{2} + q\mathbf{v} \cdot \mathbf{A} - q\phi. \qquad (11)$$

Here $\mathbf{A}$ is the vector potential: $\mathbf{B} = \nabla \times \mathbf{A}$ and $\phi$ is the scalar potential: $\mathbf{E} = \nabla \phi$, and $\mathbf{q}$ is the coordinate of the charged particle. In a toroidally symmetric configuration ($\partial/\partial\theta = 0$), the canonical angular momentum of the charged particle satisfies

$$P_\theta \equiv \frac{\partial L}{\partial \dot{\theta}} = mr^2\dot{\theta} + qrA_\theta = const. \qquad (12)$$

The variation of $\psi$ at a distance $d$ from $\psi$ at the initial point is given by

$$|\delta \psi| = d|\nabla \psi|, \qquad (13)$$

and by using Eq. (12), $d$ is calculated as

$$d = \left|\frac{m\delta(rv_\theta)}{q|\nabla \psi|}\right|. \qquad (14)$$

From Eqs. (8) and (9), $|\nabla \psi| = r(B_r^2 + B_z^2)^{1/2} = rB_p$ and also assuming that $\delta(rv_\theta) <\sim rv_\theta$, the maximum value of $d$ is given by

$$d \leq \left|\frac{mr\dot{\theta}}{qB_p}\right| \equiv r_{Lp}, \qquad (15)$$

indicating that the deviation of the charged particle from one magnetic surface is less than the poloidal Larmor radius $r_{Lp}$. When the magnetic field is strong enough and the mechanical momentum of a particle is ignored, the canonical angular momentum of the particle is approximated as

$$P_\theta \sim qrA_\theta, \tag{16}$$

and the orbits of charged particles are limited on magnetic surfaces.

When the Hamiltonian of the charged particle

$$H = -L + \sum_i p_i \dot{q}_i = -\frac{1}{2m}(\mathbf{p} - q\mathbf{A})^2 + q\phi \tag{17}$$

is not explicitly time dependent, the total energy of the charged particles is conserved:

$$W = \frac{mv^2}{2} + q\phi = const. \tag{18}$$

and $v$ has an upper limit of $v \leq (2(W - q\phi)/m)^{1/2}$, and thus, as long as the poloidal field lines form closed magnetic surfaces in a finite region, the orbits of charged particles are also limited to a finite region.

The force balance equation of single component non-neutral plasma is

$$mn\left(\frac{\partial \mathbf{v}}{\partial t} + \mathbf{v} \cdot \nabla \mathbf{v}\right) = qn(\mathbf{v} \times \mathbf{B} - \nabla \phi) - \nabla p. \tag{19}$$

When the density of non-neutral plasmas is far below the Brillouin limit

$$n_B = \frac{\varepsilon_0 B^2}{2m}, \tag{20}$$

and the pressure term is negligibly smaller than the electromagnetic forces, the equation of motion reduces to

$$qn(\mathbf{v} \times \mathbf{B} - \nabla \phi) = 0. \tag{21}$$

By taking the scalar product of Eq. (21) and $\mathbf{B}$,

$$\mathbf{B} \cdot \nabla \phi = 0, \tag{22}$$

indicating that the electrostatic potential is constant on a magnetic field line. From Eq. (21), the perpendicular velocity of the fluid is given by the $\mathbf{E} \times \mathbf{B}$ drift speed

$$\mathbf{v} = -\frac{\nabla \phi \times \mathbf{B}}{B^2}, \tag{23}$$

and the strong self electric field of non-neutral plasmas induces fast cross-field transport of charged particles on a magnetic surface. Thus in cold non-neutral plasmas, electrostatic potential $\phi$ is a function of magnetic flux $\psi$

$$\phi = \phi(\psi). \tag{24}$$

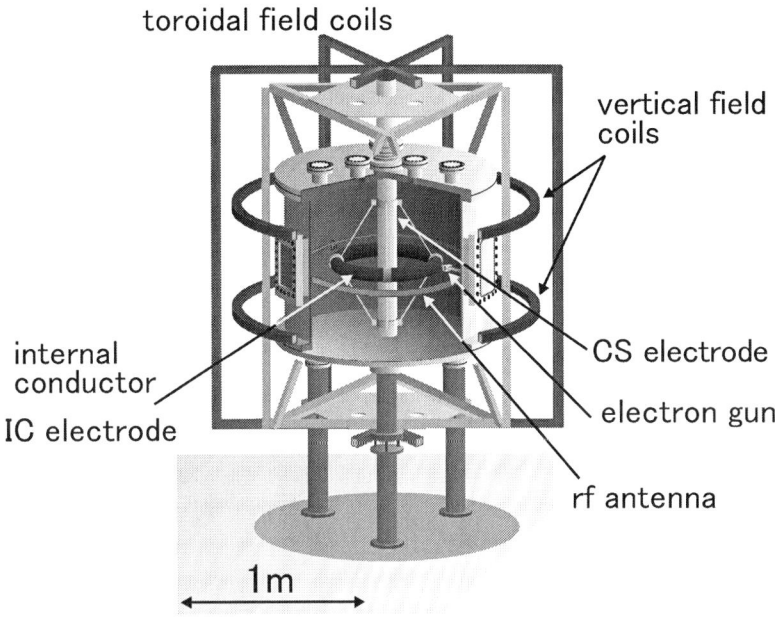

toroidal field coils

vertical field coils

internal conductor

IC electrode

CS electrode

electron gun

rf antenna

1m

**FIGURE 1.** Bird eye view of the Proto-RT device.

When the temperature of the plasma is also a function of magnetic flux $\psi$, the density in the equilibrium state is given by

$$n = N(\psi)\exp\frac{e\phi}{k_B T_e(\psi)}, \tag{25}$$

and one may solve Poisson's equation

$$\nabla^2\phi = \frac{1}{\varepsilon_0}N(\psi)\exp\frac{e\phi}{k_B T_e(\psi)} \tag{26}$$

to obtain equilibrium potential and density profiles of toroidal electron plasmas[2].

## Magnetic surface devices for non-neutral plasmas

Until very recently, little attention has been given to the confinement of non-neutral plasmas on magnetic surfaces. As well as the improvement of the confinement properties of toroidal non-neutral plasmas, the use of magnetic surface configuration is essentially important for the trap of multi-component plasmas (e.g. antihydrogen plasma or positron-electron plasma) in a toroidal geometry, in order to avoid the above mentioned charge separation and the resultant rapid loss of particles.

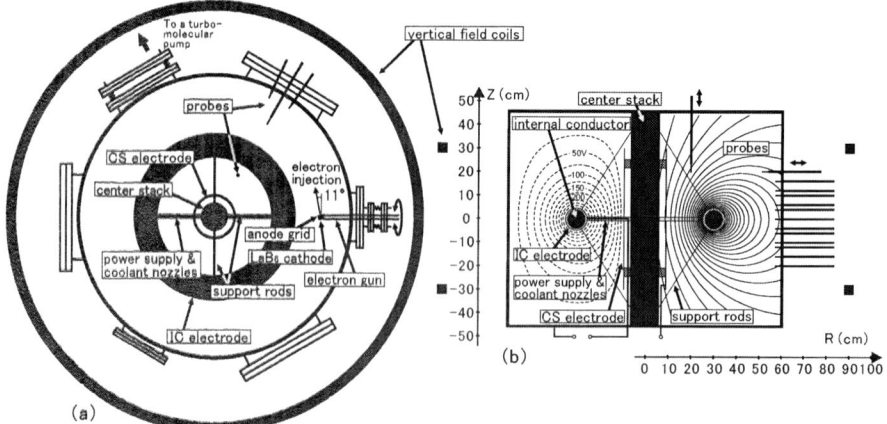

**FIGURE 2.** (a) Top view and (b) poloidal cross section of Proto-RT. Magnetic surfaces of a dipole field and equi-potential contours when the IC electrode is biased are also shown.

For the confinement of low density (below the Brillouin limit $n_B$) non-neutral plasmas, current-free configurations are suitable for the generation of magnetic surfaces. Closed magnetic surfaces of a dipole field can be produced by a ring conductor hung in a vacuum vessel. For the reduction of disturbance due to the mechanical structures of the coil support, one may use a super-conducting levitated coil ring [1]. As a first step toward the realization of an antiproton plasma, a positron plasma, and other multi-component plasmas with a radial electric field in a toroidal geometry, a pure electron plasma is relatively easy to generate, and is suitable for the fundamental test on the various toroidal non-neutral plasmas.

Proto-RT is a prototype ring trap device constructed for the investigation of various topics in non-neutral plasma physics, and experiments on a toroidal pure electron plasma has been carried out. In the following sections, recent experimental investigation in Proto-RT, including the plasma formation, confinement properties, and electrostatic potential structures of a toroidal electron plasma are described.

## EXPERIMENTAL SETUP IN PROTO-RT

### Device and magnetic field configuration

The Proto-RT device is an internal conductor device constructed in 1998 for the study of physics related to non-neutral plasmas [1], chaos-induced resistivity of collisionless plasma and its application to low-gas-pressure plasma sources [25, 26], and experimental investigation of ultra-high $\beta$ plasma with fast flow [3, 4]. The bird-eye view and cross sections of Proto-RT are described in Figs. 1 and 2.

Proto-RT consists of a cylindrical chamber, three kinds of magnetic field coils, vacuum pumps, and other diagnostic apparatus connected to the main chamber, as shown in the figures. The chamber has a rectangular poloidal cross section of 90 cm × 53.3 cm and it is evacuated to the base pressure of $4 \times 10^{-7}$ Torr by a turbomolecular pump. Proto-RT has three kinds of magnetic field coils: internal conductor (IC), vertical field (VF) coils, and toroidal field (TF) coils, and a variety of field configurations can be produced by the combination of these coils. Inside the chamber, a ring-shaped internal conductor for the production of dipole magnetic field is hung by support rods that are connected to the center stack (CS) of 11.4 cm diameter. The major radius and minor radius of the IC case is 30 cm and 4.3 cm, respectively. Electric current and coolant for the IC are fed through a pair of tube structures via the CS. All these structures, including the support rods and coolant or feeder tubes, are covered by ceramic tubes for the electric insulation from plasmas. Outside the chamber, a pair of the vertical field coils are installed at $R = 90$ cm with a vertical interval of 60 cm. It provides almost parallel magnetic field lines inside the chamber, and used for deforming the shape of magnetic surfaces. For the production of magnetic shear, toroidal field coils are installed in the CS. All the coils are power fed by DC power sources and the maximum coil currents are 10.5 kAT (IC), 5.25 kAT (VF), and 30 kAT (TF), respectively. The typical strength of the magnetic field in the confinement region of the chamber is of the order of $10^{-2}$ T. Magnetic surfaces in the poloidal cross section of the Proto-RT chamber are shown in Fig. 2 as solid lines.

A toroidal non-neutral plasma relaxes to an equilibrium state under the effects of external electric fields from outside of the space charges [20]. As far as an equilibrium state is found, this external fields are automatically generated by the induced image charges on the chamber. The equilibrium can also be externally controlled by the applied electric fields generated by the electrodes. The electrodes are also effective for the potential optimization and formation of radial electric field inside neutral plasmas. In this study, the effects of electrode biasing were examined. For the potential optimization and formation of radial electric field inside the plasma, a pair of electrode are installed on the IC and the CS. The dotted lines in Fig. 2 show the equi-potential contours when the IC electrode is biased to $V_{IC} = 300$ V.

## Electron injection

An electron gun is installed in Proto-RT at the peripheral location of the confinement region as shown in Fig. 3. A lanthanum hexaboride (LaB$_6$, work function is $\phi_W = 2.6$ eV) sintered compact is employed as a cathode of the electron gun. The electron gun is located near $Z = 0$ plane, and it is movable in the radial direction. Injection angle of the gun is also variable using a rotating flange. Acceleration voltage of up to $V_{acc} = 1.3$ kV is applied between the LaB$_6$ cathode and a molybdenum anode grid of 65 % transparency, located 2 mm in front of the cathode. The cathode is heated with a current of up to 28 A by a regulated DC power source and operated at temperature of $\sim 1800$ K. Acceleration voltage $V_{acc}$ is controlled by a MOS-FET open-drain semiconductor switch. The obtained drain current is $I_{drain} \sim 0.8$ A when $V_{acc} = 1.2$ kV and no magnetic field. With $V_{acc} = 300$ V, the electron injection beam current $I_{beam}$ is 26 mA when $V_{IC}$ (IC electrode bias voltage) is 0, and 5.9 mA when $V_{IC} = -300$ V. The gun

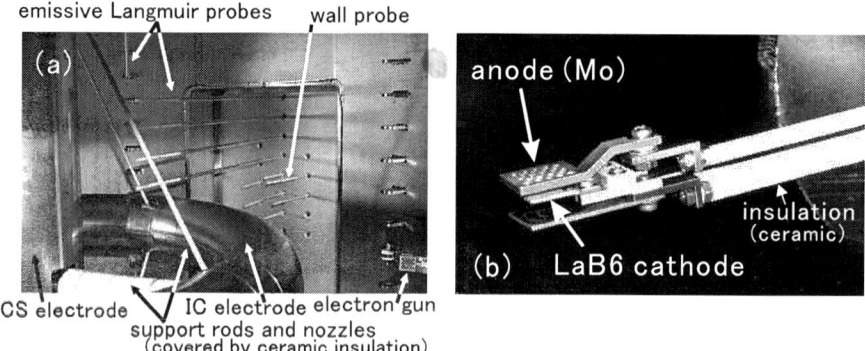

emissive Langmuir probes    wall probe

(a)

anode (Mo)

insulation
(ceramic)

(b)    LaB6 cathode

CS electrode    IC electrode electron gun
support rods and nozzles
(covered by ceramic insulation)

**FIGURE 3.**    (a) Internal view of Proto-RT and (b) electron gun structure.

is usually operated at a base pressure of below $1 \times 10^{-5}$ Torr, in order to avoid rapid evaporation and evaporation coating due to the oxidation of the cathode.

The injection angle of the electron gun is decided according to a numerical calculation [6] so that the electrons take sufficiently long orbit lengths. Because of the large gyration radius, electrons obey a strongly nonlinear equation of motion in the inhomogeneous magnetic field, resulting in chaotic (non-periodic) motion with long orbit lengths.A numerical calculation shows that electrons are effectively injected from the edge of the confinement region [6]. The typical orbit of a single electron injected into a dipole magnetic field is shown in Fig. 4 (bird-eye view where $X = Y = 0$ is the center axis of the device). When the pitch angle between the dipole magnetic field lines and initial injection velocity is large, electrons take transit orbit around the IC as shown in Fig. 4 (a). Electrons with small parallel velocities are mirror trapped in the bad curvature region (Fig. 4 (b)). Electrons also undergo the Larmor rotation around the magnetic field lines, and drift motion in the toroidal direction due to the $\mathbf{E} \times \mathbf{B}$ and $\nabla B$/curvature drifts.

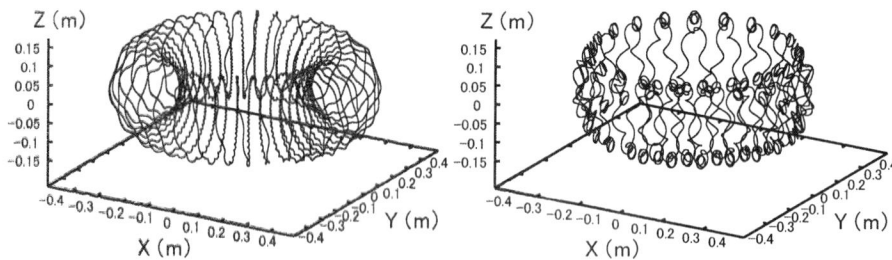

**FIGURE 4.**    Typical electron trajectories of (a) "passing" and (b) "trapped" orbits.

380

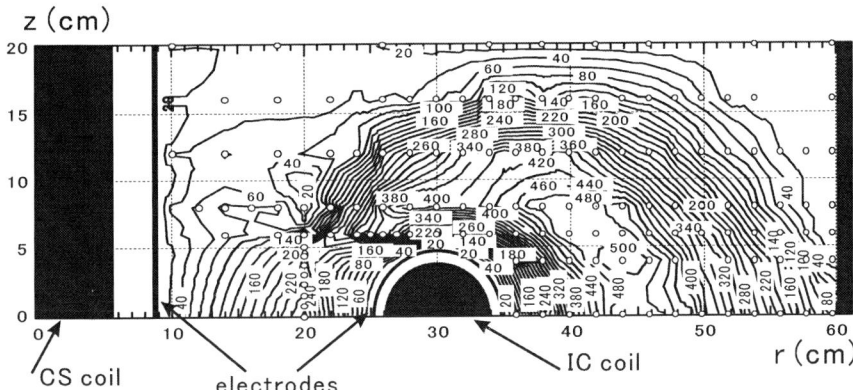

**FIGURE 5.** Measuring points of Langmuir probes and an example of the reconstructed potential profile.

## Diagnostics on toroidal electron plasma

Among other parameters, space potential $\phi_s$ is one of the essential parameters for the measurements of the structures of pure electron plasmas. The electrostatic potential of the plasma is measured by emissive Langmuir probes and it provides the information of potential profiles with fine spatial resolution. Basically, the probe is a simple metallic tip inserted into a plasma, and biased to some potential. Analysis of the relation between the bias voltage and the obtained current provides the information of electron parameters. For pure electron plasmas, structures like Langmuir probes inside the plasma are serious obstacles for the realization of stable confinement, but it can be used for the measurements of space potential during the electron injection phase.

The probes are arranged as an array located between $Z = -20$ cm and $+20$ cm (the configuration of the probes is shown in Fig. 2) and two dimensional potential structures are measured in the confinement region. Schematic and photographic views of emissive Langmuir probe configuration are shown in Figs. 3. The probes were inserted from gauge ports (*r-probes*) on the diagnostic flange and one gauge port (*z-probe*) on a diagnostic port. The probes are movable along **r** (*r-probe*) or **z** (*z-probe*) direction, and we can obtain two-dimensional potential profiles. Measuring points and reconstructed 2-d profiles are shown in Fig. 5.

After the stop of the electron supply, Langmuir probes placed inside the confinement region are serious obstacles for the realization of stable equilibrium. For the measurement of electrostatic fluctuations and determination of the confinement time of the plasma without such perturbations, $5 \times 15$ mm and $5 \times 250$ mm copper foils are covered by insulating quartz tubes and installed in the chamber, and used as wall probes [17]. The sensor foil is connected to the vessel wall through a current amplifier for the observation of image current, which indicates the motion of the plasma. The longer wall is placed over the entire confinement region and used for the estimation of the remaining charge of the plasma by integrating the passing image current.

During the electron injection, a current collected by the high-impedance emissive

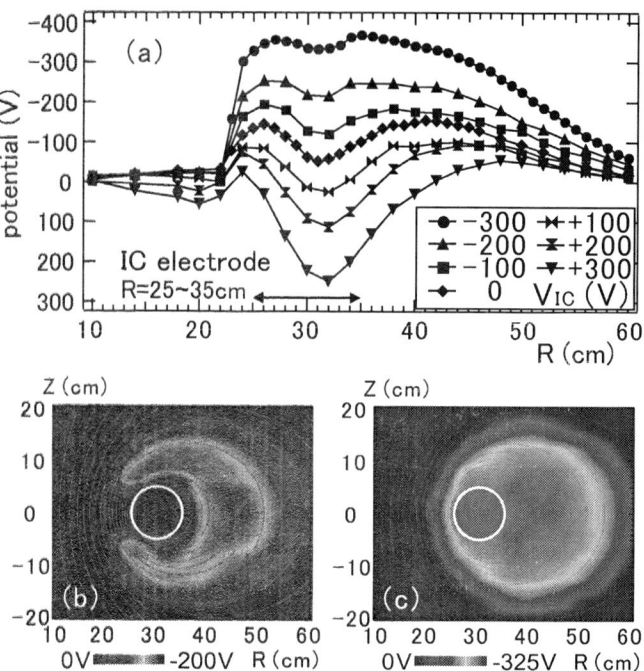

**FIGURE 6.** (a) Radial potential profiles of electron plasma at $Z = 6$ cm in the variation of the electrode bias voltage $V_{IC}$. Two dimensional potential distribution (b) when $V_{IC} = 0$ V and (c) when $V_{IC} = -300V$. Thin lines show the magnetic surfaces of the dipole field.

Langmuir probe is less than 10 $\mu$A. It is much smaller than the injected beam current of the electron gun of $2 \sim 30$ mA, and the resultant disturbance is supposed to be small. In fact, additional insertion of another Langmuir probe does not significantly change the measured $\Phi_H$, and we therefore conclude that during the electron injection, the Langmuir probes do not seriously perturb the plasma.

## TOROIDAL ELECTRON PLASMA IN PROTO-RT

### Electron injection and potential formation

Figure 6 shows the structures of the internal electric potential in the poloidal cross section of the device during the electron injection (the injection period was 100 $\mu$s). Potential profiles when the IC electrode is grounded is shown in Fig. 6 (b). Poisson equation was solved in the geometry of Proto-RT [7], and we may reproduce the measured potential profile, assuming an electron cloud with a peak number density of $1 \times 10^{13}$ m$^{-3}$ and a total space charge of $3 \times 10^{-7}$ C. When $V_{IC} = -300$ V, the peak density and total charge are estimated to be $1 \times 10^{13}$ m$^{-3}$ and $4 \times 10^{-7}$ C, respectively. The high density region of the electron plasma is relatively remote from the support rods of the internal

382

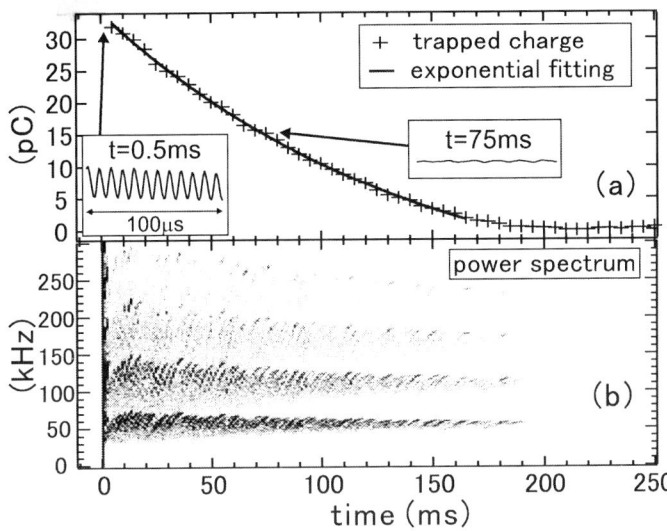

**FIGURE 7.** (a) Trapped charge of electron plasma after the stop of electron supply at $t = 0$ and (b) temporal evolution of the power spectrum of the electrostatic fluctuation.

conductor throughout the experiments, and these structure does not seriously perturb the electron plasmas so long as the electrons rotates in the toroidal direction.

When the potential is not externally controlled (the IC electrode is grounded) in Fig 6 (b), the distribution has a reversed "C" shape and the potential contours show strong disagreement with the magnetic surfaces of the dipole field configuration. Thus the potential contours without electrode bias do not coincide with the magnetic surfaces, implying a rapid particle loss across the field lines. Because the electrostatic potential $\phi$ is not constant on a magnetic surface, Eq. (22) is not satisfied, implying that thermalized (low energy) electrons are not confined when the potential is not externally controlled. The injected electron beam current when $V_{IC} = 0$ V is $I_{beam} = 26$ mA, and it is an order of magnitude larger than when $V_{IC} = -300$ V, also may suggesting that stable confinement is not realized unless the IC electrode is negatively biased.

In contrast, when the IC electrode is negatively biased, the potential profile was successfully modified. Figure 6 (c) shows the potential profile when $V_{IC} = -300$ V, the same potential as the acceleration voltage of the electron gun. The potential hall near the IC electrode is eliminated, and the potential contours surround the IC, which is topologically close to the shape of the magnetic surfaces.

Figure 6 (a) shows the radial potential profiles at $Z = +6$ cm, just above the surface of the IC electrode. The grounded or positively biased IC electrode yields a hollow potential profiles around the IC. These concave potential distributions are not favorable for the stable confinement, because the resultant sheared $\mathbf{E} \times \mathbf{B}$ flow can be a energy source for the destabilization of the Kelvin-Helmholtz (diocotron) modes. By negatively biasing the IC electrode, the hollow structures are cancelled as illustrated in the figure, and it might lead to stabilize the diocotron instability of non-neutral plasmas. As described

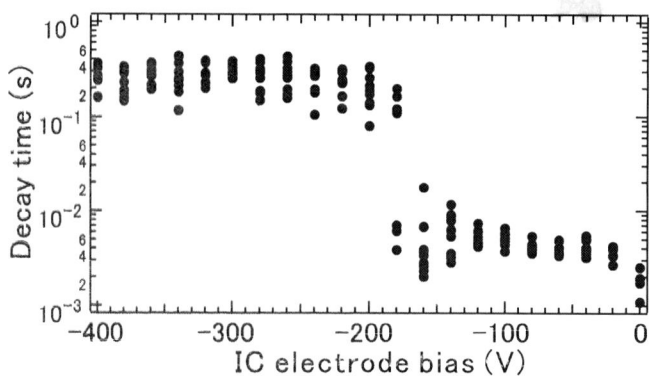

**FIGURE 8.** Confinement time of an electron plasma in the variation of $V_{IC}$.

in the next subsection, the measurements of electrostatic fluctuation show that the confinement time is enhanced for more than an order of magnitude by the negative bias of the IC electrode.

## Fluctuation and trapped charge decay

The typical temporal evolution of the electron plasma, the trapped charge and the power spectrum of the electrostatic fluctuation are shown in Fig. 7. Electrons are injected with an acceleration voltage of 300 V from $t = -100$ to 0 $\mu$s, into DC magnetic and electric fields generated by a dipole filed coil current of $I_{IC} = 7$ kAT (magnetic field strength at $R = 40$ cm is 70 G) and electrode bias voltages of $V_{IC} = -300$ V and $V_{CS} = 0$ V. After the electron gun is turn off at $t = 0$ s, a large oscillation during the electron injection phase decays typically in a time constant of $\sim 1$ ms. Subsequently, a quiet oscillation mode is realized only when the IC electrode is negatively biased.

The wall signals before and after the stabilization of the initial large fluctuation are shown in the small boxes in Fig. 7 (a). The magnitude of the electrostatic fluctuation before and during the quiet mode normalized by the DC electrostatic potential in the plasma is $\tilde{\phi}/\phi = 12$ % and 0.6 %, respectively. In the quiet confinement phase, the remaining charge drops approximately exponentially as indicated in Fig. 7 (a). As the plasma enters the quiet confinement mode, the frequency of oscillation, illustrated in Fig. 7 (b), decreased from 240 kHz to 62 kHz. In the quiet confinement phase, the decrease of the frequency is relatively small, and the peak of the fundamental frequencies in the power spectrum are 62 kHz at $t = 5$ ms and 57 kHz at $t = 180$ ms, indicating that both self and external electric fields decide the frequencies of the electrostatic oscillation.

The frequencies of the observed electrostatic oscillation is inversely proportional to the magnetic field strength and it shows approximately linear dependence on the external electric fields. As well as the characteristics of the fluctuation frequencies, the direction of the wave propagation also agrees with the properties of diocotron oscillation mode.

The confinement time of the particles (time constants of the exponential fitting curves) as a function of $V_{IC}$ is shown in Fig. 8. When $V_{IC}$ is above $-180$ V, the initial fluctuation of the plasma does not enter a quiet state and the charge decays within some milliseconds. The amplitude of the initial fluctuation when $V_{IC} = -80$ V is $\tilde{\phi}/\phi = 32$ % and it decays without entering the quiet confinement mode. As indicated in Fig. 6, the drastic improvement of the confinement is observed when the hollow potential structure is eliminated by the sufficient bias voltage of the IC electrode.

## Confinement time scalings

Figure 9 (a) shows the confinement time $\tau$ of electrons as a function of background neutral gas (hydrogen) pressure $P$. At the base pressure of $4 \times 10^{-7}$ Torr and the maximum dipole field coil current of 10.5 kAT, the obtained confinement time is $\tau = 0.5$ s. Assuming that the loss of electrons is caused by the collisions with neutral atoms, the force balance equation of electrons in the confinement phase is given by

$$qn(\mathbf{E} + \mathbf{v}_e \times \mathbf{B}) - m_e n_e v_{en} \mathbf{v}_e = 0. \tag{27}$$

Here the small inertia term is neglected. When the confinement time of electrons is determined by the loss of the momentum of toroidal $\mathbf{E} \times \mathbf{B}$ motion due to the collisions with neutral gas molecules, the toroidal viscous force is balanced by the Lorentz force caused by the outgoing radial velocity of electrons:

$$0 = -e n_e v_r B - m_e n_e v_{en}(v_t - v_n), \tag{28}$$

where $e$ is charge, $v_r$ and $v_t$ are radial and toroidal speed of electrons, $m_e$ is electron mass, $n_e$ is electron density, $v_n$ is the velocity of neutral atom, $v_{en} = n_n \sigma v_t$ is the electron-neutral atom mean collision frequency, $n_n$ is neutral gas density, and $\sigma$ is the collision cross section. Assuming that $v_n = 0$, the radial outgoing velocity of the electrons is given by

$$v_r = \frac{m_e n_n \sigma E^2}{eB^3}. \tag{29}$$

Taking the minor radius of the electron plasma $a$ as the typical length, the typical diffusion time of the electrons is given by

$$\tau_D \sim a/v_r = \frac{eaB^3}{m_e n_n \sigma E^2} \propto P^{-1} B^3. \tag{30}$$

Substituting the experimental parameters of $B \sim 0.005$ T, $P = 10^{-6}$ Torr, $E = 3 \times 10^2$ Vm$^{-1}$, $\sigma \sim 10^{-19}$ m$^2$, and $a = 0.1$ m, the typical confinement time $\tau_D$ is in the order of 1 s, and it is comparable to the observed confinement time. For the pressure range of $10^{-6}$ to $10^{-4}$ Torr, $\tau$ is scaled as $\propto P^{-1}$, as shown in Fig. 9 (a), indicating that the effects of residual neutral gas set the confinement time. When $P$ is below $\sim 10^{-6}$ Torr, $\tau$ deviates from the $P^{-1}$ line and it saturates near 0.5 s. The confinement time at the base pressure of $4 \times 10^{-7}$ Torr in the variation of the dipole field coil current is shown in Fig. 9

385

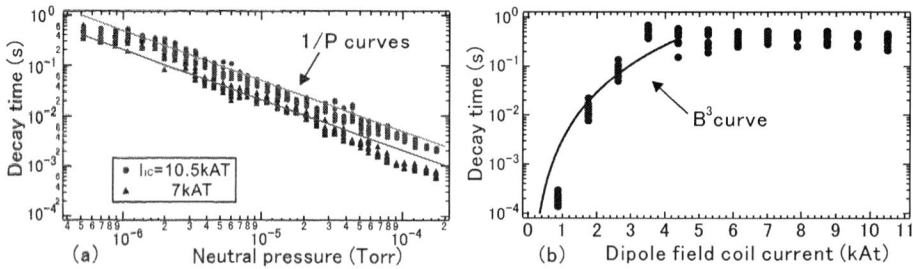

**FIGURE 9.** Confinement time of toroidal electron plasma in the variation of (a) base pressure and (b) magnetic field strength.

(b). Before entering the saturation region (when $I_{IC}$ is below $\sim 5$ kAT), the observed $\tau$ is approximately proportional to $B^3$ and it also agrees with the parameter dependence of the calculated $\tau_D$, but $\tau$ has an upper limit in spite of the increase of the magnetic field strength above $I_{IC} \sim 5$ kAT. As indicated in the change of the electron charge in Fig. 9 (b), the amount of the trapped charge as well as $\tau$ saturate above $I_{IC} \sim 5$ kAT, although the obtained electron density ($\sim 10^{12}$ m$^{-3}$) is far below the Brillouin density limit ($\sim 10^{14}$ m$^{-3}$), suggesting the existence of some anomalous loss of the electrons.

In the initial large fluctuation phase just after the stop of the electron injection, charge on the wall decreased from $\sim 200$ pC to 32 pC, and only small fraction of electrons shows good confinement properties. In comparison with the trapped charge during the electron injection calculated from the Poisson's equation, stably confined electron charge when $I_{IC} = 10.5$ kAT, $P = 4 \times 10^{-7}$ Torr, and $V_{IC} = -300$ V is estimated to be $\sim 5 \times 10^{-8}$ C.

## Stabilizing effects of magnetic shear

As described in the previous section, the reduction of electrostatic fluctuation was realized by properly adjusting the potential profiles. The addition of toroidal magnetic field (sheared magnetic field) also can stabilize the diocotron instability. The effects of magnetic shear is applicable for the stabilization of MHD instability of two-fluid plasmas, and thus it might be potentially effective for the stable confinement of multi-component plasmas in future experiments.

Electrostatic fluctuations during the electron injection, measured by a wall probe, are shown in Fig. 10 in the variation of added toroidal field. Dipole field coil current $I_{IC} = 7$ kAT was kept constant. When $I_{TR}$ is larger than $\sim 4$ kAT (the strength of a toroidal field is close to the dipole field strength), the amplitude of the total fluctuation decreases by a factor of 10, compared with when in pure dipole field configuration. During these experiments, the generated space potential is kept approximately constant ($\phi_H$ of emissive Langmuir probe at $R = 46$ cm is between 230 V and 290 V, for the variation of $I_{IC}$ from 0 kAT to 9 kAT), and stabilization of diocotron instability by magnetic shear [28] is experimentally demonstrated.

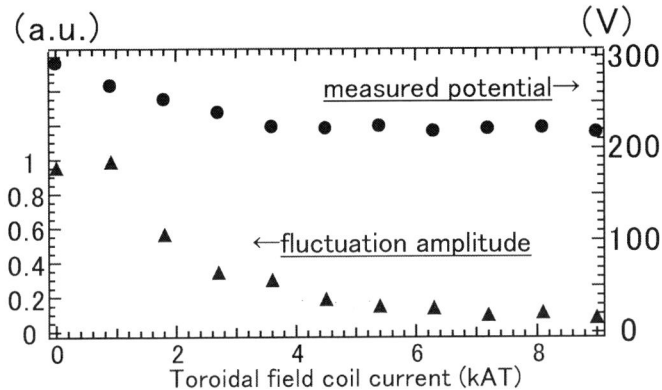

**FIGURE 10.** Electrostatic fluctuations and space potentials at $R = 46$ cm in the variation of added toroidal field $I_{TF}$. Fluctuation amplitude is normalized by the absolute value of space potential when $I_{TF} = 0$.

Although the addition of a toroidal magnetic field is effective for the realization of quiet plasma during the electron injection, it induces the earlier onset of the rapid growth of instability after the stop of the electron injection. Figure 11 (a) shows the wall probe signals in the variation of added toroidal field. The stable charge decay, as observed in pure dipole field configuration (Fig. 7), is not realized when the magnetic shear exists. As shown in the life time of the electron plasma as function of toroidal field strength in Fig. 11 (b), the plasma tends to disrupt earlier in the stable confinement phase as the stronger toroidal field is added.

Reduction of the stable confinement time is similarly observed even in the pure dipole field when structures like Langmuir probes are inserted into the confinement region. One of the possible interpretations is that when electrons hit the surfaces of the obstacles, accumulated neutral gas molecules are released and ionized, and it causes the ion resonance instability.When toroidal field is added to the pure dipole field configuration, the $\mathbf{E} \times \mathbf{B}$ drift motions of electrons take spiral orbits around the internal conductor, and the trajectories may intersect the coolant and feeder structures of the internal conductor. Then these effects could lead to the collisions of electrons with the obstacles. In contrast, when electrons are confined in pure poloidal field, the charged particles do not hit the coolant structures as long as they rotate in the toroidal direction due to the $\mathbf{E} \times \mathbf{B}$ drift motion. The addition of toroidal field lead to the increase of space potentials near the bar structure of the IC electrode, indicating that electrons are transported inwardly due to the spiral motion.

# PROSPECT TOWARD ANTIMATTER PLASMAS AND MULTI-COMPONENT PLASMAS

For the trap of antimatter plasmas and its mixtures, where the particles are supplied from relatively weak radiation sources compared with conventional plasma sources, excel-

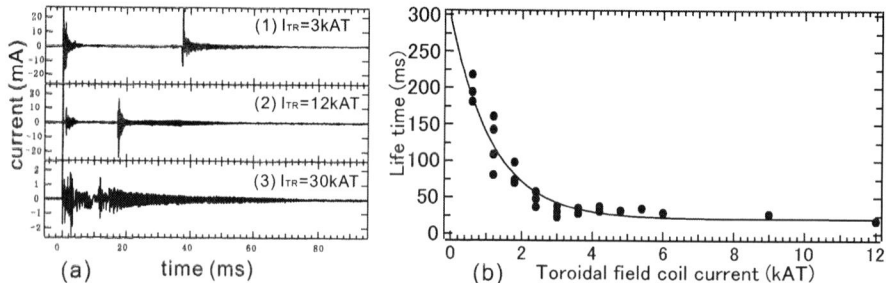

**FIGURE 11.** (a) Wall probe signals in the variation of added toroidal field. Electrons are injected from $t = -100$ to $0$ $\mu$s. (b) Stable confinement time before the rapid growth of the instability.

lent confinement properties are important for trap devices. The Proto-RT experiments demonstrated the fine confinement properties of non-neutral plasmas in the magnetic surface configuration, and the next task is the injection and simultaneous confinement of another particles to form multi-component plasmas. At present, as well as the experiments of toroidal non-neutral plasmas on internal conductor devices [1], a stellarator [2], and a helical system [27] are in progress or under construction. A helical device can also produce closed magnetic surfaces [2] with only "external" field coils located outside of the plasma.

For the creation of multi-component plasmas, we can inject another species of particles into a magnetic surface configuration by means of several methods. By the combination of dipole and vertical magnetic fields, a magnetic neutral loop is generated between two kinds of coils along the confinement region of plasmas. One can inject charged particles from peripheral region near the magnetic null line [6]. Because the magnetic moment $\mu$ is not conserved near $X$ point of the magnetic field lines, the particles take chaotic trajectories with long orbit length. When an electron plasma is generated and stably confined, the electrostatic force caused by the self-space potential can be used for the trap of positrons until the plasma is neutralized. The use of potential well of the initially generated plasma is also effective for other opposite sign of charged particles like positrons and antiprotons. For a device like a helical system, the inductive charging method [14] is an another solution to inject and compress both signs of magnetized charged particles.

## SUMMARY

As a first step toward the multi-component non-neutral plasmas, experimental investigation into the equilibrium and confinement properties of pure electron plasmas was carried out in a toroidal magnetic surface configuration. The obtained results are summarized as follows. Long-term stable confinement of a toroidal electron plasma was demonstrated in a magnetic surface configuration. In the initial fluctuating phase, the trapped charge adjusts (diminishes) to enter a quiescent phase, when the potential contours are externally adjusted to the structure of magnetic surfaces. In the present device, we confined electrons with a peak density of an order of $10^{12}$ m$^{-3}$ and total charge of an order of $\sim 10^{-8}$ C for more than 0.1 s. When $\tau$ is below $\sim 0.1$ s, it is scaled as $\tau \propto P^{-1}B^3$ for the pressure range between $10^{-6}$ and $10^{-4}$ Torr, indicating that the collisions with remaining neutral gas limit the confinement of electrons. In the lower pressure region of below $\sim 10^{-6}$ Torr, $\tau$ saturates above $\sim 0.1$ s, suggesting the anomalous loss of electrons. Although the stabilizing effects of magnetic shear is observed, the addition of toroidal field shortens the stable confinement time, possibly because of the obstacles of the support structure for the internal conductor. The adverse effects of the internal structures are inevitable in the present device, and this problem will be solved in a superconducting levitated ring device [1] in future experiments.

## ACKNOWLEDGMENTS

The authors would like to acknowledge the useful discussions and advices of Mr. J. Morikawa, Prof. H. Himura, and Mr. H. Wakabayashi. This work was funded by a Grant-in-Aid for Scientific Research No. 14102033 from Japanese Ministry of education, culture, sports, science and technology. The work of H. S. was supported in part by the Japan Society for the Promotion of Science.

## REFERENCES

1. Z. Yoshida et al., Y. Ogawa et al., H. Himura et al., C. Nakashima et al., in *Nonneutral Plasma Physics III*, edited by J. J. Bollinger, R. L. Spencer, and R. C. Davidson, AIP Conference Proceedings 498, New York, 1999, pp. 397–422.
2. T. S. Pedersen and A. H. Boozer, *Phys. Rev. Lett.*, **88**, 205002-1–205002-4 (2002).
3. S. M. Mahajan and Z. Yoshida, *Phys. Rev. Lett.*, **81**, 4863–4866 (1998).
4. Z. Yoshida and S. M. Mahajan, *Phys. Rev. Lett.*, **88**, 095001-1–095001-4 (2002).
5. L. C. Steinhauer and A. Ishida, *Phys. Rev. Lett.*, **79**, 3423–3426 (1997).
6. C. Nakashima, Z. Yoshida, H. Himura et al., *Phys. Rev. E*, **65**, 036409-1–036409-6 (2002).
7. H. Saitoh, Z. Yoshida, C. Nakashima et al., *Phys. Rev. Lett.*, **92**, 255005-1–255005-4 (2004).
8. D. H. E. Dubin and T. M. O'Neil, *Rev. Mod. Phys.*, **71**, 87–172 (1999).
9. M. Amoretti, C. Amsler, G. Bonomi et al., *Nature*, London, 2002, **419**, 456–459.
10. G. Gabrielse, N. S. Bowden, P. Oxley et al., *Phys. Rev. Lett.*, **89**, 213401-1–213401-4 (2002).
11. A. Mohri and Y. Yamazaki, *EuroPhys. Lett.*, **63**, 207–213 (2003).
12. R. G. Greaves and C. M. Surko, in *Nonneutral Plasma Physics IV*, edited by F. Anderegg, L. Schweikhard, and C. F. Driscoll, AIP Conference Proceedings 606, New York, 2002, pp. 10–23.
13. G. P. Zank and R. G. Greaves, *Phys. Rev. E*, **51**, 6079–6090 (1995).
14. G. S. Janes, *Phys. Rev. Lett.*, **15**, 135–138 (1965).
15. J. D. Daugherty and R. H. Levy, *Phys. Fluids*, **10**, 155–161 (1967).
16. J. D. Daugherty, L. Grodzins, G. S. Janes, and R. H. Levy, *Phys. Rev. Lett.*, **20**, 369–371 (1968).
17. J. D. Daugherty, J. E. Eninger, and G. S. Janes, *Phys. Fluids*, **12**, 2677–2693 (1969).
18. A. Mohri, M. Masuzaki, T. Tsuzuki, and K. Ikuta, *Phys. Rev. Lett.*, **34**, 574–577 (1975).
19. W. Clark, P. Korn, A. Mondelli, and N. Rostoker, *Phys. Rev. Lett.*, **37**, 592–595 (1976).
20. K. Avinash: *Phys. Fluids B*, **3**, 3226–3231 (1991).
21. P. Zaveri, P. I. John, K. Avinash, and P. K. Kaw, *Phys. Rev. Lett.*, **68**, 3295–3298 (1992).
22. S. S. Khirwadkar, P. I. John, K. Avinash et al., *Phys. Rev. Lett.*, **71**, 4334–4337 (1993).
23. M. R. Stoneking, P. W. Fontana, R. L. Sampson, and D. J. Thuecks, *Phys. Plasmas*, **9**, 766–771 (2002).
24. M. R. Stoneking, M. A. Growdon, M. L. Milne, and R. T. Peterson, *Phys. Rev. Lett.*, **92**, 095003-1–095003-4 (2004).

25. T. Uchida, *Jpn. J. Appl. Phys.*, **33**, L43–L44 (1994).
26. Z. Yoshida, H. Asakura, H. Kakuno, et al., *Phys. Rev. Lett.*, **81**, 2458–2461 (1998).
27. H. Himura, H. Wakabayashi, M. Fukao et al., *Phys. Plasmas*, **11**, 492–495 (2004).
28. S. Kondoh, T. Tatsuno, and Z. Yoshida, *Phys. Plasmas*, **8**, 2635–2640 (2001).

# Workshop on the Physics with Ultra Slow Antiproton Beams

March 14-16, 2005

O-kouchi Hall, RIKEN, Wako, Saitama, Japan

Sponsored by RIKEN & The Society for Atomic Collision Research

**March 14 (Monday)**

9:30-9:40 Opening/Business announcements                           Y.Yamazaki

(chair:Y. Yamazaki)
9:40-10:20 Sub-Femtosecond Correlated Dynamics Probed with Antiprotons                J. Ullrich
10:20-10:50 Capture of Slow Antiprotons by Atoms, Molecules, and Ions              J. Cohen

10:50-11:10 Coffee break

(chair: I. Shimamura )
11:10-11:40 Ionization Dynamics by p and pbar                    H. Schmidt-Boecking
11:40-12:10 Stopping and Ionization at few keV                       U. Uggerhoj
12:10-12:40 Antiproton-nucleus Annihilations at Very Low Energies Down to Capture

E. Lodi-Rizzini

12:40-14:10 Lunch

(chair: J. Eades )
14:10-14:50 Low Energy Antiproton Experiments- A review                  K. Jungmann
14:50-15:20 Project ALPHA - A New Experiment for Antihydrogen Production and Trapping

J. Hangst
15:20-15:50 ATRAP - on the way to trapped Antihydrogen                 D. Grzonka

15:50-16:20 coffee break

(chair: M. Fujiwara)
16:20-17:00 Lorentz and CPT Tests Involving Antiprotons                   R. Lehnert
17:00-17:30 Anti-hydrogen Production Conditions in ATHENA: What we (don't) Know   P. Bowe
17:30-18:00 A New Path Toward Gravity Experiments with Anti-hydrogen          P. Perez

**March 15 (Tuesday)**

(Chair:T. Kambara)

| | |
|---|---|
| 9:00-9:30 Production of Ultra-slow Antiproton Beams | H.Torii |
| 9:30-9:50 The Anticyclotron Project | D.Horvath |
| 9:50-10:20 Intense Source of Slow Positrons | A.Rosowsky |
| 10:20-10:50 ASACUSA Gaas-jet Target: Present Status and Future Development | V.Varentsov |

10:50-11:20 coffee break

(Chair: H. Schmidt-Boecking)

11:20-12:00 Prospects of CPT Tests Using Antiprotonic Helium and Antihydrogen     R.Hayano

12:00-12:30 Non-Neutral Plasma Confinement in a Cusp -Trap and   Possible Application to Anti-Hydrogen Beam Generation     A.Mohri

12:30-13:00 A Compact Setup of Fast pnCCDs for Exotic Atom Measurements     H.Gorke

13:00-14:30 Lunch

(Chair: J. Ullrich)

14:30-15:00 Imaging Antimatter: The Challenges and Applications     M.Fujiwara

15:00-15:30 Confinement of Toroidal Non-neutral Plasma in Proto-RT     H.Saito

15:30-15:50 On the Possibility of Nonneutral Antiproton Plasmas and Anti-hydrogen Plasmas     H.Higaki

15:50-16:20 Control of Plasmas for Production of Ultraslow Antiproton Beams     N.Kuroda

16:20-16:50 Coffee break

(Chair: K. Jungmann)

16:50-17:30 The Deepest Symmetries of Nature     D.Horvath

17:30-18:10 The Antiproton and How It Was Discovered     J.Eades

18:10-20:00:   Banquet

**March 16 (Wednesday)**


(chair: P. Kienle)

9:00-9:40: Nuclear Structure Studies with the Low Energy Antiprotons          S. Wycech

9:40-10:10: Antiprotonic Atoms - A Tool for the Investigation of the Nuclear Periphery

A. Trzcinska

10:10-10:40: Light Antiprotonic Atoms          D. Gotta

10:40-11:00  Study of S=-2 Baryonic States at FLAIR          D. Grzonka


11:00-11:20 Coffee break


(Chair: H.Wollnik)

11:20-11:50 Muonic Atoms of Unstable Nuclei          P. Strasser

11:50-12:20 Neutron Density Distributions of the Sn Isotopes and Ca Isotopes Extracted from the

Proton Elastic Scattering          S. Terashima

12:20-12:50 Highly Charged Ions at Rest: The HITRAP Project at GSI          F. Herfurth


12:50-14:10 Lunch


(chair: T. Motobayashi)

14:10-14:40 An Antiprotoni Ion Collider (AIC) for Measuring Neutron and Proton Distributions of

Stable and Radioactive Nuclei          P. Kienle

14:40-15:10 Nuclear Matter Radii Determined by Interaction Cross Sections          A. Ozawa

15:10-15:40 Structure Studies of Unstable Nuclei by Electron Scattering          T. Suda


15:40-16:00 Coffee Break


(chair: J. Zmescal)

16:00-16:30 Antiprotonic Radioactive Atoms for Nuclear Structure Studies          M.Wada

16:30-17:00 Muonic Antihydrogen, Production and Test of CPT Theorem          K. Nagamine

17:00-17:10 Do Electrons and Ions Coexist in an EBIT?          J. Ullrich

17:10-17:40 Toward the Ultra Slow Antiproton Physics: General Discussion          Y. Yamazaki

# List of Participants

**Bowe** Paul
University of Aarhus
Paul.Bowe@cern.ch

**Cohen** James S.
Los Alamos National Laboratory
cohen@lanl.gov

**Eades** John
Univ. Tokyo
john.eades@cern.ch

**Fujiwara** Makoto
TRIUMF
Makoto.Fujiwara@triumf.ca

**Gorke** Hubert
Forschungszentrum Jülich
h.gorke@fz-juelich.de

**Gotta** Detlev
Forschungszentrum Jülich
d.gotta@fz-juelich.de

**Grzonka** Dieter
Forschungszentrum Jülich
d.grzonka@fz-juelich.de

**Hangst** Jeffrey
University of Aarhus
jeffrey.hangst@cern.ch
hangst@phys.au.dk

**Hatakeyama** Atsushi
University of Tokyo
hatakeya@phys.c.u-tokyo.ac.jp

**Hayano** Ryugo
University of Tokyo
hayano@phys.s.u-tokyo.ac.jp

**Herfurth** Frank
GSI
F.Herfurth@gsi.de

**Higaki** Hiroyuki
Tsukuba University
higaki@prc.tsukuba.ac.jp

**Horvath** Denzo
KFKI
horvath@sunserv.kfki.hu

**Ikeda** Tokihiro
RIKEN
tokihiro@riken.jp

**Ishida** Yoshihisa
RIKEN
yishida@riken.jp

**Ishida** Katsuhiko
RIKEN
ishida@riken.jp

**Ishikawa** Takashi
University of Tokyo
ishikawa@phys.s.u-tokyo.ac.jp

**Jungmann**Klaus-Peter
KVI
jungmann@KVI.nl

**Kambara** Tadashi
RIKEN
kambara@rarfaxp.riken.jp

**Kanai** Yasuyuki
RIKEN
kanai@rarfaxp.riken.go.jp

**Katayama** Ichiro
KEK
ichiro.katayama@kek.jp

**Kienle** Paul
Technische Universität München
Paul.Kienle@Physik.TU-Muenchen.DE

**Komaki** Ken-ichiro
University of Tokyo
komaki@phys.c.u-tokyo.ac.jp

**Kuroda** Naofumi
RIKEN
nkuroda@riken.jp

**Lehnert** Ralf
Vanderbilt University
ralf.lehnert@vanderbilt.edu

**Lodi-Rizzini** Evandro
Brescia University
lodi@bs.infn.it

395

**Mohri** Akihiro
RIKEN
akimohri@crux.ocn.ne.jp

**Motobayashi** Toru
RIKEN
motobaya@riken.jp

**Nagamine** Kanetada
KEK
kanetada.nagamine@kek.jp

**Nagata** Yugo
University of Tokyo
nagata@radphys4.c.u-tokyo.ac.jp

**Nakai** Yoichi
RIKEN
nakaiy@postman.riken.jp

**Nakamura** Takashi
KEK
t-nkmr@riken.jp

**Narihara** Kazumichi
NIFS
narihara@nifs.ac.jp

**Ogata** Koremitsu
University of Tokyo
ogata@radphys4.c.u-tokyo.ac.jp

**Okada** Kunihiro
Sophia University
okada-k@sophia.ac.jp

**Ono** Naoya
University of Tokyo
ono@nucl.phys.s.u-tokyo.ac.jp

**Ozawa** Akira
University of Tsukuba
ozawa@tac.tsukuba.ac.jp

**Perez** Patrice
CEA/Saclay
patrice.perez@hep.saclay.cea.fr

**Posada** Lawrence
University of Tokyo
lposada@nucl.phys.s.u-tokyo.ac.jp

**Rosowsky** Andre
CEA/Saclay
andre.rosowsky@cern.ch

**Saitoh** Haruhiko
RIKEN
hsaitoh@riken.jp

**Schmidt-Böcking** Horst
University Frankfurt
schmidtb@atom.uni-frankfurt.de

**Shibata** Masahiro
University of Tokyo
mshibata@riken.jp

**Shimada** Hiroyuki
RIKEN
hshimada@postman.riken.jp

**Shimamura** Isao
RIKEN
shimamur@rarfaxp.riken.jp

**Strasser** Patrick
KEK
patrick.strasser@kek.jp

**Suda** Toshimi
RIKEN
suda@rarfaxp.riken.go.jp

**Takamine** Aiko
University of Tokyo
icot@riken.jp

**Tatsuno** Hideyuki
University of Tokyo
s31541@mail.ecc.u-tokyo.ac.jp

**Terasima** Satoru
Kyoto University
tera@nh.scphys.kyoto-u.ac.jp

**Torii** Hiroyuki
University of Tokyo
torii@radphys4.c.u-tokyo.ac.jp

**Trzcinska** Agnieszka
Warsaw University
agniecha@slcj.uw.edu.pl

**Uggerhoj** Ulrik
University of Aarhus
ulrik@phys.au.dk

**Ullrich** Joachim
MPI Heidelberg
joachim.ullrich@mpi-hd.mpg.de

**Varentsov** Victor
RIKEN
v.varentsov@riken.jp

**Wada** Michiharu
RIKEN
mw@riken.go.jp

**Watanabe** Yasushi
RIKEN
watanaby@riken.jp

**Wollnik** Hermann
Giessen University
Hermann.Wollnik@exp2.physik.uni-giessen.de

**Wycech** Slawomir
Soltan Institute for Nuclear Studies, Warsaw
Slawomir.Wycech@fuw.edu.pl

**Yamanaka** Nobuhiro
RIKEN
yam@postman.riken.go.jp

**Yamazaki** Yasunori
RIKEN/Univ.Tokyo
yasunori@riken.jp

**Zmeskal** Johann
Austrian Academy of Sciences
Stefan Meyer Institute for subatomic Physics
johann.zmeskal@oeaw.ac.at

# Author Index

## B

Barna, D., 293, 307
Bassalleck, B., 190
Beier, T., 278
Bianconi, A., 183
Braun, J., 361
Brenner, G., 361
Bruhns, H., 361

## C

Cohen, J. S., 98
Comeau, D., 122
Corradini, M., 183
Crespo López-Urrutia, J. R., 361

## D

Dahl, L., 278
Dorn, A., 71
Dörner, R., 89

## E

Eades, J., 3, 293, 307
Eliseev, S., 278
Erven, W., 341

## F

Fischer, D., 71, 361
Fujiwara, M. C., 111

## G

Gabrielse, G., 122
Gillitzer, A., 190
Godunov, A. L., 89
Goldenbaum, F., 122
González Martínez, A. J., 361
Gorke, H., 341

Gotta, D., 169, 341
Grzonka, D., 122, 190

## H

Hänsch, T. W., 122
Hartmann, F. J., 214
Hartmann, R., 341
Hayano, R. S., 136
Heinz, S., 278
Herfurth, F., 278
Hessels , E. A., 122
Higaki, H., 351
Hori, M., 293, 307
Horváth, D., 54, 307, 318

## I

Ishida, K., 242
Iwasaki, M., 242

## J

Jagutzki, O., 89
Jastrzębski, J., 214
Jungmann, K. P., 18

## K

Kanai, Y., 147
Kester, O., 278
Khayyat, K., 89
Kienle, P., 222
Kilian, K., 190
Kingsberry, P., 190
Kłos, B., 214
Kluge, H.-J., 278
Komaki, K., 293, 307
Kozhuharov, C., 278
Kuroda, N., 293, 307, 328

399

Walz, J., 122
Watanabe, S., 372
Weber, Th., 89
Welsch, C. P., 71
Whelan, C. T., 89
Winter, P., 190
Wycech, S., 201, 214

Yoshida, Z., 372

## Z

Zhang, Z., 122
Zurlo, N., 183

## Y

Yamazaki, Y., 147, 233, 293, 307, 328, 361